汉英英汉
服装分类词汇

Chinese-English/English-Chinese
Classified Clothing Dictionary

（第4版）

周叔安 编

中国纺织出版社有限公司

图书在版编目(CIP)数据

汉英英汉服装分类词汇 / 周叔安编. -- 4版. -- 北京：中国纺织出版社有限公司, 2023.3
ISBN 978-7-5180-9922-1

Ⅰ.①汉… Ⅱ.①周… Ⅲ.①服装—词汇—英、汉 Ⅳ.①TS941.7-61

中国版本图书馆 CIP 数据核字(2022)第 189282 号

责任编辑：苗　苗　　特约编辑：陈静杰　　责任校对：寇晨晨
责任印制：王艳丽

中国纺织出版社有限公司出版发行
地址：北京市朝阳区百子湾东里 A407 号楼　邮政编码：100124
销售电话：010—67004422　传真：010—87155801
http://www.c-textilep.com
中国纺织出版社天猫旗舰店
官方微博 http://weibo.com/2119887771
唐山玺诚印务有限公司印刷　　各地新华书店经销
2001 年 5 月第 1 版　2009 年 3 月第 2 版
2012 年 5 月第 3 版　2023 年 3 月第 4 版第 1 次印刷
开本：880×1230　1/32　印张：15.375
字数：603 千字　定价：68.00 元

凡购本书，如有缺页、倒页、脱页，由本社图书营销中心调换

编者的话（第4版）

当中国纺织出版社有限公司通知我，决定再版这本《汉英英汉服装分类词汇》时，我着实感到有点意外。众所周知，这些年来传统阅读被电子阅读冲击得越来越猛，而持续的疫情更是让出版业雪上加霜。因此，出版社在如此困境下的信念和坚持让我十分感动！

这次新版本对第3版进行了少量的删改和勘误，并顺应时尚潮流，增加了1300多条新词目，还对部分音标进行了修正。此外，在附录部分新添了摘编的"服装英语100例"，内容涉及服装时尚、服饰文化、智能服装，以及服装设计、款式、材料、配饰、工艺、检验、设备、管理等方面，权当服装词汇的延展。

值此第4版面世之际，我思绪飞扬，心中充满感激！感恩改革开放的伟大时代，感恩培养提携我的中国服装研究设计中心和原湖南省服装工业公司，感恩中国纺织出版社有限公司先后为本书辛勤付出的各位老师和所有工作人员，包括包含芳、郭慧娟、陈静杰、魏萌、苗苗等。我还要感恩支持和激励我的家人和朋友们，并特别感恩21年来始终陪伴和给予厚爱的广大读者！

科技飞速发展，新词不断涌现。自当竭力而为，惶愧学识有限。拜请行家读者，惠予指正高见！

周叔安
2022年7月

编者的话（第1版）

中国服装业日益发展，拥有产量和出口额两个世界第一的中国服装，必将在21世纪独领风骚。

为了满足广大服装研究者、设计者、服装院校师生、外贸、外事及工业、商业等有关人员工作和学习的需要，笔者编写了这本《汉英英汉服装分类词汇》，以供查阅参考。

本书的编写参阅了国家标准中服装工业名词术语的有关内容和国内外有关出版物，同时还收集了近年来新出现的一些词汇，并注意联系国内服装业的实际，力求简明、实用。

本书以服装汉英分类为特点，同时提供英汉对照以便查阅，共收实用服装词目约12000条，有附录六则。

由于服装词汇按汉英分类编写是一项全新的工作，同时服装业发展日新月异，新名词层出不穷，再加之编者学识有限，故在收集、选词、分类诸方面疏漏在所难免，恳请广大读者批评指正。

本书承蒙中国服装研究设计中心和湖南省服装工业公司的支持和指导，并得到中国香港和内地诸多同行、好友的赐教和鼓励，谨此衷心致谢。

编　者
2000年12月

第 2 版说明

《汉英英汉服装分类词汇》是国内首本按服装部位分类编写的词汇，自 2001 年 5 月正式出版以来，为广大服装研究者、设计者、服装院校师生、外贸、外事及工商业有关人员的工作和学习提供了有益的帮助，为我国的服装文化发展做出了重要贡献，深受读者好评，迄今为止第 1 版已印刷 12 次，印量达 70000 册。

随着世界经济一体化进程的加速，我国服装设计水准和工业技术也得到了长足的发展与提高。为了顺应时代的发展，更好地满足读者使用需求，编者花了两年多的时间对《汉英英汉服装分类词汇》进行了修订。本次修订的重点有两个方面：一是调整收词，增加新词新义，订正原来的错误疏漏之处；二是调整英汉服装词汇部分的编排方式，以便读者更加方便地查找使用。

这次修订除了订正原来错误疏漏之处，另在原有词目基础上增加了 5500 余条，全书共收词约 17500 条，基本反映了目前服装词汇的面貌，能够满足广大读者的查阅需要。

这次修订在排版方式上采用了多处更贴近读者使用的格式，如：将第 1 版英汉部分中用"~"省略的英语中心词目恢复并以斜体表示；将英汉部分的书眉字以该页的首词或尾词表示。

由于水平有限，本次修订工作肯定会有一些不足甚至失误的地方，希望读者一如既往地提出宝贵意见，通过不断修订，使《汉英英汉服装分类词汇》更加臻于完善。

编 者
2009 年 3 月 10 日

编者的话（第3版）

  2001年5月，《汉英英汉服装分类词汇》（第1版）由中国纺织出版社首次出版发行。2009年3月，本书出了第2版。今天，第3版又与读者见面了。

  与第2版相比，新版主要的变化是：在英汉部分为所有的英语中心词目添注了音标，以更加方便读者的学习。此外，在参阅了大量最新资料的基础上，又选加了约2500条新词目，使全书收词达到20000余条。同时，还对部分词汇再次进行了调整或修正，力求更加准确。总之，本书努力追逐时尚潮流，积极扩充信息容量，以尽量满足读者的需求。

  今年，《汉英英汉服装分类词汇》10岁了！10年来，承蒙广大读者，特别是服装业内人士的厚爱，本书取得了较好的成绩。第1版曾获得"2003年全国优秀畅销书奖"，第2版亦荣获"2009年中国纺织工业协会优秀图书奖"。借此机会，编者衷心感谢广大读者多年的支持，是你们给了我莫大的信心和持续的动力。感谢中国服装研究设计中心和原湖南省服装工业公司的培养。感谢中国纺织出版社包含芳、郭慧娟、陈静杰等编辑及其他有关人员的辛勤工作。还要谢谢我定居澳大利亚的女儿及我香港的老朋友王巧玲女士，是她们持续多年帮我收集资料。最后，我要特别感谢已经辞世20年的父亲，他是我自学英语的老师，也是我跋涉人生的导师。没有他当年的教诲和激励，我难以坚持到今天。

  山外有山，天外有天，虽不能及，心向往之。这些年来，虽然为服装行业做了点实事，但本人自知学识和能力十分有限，唯有学习，学习，再学习！也恳请广大读者一如既往，多多关注本书。对新版的疏漏之处，还望不吝赐教并及时给予批评指正。

  谢谢大家！

编 者
2011年10月20日于墨尔本

用法说明

一、本词汇主要有两部分内容

第一部分为汉英服装分类词汇，按服装成品、设计、材料、工艺、服饰品、设备等分类编排。

第二部分为英汉服装词汇，按英文字母顺序排列，并根据需要以英语词目为中心，构成相应的词组或短语；另外动词词汇则在 to 后按英文字母顺序集中列出。

此外，还附有一些对服装工作者有实用价值的汇编资料。

二、符号含义

1. 圆括号"（ ）"中的词语为可替换词语或解释性说明。
2. 方括号"〔 〕"中的字母为国际音标。
3. 六角符号"〔 〕"中的字为略语，如：〔英〕英语，〔美〕美语，〔法〕法语，〔日〕日语，〔粤〕粤语，〔设〕设计，〔工〕工艺，〔检〕检验，〔复〕复数，其余类同。
4. 同一汉语词目有两个或两个以上的英语释义时，用"；"隔开。
5. 同一英语词目有多个汉语释义时，词义相同或相近的译名用"，"隔开，词义不同的则用"；"隔开。
6. 下位词目中的斜体部分为英语中心词目。

三、词组或短语中的冠词 the，a 一般省略

关于本书中涉及的保护动物说明

第 6 页涉及的保护动物

貂:《中国国家野生动物保护法》已将紫貂列为国家一级重点保护动物。

海豹:中国的环保团体把 3 月 1 日作为国际海豹日。

第 78~80 页涉及的保护动物

海狸:列入《世界自然保护联盟》(IUCN) 2008 年濒危物种红色名录 ver 3.1——无危 (LC)。

赤狐:列入中国国家林业局 2000 年 8 月 1 日发布的《国家保护的有益的或者有重要经济、科学研究价值的陆生野生动物名录》。

沙狐:已列入国家林业局 2000 年 8 月 1 日发布的《国家保护的有益的或者有重要经济、科学研究价值的陆生野生动物名录》。

貂:野生貂为十分濒危的动物。

黑貂:即紫貂,《中国国家野生动物保护法》已将紫貂列为国家一级重点保护动物。

扫雪貂:国家二级保护动物。

海獭:列入《海洋哺乳类保护条例》,为"枯竭种"。

水獭:列入《世界自然保护联盟》(IUCN) 2015 年濒危物种红色名录 ver 3.1——近危 (NT)。

虎:列入《华盛顿公约》CITES I 级保护动物。

豹:列入《国家重点保护野生动物名录》:国家一级保护动物 (1988 年 12 月 10 日生效)。

浣熊:列入《世界自然保护联盟》(IUCN) 2012 年濒危物种红色名录 ver 3.1——无危。

貂熊:已列入中国国家一级重点保护动物 (1989)。

臭鼬:列入《世界自然保护联盟》(IUCN) 2008 年濒危物种红色名录 ver 3.1——低危 (LC)。

犀牛:列入《世界自然保护联盟濒危物种红色名录》(IUCN) 2008

年 ver 3.1。

麂：列入中国生物多样性红色名录——脊椎动物卷，评估级别为濒危（EN）。

鹿：梅花鹿、黑麂为国家一级保护动物。

鳄鱼：扬子鳄、中美短吻鳄、南美短吻鳄、窄吻鳄、尖吻鳄、中介鳄、菲律宾鳄、佩滕鳄、尼罗鳄、恒河鳄、湾鳄、菱斑鳄、暹罗鳄、短吻鳄、马来鳄、食鱼鳄等属于濒危野生动植物种，是国际性重要保护物种，被《华盛顿公约》CITES（濒危野生动植物种国际贸易公约）列入附录Ⅰ名单，禁止其国际贸易。另外，爬行纲鳄目除列入附录Ⅰ的以外所有物种均被列入 CITES 附录Ⅱ名单，管制国际贸易。

鲨鱼：1999 年已将鲸鲨、姥鲨（又叫象鲛）和噬人鲨三种列入《濒危野生动植物种国际贸易公约》（CITES）附录Ⅱ中，严禁贸易交往活动。我国南海鲨鱼资源丰富，种类又多，广东省有关部门于 2001 年将鲸鲨、姥鲨列入水生野生动物重点保护的物种。

鲸鱼：我国的鲸类动物除白鳍豚和中华白海豚两种被列为国家一级保护动物外，其他所有种均被列为二级保护动物。

海象：列入《世界自然保护联盟濒危物种红色名录》。

鸵鸟：列入《濒危野生动植物种国际贸易公约》。

目　录

汉英服装分类词汇

一、服装成品 …………………… 2
 （一）一般名称 ………………… 2
 （二）款式名称 ………………… 7
  1. 上衣 …………………………… 7
   ① 上衣 ………………………… 7
   ② 夹克衫 ……………………… 8
   ③ 衬衫 ………………………… 10
   ④ 背心、马甲 ………………… 12
   ⑤ 袄 …………………………… 12
  2. 裤子 …………………………… 13
   ① 长裤 ………………………… 13
   ② 牛仔裤 ……………………… 15
   ③ 短裤 ………………………… 16
  3. 裙 ……………………………… 17
   ① 裙 …………………………… 17
   ② 连衣裙 ……………………… 19
  4. 套装，制服，礼服 …………… 20
   ① 套装 ………………………… 20
   ② 西装 ………………………… 21
   ③ 运动套装 …………………… 22
   ④ 制服 ………………………… 22
   ⑤ 防护服 ……………………… 23
   ⑥ 礼服 ………………………… 24
  5. 外套，大衣 …………………… 25
   ① 外套 ………………………… 25
   ② 大衣 ………………………… 26
   ③ 长袍 ………………………… 27
   ④ 风雨衣，披风 ……………… 28
  6. 内衣 …………………………… 28
   ① 内衣裤 ……………………… 28
   ② 胸衣 ………………………… 29
   ③ 内裤，衬裤 ………………… 31
   ④ 睡衣裤 ……………………… 32
  7. 针织服装 ……………………… 32
   ① 毛衣，毛衫 ………………… 32
   ② T恤衫 ……………………… 34
   ③ 针织套装 …………………… 35
   ④ 套头衫 ……………………… 35
   ⑤ 泳装，泳衣 ………………… 36
 （三）部位、部件名称 ………… 36
  1. 衣前身 ………………………… 36
  2. 衣后身 ………………………… 38
  3. 领子 …………………………… 39
  4. 领口，领圈 …………………… 41
  5. 袖子 …………………………… 42
  6. 裤、裙的部位及部件 ………… 44
  7. 口袋，衬布，褶裥，襻 ……… 45
   ① 口袋 ………………………… 45
   ② 衬布 ………………………… 47
   ③ 褶裥，襻 …………………… 47

二、服装设计制图 ········· 49
- （一）人体主要部位 ········· 49
- （二）量身 ··············· 50
- （三）设计和款式 ········· 53
- （四）图样和符号 ········· 56
- （五）线条，线形 ········· 58
- （六）颜色，色彩 ········· 60

三、服装材料 ············· 61
- （一）原料 ··············· 61
  - 1. 纺织纤维 ············· 61
  - 2. 纺织用纱 ············· 64
- （二）面料 ··············· 66
  - 1. 一般名称 ············· 66
  - 2. 面料工艺 ············· 67
  - 3. 面料性能 ············· 68
  - 4. 棉织物 ··············· 69
  - 5. 麻织物 ··············· 72
  - 6. 丝织物 ··············· 72
  - 7. 毛织物 ··············· 74
  - 8. 化纤和混纺织物 ······· 75
  - 9. 毛皮，皮革 ··········· 78
  - 10. 针织物及其他布料 ···· 80
- （三）辅料 ··············· 81
  - 1. 里料，衬里 ··········· 81
  - 2. 衬料，衬布 ··········· 82
  - 3. 填料 ················· 83
  - 4. 线，纽扣，拉链 ······· 84
    - ① 线 ················· 84
    - ② 纽扣 ··············· 84
    - ③ 拉链 ··············· 87
  - 5. 绳带，唛头，徽章，衬垫 ··············· 87
    - ① 绳带 ··············· 87
    - ② 唛头，商标 ········· 88
    - ③ 徽章 ··············· 88
    - ④ 衬垫，等 ··········· 89
- （四）包装材料 ··········· 89

四、服装工艺操作 ········· 90
- （一）技术室打样 ········· 90
- （二）原辅料检验 ········· 92
- （三）裁剪 ··············· 92
- （四）缝纫，缝制 ········· 92
  - 1. 一般名称 ············· 92
  - 2. 缝纫工艺 ············· 93
  - 3. 印花，绣花，等 ······· 96
- （五）锁钉，水洗，熨烫，包装 ··············· 97
  - 1. 锁钉 ················· 97
  - 2. 水洗 ················· 97
  - 3. 熨烫 ················· 99
  - 4. 包装 ················· 99
- （六）成品检验 ··········· 99
- （七）缝 ················· 103
- （八）线迹，针法 ········· 105
  - 1. 基本线迹 ············· 105
  - 2. 手缝针法 ············· 106
- （九）线缝 ··············· 106

五、服饰品 ··············· 107
- （一）帽 ················· 107
  - 1. 品种和款式 ··········· 107
  - 2. 帽部件 ··············· 110
- （二）鞋 ················· 110
  - 1. 品种和款式 ··········· 110
  - 2. 运动鞋 ··············· 112
  - 3. 靴子 ················· 113
  - 4. 凉鞋 ················· 114
  - 5. 拖鞋 ················· 115
  - 6. 鞋部件 ··············· 115

（三）袜 …………………… 116
（四）手套 ………………… 118
（五）领带，围巾 ………… 119
　1. 领带，领结 …………… 119
　2. 围巾 …………………… 120
（六）腰带，手袋 ………… 121
　1. 腰带 …………………… 121
　2. 手袋，手提包 ………… 122
（七）首饰，饰物，花边 … 123
　1. 首饰，珠宝 …………… 123
　2. 饰物，装饰品 ………… 125
　3. 花边 …………………… 127

六、服装设备和工具 ……… 128
（一）一般名称 …………… 128
（二）设计打样设备和
　　　工具 ………………… 129
（三）裁剪设备和工具 …… 130
（四）缝纫设备和工具 …… 131
（五）包缝，锁钉，绣花，
　　　针织等设备和工具 … 132

（六）缝纫机部件 ………… 135
（七）熨烫设备和工具 …… 135
　1. 熨烫机，整烫机，压
　　烫机 …………………… 135
　2. 部位熨烫机 …………… 136
　3. 熨斗 …………………… 137
　4. 黏合机，洗衣机，等 ……
　　　………………………… 137
（八）包装设备和工具 …… 137

七、其他 …………………… 138
（一）名词和词组 ………… 138
　1. 一般名称 ……………… 138
　2. 服装团体及组织 ……… 139
　3. 服装从业人员 ………… 140
　4. 文化，艺术与美 ……… 141
　5. 美容与妆扮 …………… 142
　6. 居家用品 ……………… 143
　7. 管理，质量，市场 …… 144
（二）动词和词组 ………… 147

英汉服装词汇

A	…………… 152		L	…………… 252
B	…………… 155		M	…………… 263
C	…………… 174		N	…………… 274
D	…………… 200		O	…………… 278
E	…………… 209		P	…………… 281
F	…………… 213		Q	…………… 299
G	…………… 227		R	…………… 300
H	…………… 234		S	…………… 305
I	…………… 241		T	…………… 344
J	…………… 244		U	…………… 366
K	…………… 250		V	…………… 368

W ················· 372　　Y ················· 381
X ················· 380　　Z ················· 384

附　录

一、颜色 ················· 388
　　1. 汉英顺序 ············ 388
　　　① 红色类 ··········· 388
　　　② 橙色类 ··········· 390
　　　③ 黄色类 ··········· 390
　　　④ 绿色类 ··········· 392
　　　⑤ 青色类 ··········· 394
　　　⑥ 蓝色类 ··········· 395
　　　⑦ 紫色类 ··········· 396
　　　⑧ 黑色类 ··········· 397
　　　⑨ 白色类 ··········· 398
　　　⑩ 灰色类 ··········· 398
　　　⑪ 棕色类及其他 ····· 400
　　2. 英汉顺序 ············ 402
二、常用缩写词 ············ 422
　　1. 制图缩写词 ·········· 422
　　2. 色相（hue）缩
　　　写词 ················ 425
　　3. 色调（tone）缩
　　　写词 ················ 425
　　4. 纤维（fibre）缩写词
　　　（或代码）············ 426
　　5. 贸易和其他有关缩
　　　写词 ················ 428
三、粤语、普通话、英语对照词
　　200 例 ················ 442
四、公、市、英制长度单位及其
　　换算 ·················· 450
　　1. 长度单位 ············ 450
　　　① 公制 ············· 450
　　　② 市制 ············· 450
　　　③ 英制 ············· 451
　　2. 长度换算 ············ 451
　　　① 公制折合市制或
　　　　英制 ·············· 451
　　　② 市制折合公制或
　　　　英制 ·············· 451
　　　③ 英制折合公制或
　　　　市制 ·············· 452
五、国际通用服装洗涤、熨烫
　　符号及含义 ············ 452
　　1. 基本符号及其附加
　　　符号 ················ 453
　　2. 洗涤符号 ············ 453
　　3. 氯漂符号 ············ 454
　　4. 熨烫符号 ············ 454
　　5. 干洗符号 ············ 454
　　6. 滚筒干燥符号（不适于
　　　真皮和毛皮制品）····· 455
六、部分词汇图释 ·········· 455
　　1. 上衣 ················ 455
　　2. 裤 ·················· 456
　　3. 裙 ·················· 457
　　4. 背心 ················ 458
　　5. 领子 ················ 458
　　6. 领口 ················ 460
　　7. 袖 ·················· 461
　　8. 口袋 ················ 462
　　9. 廓型 ················ 463
七、服装英语 100 例 ········ 464

汉英服装分类词汇
Chinese-English Classified Clothing Dictionary

一、服装成品
Finished Products of Clothes

(一) 一般名称 General Terms

服装 garments; apparel; clothing; wear; dress; costume; costumery; garb; wearing apparel; habit〔法〕

时装 fashion; the latest fashion; fashion clothing; fashionable dress; stylish garments; habit〔法〕

衣服 clothes; clothing; apparel; attire; gear; fig; toggery; togs; vestments; vesture; wearables; wearings; array; dud

衣裳 costume; clothing; clothes

衣着 garments; clothes; clothing

衣物 articles of clothing

装束 attire; garb; dress; costume; costumery; rig

穿着 dressing; apparel; fig

服饰 apparel and accessories; clothing; costume; dress; housings; garniture; trappings; finery; toilet; dress and personal adornment

中西服饰, 华洋服饰 Chinese and Western clothing

历代服饰 clothes and ornaments of every dynasty

千禧年服饰 millennium fashion and accessories

衣冠 clothes and hat; dress

帽类 headgear

鞋袜类 footwear; footgear

鞋帽 footwear and headgear

可穿戴智能设备 intelligent wearable equipment; smart wearable device

三维虚拟服装 3D virtual clothing; 3D virtual garments

现成服装 ready-to-wear; ready-made clothes; prêt-à-porter〔法〕; confection〔法〕

成衣 ready-to-wear; ready-made clothes; ready-made goods; tailoring, off the peg dress; off-the-peg; off-the-rack; reach-me-down; store clothes

高级成衣 prêt-à-porter〔法〕

一般成衣 confection〔法〕

定制衣服 custom-made clothes; customized apparel; tailored clothes; bespoke suit; tailleur〔法〕; couture〔法〕

高级定制女装 haute couture〔法〕

外衣 outerwear; outergarment; overclothes; overdress; overgarment; upper clothes; garment

内衣 underwear; underclothes; undergarment; underdress; inner wear

上装 top

下装 bottom

中国服装 China garments; Chinese wear

中式服装, 中装 Chinese-style costume; Chinese-style suit; Chinese-style clothes; Chinese costume; Chinese wear

西装 Western-style clothes; tailored suit; Western suit; suit of clothes; sack suit; suit

一般名称

西装型服装　tailored clothes
中西式服装　Chinese and Western-style clothes
民族服装　national costume; ethnical garments; ethnical costume
汉服　Hanfu; Han Chinese clothing; Han Chinese costume
唐装　Chinese-style costume; Tang-style suit; Tang suit
少数民族服装　costume of national minorities; minority costume; minority clothing
民俗服装　folk costume; ethnical costume; ethnical garments
经典服装　classic clothing
古典服装　classical costume; classic clothing
古代服装　ancient costume; historical costume
传统服装　traditional costume
现代服装　modern dress
春装　spring wear
夏装　summer wear
秋装　autumn wear
冬装　winter wear
春秋装　between season wear
男装　men's wear; menswear; men's clothing; men's suit
女装　women's wear; ladies' wear; womenswear; ladieswear; women's clothing; women's suit; dress; costume
童装　children's wear; kid's wear; childrenswear; kidswear; kids' clothing
男童装　boy's clothes; boyswear
女童装　girl's dress
中老年服装　middle and old aged people's wear
青少年服装　teen; teen's wear
少年服装　school age's wear; preteen's dress
学龄前儿童服装　preschooler's wear
幼儿服装　toddler's wear; infant's wear
娃娃装　babydoll dress
婴儿服装　baby's wear; babywear; infant's wear
婴儿衣服　swaddling clouts; swaddling clothes; long clothes
新生婴儿洗礼服　chrisom
新生婴儿服装　newborn clothing
新生婴儿全套衣物　layette
机织服装　woven garments; woven wear
针织服装　knitted garments; knitted wear; knit dress; jersey dress
系列服装　series garments; collection
组合服装　combined garments
全套服装　integral garments
配套服装，套装　coordinates
不配套服装　separates
基本型服装　basic dress
特种服装　special garments
特许服装　licensed clothing
名牌服装，品牌服装　brand clothing; designer's clothes; famous brand fashion; licensed clothing
生态服装，环保服装　ecological clothing; ecological fashion; Eco-Tex; environmental garments; organic garments; eco-friendly apparel
低碳服装　low carbon clothing; carbon neutral apparel
素食服装　vegan clothing
都市服装　urban wear
郊外服装　suburban wear
生活服装　casual wear

正式服装　official wear; dressy clothes
职业服装　professional garments; business wear; occupational clothing; career apparel
运动服装　sportswear; active wear; activewear; gym wear; playwear; warmup jumpers and pants
舞台服装，戏装　stage costume; theatrical costume; costume; costumery
军用服装　military garments
民用服装　civil garments
实用型服装　utility clothing
宗教服装　religious clothing
社交服　social wear; party dress; party gear
礼服　formal dress; formal wear; full dress
便服　leisure wear; casual wear; casuals; everyday clothes; informal dress; mufti; dishabille; déshabillé〔法〕; undress; negligee; civil clothes; plain clothes; civvies
军便服　undress uniform; service dress; undress
户外服　outdoor apparel; outerwear
室内服　indoor apparel
家居服　home wear; lounge wear; loungewear; family uniform; dressing gown; housecoat; house dress
主妇服　hostess gown
孕妇服　maternity dress; maternity wear
防辐射孕妇服　anti-radiation maternity dress
阿婆服　granny dress
化装舞会服　fancy dress
盛装　splendid attire; rich dress; full fig; fine array
华丽衣服，华服　gorgeous clothes; caparison; finery

休闲服　leisure wear; lounge wear; loungewear; resort wear; casual suit; casual attire; spare attire; sports wear
节日服　sunday clothes; best clothes
外出服　town wear
游戏服　play wear
旅游服　resort wear
郊游服　country wear
海滨服　beachwear; cover-up
海洋服　sea gear
猎装　hunting wear; hunting dress; habit de chasse〔法〕
寿服　graveclothes
丧服　mourning; sables; weeds
囚服　prisoner's wear
工艺服装　craft dress
钩编服装　crochet wear; crocheted dress
抽纱服装　drawnwork garments
绣花服装　embroidered dress
珠绣服装　embroidered dress with paillettes and beads; beaded embroidery garments
真丝手绘服装　hand painting silk clothing
贝珠服装　apparel with shell sets
扎染服装　tie-dyed clothing
彩装，彩衣　coloured clothes; motley
迷彩装　camouflage clothes; camo
嫁衣　trousseau
风景花样衣　robe-paysage〔法〕
街头服装　street wear; streetwear
奇装异服　fanciful costume; weird dress; bizarre dress
奇巧服装　tricky clothes; fancy dress
个性服装　individual clothing
应时服装　occasional wear
面试服装　interview clothing

一般名称

二手服装　second hand clothes
百衲衣，乞丐装　ragged clothes
破旧衣　rags
粗布衣　rough clothes
（宽松）布袋装　sack dress
（细长、直筒）铅笔装　pencil dress
四季流行组合装　year-round fashion
创新组合装　robe gag［法］
幻想型服装　fantasy clothing
一件头服装　one-piece clothes；one-piece
一次性服装　disposable dress
无构造服　unsuits
无性别服装　ambisextrou clothing；unisex fashion；unisex
高（中、低）档服装　high(mid, low)-grade clothing
值得投资的服装　investment clothes
命名服装，标名服装　designer's clothes
高科技服装　high-tech clothing；high-tech clothes
智能服装　intelligent garments；smart clothing；smartwear
智能医疗服装　intelligent medical apparel
保健服装　health care clothing
展示服装　show-off clothes
功能服装　functional clothing
防辐射服　anti-radiation suit
防静电服　antistatic clothing
防创伤服　anti-trauma clothing
抗菌防臭服　antibacterial deodorant clothing
吸湿排汗服　moisture wicking clothes
远红外发热服　far-infrared heating clothing
保暖服　thermal insulation clothes；thermowear；thermal(s)
空调服　air-conditioning clothes

光能服　solar-energy clothes
发光服　luminous clothes
储能闪光服　energy storage flash clothes
变色服　colour-changeable clothes
可食服　eatable clothes
香味服　fragrant clothes
防污服　soil-release clothes
防盗服　clothes against theft
抗皱服，免熨烫服　wrinkle-free clothing
适应性服装　adaptive clothing
锁扣衣　locking clothing
驱蚊衣　mosquito-repellent clothes
两面穿衣服　reversible；AB wear
可调节的衣服　adjustable clothes
无缝衣　closed clothes
无缝服装　seamless garments
轻便服装　light clothing
巴黎时装　Paris mode
高级定制时装　haute couture［法］；haute couture fashion
职业时装　professional fashion；occupational fashion
军服式时装　militant fashion
大众时装　street fashion；mass fashion
怀旧时装　retro-fashion
加一时装　plus one fashion
性感时装　sexy fashion
清爽时装　neat fashion
环保时装　ecological fashion
道德时装　ethical fashion
垃圾时装　fashion made of refuse
竹制时装　fashion made of bamboo
缀窝钉装　studded clothes
豹纹装　leopard print clothes
吉卜赛装　gypsy dress；flamenco dress；robe gitane［法］
冈多拉装　gandoura

（印度）莎丽装　saree; sari; sarrie
传统韩装　Hanbok
哥萨克装　Cossack
忽必烈汗装　Kublai Khan clothes
古代武士装　tabard
铠甲装　coat of mail; mail
圣经时代装　robe biblique〔法〕
特体服装　special measurement clothes
特大号型服装　outsize; oversize; plus clothing
特长服装　long-long dress
超长服装, 迷喜装　mixi
中长服装, 迷地装　midi
特短服装　shorty; shorties
短迷你服装　skimp; mini
超迷你服装, 迷哥装　micro
高腰装　high rise dress
紧身服装　tights; skintight; skin dress; clinging clothes; sheath; skinwear
塑形服装　shaped wear; shapewear; shaper
健美服　body suit
情侣装　lovers clothes; couple dress
性感装　sexy clothes
暴露装　bare look; décolletage〔法〕
上空装, 无上装　topless
中空装　midriff
下空装　bottomless
露上装　bare top
露背装　backless; bare-back dress
露肩装　off the shoulder dress
露脐装　bare midriff
露股装　micro
透明装　see-through clothing; see-through
棉布服装　cotton garments
环保棉布服装　ecological cotton garments
麻布服装　bast garments

麻棉服装　ramie/cotton garments
毛呢服装　woolen garments
丝绸服装　silk garments; silk wear
砂洗丝绸服装　sand-washed silk dress
化纤服装　chemical fibre garments
皮革服装　leather garments; leatherwear
裘皮服装　fur garments
羊皮服装　goatskin wear; sheepskin garments
猪皮服装　pigskin wear; pigskin garments
牛皮服装　cowhide garments; cattlehide wear
麂皮服装　chamois garments
鹿皮服装　deerskin
貂皮服装　marten dress
狐皮服装　fox dress
海豹皮服装　sealskin
羊毛皮服装　goat fur wear
兔毛皮服装　rabbit wear
人造革服装　artificial leather wear
人造毛皮服装　artificial fur wear
羽绒服装　down wear; down coat
防寒服装　warm wear; snow suit; thermal wear
牛仔服装　denim garments; jeanswear
毛巾服装　towelling clothes
弹力毛巾服装　stretch towelling sheath
手帕服装　handkerchief dress
漆布服装　varnished cloth garments
石棉服装　asbestos clothes
乳胶服装　latex clothing
泡沫塑料服装　foamed plastic
塑料服装　plastic garments
纸服装　paper dress
桑纸服装　mulberry paper clothes

(二) 款式名称 Terms of Style

1. 上衣 upper outer garment

①上衣 upper garment; jacket; top; top coat; coat; blouse; blouson

西装上衣 lounge jacket; tailored jacket; suit jacket
中式上衣 Chinese blouse
中西式上衣 Chinese and Western style blouse
单（双）襟式上衣 single (double)-breasted jacket
外上衣 over blouse
内上衣 tuck in blouse
套装上衣 suit jacket
中山装上衣 Zhongshan coat; Chinese style jacket; Sun-Yat-Sen style jacket
（立领）毛式上衣 Mao jacket
列宁装上衣 Lenin coat
尼赫鲁上衣 Nehru jacket; Nehru coat
青年服上衣 young men's coat
男式短上衣 mannish jacket
男便装上衣 sack coat; blazer coat
男紧身上衣 doublet; justaucorps〔法〕
短西装上衣 minisuit
猎装上衣 hunter jacket; shooting coat
军上衣 blouse; tunic
轻便型上衣，运动上衣 sports coat; sports-coat
运动上衣 blazer
田径上衣 track jacket
滑雪上衣 ski jacket
束带滑雪上衣 drawstring anorak
运动短上衣 stadium jumper
日式柔道上衣 vest de judo-ka〔法〕
制服上衣 tunique〔法〕
袋型上衣 sack
双色短上衣 two-tone jumper
三角形短上衣 type jumper
双排扣短上衣 pea coat; spencer croise〔法〕
双排扣截筒（短身）上衣 cropped peacoat
水手短上衣 jumper; monkey jacket; sailor blouse
水手粗呢上衣 pea coat; peacoat; pea jacket; caban〔法〕
双排扣水手上衣 reefer; reefing jacket; pea jacket
水兵服上衣 marinière〔法〕
水手领上衣 middy
高领上衣 turtleneck top
V形领口套头上衣 pull en V〔法〕
驳领上衣 habità revers〔法〕
防风厚上衣 donkey jacket
风帽上衣 hood jacket; anorak; hooded jack
束腰带上衣 belted jacket
长线型上衣 long line jacket
蓬松（紧身）上衣 voluminous (tight) top
巴尔干宽松上衣 Balkan blouse
套头宽上衣 pull blouson
（旧时）紧身皮制无袖短上衣 jerkin
宽大短上衣 talma
紧身短上衣 tunic; coatee
超短上衣 minicoat; minisack
救生衣 life jacket
防弹衣 body armor; bulletproof jacket
女上衣 frock; blouse; feminine frock; blazer

女宽松上衣　blousing blouse
女束腰上衣　blouson;tunic
女束腰宽松上衣　blouson tunic
女皱褶上衣　farmer blouse
女饰带上衣　sash blouse
女紧身短上衣　basque;coatee;polka
女胸围式紧身短上衣　bustier
女绣花短上衣　Zouave
和服袖短上衣　bolero kimono〔法〕
女防风厚上衣　donkey jacket
女便装上衣　house jacket
女军式上衣　military coat
男式女上衣　boyfriend blazer
白来罗女上衣　bolero
女无袖上衣　blousette
女宽松无袖上衣　cymar
女贝壳衫　shell
女卡拉扣上衣　caraco
女单排扣短上衣　shrug
女双排扣紧身上衣　reefer
女衬衫上衣　sur chemise;chemisier
女盒型短上衣　short boxy jacket;short square jacket
朱阿夫型女短上衣　Zouave jacket
袒胸衣　topless;decolletage;low dress;bare top
女露腰上衣　midriff;bare midriff
女露背上衣　bare-back blouse;bare back
春秋衫　spring and autumn coat
两用衫　sport coat;dual coat
罩衣　over jacket;outer jacket
罩衫　smock;blouse;blouson
长罩衫　smock frock;overshirt;overblouse
两片连肩袖罩衫　two-piece raglan smock
宽大罩衫　slop;overall
工作罩衫　smock;slop
防尘罩衫　dust coat;duster

画家工作罩衫,油漆工罩衫　painter smock
(19世纪俄国)军队罩衫(立领,双扣,双胸袋的套头衫)　gymnasterka
运动罩衫　sport blouse
女水手领罩衫　middy blouse
女无袖套领罩衫　jumper
孕妇罩衫　maternity blouse;maternity top
农夫罩衫　farmer blouse
束腰外衣　tunic
土耳其式束腰外衣　Turkish tunic
吊带束腰外衣　camisole tunic
V领束腰外衣　tunic with V-neckline
(土耳其)阿巴依外衣　Abai
(低腰)古鲁衫　gurus
(旧时)大襟衫　large front clothes
马褂　mandarin jacket;magua
黄马褂　yellow jacket
金缕玉衣　jade clothes sewn with gold thread
童罩衫　smock
童外衣　frock
婴幼短上衣　sacque
娃娃上衣　babydoll top

②夹克衫　jacket;blouson〔法〕
长夹克衫　maxijacket
宽松短夹克　mini sack
短外套式夹克　spencer jacket
衬衫式夹克　shirt jumper;shirt jacket
便装夹克　sport jacket;casual jacket
休闲夹克　casual jacket;lounge jacket;sack jacket
应时夹克　occasion jacket
胸式短夹克　wind jacket
截短夹克　cropped jacket;crop jacket
无领开襟夹克　cardigan jacket

无领夹克 collarless jacket
无里夹克 unlined jacket
全（半）里夹克 fully(half)-lined jacket
女式夹克 costume jacket
女紧身夹克 polka jacket
白来罗短夹克 bolero jacket
双面夹克 reversible jacket; double-faced jacket
替换夹克 sport jacket; odd jacket; spare jacket
奇特夹克 fancy jacket
（拼接）花式夹克 detailed jacket
可拆卸夹克 detachable jacket
箱型夹克 box jacket; beer jacket
圆形夹克 round jacket
（门襟）拉链夹克 zip-front jacket
（美国）西部夹克 Western jacket
爱德华夹克 Edwardian jacket
艾森豪威尔夹克 Eisenhower jacket
伊顿夹克 Eton jacket
香奈儿女装夹克 Chanel jacket
军装夹克 military jacket; battle jacket
军用紧身夹克 shell jacket
步兵夹克 doughboy jacket
堑壕夹克 trench jacket
迷彩夹克 camo jacket
射击用夹克 shooting jacket
防弹夹克 bulletproof jacket
飞行员夹克 flight jacket; aviator's jacket; bomber jacket; pilot jacket
水手夹克 pea jacket
法式水兵夹克 French sailor jacket
海员厚夹克 pilot jacket
运动夹克 track jacket; blazer
赛船夹克 regatta blazer
巡航夹克 cruising blazer
骑士夹克 rider's jacket; hacking jacket; riding jacket
猎装夹克 safari jacket; bush jacket; hunting jacket; vest de chasse〔法〕
伐木人夹克 lumber jacket; lumberjack
自加热式智能夹克 self-heating smart jacket
防寒夹克 anorak jacket
防雨夹克 thunder jacket; rain jacket; spray jacket
防风夹克 wind jacket; windcheater
毛毛雨夹克 drizzle jacket
登山夹克 alpine jacket
滑雪夹克 snowboard jacket; ski jacket
摩托夹克 motorcycle jacket; motor jacket
高尔夫短夹克 swing top
棒球夹克 baseball jacket
（男士）吸烟夹克 smoking jacket
二合一夹克，两用夹克 2 in 1(3 in 1) jacket
（披在夹克上面的）宽夹克外衣 surveste〔法〕
牛仔夹克 cowboy's jacket; denim jacket
斜纹布夹克 twill jacket
灯芯绒夹克 corduroy jacket; cord jacket
皮夹克 leather jacket
短皮夹克 dumb jacket
羊皮夹克 goatskin jacket
麂皮夹克 chamois jacket
仿麂皮夹克，绒面革夹克 suede jacket; suedette jacket
猪皮夹克 pigskin jacket
人造皮革夹克 leatheret jacket; faux leather jacket
弹力绒面革夹克 stretch suedette jacket
涤/棉涂层布夹克 T/C poplin coated jacket
涤/棉水洗布夹克 T/C washer jacket

桃皮绒夹克　micro-fibre jacket
毛呢夹克　woolen jacket
（拉毛绒）津布夹克　chimp jacket
填棉夹克　quilted jacket; padded jacket; puffer jacket
牛仔填棉夹克　denim padded jacket
锦纶/PVC 填棉夹克　nylon/PVC padded jacket
填棉刺绣夹克　quilted and embroidered jacket
羽绒夹克　down jacket
毛皮夹克　fur jacket
毛领夹克　fur collar jacket
毛皮衬里夹克　fur-lined jacket
仿羔皮里夹克　sherpa-lined jacket
丝绸夹克　silk jacket
锦纶（涤纶）/PVC 涂层夹克　nylon (polyester)/PVC coated jacket

③衬衫　shirt; blouse

男衬衫　shirt; chemise〔法〕
男用外衬衫　liquette〔法〕
美式男用衬衫　chemise American〔法〕
美国士官衬衫　CPO (Chief Petty Officer) shirt
（高档）法式衬衫　French shirt
礼服衬衫　dress shirt; formal shirt; tuxedo shirt
礼仪衬衫　ceremonial shirt; formal shirt
制服衬衫　formal shirt; uniform shirt
商务衬衫　business shirt
智能衬衫　smart shirt
外套衬衫　coat shirt
夹克式衬衫　jacket shirt
袋形衬衫　baggy shirt
两用衬衫　flexible shirt
办公衬衫　business shirt

工作衬衫　work shirt; business shirt
学生衬衫　school shirt
大学生衬衫　college shirt
飞行员衬衫　pilot shirt; aviator shirt
牧师衬衫　cleric shirt
海盗衬衫　pirate shirt
西式衬衫　Western shirt
军式衬衫　army shirt
休闲衬衫　leisure shirt; casual shirt; sports shirt
牛仔衬衫　cowboy shirt
夏威夷衬衫　Hawaii shirt; Aloha shirt
加拿大衬衫　Canadian shirt
阿拉伯衬衫　Arabian shirt
阿斯科衬衫　Ascot shirt
墨西哥婚礼男衬衫　Mexican wedding shirt
（伊斯兰）长宽衬衫　kamis
（印度）宽松衬衫　banian
（印度）无领长衬衫　kurtah shirt; kurtah
亨利衬衫　Henley shirt
阻特衬衫　zoot shirt
常春藤风格衬衫　Ivy shirt
袋鼠衬衫　kangaroo shirt
女衬衫　blouse; shimmy
男式女衬衫　shirt blouse; shirt waist
女束腰衬衫　tunic blouse
女露腰衬衫　midriff blouse
女卷裹衬衫　wrap blouse
女长下摆衬衫　over blouse; under blouse
女前下摆打结衬衫　tie front blouse
女打结领衬衫　tie-bow blouse
女小披肩领衬衫　capelet blouse
女时髦领衬衫　confection
女无领衬衫　chemise; collarless blouse
女无袖衬衫　sleeveless blouse; chemise
女七分袖衬衫　3/4 sleeve shirt

衬衫

吉卜赛女衫　gypsy blouse
蝙蝠衫　batwing blouse
嘻哈衬衫　Hip Hop blouse
女童衫　girl's blouse
宽大女衬衫　maxi-chemise
套头宽大女衬衫　marinière〔法〕
孕妇衬衫　maternity shirt
花样衬衫　patterned shirt(blouse)
印花衬衫　printed blouse(shirt)
绣花衬衫　embroidered blouse
抽绣衬衫　drawnwork blouse
小提花衬衫　dobby shirt
通花衬衫　crocheted blouse
网眼衬衫　mesh shirt
珠饰衬衫　beaded blouse
胸饰衬衫　bosom blouse; jabot blouse
胸褶衬衫　pleated bosom blouse(shirt)
皱褶衬衫　ruffle blouse
泡泡袖衬衫　pull sleeve blouse
硬胸衬衫　boiled shirt
装领衬衫　collar attached shirt
翻领衬衫　open-neck shirt
套头衬衫　pullover shirt
（错色）双层衬衫　twofer shirt
燕尾衬衫　swallowtail shirt
长身线形衬衫　longline shirt
细长衬衫　tapered shirt
合体衬衫　fitted shirt
紧身衬衫　body shirt; fitted shirt
弹力紧身衬衫　stretch shirt; lycra shirt
半紧身衬衫　semi-fitted shirt
无领衬衫　collarless shirt
无袖衬衫　sleeveless shirt
饰狭窄领带的无领衬衫　neckband shirt
饰金银丝的细条衬衫　lurex pinstripe shirt
中式小褂　Chinese-style shirt

长（短）袖衬衫　long(short) sleeved shirt
深（浅）色衬衫　dark(light)-coloured shirt
条（格）子衬衫　stripe(check)shirt
平（圆）下摆衬衫　straight(tail) bottomed shirt
花点衬衫　coloured spots shirt
免烫衬衫　memory shirt
填棉衬衫　padded shirt; quilted shirt; lumber shirt; quilted lumber jack shirt
全棉衬衫　cotton shirt
牛仔布衬衫　denim shirt
牛津布衬衫　oxford shirt
青年布衬衫　chambray shirt
灯芯绒衬衫　corduroy shirt
经条灯芯绒衬衫　bedford cord shirt
绒布衬衫　flannel shirt
格绒衬衫　brushed check shirt; flannelette check shirt
植绒布衬衫　flocking shirt
泡泡纱衬衫　seersucker shirt
亚麻衬衫　linen shirt
涤/棉衬衫　polyester cotton shirt; T/C shirt
麻/棉衬衫　ramie/cotton shirt; linen/cotton shirt
人造棉衬衫　rayon shirt
绒面革衬衫　suedette shirt
丝绸衬衫　silk shirt
砂洗丝绸衬衫　sand-washed silk shirt (blouse)
柔姿纱衬衫　polyester georgette blouse
巴厘纱衬衫　voile blouse
毛巾衫　towelling blouse
纸衬衫　paper shirt

④背心，马甲 vest; waistcoat
礼服背心 dress vest; evening vest; dress waistcoat; formal vest
（礼服用）无背背心 backless vest
西服背心 waistcoat; vest; weskit
套装背心 suit vest
夹克式背心 vest in jacket style
变化背心 fancy vest
两面穿背心 reversible vest
多功能背心 multi-functional vest
智能背心 smart vest
远红外加热背心 far-infrared heating vest
防辐射孕妇背心 anti-radiation maternity vest
替换背心 odd vest
户外背心 outer vest
休闲背心 leisure vest
运动背心 sport vest; gym vest
钓鱼背心 angler's vest; fishing vest
牛仔背心 cowboy vest
邮递员背心 postboy waistcoat; postboy vest
摄影师背心 photographer's vest
警察背心 police vest
作（战）训（练）背心 field training vest
救生背心 life vest; survival vest
战术背心 tactical vest
防弹背心 armoured vest; bullet proof vest
反光安全背心 reflective safety vest
单（双）襟背心 single(double)-breasted vest
两（三）扣背心 two (three)-button waistcoat
拉链（按扣）背心 zipper (snap) front vest
长身线形背心 longline vest
燕尾背心 swallowtail vest
套头背心 pullover vest
带帽背心 vest with hood
有领背心 collar vest; lapeled vest
紧身背心 doublet; weskit
卡马背心 cummervest
女用宽大背心 bodice
女用三角背心 halter
女式短马甲 gilet〔法〕
女紧身短马甲 hug-me-tight
填棉背心 cotton-padded vest; padded vest; puff vest; bodywarmer
羽绒背心 down vest
（泡泡面）蓬松背心 puff vest
皮背心 leather vest
反绒皮（猄皮）背心 suede vest
绒面猪皮背心 pig suede vest
二层猪皮背心 pig split vest
鹿皮背心 buckskin vest
仿鹿皮背心 suede vest
毛皮背心 fur vest
羊剪绒背心 shorn sheepskin vest
毛羊皮衬里背心 shearling vest
牛仔布背心 denim vest
丝绒背心 velvet vest
绒布背心 flannelet vest
摇粒绒背心 polar fleece vest
网眼背心 mesh vest

⑤袄 Chinese-style coat
单衣 unlined dress
夹衣 lined dress
棉衣 wadded clothes; cotton-padded coat
中式棉袄 Chinese-style padded coat
紧身棉袄 tight quilted jacket

长裤

丝绵袄　floss-padded clothes
毛皮袄　fur-lined jacket
羊皮袄　zamarra
毛皮衣　fur clothes

2. 裤子　trousers; pants; slacks; pegs; pantaloons〔美〕; pantalon〔法〕; jupon〔法〕

①长裤　trousers; pants; slacks; longs

西装长裤　tailored trousers; tailored pants; trousers
英式西裤　British trousers
高尔夫西裤　golf jupon
无腰带西裤　trousers without belt
半长西裤　breeches
礼服裤　dress trousers; dress pants; formal trousers; dress slacks; court breeches
礼服条纹西裤　striped trousers
套装裤　suit pants
（大腰头）中裤　Chinese-style pants; Chinese-style slacks
外裤，罩裤　over pants
直筒裤　straight trousers; stem-pipe pants; cigarette pants; straight-leg slacks
铅笔裤　pencil trousers
烟囱裤　stovepipe pants
喇叭裤　flare trousers; bell-bottom pants; bell bottoms; flares; bells
低腰喇叭裤　low-waisted bells
宽裤脚九分裤　7/8 pants with wide leg
窄脚裤　tapered trousers; tapered pants; hose-bottom pants; shim slacks
细窄裤　pantalon mince〔法〕
高腰细脚裤　high-waist tapered trousers
拉链细脚裤　zip skinny leg pants
卷脚口裤　pleat trousers; roll-leg pants;

cuffed pants; pantalonà revers〔法〕
平脚口裤　pleatless trousers; cuffless pants; cuffless
紧身裤　body fitting trousers; skinny pants; leggings; pantalon collant〔法〕
全棉弹力紧身裤　cotton lycra leggings
缀亮片紧身裤　sequined leggings
低腰裤　low-waist pants; hipsters
低腰紧身长裤　hip-huggers
低裆裤　low-cut crotch pants
挂臀裤　hip hangers; hipsters
休闲裤　leisure pants; casual pants; casual trousers
大贴袋休闲裤　cargo pants
懒散裤　slouchy pants
宽松长裤，便裤　slacks; easy care trousers; bags〔英〕
无腰带便裤　beltless slacks
宽松舒适裤　easy pants
（南亚）宽松女长裤　shalwar
（板状）宽大裤　boardies
宽腰头裤　wide waistband pants
腰头前打褶裥裤　pleat front pants
腰头前无褶裥裤　flat front pants
松紧腰头裤　pull on pants
大脚口宽松女裤　palazzo pants
灯笼裤　knickerbockers; knickers; bloomers; pantalets; golf pants; plus-fours
袋型裤　full pants; baggy pants; Oxford bag trousers; Oxford bags
半袋型裤　semi-baggy pants
降落伞裤　parachute pants
陀螺型裤　peg-top pants; peg tops
木钉裤　pegged trousers; pegged leggers
楔型裤　wedge trousers
贴身型裤　tunic trousers
（细窄贴脚）摇滚裤　rock pants

截筒裤　cut-off pants; chop pants
束脚裤　ankle tied pants; ankle banded pants
佐阿夫女束脚裤　Zouave
女式牧裤　gauchos
裙裤　culottes; culotte pants; jube-culotte〔法〕
卷裹裤　wrap pants
闺阁裤，后宫裤　harem trousers; harem pants
宫殿裤　palazzo pants
街市裤　city pants
卡普里裤　capri pant; Capris
猎装裤　hunting breeches
登山裤　climbing breeches
滑雪裤　ski pants; ski jupon; snowboard pants
滑雪板裤　snowboard pants
(美国) 西部风格裤　Western pants
常春藤风格裤　Ivy slacks
印度风格裤　pantalon Indian〔法〕
男式风味裤　pantalon masculin〔法〕
祖父裤　pantalon de grandpére〔法〕
君王裤　pantalon de maharajah〔法〕
老板裤　boss pants
军裤　military pants; army pants
迷彩裤　camouflage pants; camo pants
水兵裤　sailor jupon; sailor pants
海盗裤　pirate pants; corsair pants
小丑裤　clown pantaloons
舞蹈裤　dance pants
(配) 靴裤　bootleg pants; bootscut pants
马裤　riding breeches; show breeches; breeches; jodhpur pants; jodhpurs; pantaloons; culotte de cheval〔法〕
鹿皮马裤　buckskins
裙式马裤　jupe pantaloons

(赛马) 骑师裤　jockey pants
斗牛裤　toreador pants
吊带裤　suspender pants
背带裤　suspender pants; bib pants
连体长裤　jumpsuit
工作裤　working trousers; houseboy pants; corduroys; dungarees
办公裤　business pants
背带工装裤　overalls; salopettes
沙罗佩工装裤　salopette pants; salopettes
木工裤　carpenter pants
货船工作裤　cargo pants
画家工作裤，油漆工裤　painter pants
瑜伽裤　yoga pants
远红外棉裤　far-infrared cotton pants
按摩裤　massotherapy pants
空调裤　turned air pants
防皱裤　wrinkle-free pants; wrinkle resistant trousers
起绒裤　brushed pants
雨裤　rain trousers
替换裤　spare trousers
三用可变裤　3-way convertible pants
裤腿可脱卸裤　zip-off leg pants
多袋裤　multi-pocket pants
快干裤　quick dry pants
八分之七裤，九分裤　seven eighths pants; 7/8 length pants; ankle-length trousers
四分之三裤，七分裤　3/4 length pants
半长裤　clamdiggers
齐膝裤　knee pants; knee-hi
女长裤　women's slacks; ladies' slacks
孕妇裤　maternity pants
童裤　children pants
(女) 童吊带裤　girl's jumpsuit
(男) 童牛仔裤　boy's jeans
童套裤　leggings

童开裆裤 open-seat pants

② **牛仔裤** jeans; jean pants; Levis; cowboy pants; denim jeans; denim trousers; denim pants; denims; gaucho pants; dungarees

正装牛仔裤 formal jeans
古典牛仔裤 vintage jeans
五袋款牛仔裤 five-pocket jeans
石磨水洗牛仔裤 stone-washed jeans
黑洗水牛仔裤 black wash jeans
喷砂牛仔裤 sandblasted denim jeans
漂白牛仔裤 bleached jeans
彩色牛仔裤 coloured jeans
双色牛仔裤 two tone jeans
褪色牛仔裤 fade-out jeans
马裤式牛仔裤 jodhpurs-style jeans
配靴穿牛仔裤 bootleg jeans; bootscut jeans
压皱牛仔裤 goffered jeans
三褶饰牛仔裤 triple pintuck front jeans
皱折效果牛仔裤 crinkle effect jeans
古旧效果牛仔裤 vintage jeans
时款牛仔裤 fashion jeans
窝钉牛仔裤 studded jeans
豹纹牛仔裤 leopard print jeans
标签牛仔裤 label jeans
激光雕刻牛仔裤 laser engraving jeans
烂花牛仔裤 burnt-out jeans
绣花牛仔裤 embroidered jeans
贴绣牛仔裤 appliqué jeans; patchwork jeans
拷花牛仔裤 embossed jeans
印花牛仔裤 printed jeans
涂层牛仔裤 coated jeans
闪光牛仔裤 glitter denim jeans
工作牛仔裤 fatigue jeans
吊带牛仔裤 overall jeans

截筒牛仔裤 cropped jeans; crop jeans
卷边脚口牛仔裤 cuffed jeans; turn up cuff jeans
弹力牛仔裤 stretch denim pants; lycra jeans
弹力牛仔靴裤 stretch bootleg jeans
弹力紧身牛仔裤 stretch skinny jeans
紧身牛仔裤 skinny jeans; tight jeans
卡普里（紧身）牛仔裤 capri jeans
轻薄牛仔裤 light jeans
厚重牛仔裤 heavy weight jeans
硬挺牛仔裤 rigid jeans
再生牛仔裤 recycled jeans
软身牛仔裤 soft jeans
仿麂皮牛仔裤 suede jeans
灯芯绒牛仔裤 cord jeans
宽腰头牛仔裤 wide waistband jeans
搭襻腰头牛仔裤 tab waistband jeans
无腰头牛仔裤 waistless jeans
低腰牛仔裤 hipster jeans
低裆牛仔裤 low-cut crotch jeans
（低裆）松垮牛仔裤 loose jeans
喇叭牛仔裤 flare jeans
微喇牛仔裤 little flared-leg jeans; bootleg jeans; boots cut jeans
拉链（纽扣）门襟牛仔裤 zipper (button) fly jeans
宽大（紧窄）脚口牛仔裤 wide (skinny)-leg jeans
宽适型牛仔裤 easyfit jeans
宽松牛仔裤 oversize jeans
实用型牛仔裤 utility jeans
直筒牛仔裤 straight-leg jeans; cigarette jeans
时尚破烂型牛仔裤 ripped to fashionable shreds jeans
摇滚爵士牛仔裤 jitterbug

宽裤腰袋状牛仔裤　rindy
不像牛仔裤的牛仔裤　non jeans
舒适修身牛仔裤　jeggings
棉斜纹裤　cotton twill trousers; moleskins
丝光卡其布裤　drill chinos; chinos
灯芯绒裤　corduroys; corduroy pants
帆布裤，珠帆裤　canvas pants
防水帆布裤　tin pants
麻/棉裤　ramie/cotton pants
人棉裤　rayon trousers
涤/棉裤　truean trousers; T/C pants
涤纶裤　polyester trousers
涤纶桃皮绒裤　polyester microfibre pants
锦纶裤　nylon pants
腈纶裤　acrylic trousers
丝绸裤　silk trousers
䌷丝绸裤　silk noil pants
夹裤　lined trousers
棉裤　cotton wadded trousers; cotton padded pants
羽绒裤　down wadded trousers
长毛绒裤　plush trousers; plushes
皮裤　leather trousers
人造毛皮裤　fur cloth trousers
毛呢裤　woolen trousers

③ 短裤　shorts; breeches; breeks

西装短裤　tailored shorts; shorts
便装短裤　casual shorts
工作短裤　work shorts
百慕大短裤　Bermuda shorts
牙买加短裤　Jamaica shorts
(美国)西部短裤　Western shorts
沙滩短裤　beach shorts; board shorts
运动短裤　trunks; sports shorts; gym shorts
田径短裤　Jamaica shorts; track shorts
赛跑短裤　running pants
篮球短裤　basketball shorts
网球短裤　tennis shorts
橄榄球短裤　rugger shorts
(女)骑车短裤　pedal pushers; bike shorts
护裆骑车短裤　cycling bib shorts
拳击短裤　boxer shorts; boxing trunks; boxers
登山短裤　climbing shorts
溜冰短裤　skate shorts
游泳短裤　swim pants; swim shorts; swimming trunks; slips
冲浪短裤　jams; surfer's shorts; board shorts; boardshorts; boardies
快干冲浪短裤　quickdry boardshorts
印花冲浪短裤　floral boardshorts
甲板短裤　deck pants
散步短裤　walking shorts
休闲短裤　leisure shorts; casual shorts
大贴袋短裤　cargo pocket shorts; cargo shorts
松紧腰头短裤　pull-on shorts
高腰短裤　high-rise shorts
都市短裤　city shorts
热裤　hot pants; hotpants; city shorts
(比热裤稍长的)暖裤　warm pants
宽松短裤　loose cut shorts; baggys
(拉绳)束带短裤　drawstring shorts
背带短裤　bib shorts
连身短裤　shortalls; jumpshorts; rompers
宽脚口短裤　flared shorts
平脚短裤　boxers
卷脚口短裤　cuffed shorts
灯笼短裤　bloomer shorts; knickerbockers
紧身短裤　culottes anglaise[法]
剪截式短裤　cut-offs

毛边脚口的牛仔短裤　cut-off jeans
破烂型卷脚（牛仔）短裤　ripped cuffed shorts
长短裤　maxi-shorts;pantacourt〔法〕
齐膝短裤　knee pants;knee breeches
膝上短裤　above-the-knee pants
超短裤　minipants; minishorts; short shorts;hot pants;shorty;shortie
全棉斜纹短裤　cotton twill shorts
牛仔短裤　denim shorts
烂花牛仔短裤　burnt-out denim shorts
涂料染色帆布短裤　pigment dyed canvas shorts
锦纶塔斯纶短裤　nylon taslan shorts
麻/棉短裤　ramie/cotton shorts
丝绸短裤　silk shorts

3. 裙　skirt

①裙　skirt;jupe〔法〕

标准样裙　standard skirt
西服裙　tailored skirt;costume skirt;skirt
礼服裙　tuxedo skirt;formal skirt
A 型裙　A-line skirt
喇叭裙　flare skirt; trumpet skirt; bell-type skirt
钟型裙　bell skirt;bell-type skirt
梯形短裙　short trapezium skirt
伞裙　umbrella skirt;parachute skirt
圆锥形裙　cone skirt
蓬松裙　bouffant skirt
气球裙　balloon skirt
泡泡裙　bubble skirt
花瓣裙　petal skirt;jupe petale〔法〕
圆裙　circular skirt
半圆裙　semi-flared skirt; semi-circular skirt;half-circle skirt

直筒裙　straight skirt
筒形裙　barrel skirt;tube skirt
笔杆裙　pencil skirt
沙漏形裙　hourglass-shape skirt
斜裙　bias skirt;jupe en bias〔法〕
斜裁裙　bias-cut skirt
褶裥裙　pleated skirt
永久褶裥裙　permanently pleated skirt
箱褶裙　box pleated skirt
阴裥裙　inverted pleat skirt
倒褶裙　kick-pleat skirt;inverted skirt
百褶裙　accordion-pleat skirt; all-around pleated skirt;pleated skirt
碎褶裙　gathered skirt
细褶裙　square skirt
片褶裙　pleat-gore skirt
边褶裙　side-pleat skirt
襞褶裙　hip pleated skirt; back pleated skirt;jupe plissée〔法〕
皱褶裙　ruffle skirt;jupe fronce〔法〕
叠褶裙　tucked skirt
（苏格兰式男用）褶叠短裙　kilt skirt; kilt
笼基（缅甸传统筒裙）　longyi; lungyi; lungee;lungi
超短叠裙　minikilt
包叠裙，围裹裙　wrap-around skirt; wrapped skirt;wrap around;wrapover; wrapround
腰布卷身裙　pagne〔法〕
螺旋裙，旋涡裙　spiral skirt;swirl skirt
陀螺裙　pegged skirt;peg-top skirt
蜗牛裙　escargot skirt
双层裙　double skirt
多层波浪裙　flare tiers skirt
波边下摆裙　fluted hem skirt
宝塔裙　tiered skirt

伊斯兰教托钵僧裙	dervish skirt
旦多尔裙	dirndl skirt
塔尼克裙	tunic skirt
吉卜赛裙	gypsy skirt
波希米亚裙	Bohemian skirt
宽摆裙	flare skirt; expansion skirt
（高腰大摆）摇滚裙	rock skirt
摇曳裙	swing skirt
展翅裙	robe ailee〔法〕
翘摆裙	flippy skirt
窄摆裙	hobble skirt; slim skirt
旗袍裙	cheongsam skirt; Qipao style skirt; slim skirt
蹒跚裙	hobble skirt
紧身裙	tight skirt; sheath skirt
合身裙	fitted skirt
围裙式裙	apron skirt
片裙	gored skirt
一片裙，单片裙	one-piece skirt
两（四、六、八）幅裙	two (four, six, eight)-gore skirt
连接裙	joint skirt
掩块裙，拼幅裙	panel skirt
腰骨裙	hip-bone skirt
拼腰裙，剪接裙	yoke skirt
垂饰裙	draped skirt
鱼尾裙	fish-tail skirt
吊带裙，背带裙	suspender skirt; braces skirt; suspender jumper
花边裙	lace skirt; skirt with laces
荷叶边裙	flounce skirt
开门襟裙	button through skirt; button-down skirt
女骑装裙	astride overskirt; riding skirt
工装裙	salopette skirt
农妇裙	peasant skirt
网球裙	tennis skirt
芭蕾裙	ballet skirt; tutu skirt; tutu
纱笼裙	sarong skirt; sarong
闺阁裙	harem skirt
印第安女裙	squaw skirt
高腰裙	high-waist skirt; empire skirt
低腰裙	torso skirt; hip bone skirt; hipster skirt; hip hangers
低吊裙	low-slung skirt
长（短）裙	long (short) skirt
齐地长裙	full length skirt
齐膝裙	knee skirt
迷喜裙，超长裙	maxiskirt; maxi
迷地裙，半长裙	midiskirt; mid length skirt; midi
迷你裙，超短裙	miniskirt; mini
迷哥裙，超迷你裙	micro skirt; microskirt; micro-mini skirt; micromini; mini-mini skirt
套裙，罩裙	overskirt; slip-on; slip
裤裙	culotte skirt; culottes; pantskirt; divided skirt; separate skirt; jupe-culotte〔法〕; robe culotte〔法〕
（旧时）撑架裙	hoopskirt; crinoline
可拆卸裙	detachable skirt
不对称裙	asymmetrical skirt
大贴袋裙	cargo skirt
扎染裙	tie-dyed skirt
图案花裙	figured skirt
绣花裙	embroidered skirt
印花布裙	print skirt
牛仔裙	denim skirt
毛巾裙	towelling skirt
平绒裙	panne skirt
丝绸裙	silk skirt
绵绸裙	silk noil skirt
填棉裙	quilted skirt
毛呢裙	woolen skirt; stuff skirt

粗毛花呢裙　tweed skirt
皮裙　leather skirt
草裙　grass skirt

②连衣裙　one-piece dress; dress

A 型连衣裙　A-line dress
紧身连衣裙　sheath dress; body dress
公主线紧腰连衣裙　princess line dress
衬衫连衣裙　shirt dress; shirtdress; shirt-waist dress; shirtwaister
套衫连衣裙　jumper dress
毛衫连衣裙　sweater-top dress
猎装式连衣裙　safari dress
披肩连衣裙　dress with cape
有裙裤的连衣裙　pant-dress
泳装式连衣裙　swim dress
溜冰连衣裙　skate dress
背心连衣裙　jumper dress; chemise skirt
长（中，短）袖连衣裙　long (half, short) sleeve dress
无袖连衣裙　chemise frock; jumper skirt; jumper dress
无袖紧身连衣裙　sleeveless slim-fit dress
小圆领连衣裙　Peter Pan collar dress
U 形领口连衣裙　U-neck dress
V 形叠领口连衣裙　crossover V-neck dress
无领短袖连衣裙　collarless short-sleeved dress
露背连衣裙　bare-back dress; low cut back dress
超短连衣裙　minidress
太阳连衣裙　sun dress
吊带连衣裙　shoestring strap dress
花边吊带连衣裙　camisole dress
蝴蝶结带连衣裙　bow-tie dress
挂脖式露背连衣裙　halter dress
单肩连衣裙　one shoulder dress
女学生裙　jumper skirt; school skirt; school jumper; school dress
"吉吉"少女连衣裙　Gigi dress
（香奈儿）小黑裙　little black dress
彼得·汤姆森（水手领）连衣裙　Peter Thomson dress
弗拉门戈舞连衣裙　flamenco dress
波希米亚连衣裙　Bohemian dress
村姑连衣裙　peasant dress
直筒式连衣裙　shift dress; tube dress
大摆连衣裙　flared bottom dress
蓬松连衣裙　tent dress
泡泡连衣裙　bubble dress
气球连衣裙　balloon dress
裹身连衣裙　wrap dress; robe housse〔法〕
郁金香裹身连衣裙　tulip dress
派内尔（镶饰）连衣裙　panel dress
牛仔缩褶连衣裙　denim smock dress
珠饰连衣裙　beaded dress
三色连衣裙　tri-tone dress
双色连衣裙　bicoloured dress; two-tone dress
（错色）双层连衣裙　twofer dress
可变化的连衣裙　shift dress
单肩带连衣裙　one shoulder dress
不对称连衣裙　asymmetric dress
不规则连衣裙　irregular dress
豹纹连衣裙　leopard print dress
亮片连衣裙　sequined dress
和服袖连衣裙　robe kimono〔法〕
圆领套裙　round-neckline slip-on
挂肩套裙　built-up shoulder slip-on
旗袍　Qipao; chi-pao; Chinese dress; mandarin dress; cheongsam; cheungsam; Chinese gown; Chinese-style frock
珠绣旗袍　embroidered Qipao with paillette and beads

软缎绣花旗袍　satin embroidered cheongsam
红色婚礼旗袍　red bridal dress
婆婆裙　granny dress
新娘婚纱装　bridal gown; bridalwear; wedding gown
结婚礼服拖裙　train
拖地裙装　maxilength
超长连衣裙　maxi dress
连衫围裙装　apron dress; pinafore dress
童裙衫　girl's frock
童围裙　apron dress; slip; pinafore; pinny

4. 套装，制服，礼服　suit, uniform & formal dress

①套装　suit; outfit; tailleur〔法〕

正式套装　classic suit
正统套装　suit-suits
礼服套装　dress suit; tuxido suit
商务套装　business suit
男套装　men's suit; tailleur masculin〔法〕
女套装　lady's suit; ensemble; costume; separate dress
男式常服　lounge suit; business suit
女式常服　women's suit
春秋套装　spring and autumn suit
夏季套装　summer outfit, summer suit; light suit
全年通用套装　year-round suit
组合套装　package suit; setup suit
混合套装　mixed suit
调和套装　coordinate suit; coordinated set
不同布料组合的调和套装　duos; unmatched suit
上下套装　top and bottom set
上下相连的套装　combinaison〔法〕

（上衣、背心、裤子不同布料组合）分开套装　split suit
变化套装　non suit
两件套装　two-piece suit; two-piece set; twin set
三件套装　three-piece suit; ensemble; trios
三件套裙　three-piece skirt set
四件套装　four-piece suit
西装式套装　tailored suit
三件套西装　vested suit
外套套装　coat suit
风衣套装　windbreaker suit
罩衫套装　smock suit
短外衣套装　jumper suit
束腰外衣套装　tunic suit
短夹克套装　jack suit; blouson suit
衬衫套装　shirt suit
衬衫T恤套装　shirt and tee set
背心套装　vested suit
背心热裤套装　panties-slip
连衣裙套装　one-piece dress suit
西服裙套装　tailored skirt suit
针织领袖裙套装　skirt suit with knitted collar and sleeves
褶裥短裙套装　peplum suit
裙裤套装　culotte suit; pantdress
裤套装　trouser suit; pant suit; pantalon suit
连衣裤套装　jumpsuit
紧身连衣裤套装　cat suit; catsuit; body suit
连体装　siamese clothes; jumpsuit; bodysuit; catsuit
外套装　oversuit
奥黛（越南传统女套装）　Ao Dai
旁遮普服（印度传统女套装）　salwar ka-

meez
布卡罩袍装 burka;burqa;burqua;burkha;bourkha
厚套装 beefy suit
宽肥套装 zoo-suit
雨衣套装 rain suit
牛仔套装 cowboy suit;denim suit;denim tops and bottoms
男（女）童套装 boy's(girl's)suit;boy's(girl's)set
水手童套装 boy's sailsuit
游戏童套装 playsuit
童连衫裤 jumpers;pantywaist;rompers;creepers
童填棉连衫裤 ski rompers
婴幼连衫裤 infant's rompers;grosuit;easysuit;wondersuit;bodysuit;unitard;coveralls
新生婴儿连衫裤 newborn rompers
婴儿背心连裤装 singletsuit
婴儿舒适连身装 snugglesuit
婴儿连袜衫裤 footies
婴儿套装 baby's suit;infant's suit;bodysuit
婴儿填棉套装 baby's padded set
婴儿成长套装 infant's grow suit

② 西装 Western-style clothes;Western dress;suit

女西装,妇女服装 women's suit
正统男西装 classic suit;suit-suits
定制西装 bespoke suit
普通西装 lounge suit;sack suit
休闲西装 leisure suit;casual suit
古典西装 classic model suit
意大利欧式西装 Italian continental suit
英式西装 British-style suit
两（三）件套西装 two(three)-piece suit
单（双）襟西装 single(double)-breasted suit
单纽西装 one-button suit
窄驳领西装 small lapel suit
宽驳领西装 wide lapel suit
尖驳领西装 L-peak lapel suit
Y型西装 Y-shaped suit
防雨西装 rainproof suit
吸烟（套）装 Le smoking〔法〕
欧洲大陆型套装 continental suit
尼赫鲁套装 Nehru suit
卓别林套装 Charlie Chaplin suit
斯宾塞套装 Spencer suit
诺福克套装 Norfolk suit
百老汇套装 tailleur Broadway〔法〕
尼卡套装 knicker suit
爵士套装 jazz suit
阻特套装 zoot suit
香奈儿套装 Chanel suit
白来罗套装 belero suit
卡帝冈套装 cardigan suit
战斗套装 battle suit
坦克套装 tank suit
伪装迷彩套装 camouflage pattern suit;camo suit
变色龙套装 chameleon suit
跳跃套装 jump suit;cat suit
短裤型跳跃套装 combinaison short〔法〕;combishort
贴身套装 fitting suit
拉链套装 zip suit
日常套装 afternoon suit
休闲套装 leisure suit;casual suit
简便套装 easy suit;easysuit;slack suit;un-constructed suit;unsuits
中式套装 Chinese style suit

中山服　Chinese tunic suit; Mao suit; Sun Yat sen's uniform
民初装　Suit at the early Republic of China
客家装　Hakka suit
妈祖装　maid's suit; amah's suit
衫裤装　samfoo; samfu

③运动套装　sweat suit; sports suit; gym outfit; play-suit

锦纶运动套装　nylon jogging suit; nylon trilobal suit
拉拉队套装　rah-rah suit
竞赛套装　racing suit
登山服　mountaineering suit
滑雪服　ski suit
溜冰服　skating suit
田径服　athlete's wear; track suit
体操服　gymnastic suit; gym suit; gym outfit
乒乓球服　table tennis suit
羽毛球服　badminton suit
垒球服　softball suit
曲棍球服　hockey suit
网球服　tennis dress
高尔夫球服　golf wear
柔道服　judo suit; judo wear
跆拳道服　taekwondo suit
空手道服　karate suit
摔跤服　wrestling suit
举重服　weight lifter's suit
击剑服　fencing suit
射箭服　archery suit
游泳服　swimming suit; swimsuit; bathing suit; swimwear
冲浪服　surfing suit; surfwear
骑自行车服　cycling suit; cycle wear
功夫服　Gongfu suit

健身服　fitness suit; exercise wear
瑜伽服　yoga suit
练功服　practice clothes
竞技表演服　rodeo suit
训练服　training wear; trainer; exercise wear
裁判员服　referee's suit
杂技服　acrobatic wear
舞蹈服　dancing dress; dance wear
芭蕾舞服　ballet wear
狩猎服　safari suit; shooting suit; safari set
户外活动套装　blazer suit
慢跑运动套装　jogging suit; jogging set
宽松运动衣裤　trackie
出街套装　town suit
散步套装　walking suit
周末套装　weekend suit
郊外套装　country suit
轻松套装　luncheon suit
轻快套装　tracksuit; sports suit
奥特曼套装　ultraman suit
沙滩装　beach suit; beach wear
男海滩装　cabana set
太阳装　sun clothes; sun suit; sunwear
骑装　hacking outfit; riding suit; riding wear
斗牛士套装　toreador suit
女骑装　riding habit; habit

④制服　uniform; livery; garb; habilime; vestment

职业服　professional garments; business wear; business clothing; business attire; occupational clothing; career apparel
军服　military uniform; army uniform; uniform; battle dress; regimentals; monkey clothes

服役军服　army service uniform
陆军作战服　army combat uniform
野战军服　field uniform
变色龙军服　chameleon suit
军便服　service dress; military jacket
英国军服　King's(Queen's)coat
(旧时)雇佣兵服　mercenary dress
作(战)训(练)服　field training uniform
警服　policemen's uniform
司法服　judicial clothing
民航服　civil aviation uniform
办公服　office uniform; business suit; dark suit
行政服　executive wear
商务服　business garments
商务休闲服　business casual attire
(职场)精致休闲服　smart casual attire; smart casuals
海关服　customs uniform
邮电服　post and telecom uniform
银行服　bank uniform
税务服　tax collectors' uniform
工商服　industrial and commercial managers' uniform
医务服　medical uniform; medical wear; hospital uniform
护士服　nurse uniform; nurse dress
铁道服　railwaymen's uniform
列车员制服　conductors' uniform
机组人员制服　aircrew uniform
空姐制服　flight attendants' uniform
仪仗队制服　dress uniform
保安制服　security personnel's uniform
服务员制服　waiters' uniform
酒店服　hotel stuff uniform
炊事服　cooking uniform; cooking wear

校服　school uniform; school wear; schoolwear
学生服　students wear; school uniform; campus wear; pupil's wear
伊顿服　Eton suit
幼儿园服　kindergarten uniform
团队制服　team uniform

⑤防护服　protecting apparel; protective clothing; protective gear; safety clothing

危险品防护服　hazmat suit
纳米防护服　nano-protective uniform
劳保服　labour protective coveralls; working wear; safety wear
作业服　career suit
工作服　working uniform; business clothing; overalls; jumper; fatigue clothes; fatigues; working dress; working wear; workwear; habits de travail
连衣裤工作服　coveralls; boiler suit; combinaison d'homme〔法〕
特种工作服　special working uniform
宇航服　space suit; spacesuit; spacewear; penguin suit; astronaut's apparel
舱内宇航服　IVA spacesuit
舱外宇航服　EVA spacesuit
飞行服　flying suit; flying wear; flight gear
防弹飞行服　flyer's armour
抗超重飞行服　G suit; gravity suit; anti-G suit
飞行员救生服　Mae West
海洋救生服　ocean survival suit
搜救服　search and rescue wear
增压服　pressure suit
伞兵服　jump suit
潜水服　diving suit; diver's coveralls; sub-

marine armour; drysuit
保暖紧身潜水服　wet suit
干式潜水服　dry suit
远洋巡航服　cruising wear
航海服　sailing wear
水手服　sailors suit
消防服　firemen's uniform
炼钢服　steelworker's suit
石油工人服　oil worker's uniform
养路工人服　road maintenance worker's uniform
环卫工人服　dustmen's uniform
矿工服　miner's suit
实验室工作服　lab coat
超净防尘服　super-excellent clean uniform
防静电服　antistatic uniform
绝缘服　insulated wear
防感染服　anti-infection apparel
防蚊服　mosquito prevent suit
抗菌防臭服　antibacterial deodorant clothing
吸湿排汗服　moisture wicking clothes
防水透气服　breathable-waterproof clothing
阻燃服装　flame retardant clothing
防火服　flame-resistant clothes
防污染服　anti-pollution garments
防生化服　chemical-biological protective garments
防毒气服　antigas clothes
防辐射服　radiation-proof suit
防微波服　microwave shielding clothes
防紫外线服　ultraviolet resistance suit
防核辐射服　nuclear protective clothing
防弹服　ballproof clothes
芳纶防弹服　aramid body armour
道路安全服　road safety garments
反光服　reflective clothes

（荧光）信号服　signal clothing

⑥礼服　formal dress; formal wear; formal attire; dress coat; dresscoat; dress; official uniform; ceremonial dress; ceremonial robe; robes; trappings; official wear; monkey clothes; monkey suit

礼仪服　ceremonial wear; ceremony clothing; clothes of ceremony
正式礼服　full dress; formal dress; most formal wear; toilette
晚礼服，夜礼服　evening dress; evening attire; evening suit; evening coat; dress suit; dress coat; dinner dress; formal evening wear; evening gown; after dark; after six; glad rags; bobtail; formal; habit de soirée〔法〕
法国古典晚礼服　Creole evening dress
早礼服　morning coat; cutaway
日礼服　formal day wear; morning coat; afternoon dress
大礼服　full dress; full dress coat; full dress uniform; court dress; grande toilette〔法〕; dress suit; frock coat
小礼服　tuxedo; tux; dinner jacket; evening jacket; mess jacket; smoker; dinner coat
常礼服　black suit; tuxedo; morning dress
简略礼服　semi-formal wear; semi-evening dress
准礼服　informal wear
燕尾服　swallow-tailed coat; swallowtail; tails; tailcoat; dress coat; evening dress; full dress; dress suit; enbifuku〔日〕
鸡尾酒会服　cocktail dress; cocktail suit; cocktail attire

外套

宴会服 party dress; pageant dress
庆典礼服 pageant dress
（女）舞会礼服 ball-dress; ball gown
军礼服 dress uniform; monkey clothes
婚礼服 wedding dress; wedding gown
婚纱服 bridalwear; bridal gown; wedding gown
新娘礼服 wedding gown; bridal dress; bridal gown; robe de mariée〔法〕
大学礼服 academical dress; academicals
学位服 academic costume; academicals
博（硕、学）士服 doctor's (master's, bachelor's) gown and hood
变化式礼服 fancy dinner suit; fancy tuxedo
缀亮片礼服 sequined dress
（旧时上朝穿的）朝服 court dress; official dress

5. 外套, 大衣 coat; overcoat; manteau〔法〕

① 外套 coat; overcoat; overdress; loose coat; jacket; outer garments; wraps

长外套 long coat; maxicoat
中长外套 midi coat
短外套 short coat; car coat; carcoat; covert coat; jacket; over sack; spencer; cardinal
齐腰短外套 half coat
超短外套 mini coat
单（双）排扣外套 single (double)-breasted coat
男女通用宽外套 paletot〔法〕
宽短外套 petit paletot〔法〕
开襟宽腰直筒外套 coatdress

无袖外套 cloak
无袖长外套 inverness
外用外套 outer coat
填棉外套 padded coat; padded jacket; quilted coat
厚外套 greatcoat
薄型外套 summer coat; thin jacket
春秋外套 top coat
冬季外套 winter coat
三季通用外套 three-season coat
四季通用外套 all year-round coat
晴雨外套，全天候外套 all-weather coat
合身外套 shaped coat
细长外套 slim coat
箱型外套 box coat
A 型外套 A-line coat
直筒型外套 straight coat
茧型外套 cocoon coat
连肩袖外套 raglan coat; balmacan
公主线外套 princess coat
宽摆外套 flared coat
宽胸外套，围裹式外套 wrap coat; clutch coat
女宽松外套 camisole
阔步外套，潇洒外套 swagger coat
女西装外套 dress coat
女骑装外套 riding coat; redingote〔法〕
女长皮外套 pelisse
轻短皮外套 covert coat
束带长外套 sleepcoat
束腰带外套 belted coat; tunic coat
无腰带外套 tuxedo coat
背部紧身外套 pinch back coat
衬衫型外套 shirtcoat; pardessus chemise〔法〕
和服式外套 kimono slip-on
带风帽外套 hooded coat

达夫尔（连帽）外套 duffel coat
绅士外套 gentleman coat
柴斯特外套 chesterfield
巴尔玛外套 balmacaan coat;balmacaan
堑壕外套 trench coat
沙沙（堑壕）外套 sassard coat
军装式外套 military coat
丛林外套 bush coat
水手外套 reefer coat;pea jacket;pea coat
飞行员外套 pilot coat
运动型外套 sport coat
网球外套 tennis wrapper
马球外套，波鲁外套 polo coat
（球队）替补队员外套 bench warmer
运动场（观赛）外套 stadium coat
骑装外套 riding coat;riding jacket
（赛马）骑师外套 jockey coat
跑车外套 car coat
猎装短外套 hunting coat;covert coat; safari blouse
旅行外套 travel coat
休闲外套 casual coat;casual jacket; sportscoat
时尚外套 fashion coat
窝钉皮外套 studded leather coat
晚会外套 evening wraps
摄影师外套 cameraman coat
马车夫外套 coachman's coat
伐木人外套 lumber coat;lumber jacket
苦力外套 coolie coat
牧场外套 ranch coat
防尘外套 duster coat;dust free coat; dustcoat;duster
驴子外套 donkey coat
伊顿外套 Eton coat;Eton jacket
钮贝里外套 Newbury coat

马基诺外套 Mackinaw coat;Mackinaw
阿拉斯加外套 Alaska coat
阿尔斯特长外套 ulster coat;ulster
哈德孙湾外套 Hudson Bay coat
维也纳外套 Vienna coat
西班牙外套 Spanish coat
阿富汗外套 Afghan coat
带头巾的阿拉伯外套 burnous;burnoose
（配面罩的）多米诺外套 manteau domino
粗绒大衣呢外套 feamaught;feamought
牛仔（布）外套 jean coat;denim jacket
皮外套 leather coat;leather jacket
PVC 长外套 PVC long coat

②大衣 overcoat;topcoat;coat;cloak
礼服大衣 frock coat;frock;paletot〔法〕
柴斯特礼服大衣 chesterfield
西装大衣 manteau veston〔法〕
军大衣 military coat
军用短大衣 British warm;warm〔英〕
派克大衣 parka
带帽派克短大衣 anorak
带帽达尔夫粗呢大衣 duffel coat
带帽长大衣 capote
单（双）排扣软领长大衣 chesterfield
无纽大衣 clutch coat
轻便大衣 top coat;covert topcoat;topcoat;topper
快艇防寒大衣 yacht parka
英式厚冬大衣 British warmer;British warm coat
春秋大衣 manteau de demi-saison〔法〕
巴尔马干大衣 balmacaan coat;balmacaan
男大衣 pardessus〔法〕
男紧身大衣 benjamin
女抱合式大衣 clasp coat
女宽松短大衣 wrap topper;topper

束带短大衣	belted topper
童大衣	children's coat
童风雪大衣	snowsuit
风雪大衣	parka
空军风雪大衣	airforce parka
双面大衣	reversible coat
羽袖大衣，披风大衣	inverness coat
披肩大衣	caped coat
斗篷大衣	cloak
套袖大衣，连肩袖大衣	raglan
半套袖大衣	semi-raglan coat
三片套袖大衣	three-piece raglan coat
棉大衣	cotton-padded coat
（涤纶）喷胶棉大衣	polyfilled coat
羊绒大衣	cashmere coat
毛呢大衣	woolen coat
粗呢大衣	duffel coat
厚呢大衣	dreadnaught; dreadnought
洛登缩绒（防水）厚呢大衣	loden coat; loden
皮大衣	leather coat
镶饰毛皮的皮大衣	leather coat inserted with fur
裘皮大衣，毛皮大衣	fur coat; pelisse〔法〕
羊皮大衣	sheepskin coat
貂皮大衣	marten coat; mink coat
狐皮大衣，狐裘	fox coat; fox robe
羽绒大衣	down coat
长毛绒大衣	plush coat
人造毛皮大衣	artificial fur coat
摇粒绒大衣	polar fleece coat
绒毯短大衣	blanket coat
毛皮衬里大衣	fur-lined coat
驼绒衬里大衣	fleece-lined coat
（防水）油布大衣	oilskin coat
夹大衣	spring coat; light coat

③长袍 gown; robe

阿拉伯大袍	aba
非洲短袖宽袍	dashiki
（西非）宽大长袍	boubou
土耳其长袍	caftan; kaftan; dolman
（古希腊、罗马）短袖或无袖齐膝宽袍	tunic dress; tunic
古罗马宽袍	toga
南美斗篷装	poncho
日本和服	kimono〔日〕; wafuku〔日〕
英国国王加冕服	dalmatic
龙袍	dragon robe; imperial robe
学位袍	academic gown; degree gown
博（硕、学）士学位袍	doctor's (master's, bachelor's) gown
法袍	surplice; vestment; judge's robe; lawyer's robe
法官袍	judge's robe; long robe
教士袍	cassock; cope; zimarra; long robe
神父（牧师）白袍	alb
牧师法衣	canonicals; clericals
主教法衣	chimere; dalmatic; cassock tunicle; tunic; tunique〔法〕
祭司法衣	soutane
麻布僧衣	alb
僧袍，道袍	cowl; frock; monk's habit; priest robe
袈裟	pallium
十字褡	chasuble
穆斯林男用长布袍	jibba; jibbah
穆斯林男女穿的开襟长布袍	jubbah
阿拉伯白布袍	haik
中式长袍	Chinese gown
棉袍	cotton wadded robe
衬绒袍	wool lined robe
风帽绒布袍	hooded fleece robe

（民初时男人穿）长衫　Chinese robe; Republican robe
女便袍　cymar
主妇长袍　hostess gown
家居餐袍　brunchcoat
蓝布大褂　blue cotton gown
毛皮袍　fur robe
丝绵袍　floss-padded robe
印花摇粒绒袍　printed polar fleece gown
纽扣全开襟袍　button through gown
罩袍　overgown, dust gown; over robe
宽袍　wrapper
（泳后）海滨袍　beach gown
手术袍　surgical gown
产妇育婴袍　maternity nursing gown

④风雨衣，披风　weather-all coat & cloak

风雨衣　weather-all coat; spring coat; ulster; wind coat; reversible raincoat; weatherproof clothing
防风衣，风衣　windbreaker; windcheater; wind coat; wind jumper; trench coat; duster coat; dustcoat; duster; coupe-vent〔法〕
防风暴衣　stormbreaker
雨衣　raincoat; raindress; rainsuit; rainwear; rainproof; waterproof; mackintosh
manteaude pluie〔法〕
工业雨衣　industrial rainsuit
劳工雨衣　labour raincoat
摩托雨衣　rainsuit for motorcycle rider
连身雨衣　coverall rainsuit
迷彩雨衣　camouflage rainsuit
毛呢雨衣　barret
博柏利雨衣　Burberry
尼龙雨衣，锦纶雨衣　nylon raincoat;
coated nylon jacket
塑料雨衣　plastic raincoat
PVC 雨衣　PVC raincoat
PVC 长雨衣　PVC long coat
PE 一次性雨衣　PE disposable raincoat
橡胶雨衣　rubber raincoat; neoprene raincoat
单面胶雨衣　PVC/polyester raincoat
双面胶雨衣　PVC/polyester/PVC raincoat
油布雨衣　oilskin; oils; tarpaulin
宽大油布雨衣　slicker
无纽束腰男雨衣　tielocken
斗篷雨衣　raincape; rainclock
蓑衣　coir raincoat
雨披　poncho; rain cape; raincape; rainproof; rainwear
军用雨披　military poncho cloth
PVC 雨披　PVC poncho
披肩，斗篷　cape; cloak; manteau; mantle; pelerine; pelisse; tippet; poncho; dress smock
大氅　cape; cloak
礼服斗篷　dress cape; mantilla
带帽斗篷　capuchin
斗牛士红色斗篷　capa
人造毛皮披肩　fake fur cape
（古希腊、古罗马）大披肩　pallium
阿拉伯披肩　aba mantle
小披肩　capelet; mantelet
长披风　kirtle; inverness
童披风　wrapper; mantle

6. 内衣　underwear; undergarment; lingerie〔法〕

①内衣裤　body clothes; underclothes; underclothing; skivvies
基本内衣　foundation garment; foundation

针织内衣 knitted underwear
弹力内衣 elastic underwear
女内衣裤 underthings; undies; body briefer
长内衣裤 longies; long johns〔美〕
女内衣裤套装 camisole and tap pants
女内衣 chemise; undies; lingerie〔法〕
童内衣 underwaist; waist
情侣内衣 lovers underwear
情趣内衣,性感内衣 sexy underwear; sexy lingerie
贴身内衣 undershirt; undervest
无缝内衣 seamless underwear
(仿女式)男用内衣 lingerie for men
运动内衣 sports underwear
环保内衣 ecological underwear
智能内衣 smart underwear
功能内衣 functional underwear
航天内衣 space underwear
防X射线内衣 X-ray proof underwear
吸湿排汗内衣 moisture wicking underwear; moisture absorbing underwear
塑形内衣 shaping underwear; foundation; all-in-one
护肤内衣 skin care underwear
胶原蛋白内衣 collagen underwear
维生素E内衣 vitamin E underwear
香味内衣 fragrance underwear
远红外内衣 far-infrared underwear
保健内衣 health care underwear
保暖内衣 thermal underwear; thermals
德绒(保暖)内衣 dralon underwear
全棉弹力保暖内衣 cotton lycra thermals
真丝(全棉)内衣 silk (cotton) underwear
木(竹)纤维内衣 wood (bamboo) fibre underwear

芦荟纤维内衣 aloe fibre underwear
贴身衣物 intimates
紧身衣 body; bodywear; maillot〔法〕; skin dress; skin tights; skintight; tights
紧身内衣 maillot de corps〔法〕
连胸紧身女内衣 all-in-one
紧身背心 body shirt; under bodice
保暖(内衣)背心 thermal vest
衬里背心 underwaist; camisole; cami
吊带女背心 camisole; cami; tank top
女连裤内衣 camiknickers; combinations; union suit; teddy suit; chemiloon
连体袜衣 body stocking; bodystocking
连体网衣 fishnet dress; fishnet bodystocking
紧身连衣裤 body stocking; bodysuit; catsuit; leotard
紧身内衣裤 skin tights

②胸衣 corsage; foundation; bustier; camisa; camisole; cuirass; corselet; corselette; plastron

紧身胸衣 corset; basque; bandeau; pompadour; bodice; waist; stays; roll-on; open-button
挂脖式紧身胸衣 halter bodice
弹力胸衣 stretch foundation
花边胸衣 camisole
花边网眼胸衣 lace cami top
吊带胸衣 body bralette
窄带式胸衣 bandeau top
套筒胸衣 tube top
无肩带胸衣 sun top
无袖胸衣 chemisette
(背心式)凉爽胸衣 coolmax bra top
衬领 plastron
乳罩,胸罩 brassiere; brassière〔法〕;

bra; bralette
普通乳罩　everyday bra
智能胸罩　smart bra
全杯型乳罩　full cup bra; full figure bra
半杯形乳罩　half cup bra; balconette bra
软杯高弹乳罩　soft cup bra
廓型乳罩　contour bra
低胸深V形乳罩　plunge bra
低胸大U形乳罩　superscoop bra
模压乳罩　moulded bra
定形（免熨）乳罩　memory bra
钢丝托乳罩　underwire bra
无钢丝托乳罩　wirefree bra; wireless bra
硬托乳罩　form support bra
上托乳罩　push up bra; booster bra; uplift bra
束带乳罩　band brassiere
吊带乳罩　garter brassiere
无吊带乳罩　strapless brassiere; strapless bra; off shoulder brassiere
前扣乳罩　front hook brassiere; front opening bra; front closure bra
三角乳罩　triangular bra
羊皮乳罩　nappa bra
花边乳罩　lace brassiere
加垫乳罩　padded brassiere
轻薄乳罩　lightness bra
网眼乳罩　eyelet brassiere; lace brassiere
弹力乳罩　elastic brassiere
弹力T恤乳罩　stretch T-shirt bra
弹力塑形乳罩　shapers flexi bra
木（竹）纤维乳罩　wood (bamboo) fibre bra
硅胶乳罩　silicone bra
隐形乳罩　Nude bra; Nu bra; silicone bra
无限制乳罩　limitless bra
（无吊带）套筒胸罩　tube bra

比基尼胸罩　bikini bra
运动胸罩　sports bra
交叠形胸罩　crossover bralette
立即托高胸罩　instant uplift bra
空气胸罩，充气乳罩　air bra
多功能胸罩，多变乳罩　multiway bra
乳头胸罩　nipple bra
透明乳罩　transparent bra
魅力乳罩　glamour bra
男性胸罩　men's bra
长身线形乳罩　longline bra
可调节乳罩　convertible bra
按摩乳罩　massatherapy bra
孕妇乳罩　maternity bra
哺乳乳罩　nursing bra
少女乳罩　teen bra
小（中、大）号乳罩　a (b, c)-cup brassiere
迷你乳罩　minimal bra
泪滴形迷你乳罩　teardrop bra
绣花（印花）乳罩　embroidered (printed) bra
巴西乳罩　Brazillian bra
法式乳罩　French bra
乳罩罩杯　cup
罩杯托　cup form
罩杯尺寸　cup size
胸罩内垫，胸垫　bra pads; bust pads
吊带胸垫　strap bra pads
假胸　falsies; bust form; bust pads
无胸罩　braless
乳（头）贴　pasties
汗衫　undershirt; undervest; T-shirt; singlet; vest
弹力汗衫　elastic undershirt
网眼汗衫　mesh undershirt
圆领汗衫　skivvy shirt

圆领短袖汗衫　T-shirt
汗背心　undervest; singlet; sleeveless undershirt; chesty〔澳〕
（女用）腰封　waist nipper
兜肚衣　stomacher
③内裤，衬裤　pants; panties; underpants; under drawers; drawers
衬裤汗裤　underpants; cotton drawers
短内裤　briefs; undershorts; short panties
平脚口短内裤　boyleg briefs; boyleg
高脚口短内裤　hi-cut briefs
高腰短内裤　high waist briefs
低腰女内裤　hipster briefs
女塑形短内裤　shapers briefs
（前缝呈倒Y形的）男内裤，三角裤　Y-front briefs; Y-fronts
（男式）前开门襟短内裤　fly front briefs; Y-front briefs
运动型短内裤　sports briefs
宽松短内裤　full briefs
合体短内裤　fitted briefs
木（竹）纤维短内裤　wood (bamboo) fibre briefs
网眼花边短内裤　lace briefs
弹力窄带女内裤　string briefs
孕妇穿短内裤　maternity briefs
女无袖连衫衬裤　teddy; teddies〔复〕
女短衬裤　knickers; scanties
女月经裤　sanitary knickers
女半长内裤　pettipants
松紧短内裤　boxer shorts
一次性短内裤　disposable brief
吊袜内裤　garter panties; control briefs
三角裤　briefs; short panties; undershorts; tanga briefs; tanga; triangle
（女）窄条布三角裤　G string
绣花网眼三角裤　embroidered mesh G string
拳击短裤式三角裤　boxer briefs
丁字裤　G-string; tanga briefs; tanga
比基尼内裤　bikini briefs; bikini
低腰比基尼裤　hipster bikini; low rise bikini
弹力窄带比基尼裤　string bikini
V形布窄带女短内裤　mini V-string; V-string
收腹内裤　girdle
（女）紧身腹带　waist nipper
衬裙，内裙　petticoat; pettiskirt; underdress; underskirt; slip; chemise; placket（古时）; jupon〔法〕
长衬裙　full slip; slip
短衬裙　semi-slip; half slip; waist slip
迷你衬裙　mini petticoat
紧身衬裙　tight slip; slim slip; shaping slip
塑形衬裙　shaping slip
宽摆衬裙　flare slip
蓬松衬裙　bouffant slip
连衫衬裙　combinaison de femme〔法〕
带裙箍衬裙　hoop petticoat
带内裤衬裙　panty petticoat
带乳罩的衬裙　bra slip
有袖衬裙　sleeve slip
斜条衬裙　Balmoral
背心式吊带衬裙　camisole slip
无吊带衬裙　strapless slip
吊带丝裙　silk suspender skirt
吊袜带　garters; stocking suspenders; suspenders
内裤吊袜带　panty girdle
裙撑　pannier; panier; crinolette; crinoline
尿布裤　nappy pants

④ 睡衣裤　pajamas；pyjamas；night clothes；nightwear；sleep wear；sleep suit；sleep set；night suit；jams

针织睡衣裤　knitted pyjamas

两（三）件套睡衣裤　two(three)-piece pyjamas

绣花丝绸睡衣裤　embroidered silk pajamas

轻薄料睡衣裤　tropical pyjamas

印花睡衣裤　floral pajamas

印字图睡衣裤　character pyjamas

拼缝睡衣裤　patchwork pajamas

毛巾布睡衣裤　towelling pajamas

绒布睡衣裤　flannelette pyjamas

拉绒布睡衣裤　blanket pyjamas

木（竹）纤维睡衣裤　wood(bamboo) fibre pyjamas

珊瑚绒睡衣裤　coral fleece pyjamas

舒棉绒睡衣裤　shu velveteen pyjamas

条纹布睡衣裤　striped pajamas

格子布睡衣裤　check pajamas

翻领睡衣裤　notch collar pajamas

束带睡衣裤　pajamas with waistband

女睡衣套装　camisole and tap pant set

（女）情趣睡衣裤　sexy pyjamas

童睡衣裤　sleepsuit；sleepers

新生儿睡衣裤　newborn sleepsuit

婴幼儿连衫裤睡衣　creeper pyjamas

睡衣　pajama coat；sleep coat；sleepcoat

睡裤　pajama trousers；sleep pants

短睡裤　sleep shorts

（女、童）睡衣　nightdress；nightgown；bedgown；nighty；sleepers；nightie

女长睡衣　negligee；négligé[法]；déshabillé[法]；sleep coat

女宽大睡衣　peignoir[法]

女吊带睡衣　chemise

女背心睡衣　sleep singlet

背部交叉吊带睡衣　cross back chemise

男长睡衣　nightshirt

短睡衣　sleeping jacket；pajama jacket；bed jacket

短睡裤　bottoms

睡袍　night gown；nightgown；nightdress；nightie

衬衫式睡袍　nightshirt；sleep shirt；sleep-shirt

T恤式睡衣　sleep tee

晨衣　dressing gown；wrapper；morning gown；undress；robe[美]；robe de chambre[法]；peignor[法]

和服式女晨衣　kimono

浴衣、浴袍　bathrobe；bath gown；peignor[法]

和服式浴袍　kimono bathrobe

全棉毛巾布浴袍　cotton towelling bathrobe

珊瑚绒浴袍　coral fleece bathrobe

木（竹）纤维浴袍　wood(bamboo)fibre bathrobe

舒棉绒浴袍　shu velveteen bath robe

带肩垫浴袍　tank suit

7. 针织服装　knitted garments

针织时装　knitted fashion

针织运动装　knitted sportswear

针织外衣　knitted outerwear

针织内衣　knitted underwear

毛（丝，棉）针织服装　woolen(silk，cotton)knitted clothing

①毛衣，毛衫　woollen sweater；sweater；jumpers[英]

男、女童毛衫　sweater for men, women

毛衣，毛衫

and children
全成形毛衫　full fashioned sweater
渔人毛衫　fisherman sweater
费尔（岛式）毛衫　Fair Isle sweater
艾伦毛衫　Aran sweater
耿西毛衫　Guernsey sweater
北欧毛衫　Nordic sweater
斯堪的纳维亚毛衫　Scandinavian sweater
北美印第安人厚毛衫　Siwash sweater
阿盖尔毛衫　Argyle sweater
德尔曼毛衫　dolman sweater
羊毛衫　woollen sweater
羊绒衫　cashmere sweater
羊仔毛衫　lambswool sweater
兔毛衫　angora sweater
驼毛衫　camel wool sweater
马海毛衫　mohair sweater
雪兰毛衫　shetland wool sweater
丝织毛衫　silk sweater
棉织毛衫　cotton sweater
提花毛衫　jacquard sweater;brocade sweater
织花毛衫　fancy knitted sweater
拷花毛衫　embossed sweater
绣花毛衫　embroidered sweater; sweater with embroideries
补花毛衫　sweater with patches
珠绣毛衫　beaded embroidery sweater
珠饰毛衫　beaded sweater; sweater with beadings
手绣毛衫　hand embroidery sweater
（手工）钩编毛衫，镂空衫　hand crocheted sweater
开襟毛衫　cardigan; cardigan jacket; cardigan sweater; gilet〔法〕
开襟短毛衫　belero sweater; crop cardigan
无扣开襟衫　wrap sweater
按扣开襟衫　snap cardigan
单扣短袖开襟衫　one button short sleeve cardigan
绣字开襟衫　letter cardigan
和服式开襟衫　surplice cardigan
男式女用开襟衫　boyfriend cardigan
尖（圆）领开襟衫　V-(crew)neck cardigan
带风帽开襟衫　hooded cardigan
垂饰前襟毛衫　draped front cardigan
半开襟毛衫　half cardigan
重穿毛衫　sweater on sweater
两面穿毛衫　reversible sweater
短毛衫　cropped sweater
超短毛衫　mini sweater
背心式毛衫　vest sweater
字母毛衫　letter sweater
开肩毛衫　button-shoulder sweater
圆领毛衫　round-neck sweater
尖领毛衫　V-neck sweater
高翻领毛衫　turtle-neck sweater
水手领毛衫　middy sweater
水手领毛衫　crew sweater
滑雪毛衫　ski sweater
马球毛衫　polo sweater
网球毛衫　tennis sweater
打靶毛衫　target-shooting sweater
束带毛衫　drawstring sweater
系带领毛衫　lace-collared sweater
多层级毛衫　layered sweater
紧身毛衫　formfitting sweater;naked sweater;poorboy sweater;jersey
肥大女毛衫　Sloppy Joe;sloppy joe
毛线衫　knitted sweater;sweater;woollies
棒针毛衫　woollen hand knitted sweater

腈纶毛衫　acrylic sweater
锦纶衫　nylon sweater
棉线衫　sweater made of cotton yarn; cotton yarn sweater
麻线衫　sweater made of ramie yarn
混纺毛衫　blended sweater
毛腈混纺衫　woollen/acrylic sweater
麻棉混纺衫　ramie/cotton sweater
棉毛衫　cotton interlock jersey
卫生衫, 卫生衣, 卫衣　cotton sweater; sweater; sweat top
溜冰套头卫衣　skate sweater
带帽卫衣　sweat shirt with hood
运动衫　sport shirt; athletic shirt
厚运动衫　sweater
圆领长袖运动衫　sweat shirt; sweatshirt
低领无袖运动衫　running shirt

②T 恤衫　T-shirt; tee(s)

智能 T 恤衫　intelligent T-shirt
运动 T 恤衫　active T-shirt
快干运动 T 恤衫　active dry tee
冲浪 T 恤衫　surf T-shirt; wetshirt
吸湿排汗 T 恤衫　moisture wicking tee; moisture absorbing tee
双丝光 T 恤衫　double mercerized T-shirt
印花 T 恤衫　printed tee
印图形 T 恤衫　graphic tee
扎染 T 恤衫　tie-dyed tee
珠片 T 恤衫　sequined tee
燕尾 T 恤衫　swallowtail tee
长身 T 恤衫　longline tee
阔身 T 恤衫　oversize tee
截筒（短身）T 恤衫　cropped tee
窝钉 T 恤衫　studded tee
半开襟 T 恤衫　Henley tee
（错色）双层 T 恤衫　double layer tee

连肩袖 T 恤衫　raglan tee
层迭袖 T 恤衫　layered sleeve tee
勺形收褶领口 T 恤衫　scoop gathered tee
拼色 V 领 T 恤衫　panel V-neck tee
标签 T 恤衫　label tee
标识 T 恤衫　logo tee
个性 T 恤衫　individual tee
情侣 T 恤衫　sweethearts tee; couple tee
功能 T 恤衫　functional tee
机织面料 T 恤衫　woven fabric tee
文化衫　singlet; T-shirt
海魂衫　sailor's striped shirt
棒球衫　baseball jersey
马球衫　polo-shirt; polo top
珠底布马球衫　piqué polo-shirt〔法〕
足球衫　football jersey
橄榄球衫　rugby shirt; rugby tee; rugby top
高尔夫球衫　golf shirt
板球衫　cricket shirt
滑雪绒衫　ski fleece top
健美衫　muscle top; fitness top
健身衫　fitness top
蝙蝠衫　batwing top
广告衫　advertising T-shirt; slogan tee
彩条衫　colour-striped jersey
豹纹衫　leopard print top
圆领间条衫　round neck striped top
棉毛短袖圆领衫　interlock T-shirt
无袖圆领衫　tank top; tanktop; muscle tanker; muscle top
针织网眼衫　mesh knit top
网眼叠合衫　wrap lace top
饰边叠合衫　frill wrap top
（错色）双层衫　double layer top; twofer top
挂脖领衫　halter top

针织套装

T形运动胸衫 T-back sport top
垂坠领口紧身衫 drape neck slinky top
褶裥领口衫 gathered neck top
不对称衫 asymmetric top
单肩带弹力罗纹衫 one shoulder stretch rib top
露腰衫 midriff top

③针织套装 knitted suit;knitwear set
针织三件套 three-piece set knitted;knit ensemble
针织运动套装 knitted jogging suit
涤纶经编运动套装 polyester tricot track-suit
无缝运动装 seamless activewear
丝绸针织套装 silk jersey blouse and pants
紧身套装 body suit;unitard
毛衫套装 twin sweater set,sweater suit
开襟毛衫裙套装 cardigan suit
毛线裤套装 sweater pants
羊毛衫裤 woolen sweater and trousers
棉毛衫裤 cotton interlock singlet and trousers;cotton interlock underwear
绒布衫裤 flannelette underwear
运动衫裤 sport singlet and trousers
紧身运动衣裤 active tights
卫生衣裤 sweat shirt and pants
针织睡衣裤 knitted pajamas
毛衫连裙装 sweater-top dress

④套头衫 pullover;overpull;slipover;slip-on;pullover shirt;jumper

罗纹套衫 rib knitted pullover
拉链套衫 zipper pullover
厚针织套衫 shaker knit pullover
绒线提花套衫 chenille jacquard pullover
约翰尼领套衫 Johnny collar pullover

翻领条纹套衫 stripe roll neck jumper
圆领套衫 crew neck pullover
二合一衬衫领套衫 2 for 1 jumper and shirt
高领套衫 turtle-neck pullover;roll neck;polo neck sweater
V领套衫 V-neck pullover;pullen V〔法〕
无袖套衫 shell top
宽大套头衫 sloppy joe
套头立领衫 skivvy
针织背心 knitted vest;sweater vest
羊毛背心 wool vest;sleeveless woollen sweater
毛线背心 knitted vest;pull-under
绳绒线背心 chenille vest
膨体纱背心 bully vest
绞花编织背心 cable vest;texture vest
钩编背心 crocheted vest
网眼背心 mesh vest;mesh singlet
高尔夫套头背心 golf pullover vest
弹力背心 stretched vest;lycra singlet;toning tank
罗纹运动背心 athletic rib singlet
带胸罩的运动背心 active bra tank
（错色）双层背心 double layered singlet
女吊带背心 shoestring singlet;shoestring cami top
胸罩式吊带背心 shelf bra singlet
帝国风格吊带背心 empire vest
针织吊带短衫 crop top
针织圆领背心衫 tank
宽摆背心衫 swing tank
宽松背心衫 sloppy tank
长身线形背心衫 longline tank
T形后身背心衫 racer back tank
V领无袖针织衫 V-neck sleeveless singlet
门襟叠合式抽褶衫 cross front shirred top

（女）露肩衫　off the shoulder top
针织夹克衫　knitted jacket
带帽绒布夹克　hooded fleecy sweat jacket
针织风帽衫　hoodie
印花风帽衫　print hoodie
套头风帽衫　pull-on hoodie
条纹V领风帽衫　stripe vee hoodie
拉链开襟风帽衫　zip thru hoodie
针织短外套　knitted coat; sweater coat; knitted jacket
针织衬衫　knitted shirt; knitted blouse
高乔牧人针织衬衫　gaucho shirt
针织裙　knitted skirt; jersey skirt
针织连衣裙　sweater dress; knit dress; jersey dress
针织背心裙　tank dress
针织水手裙　tricot sailor dress
连衫衬裙　undershirt dress
针织裤，毛线裤　knitted trousers
针织紧身裤　body pants; leggings
牛仔风格紧身裤　denim look leggings
拉链脚口截筒紧身裤　zip crop leggings
（防走光）打底裤　leggings
丝弹力裤　silk spandex pants
踩脚裤　anchored pants
健身裤　fitness pants
运动裤　athletic pants; trackpants; sweatpants
长运动裤　sweat pants; sweatpants; track pants; trackpants
女半长运动裤　toreador pants
针织运动裤　jersey trackpants
绒布运动裤　fleece trackpants
两侧间条运动裤　side stripe trackpants
前开门襟运动短裤　guy front trunks; fly front trunks
前中膨鼓形运动短裤　pouch front trunks

低腰运动短裤　hipster trunks
长脚运动短裤　long leg trunks
无缝运动短裤　seamfree trunks
护裆骑车短裤　cycling bib shorts
溜冰绒裤　fleece skate pants

⑤泳装，泳衣　swimsuit; swimwear; bather; cozzie; cossie

（女）泳装　swimming costume; bathing costume
单（两）件式女泳装　one(two)-piece swimsuit
比赛用泳装　racer bather; racer swimsuit
喷气概念（赛用）泳装　jet-concept swimsuit
鲨鱼皮（赛用）泳装　LZR Racer swimsuit
比基尼泳装　bikini; apron swimsuit
可卸式比基尼泳装　detachable bikini
比基尼连裙装　bikini dress
冲浪泳装　surfer swimsuit
超短两截式女泳装　minibikini
无胸罩女泳装　monokini; toplesssuit
背心式女泳装　tankini
大圆领口连身女泳装　tank swimsuit
连衣裤女泳装　maillot〔法〕
灯笼裤式泳装　bloomer swimsuit; maillot bloomer〔法〕
弹力丝泳装　silk spandex swimwear
优质泳装　maillot nageur〔法〕

（三）部位、部件名称　Terms of Positions and Parts

1. 衣前身　front body; front; bodice front
裁片，片料　cutted pieces; cut pieces; cut

衣前身

parts
衣大身 bodice; body
大身衣片 body piece
前身里子 front lining
全(半)衬里 full(half) lining
部分衬里 partial lining
活动里 detachable lining
拉链脱卸里 zip-out lining
防缩衬里 shrink-proof lining
原身衬里 self lining
隔层 layer
单(双)层 single(double) layer(s)
吸汗层 wicking layer
衣肩 shoulder
过肩 yoke
单(双)层过肩 one(two) layer yoke
前过肩 front yoke
领嘴 notch
衣胸部 chest; breast; bosom
(衬衫)硬衬胸 front stiff
前襟，前片，前幅 forepart; front; front panel
左(右)前襟 left(right) forepart
单(双)襟 single(double) breast
开襟 opening; placket; open front; cardigan front
全开襟 full open front; full placket
长开襟 deep placket
半开襟 half placket; placket front; neckline placket
对襟 front opening
偏襟 slanting front; side opening
曲襟 crank opening
门襟 top fly; front fly; left fly; fly; placket; plaquette; button stay; storm flap; closing; closure
明门襟 front strap, front band; box pleat front; top centre plait; top centre; placket front; neckline placket
贴门襟 facing strap; taped front
暗门襟 French front; plain front; wrap over front; fly front; button panel; cover placket; concealed placket
交叉门襟，叠门襟 crossover placket; crossed-over placket
按扣门襟 snap closing
系带门襟 tied closing
扣襻门襟 tab closing
套索纽门襟 toggle closure
尼龙搭扣门襟 velcro closure
拉链门襟 zip fly; zip front; zip through front
假门襟 mock fly; false placket
门襟里搭襻 fly tongue
里襟 under fly; right; right front; catching fly; under placket; under lap; underlap; button stand
挂面 front facing; facing
前搭门 front overlap
门襟止口 front edge
衣腰部，腰身 waist
腰节 waistline
侧幅 side panel
下摆，衣裾 bottom; hem; hem opening; lap; sweep; skirt; expansion; suso[日]
平下摆，平裾 square-cut bottom; square front; plain bottom; flat bottom
斜下摆 slant-cut bottom
弧形下摆，曲裾 curve bottom
圆下摆 round bottom
扇形下摆 scalloped bottom
波浪形下摆 petal bottom
不对称下摆 asymmetric bottom
绉褶波形下摆 fluted hem

贴边下摆　faced hem
压线下摆　topstitched hem
暗缝缲边下摆　blind stitched hem
衬衣下摆　shirttail hem; shirttail
罗纹下摆　rib bottom; knitted waistband
大圆角前摆　front cut away
圆角前摆　front round cut
方角前摆　front square cut; square front
止口圆角　front cut
止口圆角点线　front cut point
衣边　edge, hem
底边，折边，吊边　hem
下摆折边　bottom hem
袖口折边　cuff hem
口袋折边　pocket hem
下卷边　roll-down hem
小卷边　baby hem
反折边，反吊边　turnup hem
贴边　facing; welt
单（双）贴边　single(double) welt
绲边　piped edge; welted edge
包边　covered edge
假缝边　tack edge
激光（雕花）衣边　laser edge
绉褶波形衣边　fluted hem
毛边　fringe; fraying
皱襞　jabot
缝份　seam allowance; seam; balance
止口　edge margin
领串口线　gorge line
领驳口线　fold line for lapel; roll line
扣眼，纽孔　buttonhole
圆头纽孔，凤眼　eyelet buttonhole
平头纽孔，平扣眼　straight end buttonhole; flat buttonhole
横扣眼　horizontal buttonhole
直扣眼　vertical buttonhole
花式扣眼　fancy buttonhole
假扣眼　mock buttonhole; dummy buttonhole; decoration buttonhole
绲边扣眼，绲眼　welt buttonhole; bound buttonhole
嵌线（绳）扣眼　corded buttonhole
皮扣眼　leather buttonhole
手工扣眼　manual buttonhole
睡扣眼　nemuriana
扣位　button stand
扣眼位　buttonhole position
扣眼档　buttonhole distance
省　dart
前肩省　front shoulder dart
前腰省　front waist dart
前身通省　front open dart
胸省　chest dart; breast dart
领省　neck dart
领口省　neckline dart; gorge dart
驳头省　lapel dart
肋省　underarm dart; side dart
侧缝省　side seam dart
横省　side dart
斜省，对角省　diagonal dart
袖窿省　armhole dart
肘省　elbow dart
肚省　stomach dart; fish dart
法式省　French dart
曲线省，刀背缝　French dart; contour dart
鱼型褶（省）　fish dart
无省　dartless

2. 衣后身　back body; back; bodice back

后片, 后幅　backpart; back; back panel
后身里　back lining
后身半衬里　half back lining

总肩　across back shoulders
小肩　shoulder line
后过肩　back yoke
后开襟　back opening
后搭门　back overlap
后半腰带　half back belt
后摆　tail; sweep; coattail
后腰省　back waist dart
后肩省　back shoulder dart
后领省　back neck dart
后身通省　back open dart
衩　vent; slit; split; slash; umanori〔法〕
面衩　top vent
背衩　back vent; vent
单衩　centre vent
明单衩　hook vent
阴衩，暗衩　inverted vent
边衩，双开衩　side vent; side split; side slit
工字衩　box vent
钩形衩　hook vent
假衩　mock vent

3. 领子　collar

开门领　open collar
关门领　closed collar
开关领，两用领　convertible collar
翻领　turn collar; roll collar; turnup collar; rever collar; turnover collar; turndown collar; notch collar; col à revers〔法〕
高翻领　high roll collar
花瓣翻领　petal collar
高（低）领　high (low) collar
软（硬）领　soft (stiff) collar
半硬领　semi-soft collar
单（双）片领　one (two)-piece collar

双层折领，二重领　double collar, stand-fall collar
三式领，三用领　three-way collar
活动领，假领　detachable collar; false collar
备用领　spare collar
保洁领　everclean collar
平领　flat collar
水平领　horizontal collar
直领　band collar
立领　stand collar; Mao collar
防寒立领　blizzard collar
玳珊领　turtle collar
项圈形高直领　dog collar
颚领　chin collar
瓶颈领　bottled collar; turtle collar
圆领　round collar
短圆领　short round collar
小圆领，彼得·潘领　Peter Pan collar
方领　square collar
小方领　shirt square collar
尖角领　pointed collar; peaked collar
大尖角领　wide pointed collar
长（短）尖角领　long (short) pointed collar
斜角拼接领　miter collar
三角领　triangle collar
长椭圆领　oblong collar
翼形领　winged collar
平翼领　open-wing collar
马蹄形领，U形领　horse-hoof collar
弓形领　arched collar
波形领　cascading collar
（阶）梯形领　step collar
带形领　band collar
棒形领　stick collar; club collar
踏板形领　pedal collar

帆形叠领　reefer collar
像形领　picture collar
漏斗领　funnel collar
隧道领　tunnel collar
八字领　spread collar; semi-cut-away collar
一字领　off collar
展开领　spread collar
半展开领　semi-spread collar
展宽领，大八字领　wide spread collar; wide collar
青果领，丝瓜领，新月领，香蕉领　shawl collar
V形丝瓜领　peaked shawl collar
交叉围巾领　cross shawl collar; cross muffler collar
长围巾领　stole collar
围巾领　shawl collar; scarf collar
束带领　drawstring collar
围兜领　bib collar
打结领　tie collar
蝴蝶结领　butterfly collar; bow collar
燕子领　swallow collar; wing collar
鲤鱼领　carp collar
带扣领　belt collar
拉襻领　tab collar
别针扣领，针孔领，鸡眼领　pinhole collar; eyelet collar
下扣领　button-down collar
缺嘴领　notch collar
披肩领，斗篷领　cape collar; bertha collar; bertha; berthe
清教徒领，小斗篷领　puritanical collar
抵肩领，约克领　yoke collar
巴莎领　purser collar
范戴克领　Vandyck collar
巴斯特·布朗领　Baster Brown collar

约翰尼领　Johnny collar
普鲁士领　Prussian collar
拿破仑领　Napoleon collar
波拿巴领　Bonaparte collar
尼赫鲁领　Nehru collar
军官领　officer collar
水手领　sailor collar; middy collar
牧师领　clerical collar
农民领　peasant collar
小丑领　pierrot collar
迷蒂领　middy collar
意大利领　Italian collar
米兰领　Milan collar
荷兰领　Dutch collar
横文领，交叉领　cross-over collar
窄条领　strap collar
垂坠领　draped collar
远离领　far-away collar
直离立领　stand away collar
偏侧领　sideway collar
褶裥领　draped collar
皱褶领　ruffle collar
波褶领　frill collar
细褶波纹领　rippled collar
拉夫领，轮状皱领　ruff collar; ruff
短缝领　darted collar
（针织）嵌花领　intarsia collar
花边领　lace collar
绲边领　piping around collar
西装领　tailored collar; step collar
西装圆上领　semi-clover collar
小礼服（宽）领　tuxedo collar
中山服领　Zhongshan coat collar
立领，学生服领　stand collar
旗袍领　Chinese collar
马褂领，中式领　mandarin collar
大衣领　coat collar

阿尔斯特大衣领，倒挂领　Ulster collar
巴尔马干大衣领　balmacaan collar; bal collar
衬衫领　shirt collar
普通衬衫领　regular collar; plain collar
活动的礼服衬衫领　poke collar; attached collar
温莎（衬衫）领　Windsor collar
加利福尼亚（衬衫）领　California collar
巴利摩尔（衬衫）领　Barrymore collar
战斗（衬衫）领　combat collar
礼服背心上的白色添加领　collar slip
法衣领，葫芦领　surplice collar
伊顿制服领　Eton collar
马球衫领　polo collar
毛皮领　fur collar
长毛绒领　plush collar
罗纹领　knitted rib collar; ribbed collar
黏合领　fused collar
胶领　coated collar
上浆领　starched collar
驳领，驳头，翻领　lapel
平驳头，菱领，缺嘴领　notch lapel
戗驳头，尖领，剑领　peak lapel
窄驳头，小翻领　narrow lapel
宽驳头，大翻领　wide lapel
（西装）L形翻领　L-shaped lapel
（西装）T形翻领　T-shaped lapel
西装圆形翻领　semi-clover lapel
苜蓿叶形翻领　clover leaf lapel
鱼嘴形翻领　fish-mouth lapel
方角驳领　regular notch lapel
小方驳领　narrow notch lapel
缎面驳领　satin lapel
弧线翻领　bellied lapel
花式翻领　flower lapel

单（双）襟驳头　single (double)-breasted lapel
驳头斜度　lapel bias
驳头口　fold line for lapel; roll line
驳领尖　lapel peak
驳领缺嘴　lapel point; lapel notch
驳领眼　lapel buttonhole; boutonniere〔法〕
领型　collar shape; collar style; collar modeling
领面　outside collar; top collar; over collar
领里　inside collar; under collar; collar lining
领座，领脚，底领　collar band; collar stand; collar step; undercollar
领尖　collar point; collar tip
领口　collar opening
领豁口　notch
领上口　fold line of collar; collar roll line
领下口　under line of collar; collar neck line
领外口　collar edge; collar style line
领里口　top collar stand; collar stand line
领止点　collar stop
领角　collar corner
领舌　collar band tab

4. 领口，领圈　neckline; neck; neck opening; neck hole

领口线，领线　neckline
一字领口　boat neckline; slit neckline; off neckline
船形领口　bateau neckline; boat neckline
水手领领口　crew neckline
圆形领口　round neckline; crew neckline
大圆形领口　cut away neckline
椭圆形领口　oval neckline
长椭圆形领口　oblong neckline

勺形领口	scoop neckline
方形领口	square neckline
矩形领口	brick neckline
条形领口	strap neckline
梯形领口	trapeze neckline
远离领口	far away neckline
直离脖领口	stand away neckline
荷兰领口	Dutch neckline
马蹄形领口	horse-hoof neckline
匙孔形领口	keyhole neckline
键盘形领口	keyboard neckline
鸡心领口	sweetheart neckline; heart shaped neckline
钻石形领口	diamond neckline
扇贝形领口	scallop neckline
U 形领口	U-neckline
V 形领口	V-neckline
宽 V 形领口	wide V-neckline
锯齿形领口	zigzag neckline
法衣领口	surplice neckline
教士袍领口	gown neckline
背心领口	camisole neckline
开襟毛衫领口	cardigan neckline
马球衫领口	polo neckline
（吊带）花边胸衫领口	camisole neckline
（风帽）头巾形领口	hooded neckline
围巾形领口	scarf neckline
披巾形领口	stole neckline
披肩领口	wrap neckline
扣襻领口	tab neckline
打结领口	tie neckline
蝴蝶结领口	bow neckline
褶皱纽结领口	draped twist neckline
褶裥领口	tucked neckline
垂褶领口	cowl neckline
垂缀领口	draped neckline
垂瀑形领口	waterfall neckline
瓶颈领口	bottle neckline
玳瑁领口	turtle neckline
中装领口	Chinese neckline
亨利领口	Henry neckline
挂脖领口	halter neckline
束带领口	drawstring neckline
无吊带领口	strapless neckline
二重领口	double neckline
不对称领口	asymmetrical neckline
斜领领口	oblique neckline
深露式领口	plunging neckline
长裁领口	slashed neckline
狭长缝领口	slot neckline
短缝领口	darted neckline
露肩式领口	off shoulder neckline
宽开领口	off neckline; degage neckline
开缝领口	slit neckline
开衩领口	open neckline
半开襟领口	placket neckline
饰边领口	frilled neckline
罗纹领口	rib neckline
高（低）领口	high(low)neckline
领圈	neck; neckline; neckband; neck piece
V 形领圈	V-neck
Y 形领圈	Y-neck
罗纹领圈	rib neck
宽罗纹领圈	wide rib neck
低开领圈	low cut neck
领圈压条	neck tape

5. 袖子 sleeve

普通袖	normal sleeve; set-in sleeve
装袖，圆袖	set-in sleeve
中缝圆袖	set-in sleeve with centre seam
连肩袖，套袖	raglan sleeve
两（三）片连肩袖	two (three)-piece

袖子

	raglan sleeve
半连肩袖	semi-raglan sleeve
前圆后连袖	split raglan sleeve
和服袖	kimono sleeve; French sleeve
公主线和服袖	princess kimono sleeve
喇叭袖	flare sleeve; trumpet sleeve; bell sleeve
短喇叭袖	cape sleeve
钟形袖	bell sleeve
披肩袖	circular cape sleeve
帽形袖	cap sleeve
盖袖	cape sleeve
灯笼袖	lantern sleeve; puff sleeve
灯泡袖	lamp sleeve
主教袖	bishop sleeve
气球袖	melon sleeve; balloon sleeve
蓬袖，泡泡袖	puff sleeve; bishop sleeve
朱丽叶袖	Juliet sleeve
上推袖（袖口可推到肘部）	push-up sleeve
马蹄袖	horse-hoof sleeve
羊腿袖	gigot sleeve; gigot
蝙蝠袖	batwing sleeve
蝴蝶袖	butterfly sleeve
风筝袖	kite sleeve
德尔曼袖	dolman sleeve
匈牙利袖	Magyar sleeve
花瓣袖	petal sleeve
宝塔袖	pagoda sleeve
双层袖	double sleeve
多层袖	tiered sleeve
垂褶袖	cowl sleeve
抽褶袖	shirred sleeve
褶饰边袖	ruffled sleeve
方形袖	square sleeve
杯形袖	cup-shape sleeve
楔形袖	wedge sleeve
变形袖	styled sleeve
肩章袖	epaulet sleeve; strap-shoulder sleeve
高肩袖	high-shoulder sleeve
落肩袖	drop-shoulder sleeve
与过肩连裁的袖	yoke sleeve
吊袖，挂袖	hanging sleeve
罩袖	over sleeve
拉链脱卸袖	zip-off sleeve
（袖口）束带袖	drawstring sleeve
翻折袖	roll-up sleeve
开衩袖	slip sleeve
有袖头袖	cuffed sleeve
荷叶边袖	flounced sleeve; ruffled sleeve
衬衫袖	shirt sleeve
宽大袖	full sleeve
宽松（紧身）袖	loose(tight) sleeve
开启（封闭）式袖	open(closed) sleeve
整片袖，大裁袖，一片袖	one-piece sleeve
两片袖	two-piece sleeve
大（小）袖	top(under) sleeve
外（内）袖	outer(inner) sleeve; top(under) sleeve
前（后）袖	front(back) sleeve
长（短）袖	long(short) sleeve
九分袖	bracelet sleeve
七分袖	three quarter sleeve
半袖，中袖，五分袖	half sleeve, elbow sleeve
超短袖	French sleeve
无袖	sleeveless
袖里	sleeve lining
大（小）袖里	top(under) sleeve lining
袖型	sleeve shape; sleeve style; sleeve modeling
袖窿，袖孔	armhole; scye
袖窿牵条	armhole stay
袖山	sleeve cap; sleeve top; sleeve crown; sleeve point

袖口，袖头　cuff; sleeve cuff; cuff opening; sleeve opening; sleeve head; wristband
单袖头　single cuff
双袖头　double cuff; French cuff; turnup-cuff; foldback cuff
全（半）卡夫袖头　full(half)cuff
单（双）扣袖头　one(two)-button cuff
两用袖头　convertible cuff
松紧袖口　elastic cuff
罗纹袖口　rib-knit cuff
防风袖头　storm cuff
活动袖头　detachable cuff
假袖头　imitation cuff; mock cuff
假翻袖头　cuff band; cuff strap
紧袖口　bracelet cuff
束带袖头　strapped cuff
扇形袖头　scalloped cuff
圆（斜）角袖头　round (slant) corner cuff
尖角袖头　three points cuff
多褶袖头　pleated cuff
褶裥泡状袖头　bead cuff
无卷边袖头　cuffless
衬衫袖头　wristband
袖头里子　cuff lining
袖口纽　sleeve button; cuff button; sleeve links; cuff links; cufflink; links
袖口链扣　cuff links; cufflinks; links
袖衩　cuff opening; cuff vent; sleeve placket; sleeve vent; sleeve slit
大袖衩　sleeve facing
小袖衩　sleeve under facing, under lap; underlap
袖衩搭边　sleeve overlap
袖衩条　sleeve placket tape; sleeve placket stay

（袖腋部）吸汗垫布　shield
腋下镶布　underarm gusset; underarm gore
肘部垫布　elbow patch

6. 裤、裙的部位及部件　Positions and parts for trousers and skirts

裤前（后）片　front(back)leg
（裤）外长　outseam
（裤）内长　inseam
腰头　waistband
褶裥腰头　draped waistband
松紧腰头　elastic waistband; knitted waistband
腰头侧　topside
左（右）腰头　left(right)topside
腰头上口　upper end of waistband
腰头下口　under end of waistband
腰头里子　waistband lining
腰头门襟　front waistband; closure of trousers
腰部拼接布　yoke
腰头宝剑头　extended tab
腰头扣襻　button catch
腰头纽　waistband button; beaver button
裤耳，裤带襻　belt loop; belt keep
裤门襟，裤门襟　left fly; fly opening; fly facing; fly placket; fly
明门襟　exposed fly
门襟里子　fly lining
裤里襟　right fly; under fly; button panel; button catch; button stand; fly shield; fly catch; catching fly
里襟里子　right facing
里襟尖嘴　button tab at right fly
裤裆，裤衩，裤底十字骨　crutch; crotch
裤裆深　crutch depth
股上长　crotch length

裤直裆，裤立裆 rise; fork to waist
深直裆 high rise
短直裆 low rise
前（直）裆，小裆 front rise
后（直）裆，落裆 back rise
横裆 thigh
上裆 seat
中裆 leg width; knee
下落裆 inside leg
小（后）裆 front(back) crutch
小（大）裤底 front(back) crutch stay
裤衩布 crutch gusset
加固裆 reinforcing crutch
裤裆里子 crutch lining
雨水布 trousers curtain
膝盖绸 reinforcement for knee; knee patch
脚口 leg opening; bottom leg; bottom
平脚口 plain bottom; single bottom
喇叭脚口 bell bottom; flared leg
拉链脚口 zipper bottom
窄裤脚 tapered bottom; tapered leg
宽松直裤脚 Oxford bottom
直筒裤脚 straight leg
卷边裤脚 turn-up bottom; bottom with cuff; cuffed bottom
无卷边裤脚 single bottom; cuffless bottom
有角度裤脚 angled bottom
K形裤脚 K legs
裤脚卷边 turnup
贴脚条 bottom binding; bottom tape; heel stay
（脚口）踩脚带 foot strap
裤腰省 waistline dart; waist dart
裤前褶，裤裥 trouser pleat; waist pleat; front pleat

裤腿折缝，裤中线 front crease; crease in trouser; crease line
裙裥 skirt pleat
裙摆 skirt hem; sweep; trail; train
裙拼腰布 yoke
裙腰头 waistband
裙腰省 waistline dart; waist dart
裙衩 placket
裙腰开口 placket hole; placket
裙开口处腰袋 placket

7. 口袋，衬布，褶裥，襻 pocket, interlining, pleat, tab

①口袋 pocket

挖袋 insert pocket; slit pocket; slash pocket; set-in pocket
贴袋 patch pocket
大贴袋 roomy pocket; cargo pocket
拉链大贴袋 zip cargo pocket
袋鼠袋 kangaroo pocket
有盖贴袋 patch pocket with flap; flap jetted side pocket
蓬松贴袋 baggy patch pocket
明贴袋 outline pocket; out pocket
开贴袋 post pocket
内贴袋 sandwich pocket
压片贴袋 out pleat patch packet
裤脚贴袋 thigh pocket
嵌袋 welt pocket; piping pocket; jetted pocket
单（双）嵌袋 single (double) welt pocket; single (double) jetted pocket
细嵌线袋 slender welt pocket
一字嵌袋 wide welt pocket
双嵌里袋 double welt inside pocket
直袋 vertical pocket; straight pocket

斜袋，插袋 slash pocket; angle pocket; insert pocket; set-in pocket; side pocket
（裤、裙）弧形插袋 cut-away pocket
有盖斜袋 hacking pocket
1/2（1/4、1/8）斜袋 1/2（1/4,1/8）top pocket
横开袋 cross pocket
有盖袋 flap pocket; lap pocket
平盖袋 flat pocket
带纽扣的有盖袋 flap and button down pocket
方形袋 square pocket
桃形袋 peach-shaped pocket
J形袋 J-shaped pocket
扇形袋 scarap pocket
弧形袋 curved pocket; shaped pocket
月牙形袋 crescent pocket
船形袋 boat pocket; poche bateau〔法〕
曲尺袋 ruler shaped pocket
微弯袋 light curved pocket
圆角底袋 rounded pocket
尖角底袋 pointed pocket; three points pocket
三尖袋 three points pocket
六角袋 hexagonal pocket
褶裥袋 pleated pocket
明（阴）裥袋 box (inverted) pleated pocket
风琴袋，立体袋 accordion pocket
风箱袋，老虎袋 bellows pocket
接档袋 gusset pocket
针纹饰袋 pin tuck pocket
缘饰袋 fringed pocket
绲边袋 piped pocket; piping pocket; tambour pocket
双绲边袋 double piped pocket; jetted pocket

单（双）贴边袋 single (double) welt pocket
夹层袋 bound pocket
实用袋 utility pocket
假口袋 mock pocket; sham pocket; imitation pocket
上（下）袋 upper (lower) pocket
外侧袋 outseam pocket
摆缝袋 side seam pocket
侧缝直袋 side pocket
侧缝横袋 horizontal side pocket
侧缝斜袋 slant side pocket
后袋 back pocket
暗袋 concealed pocket; hidden pocket
内袋，里袋 inside pocket; lining pocket
锯齿形内袋 zigzag inside pocket
衣袋 coat pocket
裤袋 trouser pocket
女裤袋 fly pocket
裙袋 skirt pocket
背心袋 waistcoat pocket
中山装袋，老虎袋 Zhongshan coat pocket
衣横袋 side pocket
（上衣左右胸下）暖手袋 muff pocket
裤后袋，臀袋 hip pocket
裤插袋，裤斜直袋 side pocket; sidekick
裤斜插袋 forward pocket; frog pocket
牛仔裤斜袋 poche jeans〔法〕
背心下袋 waist welt pocket
围裙袋 apron pocket
胸袋 breast pocket
内胸袋 inside breast pocket
（西装）饰巾袋 breast pocket
表口袋 fob; fob pocket; watch pocket; cash pocket
手机袋 mobile pocket
MP3袋 MP3 player pocket

衬布

小口袋 tiny pocket; poche gousset[法]; pochette[法]
眼镜袋 glasses pocket
卡片袋 card pocket
零钱袋 change pocket; coin pocket
票袋 ticket pocket
拉链袋 zipper pocket
袋位 pocket position
袋口 pocket opening; pocket mouth
袋尖 pocket point
袋盖 pocket flap; flap
袋盖面 top flap
袋布 pocket piece; pocket lining; pocketing
袋角加固布 reinforcement patch for pocket
垫袋料 pocket stay
袋口牵布 pocket facing
袋口牙边 pocket welt
口袋挂襻 pocket bearer

②衬布 interlining

前身衬 front interlining
胸衬 chest interlining; breast interlining
帆布胸衬 breast canvas
绒布胸衬 breast fleece
绲条，帮胸衬 bias strip
胸衬条 stay tape
门襟衬 fly interlining
肩衬 shoulder interlining
盖肩衬 interlining domett; domette
领衬 collar interlining
驳头衬 lapel interlining
挂面衬 front facing interlining
袖头衬 cuff interlining
袋口衬 pocket-mouth interlining
袋盖衬 flap interlining
腰头衬 waistband interlining

下腰节衬 interlining below waistline
下摆衬，底边衬 bottom interlining; hem interlining
裤门襟衬 left fly interlining
裤里襟衬 right fly interlining
裙边衬 skirt hem interlining

③褶裥，襻 pleat; tab

褶裥 pleat; crease; gather; drape; plait; tuck
单（双）褶 single (double) pleat; one (two) tuck
同边褶 side pleat
顺褶 one-way pleat; knife pleat
倒褶 inverted pleat; kick-pleat; reverse pleat
明褶 visible pleat; visible tuck
暗褶，阴褶 inverted pleat; blind pleat; invisible tuck
箱（盒）形褶 box pleat
倒箱（盒）形褶 inverted box pleat
工字褶 box pleat
内工字褶 inverted box pleat
风琴褶 accordion pleats
管形褶 barrel pleat
蜂窝褶 beehive pleat
子弹带褶 cartridge pleat
剑褶，刀形褶 knife pleat
群刀褶 cluster knife pleats
压刀褶 pressed knife pleat
水平褶 horizontal pleat
活动褶 action pleat; bellows pleat; pleat
半活褶 dart tuck
死褶 stayed pleat; dead fold; gather
永久褶 permanent pleat; ever pleat; durable pleat; permanent crease
犀牛褶 supercrease

热定型褶　heat-set pleat
柔和褶　soft pleat
碎褶　gathered pleat
多层碎褶　shirring
细褶　pin tuck; pintuck
压线褶　stitched pleat
无压褶　unpressed pleat
横褶　tuck
皱褶　wrinkle; ruffle; pucker; goffer; crinkle
洗水皱褶　washed crinkle
弧形褶　curved pleat
扇形褶　fan pleat
伞形褶襞　umbrella pleat
波形褶襞　peplum
阳光褶襞　sun burst pleat; sunray pleat
装饰性褶裥　decorative tuck
折叠　fold
折缝，折痕　crease
中折缝　centre crease
假折缝　false crease
领襻　collar tab; collar strap
领口襻　neck tab
肩襻　shoulder tab; shoulder loop; shoulder broad; epaulet
袖襻　sleeve tab
防风袖襻　storm tab
下摆襻　bottom tab
腰襻　waist tab
后腰襻　back strap
裤头襻　adjustable tab; button catch

扣襻　button loop; button strap
活动扣襻　adjustable tab; extended tab
脚口扣襻　bottom tab
挂衣襻　hanger loop; tag
线襻　French tack; thread eye
线链襻　thread chain loop
耳朵皮　flange
三角形拼布　godet; gore; gusset
三角结　arrowhead tack
X形结　X-shape tack
绲条　binding; rouleau; welt
斜布绲条　bias binding; bias strip; bias tape
错色绲条　contrast binding; contrast piping
牵条　tape; stay tape; tape stay
嵌条　panel; welt
中心嵌条　centre panel
左（右）嵌条　left(right) side panel
压条　stitched piping
撑条　stay
垫袋条　pocket stay
特殊支撑物　special stay
加固带　reinforced tape
定位带　stay tape
镶边带　binding tape
补缀布片　patch
加固布片　reinforcement patch
嵌补布片　inserting patch
半月形贴布　half-moon patch

二、服装设计制图
Garment Designing and Drafting

(一) 人体主要部位 Important Parts of Body

人体　human body; human figure
身材　stature; figure; build
体形　figure; bodily form; build
体型　type of body; type of figure; bodily form; form; type of build; figure
标准体型　standard form; standard figure
匀称体型　well-proportioned form
特异体型　special form; special figure
肥胖体型　fat form; fat figure
矮胖体型　stout form; stout figure
苗条体型　slender form; slender figure
腆胸体型，挺胸体型　erect figure
驼背体型　stooping figure
人体比例　proportion of human body
头　head
额　forehead
太阳穴　temple
脸，面　face
脸型　feature
颈，脖　neck
粗颈　big neckline
颈背　nape
颈根　neck base
肩　shoulder
肩型　shoulder type
正常肩　regular shoulder; natural shoulder
平肩　square shoulder
直肩　classic straight shoulder

小圆肩　small round shoulder
宽(窄)肩　wide(narrow) shoulder
耸肩　high shoulder
溜肩　low shoulder
起肩　built up shoulder
落肩　drop shoulder
斜肩　sloping shoulder; slanting shoulder
凹肩　concaved shoulder
马鞍形肩　saddle shoulder
高锁骨　high collarbone
高肩胛骨　high shoulder blade
躯干　torso; trunk
胸　chest; bosom; breast
平胸　flat chest
鸡胸　pigeon chest
乳房　breast
乳头　nipple
肚脐　navel
腰　waist
粗(细)腰　large(small) waist
高(低)腰　high(low) waist
胁　side
腹　abdomen
上(下)腹　upper(lower) abdomen
中腹部　midriff
凸腹　large abdomen
背　back
驼背　hunchback
臀　buttocks; hip; derriere
大(小)臀　large(shall) hip; large(flat) derriere
翘(垂)臀　prominent(scooped) hip

臂，上肢　arm
大臂，上臂　upper arm; top arm
小臂，前臂　forearm
手　hand
肘　elbow
手腕　wrist
手掌　palm
手背　back of the hand
手指　finger
腋窝　armpit
腿，下肢　leg
大腿　thigh
小腿　shank
腿肚　calf
膝盖　knee
脚　foot; feet[复]
脚踝　ankle
脚趾　toe
脚跟　heel
脚面　instep
脚掌　sole

(二) 量身 Measurements

基本量身法　basic measurement
正确量身法　correct measurement
人体数据　somatic data
三维人体数据　3D human body data
人体尺寸测量系统　body measurement system (BMS)
三维人体测量系统　3D somatometry system
三维人体扫描技术　3D human scanning technology
三维人体建模　3D human modeling
服装号型　garment size designation
服装号型系列　size system and designation for costume; series of apparel sizes and styles; series of garment size designation
尺码，尺寸　size; measurements; dimension
标准尺寸　standard size
基准尺寸　regular size
准确尺寸　just size
可调尺寸　free size
尺寸差异　measurement discrepancy
尺寸允许公差　measurement allowance
宽松量　tolerance
人体尺寸　body measurement
成人（中童、小童）尺码　adult's (boy's, kid's) size
横向尺码　horizontal measurement
垂直尺码　vertical measurement
围度尺码　girth measurement
裁剪尺码　cutting measurements; cutting size
成品尺码　finished measurements; finished size
普通尺码　regular size
超常尺码　outsize; oversize
加肥尺码　plus size
特体尺码　special measurement size
大（中、小）码　large (medium, small) size
特大码　extra-large size; king size
第二大码　queen size
特小码　extra small size
尺码齐全　full range of size
尺码不合　off size
放大（缩小）尺码　enlarging (reducing) of size
尺码表　size specification; size tariff; size table

规格表 specification sheet
米，公尺 metre, meter
分米 decimetre
厘米，公分 centimetre
毫米 millimetre
码 yard
英尺 foot; feet〔复〕
英寸 inch
尺 chi
寸 cun
分 fen
基本量身 basic measurements
总体高，身高 height; stature; posterior full length; standing length
胸围 chest; breastline; bust; chest around; chest girth; bust girth
胸宽 chest width; bust width; front width; breast width; across chest; cross front
胸高 breast depth
胸高点间距 bust point to bust point
门襟长 placket length; fly length
门襟宽 placket width
下摆 bottom; hem; sweep
摆围 hem around; susomawari〔日〕
（松紧）下摆拉伸量 bottom stretched
（松紧）下摆放松量 bottom relaxed
前长 front length
后长，背长 back length; neck waist length; nape to centre waist; nape to waist
后（片）中线长，后中长 centre back length
后全长 full back length
背宽 back width; across back; x-back; breast of back; posterior chest width
半背宽 half across back

腰节长 waist length
后腰节长 waist back length
头围 head size; head circumference
头长 head length
风帽高 hood height
风帽宽 hood width
颈围 neck girth; neck size
颈点 neck point
前（后）颈点 front(back) neck point
侧颈点 side neck point
领围，领圈 neck; neckline; neck around; neck opening; neck girth; neck circumference
前（后）领圈 front(back) neck
领圈宽 neck apart; neck width
领深 neck drop; neck depth
前（后）领深 front(back) neck depth
领长，领大 collar; collar circumference; collar length
领高 collar height; neck-rib height
后领高 collar width with stand
领脚长 neck opening
领脚高 collar stand
领座宽 collar band
领尖长（宽） collar point length(width)
领尖距 collar point spread
驳领宽 lapel width; lapel scope
总肩宽 across back shoulder; shoulder tip to tip; across shoulder; x-shoulder
肩宽 shoulder width; shoulder length; shoulder
小肩宽 small shoulder length
过肩长（宽） yoke length(width)
肩斜 shoulder slope
肩高 shoulder height
肩点 shoulder point
颈肩点 neck shoulder point

后肩颈点　back shoulder point
乳峰点，胸高点　bust point; breast point
乳间宽　point width; nipple breadth
乳围　bust top
臂围　biceps
袖长　sleeve length; sleeve
大（小）袖长　top(under) sleeve length; overarm(underarm) length
袖宽　sleeve width
袖肥　biceps circumference; biceps
袖深　armhole depth; scye depth
袖窿，挂肩　armhole; armhole around; full armhole
袖窿宽　armhole width
袖窿直量　armhole width straight
袖窿弯量　armhole's length of curves
袖山高　sleeve cap height
上袖　muscle
中袖　arm width
袖下点　underarm point
袖头（宽），袖口（宽）　cuff opening; cuff width; cuff
（松紧）袖头拉伸量　cuff extended
（松紧）袖头放松量　cuff relaxed
袖头高　cuff height
腋深　scye depth
肘长　elbow length
肘宽　elbow width; forearm width
肘围　elbow circumference; elbow girth
肘点　elbow point
腕围　wrist girth; wrist circumference
腕点　wrist point
掌围　palm circumference; palm size
腰围　waist; waist girth; waistline
中腰围，腰带围　centre waist
（松紧）腰围拉伸量　waist extended
（松紧）腰围放松量　waist relaxed

腰头高　waistband height
肩至胸高点距离　shoulder to bust point
肩至腰线距离　shoulder to waist line
腹围　abdomen girth
臀围，坐围　hip; hipline; hip girth; hip circumference; round hip; seat; seat around; seat circumference; saddle width
上臀围　high hip
中臀围　middle hip
下臀围　low hip
臀宽　hip width
臀高　hip length; hip depth
超臀长度　over seat length
（裤）外长　outseam; outleg; side length
（裤）内长　inseam; inleg; inside length
（裤）门襟长　fly length
（裤）门襟宽　fly width
直裆，裤浪　rise; fork to waist; waist to hip; body rise
前（直）裆，前浪〔粤〕　front rise
后（直）裆，后浪〔粤〕　back rise
裆深　crotch depth
横裆　thigh width; thigh
上裆　seat; body rise
中裆　leg width
下落裆　inside leg
低裆，垂裆　low cut crotch
后机头长　back yoke length
股上　body rise; crotch length
股下　inside length
腿围　thigh girth; thigh size
中腿　knee; knee line
膝围　knee girth; knee
齐膝长度　knee length
腰线至膝长　waist to knee
超膝长度　over knee length
腿肚围　calf girth; calf

踝围　ankle circumference; ankle girth
脚口　bottom; bottom width; leg opening; leg bottom
口袋长　pocket length
口袋宽　pocket width
袋盖长　flap length
袋盖高　flap height
袋尖高　pocket point height
袋唇高　pocket welt height

(三) 设计和款式 Design and Style

美学　aesthetics; science of beauty
东方美学　oriental aesthetics
美感　aesthetic feeling; sense of beauty
审美　taste
韵律, 节奏　rhythm
平衡, 均衡　balance
对称均衡　formal balance
不对称均衡　informal balance
动态均衡　dynamic balance
不均衡　unbalance
比例　proportion
黄金分割　golden section
协调　coordination
对称　symmetry
不对称　asymmetry
强调　emphasis; accent
调和　harmony
统一调和　harmony of identity
对比调和　harmony of contrast
类似调和　harmony of similarity
上下调和　top and bottom
不调和　disharmony discordance
反复　repetition
交替　alternation
渐变　transition
渐层　gradation
单纯化　simplification
简约　simplicity
关联　relation
支配　dominance
搭配　matching
对比　contrast
统一　unity; dominance
集中　centrality
秩序　order
元素　element
东方元素　oriental element
中国元素　Chinese element
流行元素　popular element; fashion element
时尚元素　fashion element
图腾　totem
视错　optical illusion
运动感　movement
服装设计　dress design; costume design; clothing design
时装设计　fashion design
数字化服装设计　digital apparel design
男（女）装设计　men's (women's) design
款式设计　style design
外形设计　form design
结构设计　composite design
平面造型设计　graphic design
图案设计　decoration drawing; pattern design
装饰设计　decorative design
色彩设计　colour design
面料设计　fabric design
工艺设计　technological design; process design
板型设计　stereotype design; pattern design

包装设计	packing design
细节设计	detail design
最新设计	the latest design
精心设计	elaborate design
智能服装设计系统	intelligent garment design system
计算机辅助设计	computer aided design (CAD)
计算机辅助画样	computer aided pattern (CAP)
三维虚拟设计	3D virtual design
设计主题	motif of design
设计理念	design concept
设计构思	conception of design
构思，构想	idea; conception; vision
构思新颖	novel idea; original in conception
灵感	inspiration
感觉	sense; feeling
创意	original idea
创造性	creativeness
多元化	diversity
视野	view; vision
新视点	new view
态度	attitude
风格	style
艺术风格	artistic style
民族风格	national style
中国风格	Chinese style
西方风格	Western style
古代风格	ancient style
古典风格	classical style
怀旧风格	vintage style; retro look
复古风格	retro style
复兴风格	revival style
天然模拟风格	organic style
正宗地道风格	authentic style
前卫风格	avant-garde style
中性风格	neuter style
运动风格	sports style
田园风格	idyllic style
波希米亚风格	Bohemian style
浪漫风格	romantic style
学院派风格	academic style
帝国风格	empire style
文艺复兴时期风格	renaissance style
哥特风格	Gothic style
爵士音乐风格	jazz style
垃圾（摇滚）风格	grunge style
商务正装风格	business formal look
商务休闲风格	business casual look
精致休闲风格	smart casual look
风貌	look; chic
（迪奥）新风貌	New Look
朋克风貌	punk look
欧普风貌	op art look
典雅风貌	elegant chic
休闲风貌	casual chic
街头风貌	street chic
狂野风貌	wild chic
残旧风貌	worn look
波希米亚风貌	Bohemian chic
抽象派艺术	abstractionism
涂鸦艺术	graffito; graffiti [复]
流行	fashion; vogue; mode; à la mode [法]
最新流行	top fashion; top trend
超前流行	forward fashion
前卫流行	avant-garde fashion
进步流行	progressive fashion
流行观念	fashion idea
流行主题	fashion theme
流行规律	fashion rule
流行趋势	fashion trend

设计和款式

流行要素	fashion element
流行风格	fashion style
流行预测	fashion forecast
流行报道	fashion report
流行周期	fashion cycle
时尚	trend;fashion;vogue
绿色时尚	green fashion;ethical fashion
生态时尚	ecological fashion
素食时尚	vegan fashion
顶级时尚	top trend
经典时尚	classic fashion
流行时尚	pop fashion
波希米亚时尚	Bohemial
快时尚	fast fashion;fastfashion
复古时尚	vintage fashion
最佳时尚搭配	best vogue matches
时尚风向标	fashion trend
混搭	mash-up
时髦	fashion;vogue
赶时髦	following of the fashion
时兴	modern fashion;present fashion
不时兴，过时	out of fashion;out of mode
雅致	elegance;elegancy;chic
别致，时尚风貌	chic
中国风	China chic;chinoiserie
（服装）系列	collection
春夏（秋冬）系列	spring/summer(autumn/winter) collection
款式	style;look;design;pattern
基本款式	basic style;basic fashion
日常款式	day-to-day style
中国款式	Chinese look
欧美款式	European and American style
（美国）西部款式	Western look
非洲款式	African look
经典款式	classic look
怀旧款式	retro look
邦德款式	Bond look
最新款式	the latest style;up-to-date style
流行款式	fashionable style
爆款，爆板	hot style
新潮款式	chic style
奇特款式	fancy style
成对款式	pair style
海盗款式	pirate look;style corsaire［法］
嬉皮士款式	hippie style
大胆款式	bold look
暴露款式	bare look
性感款式	sensual look
迷彩款式	camouflage look
男女通用款式	unisex look
古怪款式	kinky look
富豪款式	opulent look
庞大款式	bulky look
大贴袋款式	cargo style
30年代款式	Thirties look
整体款式	total look
固定款式	stable style
款式翻新	innovation of style
款式描述	style description
装式	look
新装式	new look
传统装式	classic look
军装装式	military look
便装装式	casual look
套装装式	suit look
运动装式	sportive look
工作服装式	working wear look
船员装式	nautical look
农民装式	peasant look
狩猎装式	safari look;style chasse［法］
内衣装式	lingerie look
视幻艺术装式	op art look
服式	costume;style of clothing

式样 pattern; fashion; style; look; costume; drape
基本式样 basic pattern; basic fashion
最新式样 the latest fashion; up-to-date style
文雅式样 gentle fashion
摩登式样 modern fashion
东方式样 oriental look
类型 type; style
自然型 natural type
古典型 classical type
浪漫型 romantic type
彰显个性型 personality type
流行前卫型 popularity type
欧美型 European and American style
花样 pattern; design
古典花样 classic pattern
（回归）自然花样 nature pattern
复合花样 multi-pattern
纹样 pattern design
花色 design and colour
动物纹 animal print
豹纹 leopard print

（四）图样和符号 Drawing and Mark

造型 forming; modeling
英国西装造型 British model
（裙）型 trousers(skirt) modeling
压褶塑形 suppression
结构 structure
轮廓, 轮廓线 silhouette; contour; profile
服装廓型 clothing silhouette
桶型轮廓 barrel silhouette
鞘型轮廓 sheath silhouette
跛行（窄摆）轮廓 hobble silhouette
沙漏型轮廓 hourglass silhouette
美腰型轮廓 Mae West silhouette
长躯（低腰）轮廓 long torso silhouette
旦多尔轮廓 dirndl silhouette
巴斯克轮廓 basque silhouette
帝国式服装轮廓 empire silhouette
视觉效果 visual effect
立体感 three-dimensional effect
外形 external form; contour; outline; profile; appearance
原型 basic pattern; block
图样 drawing; draft; pattern; design
图形 sketch; figure
图案 pattern; figure; design; graphics
装饰图案 decorative pattern; decorative design
抽象图案 abstract design; abstract pattern
奇特图案 fancy pattern
几何图案 geometrical pattern
菱形花纹图案 diaper
涡旋花纹图案 paisley pattern
绣花图案 embroidered pattern
提花图案 jacquard pattern
地毯图案 carpet pattern
手绘图案 hand-draw pattern
图案放大（缩小）design enlargement(reduction)
绘图，制图 drawing; drafting
单线绘图 single line drawing
素描 sketch; charcoal drawing
图画 picture
插图 illustration
时装画 fashion sketch; fashion illustration; fashion drawing
原型制图 pattern drafting
计算机辅助画（纸）样 computer aided

pattern(CAP)
计算机辅助设计和制图 computer aided design and drafting(CADD)
基本图 original pattern
设计图 design drawing; design sketch; pattern sketch; working sketch
效果图 effect drawing; fashion drawing; lineup
款式图 working sketch
样稿 artwork layout; artwork
草图 draft; sketch
插图，图解 illustration
净样图 net draft
轮廓图 outline drawing
结构图 structural drawing; cutting illustration
分解图 resolving drawing; detail sketch
示意图 schematic drawing; sketch map; sketch
展示图 plane figure
（按比例）缩小图 scale drawing
裁剪图 cutting drawing; cutting illustration
板型 sterotype; pattern
服装板型 clothes sterotype; clothing pattern
纸样 pattern; paper pattern; pattern sheet; marking paper; paper
原型纸样 block pattern
数字化纸样 digitizing pattern
标准纸样 classic pattern
基本纸样 basic pattern
平面纸样 flat pattern
直接起图纸样 drafting pattern
立体裁剪纸样 draping pattern; modelling pattern
立体裁剪布质板样 cloth pattern

主码纸样 master pattern
女服、童服纸样 dressmaker's pattern
硬质纸样 pasteboard pattern
软质纸样 paper pattern
裁剪纸样 card pattern
生产纸样 production pattern
裁剪样板 block pattern; pattern
基码样板 master pattern
主样板 key pattern
轮廓样板 silhouette pattern
净样板 net pattern
唛架，排料图 marker; layout chart; layout
小唛架 mini marker
漏画纸板 stencil marker
裁剪用图样 delineator
细目，细节 detail
细部扩大 big details
符号 mark
基点 starting point
基本线 basic line
实线 outline
虚线 dotted line
点画线 alternate long and short dashes line
双点画线 alternate long and two short dashes line
距离线 distance line
装饰线 decorative line
等分线 equation line
裥位线 pleat line
塔克线，缝褶线 tuck line
省道线 dart line
蜂窝形褶饰 smock
经向符号 radial mark
顺向符号 direct mark
开省符号 dart position mark

假缝符号	tacking mark	基本线，基础线	basic line
钻眼符号	drilling mark	假想线，上平线	imaginary line
碎褶符号	gather mark	结构线	structure line
剪口符号	notch mark	分割线	section line
罗纹符号	rib mark	（衣、裤、裙）长度线	length line
拼接符号	piece together mark	前（后）身中线	centre front(back) line
对称符号	symmetry mark	胸围线	chest line;bust line
重叠符号	overlapping mark	下胸围线	under bust line
直角符号	vertical mark	胸宽线	chest width line;bust width line
归缩符号	shrink mark	总肩线	across shoulder line
拉伸符号	stretch mark	肩宽线，落肩线	shoulder line
容位符号	easing mark	后肩线	back shoulder line
省略符号	ellipsis mark	肩斜线	shoulder slope line
毛样符号	cutting line mark	肩头外倾线	shoulder width line
净样符号	outline mark	前（后）过肩线	front(back) yoke line
明线符号	topstitch mark	过肩宽线	yoke width line
否定符号	negative mark	过肩上（下）口线	top (under) line of yoke
扣位符号	button position mark	前后连肩线	raglan sleeve outline
扣眼位号	buttonhole position mark	背宽线	back width line

（五）线条，线形 Line

直线	straight line	背长线	neck waist line
横线	horizontal line	背中线	centre back line
曲线	curve line	背缝线	centre back seam line;back seam line
斜线	bias line;oblique line	公主线	princess line
垂直线	vertical line	肋线	side line
水平线，基准线	level line	（衣）摆缝线，（裤、裙）侧缝线	side seam line
原型线	original pattern line	领圈线，领窝线	neck line
造型线	shaped line;modeling line	后领圈线	back neck line
轮廓线，外形线	silhouette line;silhouette;outline	领口宽线	neck width line
身体轮廓线	body line	领深线	neck depth line
尺寸线	dimension line	领嘴线	notch position line
裁剪线	cutting line	领串口线	gorge line
成形线	finish line	领尖点线	line of collar point
完成线	seam line	翻领前宽斜线	collar width line
		翻领上（外）口线	inside (outside) line

of collar; collar stand (style) line
底领前宽斜线　end line of collar band
底领上（下）口线　inside (outside) line of collar band; collar band roll (neck) line
驳头止口线　lapel edge line
驳口线　lapel roll line; roll line; breakline
（风）帽顶线　centre back line of hood
（风）帽前口线　front line of hood
（风）帽下口线　under line of hood
（风）帽嘴线　front line of hood
袖长线　sleeve length line
袖围线　sleeve girth line
袖中线　sleeve centre line
袖山线　sleeve cap line
袖口线　sleeve opening line
袖衩线　sleeve slit line
前（后）偏袖线　fold line of top (under) sleeve
小袖深弧线　under sleeve depth line
袖窿线　armhole line
袖深线　armhole depth line
袖窿翘高线　up line of armhole
袖肘线　elbow line
肘围线　elbow girth line
前袖缝线　inseam line
后袖缝线　elbow seam line
袖头上口线　top line of cuff
袖头止口线　finish line of cuff
门襟止口线　front edge line
门襟圆角点线　line of front cut point
挂面止口线　finish line of front facing
挂面里口线　inner line of front facing
搭门线　front overlap line
撇门线　front cut line; front finish line
（衣）起翘横线　front pitch line
下摆线　bottom line; bottom width line;

hem line; hemline
底边线　hem line
（马甲）底边弧线　front point line
（服装上、下身）转换线　switch line
腰围线　waist line; waist girth line
高腰线　empire line
腰头上口线　top line of waistband; waistband roll line
腰头下口线　under line of waistband; waistband neckline
腹围线　stomach line; abdominal extension line
臀围线　hip line
中臀围线　middle hip line
裤（直）裆线　rise line; crotch line; crutch depth line
横裆线　thigh line
前（直）裆线，小裆线，前裆内撇线　front rise line
后（直）裆线，落裆线　back rise line
前（后）裆弧线　curve line of front (back) rise
小裆宽线　front rise width line
落裆宽线　back rise width line
中裆线，膝围线　knee line
下裆线　inside seam line; inseam line
裤门襟外（止）口线　outside (edge) line of left fly
裤里襟外（里）口线　outside (inside) line of right fly
裤后腰缝线　back waist seam line
裤后翘线　up line of waist
脚口线　bottom line; hem line; hemline
（中装）大襟弧（斜）线　curve (bias) line of front opening
（中装）抬裉线　armhole depth line
（中装）裉缝弧线　curve line of under-

arm

拆线，裤烫线	crease line
折叠线	fold line
翻折线	fold facing line; roll line
开衩线	vent line
袋位线	pocket position line
裥位线	pleat position line
扣位线	button position line
扣眼线	buttonhole position line
自然线形	natural line
宽大线形	big line
宽松线形	ample line
箱形线形，盒形线形	box line
矩形线形	rectangle line
桶形线形	barrel line
筒形线形	tube line
梯形线形	trapeze line
锥形线形	tapered line
帐篷线形	tent line
喇叭线形	trumpet line
气球线形	balloon line
低腰线形	drop waist line
丹尼尔线形	Daniel line
上贴下散线形	fit and flare line
沙漏线形	hourglass line
长身线形	long torso line
苗条线形	slim line
公主线形	princess line
T恤线形	T-shirt line
A形线形	A-line
T形线形	T-line
X形线形	X-line
Y形线形	Y-line
流线形	streamline

（六）颜色，色彩 Colour

自然色，本色 natural colour; begin colour

原色	primary colour; fundamental colour
纯色	pure colour
互补色	complementary colour
流行色	fashion colour; trend colour; season colour
国际流行色	inter colour; intercolour
中国传统色	traditional Chinese colour
基本色	essential colour
同类色	similar colour
多种色彩	multicolour
彩色	full colour
素色	plain colour
混合色	secondary colour
对比色	contrast colour; contrasting colour; complementary colour
冷色	cold colour
暖色	warm colour
中间色	neutral colour; intermediate colour; middle tint
柔和色，嫩色	soft colour; delicate colour
浓色	deep shade; rich in colour
前进色	advancing colour
后退色	receding colour
膨胀色	expansive colour
收缩色	contractive colour
光源色	light source colour
物体色	object colour
表面色	surface colour
底色	colour of ground
透明色	transparent colour
金属色	metal colour
强调色	accent colour
点缀色	additional colour
保护色	protective colour
迷彩色	camouflage colour
配色	colour matching; colour combina-

tion; colour scheme	色系　colour scheme; colour system
计算机配色　computer colour matching	同色系配色　tone on tone
单色，独色　solid colour	色相　hue
混色，搭配色　assorted colour	色相环　hue circle
调色　colour mixing	色度　tint
色彩的调和　colour coordination	色明度　value; shade
色彩不调和　clash of colour	色阶　gamut of colours
色彩设计　colour scheme	色感　colour sense
色彩基调　colour motif	色码　colour code
色调　colour tone; tone; colour shade; shade; hue	色号　colour number; colour No.
基本色调　essential tone	色卡　colour card
冷（暖）色调　cool(warm) tone	色样　colour sample; colour guide
淡色调　subtle tone	小块色样　lab-dip
柔和色调　tender tone; pastel tone	色牢度　colour fastness
中间色调　middle tone	(美国) 潘通色彩体系　pantone
色调的调和　tone harmony	潘通配色系统　Pantone Matching System (PMS)
同色调配色　tone in tone	

三、服装材料
Clothing Materials

(一) 原料　Raw Materials

1. 纺织纤维　textile fibre / textile fiber

先进纤维材料　advanced fibre materials
天然纤维　natural fibre
绿色纤维　green fibre
植物纤维　vegetable fibre
动物纤维　animal fibre
矿物纤维　mineral fibre
种子纤维　seed fibre
棉纤维　cotton fibre

棉花　cotton
有色棉，彩棉　coloured cotton
有机棉　organic cotton
绿色棉　green cotton
长绒棉　long fibre cotton
丝光棉　mercerized cotton
远红外棉　far-infrared cotton
转基因棉　transgenic cotton
过渡棉　transition cotton
精（普）梳棉　combed(carded) cotton
(美国) 比马棉　Pima cotton
亚洲棉　Asiatic cotton
木棉　kapok

韧皮纤维，麻纤维 bast fibre
苎麻 ramie
亚麻 linen;flax
大麻 hemp
木浆纤维 wood pulp fibre
竹纤维 bamboo fibre
竹原纤维 bamboo fibril;original bamboo fibre;bamboo fibre
竹浆纤维 bamboo pulp fibre
竹碳纤维 bamboo carbon fibre
莫代尔纤维 modal fibre;modal
莱赛尔纤维 lyocell fibre;lyocell
坦西尔纤维 tencel fibre;tencel
丽赛纤维 richcel fibre;richcel
德绒纤维 dralon fibre;dralon
香蕉茎纤维 banana fibre
菠萝纤维 pineapple fibre
天丝纤维 tencel; lyocell; tencel fibre; lyocell fibre
纽富纤维 newcell fibre
蚕丝 natural silk;silk
生丝，厂丝 raw silk;filature silk
桑蚕丝 mulberry silk
柞蚕丝 tussah silk
䌷丝 noil silk
羊毛 wool;raw wool;fleece
绵羊毛 raw wool of sheep;sheep wool
山羊毛 raw wool of goat;goat wool
羔羊毛 lamb's wool;lambswool
澳大利亚羊毛 Australian wool
塔斯马尼亚羊毛 Tasmanian wool
新西兰羊毛 New Zealand wool
雪特兰羊毛 Shetland wool
美利奴羊毛 merino wool;merino
新疆细羊毛 Xinjiang fine wool
特种羊毛 special wool
特种动物毛 specialty hair

改良毛 improved wool
再生毛 recovered wool
山羊绒 cashmere wool;cashmere;kasmir
马海毛 mohair
羊驼毛 alpaca wool;alpaca
骆驼毛 camel hair
原驼毛 guanaco wool
美洲驼毛 llama wool
骆马毛 vicuna wool
兔毛 rabbit hair
安哥拉兔毛 angora
牦牛毛 yak hair
原毛 raw wool;greasy wool
洗净毛 scoured wool
精梳毛 combing wool;long wool
粗梳毛 clothing wool;carding wool;coarse wool
精梳短毛，落毛 noil
鬃毛 bristle
石棉纤维 asbestos fibre
陶瓷纤维 ceramic fibre
化学纤维 chemical fibre;man-made fibre
合成纤维 synthetic fibre
人造纤维 artificial fibre;man-made fibre
复合纤维 composite fibre
蛋白质纤维 protein fibre
酪素纤维 casein fibre
大豆蛋白纤维 soybean protein fibre
玉米蛋白纤维 corn protein fibre
牛奶蛋白纤维 milk protein fibre
白蛋白纤维 albumen fibre
纤维素纤维 cellulose fibre
无机纤维 inorganic fibre
金属纤维 metal fibre;metallic fibre
玻璃纤维 glass fibre
中空纤维 hollow fibre;macaroni fibre
再生纤维 regenerated fibre

纺织纤维

再循环纤维　recycled fibre
变性纤维　modified fibre
异形纤维　profiled fibre
特种纤维　specialty fibre
高性能纤维　high performance fibre
智能纤维　intelligent fibre; smart fibre
纳米纤维　nano-fiber; nanofiber
功能纤维　functional fibre
空调纤维　temperature regulating fibre
抗静电纤维　antistatic fibre
阻燃纤维　flame-retardant fibre
防紫外线纤维　UV-proof fibre
抗菌防臭纤维　antibacterial deodorant fibre
芳香纤维　fragrant fibre
薄荷纤维　peppermint fibre
芦荟纤维　aloe fibre
海藻纤维　seaweed fibre; alginate fibre
罗布麻纤维　apocynum fibre
铜离子纤维　copper ion fibre
高收缩纤维　high shrinkage fibre
碳纤维　carbon fibre
发热纤维　heat-gererating fibre; softwarm fibre
旭化成发热纤维　thermogear fibre
吸湿排汗纤维　moisture wicking fibre; moisture absorbing fibre
酷美丝纤维　coolmax fibre
酷帛丝纤维　coolplus fibre
新光合纤　cooltech fibre
远东合纤　topcool fibre
超细纤维　micro fibre; super fine fibre
细旦纤维　fine denier fibre
弹性纤维　elastic fibre; elastomeric fibre; elastane fibre; elastane; spendex; lycra
奥佩纶（弹性纤维）　opelon
黏胶纤维　viscose fibre; rayon fibre
铜氨纤维　cuprene fibre; cupro fibre
富强纤维，富纤，虎木棉　polinosic fibre; polinosic rayon; polinosic
醋酯纤维，醋酸纤维　cellulose acetate fibre; acetate fibre; acetate
三醋酯纤维　triacetate fibre; triacetate
聚乳酸纤维　PLA fibre; polylactide fibre
聚酯纤维，涤纶　polyester fibre; coolmax
PTT（聚酯弹性）纤维　PTT fibre
葆莱绒纤维　prolivon fibre
聚酰胺纤维，锦纶　polyamide fibre; chinlon; tactel
聚氨酯纤维　polyurethane fibre
聚乙烯纤维　polyvinyl fibre
聚氯乙烯纤维　polyvinyl chloride fibre
聚丙烯纤维　polypropylene fibre
聚丙烯腈纤维　polyacrylonitrile fibre
聚苯乙烯纤维　polystyrene fibre
力莱　lilion microfibre; lilion
涤纶　polyester; dacron
锦纶，耐纶，尼龙　nylon; polyamide
腈纶　acrylic fibre
特拉纶，德绒　dralon
维纶　polyvinyl alcohol fibre; vinylon
丙纶　polypropylene fibre
氯纶　chlorofibre
芳纶　aramid fibre
奥佩纶　opelon
奥纶　orlon
氨纶　polyurethane fibre; urethane elastic fibre; spendex; elastane
人造棉　artificial cotton; staple rayon
人造毛　artificial wool
人造丝　rayon; rayon filament; artificial silk; fibre silk
黏胶人造丝　viscose rayon

醋酯人造丝 acetate rayon
铜氨人造丝 cuprammonium rayon
长丝 filament
涤纶长丝 polyester filament
锦纶长丝 nylon filament
腈纶长丝 acrylic filament
丙纶长丝 polypropylene filament
黏胶长丝 viscose filament
醋酸长丝 acetate filament
单纤维丝，单丝 monofilament
短纤维 staple fibre
涤纶短纤 polyester staple fibre
锦纶短纤 nylon staple fibre
腈纶短纤 acrylic staple fibre
丙纶短纤 polypropylene staple fibre
黏胶短纤 viscose staple fibre
醋酸短纤 acetate staple fibre
纤维长度 fibre length
长纤维 long fibre; long staple
中长纤维 medium staple fibre; mid fibre
短纤维 short fibre; short staple; staple fibre
纤维细度 fibre fineness
纤维密度 fibre density
纤维强度 fibre strength

2. 纺织用纱 textile yarn

经纱 warp yarn; twist yarn
纬纱 weft yarn; filling yarn
原纱，本色纱 grey yarn
加工纱 finished yarn
精梳纱 combed yarn
普梳纱 carded yarn
丝光纱 mercerized yarn
粗支纱 coarse yarn
中支纱 medium yarn
细支纱 fine yarn

特细支纱 extra fine yarn
高支纱 high count yarn
60支纱 60-count yarn
漂白纱 bleached yarn
染色纱 dyed yarn
间隔染色纱，多色纱 space dyed yarn
彩虹纱 rainbow yarn
混色纱 melange yarn
单色纱 plain yarn
花式纱 fashion yarn; fancy yarn
防染纱 resist-dyed yarn
竹节纱 necked yarn; slub yarn
结子纱 knot yarn
结子彩点纱 coloured knops yarn
弹力纱 stretch yarn; spandex yarn; elastic yarn
绉缩纱 crepe yarn; livery yarn
膨体纱 bulked yarn; voluminous yarn
单纱 single yarn
合股纱 ply yarn
强捻纱 hard twist yarn
涡流纺纱 vortex yarn
棉纱 cotton yarn
牛仔纱 denim yarn
亚麻纱 flax yarn; linen yarn
苎麻纱 ramie yarn
大麻纱 hemp yarn
丝线 silk yarn
白厂丝 white steam filature yarn
绢丝 spun silk yarn
双宫丝 doupion silk yarn
䌷丝 noil silk yarn; noil yarn
柞蚕丝 tussah silk yarn
柞绢丝 tussah spun silk yarn
柞䌷丝 tussah noil yarn
天丝 tencel yarn
功能纱 functional yarn

酷美丝纱（具有吸湿排汗功能）	coolmax yarn
酷帛丝纱（具有吸湿排汗功能）	coolplus yarn
新光合纤纱	cooltech yarn
远东合纤纱	topcool yarn
纳米纤维纱	nanofiber yarn
海藻纤维纱	seaweed yarn; alginate yarn
大豆纤维纱	soybean fibre yarn
玉米纤维纱	corn fibre yarn
芦荟纤维纱	aloe fibre yarn
薄荷纤维纱	peppermint fibre yarn
竹原纤维纱	bamboo fibril yarn
竹纤维纱	bamboo fibre yarn
木纤维纱	wood fibre yarn
毛条	wool tow; wool top; top
毛纱	wool yarn
粗纺毛纱	woolen yarn
精纺毛纱	worsted yarn
羊绒纱	cashmere yarn
羔羊毛纱	lamb's wool yarn
马海毛纱	mohair yarn
羊驼毛纱	alpaca yarn
原驼毛纱	guanaco yarn
美洲驼毛纱	llama yarn
骆马毛纱	vicuna yarn
牦牛毛纱	yak hair yarn
雪特兰毛纱	Shetland yarn
兔毛纱	angora yarn
羽毛纱	feather yarn
结子精纺毛纱	chenille yarn
绒结子毛纱	zibeline; zibelline
有光毛纱	luster yarn
黏纤纱	rayon yarn
涤纶毛条	polyester top
锦纶毛条	nylon top
腈纶毛条	acrylic top
丙纶毛条	polypropylene top
黏胶毛条	viscose top
涤纶丝束	polyester tow
锦纶丝束	nylon tow
腈纶丝束	acrylic tow
丙纶丝束	polypropylene tow
黏胶丝束	viscose rayon tow
醋酯丝束	acetate tow
涤纶丝	polyester yarn
锦纶丝	polyamide yarn; nylon yarn
腈纶纱	acrylic yarn
维纶纱	vinylon yarn
氨纶纱	spendex yarn
棉氨纶包芯纱	cotton/spendex yarn
人造棉纱	spun rayon yarn
人造丝（纱）	rayon yarn
黏胶人造丝	viscose rayon yarn
醋酯人造丝	acetate rayon yarn
合成纤维纱	synthetic yarn
混纺纱	blended yarn; mixed yarn; union yarn
T/C 纱，涤/棉纱	T/C yarn; T/C blended yarn
T/R 纱，涤/黏纱	T/R blended yarn
CVC 纱，棉/涤纱	CVC blended yarn
R/C 纱，黏/棉纱	R/C blended yarn
涤/黏纱	polyester/viscose blended yarn; polyester/rayon blended yarn
麻/棉纱	ramie/cotton blended yarn; linen/cotton blended yarn; hemp/cotton blended yarn
麻/涤纱	ramie/polyester blended yarn
麻/黏纱	ramie/viscose blended yarn
麻/腈纱	ramie/acrylic blended yarn
麻/毛纱	ramie/wool blended yarn
麻/丝纱	ramie/silk blended yarn
大麻/天丝纱	hemp/tencel blended yarn

大麻/牦牛绒纱 hemp/yak hair blended yarn
大麻/大豆蛋白纱 hemp/soybean protein blended yarn
维/棉纱 vinylon/cotton blended yarn
黏/棉纱 viscose/cotton blended yarn
腈/棉纱 acrylic/cotton yarn
丝/棉纱 silk/cotton blended yarn
毛/丝纱 wool/silk blended yarn
毛/黏纱 wool/viscose blended yarn
毛/涤纱 wool/polyester blended yarn
毛/锦纱 wool/nylon blended yarn
毛/腈纱 wool/acrylic blended yarn
石棉纱 asbestos yarn
玻璃纱 glass yarn
金属纱线 metal yarn
针织用纱 knitting yarn; hosiery yarn
弹力丝 stretch yarn; elastic yarn; spandex yarn
变形丝 textured yarn
空气变形丝 air-jet texturing yarn
三角异形丝 triangle profile yarn

（二）面料 Fabrics

1. 一般名称 general terms

织物 fabric; cloth
平纹织物 plain weave fabric
斜纹织物 twill weave fabric
缎纹织物 satin weave fabric
布料 fabric; cloth; material
基本布料 basic fabric
机织布料 woven fabric
针织布料 knitted fabric
交织布料 mixed fabric; interwoven fabric; union
混纺布料 blended fabric; mixed fabric; union
衣料 dress material; garment fabric; apparel fabric; fabric; cloth; suiting
高科技衣料 high technology garment fabric
高性能衣料 high performance fabric
智能衣料 intelligent fabric
环保生态衣料 ecological fabric
怀旧衣料 vintage dress material
内衣料 underwear material
面料 shell fabric; outer fabric
上衣料 coating
西服料 suiting
衬衫料 shirting
牛津布衬衫料 Oxford shirting
轻薄衬衫料 zephyr shirting
背心料 vesting
大衣料，外套料 overcoating; coating
风雨衣料 weather cloth
裤料 trousering; panting
裙料 skirting; fabric for dress
套装料 suiting
运动衣料 sportswear fabric
休闲衣料 casual wear fabric
轻薄料 light material
厚重料 heavy material
中厚料 medium-heavy material
涂层布料 coated fabric
水洗布料 washed fabric
磨毛布料 peach fabric; sanded cloth
拉毛布料 napped fabric
阻燃布料 flame-retardant fabric; fire-retardant fabric; antiflaming fabric
复合布料 composite fabric; compound fabric
层压布料 laminated fabric

金原色布料　metallasse〔法〕
植绒织物　flocked fabric
起绒织物　pile fabric
双面织物　reversible cloth; reversibles; double-faced fabric; double-faced material
夹心织物　sandwich fabric
闪光织物　iridescent fabric
网眼织物　mesh fabric; eyelet fabric; lace; openwork
提花织物　jacquard; dobby
弹性织物　elastic fabric; stretch fabric; lycra
（环保）绿色织物　green fabric
海藻纤维织物　seaweed fabric
功能织物　functional fabric
防水织物　waterproof cloth
防水透气织物　breathable waterproof fabric; proofed breathable cloth
防绒织物　downproof fabric
防火织物　fireproof fabric; flameproof fabric
防皱织物　crepe-resisting cloth
防污织物　soil-release fabric
防霉织物　antimist cloth
耐酸织物　acid-proof fabric
抗静电织物　antistatic fabric
抗菌防臭织物　antibacterial deodorant fabric
理疗保健织物　physiotherapy health fabric
磁疗织物，磁性布　magnetotherapy fabric
电热织物　electric heat fabric
静电感应织物　electrostatic induction fabric
放射性织物　radioactive fabric
防紫外线织物　anti-UV fabric
防辐射织物　anti-radiation fabric
防创伤织物　anti-trauma fabric
芳香型织物　fragrant fabric
吸湿排汗织物　moisture wicking fabric; moisture absorbing fabric
按摩织物　massotherapy fabric
保温布　thermal insulation cloth
空调布　air-conditioning cloth
液晶布　liquid crystal cloth
催眠布　hypnotic cloth
变色布　colour-changeable cloth
发光布　luminous cloth
抗癣布　anti-tinea cloth
抗菌布　anti-bacterial cloth
香味布　fragrant fabric
湿光布　wet cloth

2. 面料工艺　fabric technology

纺纱　spinning
织布　weaving; knitting
漂白　bleaching
染色　dyeing
纱线染色　yarn dyeing
匹染　piece dyeing
扎染　tie dyeing
间隔染色　space dyeing; random dyeing
涂料染色　pigment dye
染料　dye; dyestuff
直接染料　direct dye
活性染料　reactive dye
酸（碱）性染料　acid(basic) dye
硫化染料　sulphur dye
还原染料　vat dye
偶氮（非偶氮）染料　azo(azo free) dye
阳离子染料　cation dye
分散染料　disperse dye

荧光染料　optical dye
印花　printing
织物后整理　fabric finish
防缩整理　anti-shrinking finish
防污整理　anti-soiling finish
防臭整理　anti-odour finish
防皱整理　anti-crease finish
防起球整理　anti-pilling finish
防水整理　water-proof finish
防绒整理　down-proof finish
抗菌整理　anti-bacterial finish
抗静电整理　anti-static finish
抗辐射整理　anti-radiation finish
抗紫外线整理　anti-UV finish
阻燃整理　flame-retardant finish; fire-retardant finish; antiflaming finish
亲水性整理　hydrophilic finish
吸湿排汗整理　moisture wicking finish
芳香整理　aromatic finish
生化超柔整理　super soft bio finish
仿麂皮整理　doeskin finish
褶皱整理　wrinkles finish
免烫整理　easy-care finish
耐久性整理　durable finish
轧光整理　calender finish
定型整理　stabilized finish
织物洗水　fabric washing
植绒　flocking
印花（素色）植绒　printed (plain) flocking
拷花植绒　embossed flocking
满地植绒　overall flocking
转移植绒　transfer flocking
牛仔布植绒　jean flocking
针织布植绒　knitted cloth flocking
人棉布植绒　rayon cloth flocking
人造革植绒　leatherette flocking

PVC 植绒　PVC flocking
起绒　raising; emerizing
起绉　creping; crinkling
磨毛　sanding; peach; emerizing
轧光　calendering
涂层　coating
防水涂层　waterproof coating
透气涂层　breathable coating
PVC 涂层　PVC coating
PU 涂层　PU coating
PU 彩色涂层　PU colour coating
PU 双色涂层　PU two-tone coating
PU 蜡光涂层　PU wax coating
ULY 涂层　ULY coating
TPU 涂层　TPU coating
TPE 涂层　TPE coating
PO 涂层　PO coating
PA 涂层　PA coating
PE 涂层　PE coating
防羽绒涂层　downproof coating
阻燃涂层　flame (fire)-retardant coating
珠光涂层　pearl coating
蜡涂层　paraffin coating; wax coating
水晶胶涂层　acrylic coating
植绒涂层　frock coating
乳白涂层　milky coating
银涂层　silver coating
泡沫涂层　foam coating
含油涂层　oily coating
橡胶涂层　rubber coating
海帕伦涂层　hypalon coating

3. 面料性能　texture and quality

匹头　piece goods
质地　texture
质量　quality
标准　standard

规格 specification
成分 composition
组织结构 construction; texture
纱支 count; yarn count; yarn size; yarn number
密度 density
重量 weight
厚度 thickness
长度 length
幅宽 width
内幅宽 cuttable width
匹长 piece length
布面 side
正面 right side
反面 wrong side
布底 background
布纹 grain
幅面 breadth
单幅 single breadth
双幅 double breadth
色牢度 colour fastness
色差 off colour; off shade; colour deviation
疵点 flaw
缩水 shrink
缩水率 shrinkage
防缩 shrink-proof; anti-shrinking
撕裂强度 tear strength
拉伸强度 tensile strength
耐磨强度 abrasion resistance
手感 handle; handfeel; hand; feel
手感柔软 soft handle; soft feel
手感厚实 firm handle
手感粗糙 harsh handle
手感板硬 hard handle
洗水后手感 post wash handfeel
柔软度 softness
柔韧性 flexibility
舒适性 comfort; comfortability
透气性 breathability
吸湿性 absorption
拒水性 water repellency
抗水性 water resistance
悬垂性 drapability; drape; fall
抗皱性 wrinkle resistance
回弹性 resilience
可燃性 flammability
耐用性 durability
试验，测试 test
测试标准 test standard
织物结构（成分）测试 fabric construction (composition) test
织物性能测试 fabric performance test
（织物）耐洗测试 wash test
（织物）耐磨测试 wearing test
色牢度测试 colour fastness test
缩水率测试 shrinkage test
尺寸稳定性及相关测试 dimensional stability and related test

4. 棉织物 cotton fabrics

平纹布，平布 plain cloth; plains; sheeting
细平布 cambric; fine plain; shirting
阔幅平布 sheeting; broadcloth
本色布，本白布，坯布 grey cloth; grey
本白市布 grey sheeting
细薄白布 jaconet
漂白布 bleached cotton cloth; bleached sheeting; calico
染色布 dyed cloth; dyed goods
染色细纺 dyed shirting; dyed sheeting
染色水洗布 dyed washed fabric
素色布 plains

混色布　melange cloth
花色布　fancy cloth
色织布　yarn-dyed fabric
色织细纺　yarn-dyed cambric
色织条格布　yarn-dyed gingham
印花布　chintz calico; cotton print; printed sheeting; prints; printed calico; calico
有机棉布　organic cotton fabric
丝光棉布　mercerized cotton
密织棉布　percale
粗棉布　coarse calico; cheese cloth; dungaree
帆布　canvas; canvas duck
精梳帆布　combed duck
丝光漂白帆布　mercerized and bleached canvas
涂料染色帆布　pigment-dyed canvas
印花帆布　printed canvas
磨毛帆布　suede canvas
府绸　poplin
精梳府绸　combed poplin
色织府绸　yarn-dyed poplin
织花府绸　poplin broché
提花府绸　figured poplin; dobby poplin
牛津布　oxford
青年布，学生布　chambray
靛蓝青年布　indigo chambray
黑色青年布　black chambray
斜纹布，斜布　drill; twill
人字斜纹布　herringbone twill
弹力斜纹布　stretch twill
超细斜纹布　micro twill
厚毛头斜纹布　moleskin
卡其布　khaki; drill
纱卡　single-yarn drill
线卡　ply-yarn drill; drill yarn-twisted

单面卡　single-sided khaki
双面卡　double-sided khaki; reversible khaki
丝光卡其　mercerized khaki; chino
华达呢，轧别丁　gabercord; cotton gabardine
哔叽　serge; twills
原色哔叽　beige
混色哔叽　melange serge
缎纹哔叽　satin serge
粗纹哔叽　wide wale serge
小方块纹哔叽　serge canvas
细密防水哔叽　imperial serge
哔叽呢　serge cloth
线呢　twine cloth; cotton suiting
混色线呢　beige
棉缎，横贡缎　sateen
直贡缎　satin drill
直贡呢　venetian
元贡呢　cotton venetian
泰西缎　satin drill; cotton venetians
罗缎　tussores; bengaline; faille; grosgrain
全棉弹力布　cotton lycra
全棉防羽布　cotton dyed down-proof cloth
全棉防羽缎　cotton down-proof satin drill
粗斜纹棉布，劳动布，坚固呢　denim; jean
色织防缩坚固呢（劳动布）　yarn-dyed preshrunk denim
光洁厚斜纹布　satin jean
粗斜纹布　fustian
细斜纹布　jeanette
牛仔布　denim; jean
印地可（靛蓝）牛仔布　indigo denim
超靛蓝牛仔布　ultra indigo denim
黑色牛仔布　black denim
漂白牛仔布　bleached denim

棉织物

彩色牛仔布	coloured denim	waled corduroy	
套色牛仔布	overdyed denim	经条灯芯绒	bedford cord
印花牛仔布	printed denim	染色灯芯绒	dyed corduroy
装饰牛仔布	comestic denim	印花灯芯绒	printed corduroy
功能牛仔布	functional denim	雪花灯芯绒	snowflake corduroy
环锭纺牛仔布	ring spinning denim	砂洗灯芯绒	sand-washed corduroy
转杯纺牛仔布	rotor spinning denim	平绒，棉绒	velveteen
金银丝牛仔布	lurex denim	绒布，棉法兰绒	flannel; flannelet(te)
冲孔牛仔布	punching denim	斜纹法兰绒	twilled flannel
烂花牛仔布	burnt-out denim	起绒布，抓绒布	fleece; flannelet
提花牛仔布	jacquard denim	单面绒布	single faced flannelet
小提花牛仔布	dobby denim	双面绒布	reversible flannelet; both-side-raised velveteen
针织牛仔布	knitted denim		
植绒牛仔布	flocked denim	双面厚绒布	domette
竹节牛仔布	necked denim	印花绒布	printed flannelet
光亮牛仔布	satin jean	格绒布	raising gingham; check flannelet; brushed check shirting
扎染牛仔布	tie-dyed denim		
弹力牛仔布	stretch denim; lycra denim; spandex denim; elastic denim	全棉防缩色织格绒布	cotton yarn-dyed preshrunk checked flannelet
竹节弹力牛仔布	stretch necked denim	印格绒布	print raising gingham
层压牛仔布	laminated denim	绳绒线布	chenille fabric
涂层牛仔布	coating denim	绉布，绉纱	crepe
轻身牛仔布	lightweight denim	树皮绉	crepon〔法〕
超薄牛仔布	ultrathin denim	绉纹呢	creppella
条纹牛仔布	stripe denim	折皱布	wrinkle fabric
缎纹牛仔布	satin denim	泡泡纱	seersucker; blister crepe; cloqué〔法〕
白背牛仔布	white-back denim		
深色牛仔布	dark-coloured denim	格子泡泡纱	seersucker gingham
水洗牛仔布	washed denim	巴厘纱，玻璃纱	voile
特效洗水牛仔布	specific washed denim	蝉翼纱	organdie; organdy
再生牛仔布	recycled denim	透明薄纱	sheer; voile
灯芯布	pique	珠罗纱	bobbinet
灯芯绒	corduroy; cord; fustian cord〔英〕	麻纱	hair cords
		条纹布	striped cloth; stripe
粗条灯芯绒	bold wale corduroy; wide wale corduroy; cable cord	格子布	checked fabric; check; gingham
		竹节布	slubbed fabric
中条灯芯绒	medium wale corduroy	轧纹布，凸凹轧花布	gauffered cloth
细条灯芯绒	fine-needle corduroy; pin-		

拷花布　embossed cloth;gauffered cloth
提花布　figured cloth;jacquard fabric;faconne〔法〕
烂花布　burnt-out fabric
网眼布　mesh fabric
手工土布　homespun
毛蓝花布　printed blue nankin
靛蓝花布　indigo blue printed fabric
蜡防花布　wax resist printed fabric;batik
高山族花布　Gaoshan fancy cloth
单向导湿棉布　cotton fabric single side moisture transported
环保棉布　ecological cotton

5. 麻织物　bast fabrics

苎麻布　ramie fabric
苎麻平布　ramie plains
苎麻斜纹布　ramie twill
苎麻牛仔布　ramie denim
苎麻漂白细布　ramie white sheeting
苎麻细平布　ramie shirting
苎麻色织布　ramie yarn-dyed cloth
纯苎麻爽丽绸　pure ramie sheer
夏布　grass cloth;grass linen
亚麻布　linen cloth;flax fabric
本色亚麻布　brown linen
亚麻平布　linen plains;linen cambric
细亚麻平布　boiled lawns
漂白亚麻布　bleached linen;argouges
色织亚麻布　yarn-dyed linen
亚麻格布　linen checks
亚麻花布　linen damask
亚麻绉布　linen crepe
亚麻帆布　linen canvas;linen duck
亚麻牛仔布　flax denim
亚麻细布　fine linen
细麻布　grass lawn
薄麻布　toile
法国上等细亚麻布　batiste

6. 丝织物　silk fabrics

纺绸，电力纺　habotai;habutae;habutai
杭纺　Hangzhou habotai
富春纺　fuchun habotai;taffeta fuchun rayon
华春纺　huachun habotai
交织纺绸　interweaved habotai
提花纺绸　figured habotai
绸　silk
真丝绸　natural silk;real silk
彩绸　coloured silk
花绸　figured silk
印花（染色）绸　printed(dyed)silk
印花薄绸　foulard
水洗绸　washed silk;washable silk
间隔染色绸　space dyed silk
双宫绸　doupion silk;doupioni;douppioni
蓓花绸　rayon beihua
艾德莱斯绸　Adelis silk
和服绸　kimono silk;kimono brocade
领带绸　tie silk
云纹绸　moire
波纹绸　watered silk
树皮纹绸　bark silk
光亮绸　lutestring;lustring;lustrine
闪光绸　shot silk
蜡光绸　cire silk
柞丝绸　tussah silk;tussah cloth;tussah pongee;pongee silk
山东柞丝绸　Shantung pongee;shantung
鸭江绸　Yajiang tussah pongee
霓裳绸　Nichang silk
绵绸，䌷丝绸　silk noil poplin;noil silk;noil cloth;noilcloth
波斯绸　Persian

丝织物

格子绸　silk tartan
斜纹绸　silk twill
绡　sheer silk
绫　damask silk
广绫　Guangdong satin
缎　satin
真丝缎　silk satin
软缎　mixed satin
花软缎　mixed satin brocade
素软缎　mixed satin plain
织锦缎　tapestry satin; satin brocade; brocade
修花缎　broche satin
金雕缎　kingtiao satin
库缎，贡缎　palace satin
直贡缎　satin drill; twill satin
绉缎，绉背缎　crepe satin
查米尤斯绉缎　charmeuse
素绉缎　crepe satin plain
素面缎　plain satin
全丝薄缎　satinette
花缎　figured satin; brocade; damask
闪光缎　paillette satin
女公爵缎　duchesse satin
双面缎　reversible satin
弹力缎　lycra satin
桑波缎　Sangbo satin; Sangbo crepe damask; jacquard silk satin
古香缎　Suzhou brocade
锦，锦缎　brocade; damask
宋锦　Song (dynasty) brocade
蜀锦　Shu (Sichuan) brocade
云锦　Yun (Nanjing) brocade
土家锦　Tujia (nationality) brocade
侗锦　Dong (nationality) brocade
壮锦　Zhuang (nationality) brocade
苗锦　Miao (nationality) brocade
傣锦　Dai (nationality) brocade
布依锦　Buyi (nationality) brocade
春香锦　Chunxiang brocade
绉　crepe
真丝绉　silk crepe
真丝双绉　crepe de chine〔法〕
双宫绉　doupion silk crepe
碧绉，印度绸　kabe crepe
留香绉　shameuse
冠乐绉　guanle clogué〔法〕
顺纡绉，乔其绉　crepe georgette
乔其纱　georgette
雪纺乔其纱　chiffon georgette
烂花乔其纱　etched-out georgette
罗　gauze; leno
杭罗　Hangzhou silk gauze; Hangzhou leno
桑绢罗　rib silk
纱　gauze
香云纱，莨纱　gambiered Canton gauze; watered gauze
庐山纱　Lushan gauze
西湖纱　Xihu gauze
夏夜纱　summer night gauze
素纱　plain gauze
花纱　jacquard gauze
绫，真丝绫　ghatpot; twills
美丽绸　rayon lining twill
绢，绢丝纺　spun silk; spun silk habotai; spun silk pongee; Fuji silk
塔夫绸　taffeta
闪光塔夫绸　changeable taffeta; taffeta glacé〔法〕
格子塔夫绸　quadrille taffeta
蚕丝塔夫绸　silk taffeta
绢丝塔夫绸　spun silk taffeta
双宫丝塔夫绸　doupion silk taffeta

雪纺，绡　chiffon
呢　suiting silk
新华呢　Xinhua suiting silk
西湖呢　Xihu jacquard crepon
四维呢　ottoman silk
绒，丝绒　velvet;velours
平绒　panne velvet;panne
立绒呢　cut velvet
粒粒绒　pellet fleece velvet
天鹅绒　velvet;velour
素色天鹅绒　solid velvet
混色天鹅绒　melange velvet
轧花天鹅绒　ginning velvet
抽条磨毛天鹅绒　rib fleece velvet
金丝绒　peluche velvet
乔其绒　georgette velvet;transparent velvet
凸凹绒　sculptured velvet
光明绒　bright velvet
雪纺丝绒　chiffon velvet
双绉丝绒　velours crepe de chine〔法〕
闪光丝绒　velours glacé〔法〕
透明丝绒　transparent velvet
拷花丝绒　embossed velvet
烂花丝绒　faconné velvet〔法〕
缎花丝绒　brocade velvet
绫条丝绒　rib velvet
葛　grosgrain
绨　bengaline;faille〔法〕

7. 毛织物　wool fabrics

呢绒　woolen goods; wool fabric; woolen cloth;woolens;stuff;cloth
双面呢绒　double-faced woolen goods
弹力呢绒　lycra woolen goods
精纺毛织物　worsted fabric
全毛华达呢　pure wool gabardine
全毛单面华达呢　wool one-side gabardine
全毛缎背华达呢　wool satin-backed gabardine
全毛哔叽　wool serge
凡立丁　valitin;tropical suiting
巧克丁　tricotine
马裤呢　whipcord;cavalry twill
派力司　palace
啥味呢　twill coating;worsted flannel;cheviot
女衣呢　lady's cloth and dress worsted; ladies' suiting;dress worsted
女骑装呢　habit cloth
雪特兰呢　Shetland
直贡呢，礼服呢　venetian
驼丝锦　doeskin
花呢　fancy suiting
精纺花呢　worsted fancy suiting;fancy worsted
半精纺花呢　semi-worsted fancy suiting
单面花呢　one-side fancy suiting
海力蒙，人字呢　herringbone
板司呢　basket cloth;hopsack
波拉呢　poral
薄毛哔叽　twill muslin
细薄平纹毛织物　muslin delaine
精纺平纹呢　wool poplin
精纺麦尔登　worsted melton
绉纹呢　wool crepe
绒面呢　wool broadcloth
雪纺呢　wool chiffon
细薄呢　wool batiste
开士米薄毛呢　cashmere; cassimere; kashmir
中厚呢　medium cloth
粗纺毛织物　woolen cloth;woolen fabric; woolen goods;woolens
麦尔登　melton

海军呢　navy cloth; navy coating; admiralty cloth; navy suiting
制服呢　uniform cloth; livery cloth; livery tweed; uniform coating
学生呢　school uniform cloth; students' uniform cloth; student coating
大众呢　union cloth
女式呢　ladies cloth; ladies coating
艾博茨福德女式呢　abbotsford
素呢　plain coating
薄呢　woolenet; woolenette
麦尔登薄呢　meltonette
雪克斯金细呢　sharkskin
法兰绒　flannel
（英国）约克夏全毛法兰绒　Yorkshire flannel
萨克森法兰绒　Saxony
游艇呢　yacht cloth
格子呢　tartan; woolen check
苏格兰格子呢　Scotch tartan
军服呢　military cloth; army coating
海员厚绒呢　pilot cloth
缩绒厚呢　box cloth
洛登缩绒（防水）厚呢　loden
斜纹厚呢　fleuret
起绒精呢　duffel cloth
粗毛呢　bearcloth
粗花呢　tweed; tweed flannel; fancy woolens; costume tweed
海力斯粗花呢　Harris tweed
人字粗花呢　herringbone tweed
格子粗花呢　tartan tweed; chequered woolens
结子粗花呢　knot tweed
科纳马拉粗花呢　connemara tweed
帕西米纳羊绒粗呢　pashmina tweed
火姆司本，手工织呢　homespun

羽毛呢　feather cloth
兔毛呢　rabbit hair cloth
海狸毛呢　beaver cloth
长毛绒　plush; high pile
大衣呢　overcoat suiting; overcoating
全毛大衣呢　pure wool overcoating
拷花大衣呢　embossed overcoating
雪花大衣呢　snowflake overcoating
花式大衣呢　fancy overcoating
马海毛大衣呢　mohair fleece
兔毛大衣呢　rabbit hair overcoating
驼绒大衣呢　camel hair overcoating
牦牛绒大衣呢　yak hair overcoating
羊绒大衣呢　cashmere overcoating; cashmere
兔羊绒大衣呢　angora cashmere overcoating
骆马毛大衣呢　vicuna overcoating
长毛大衣呢　woolen fleece
短绒大衣呢　kerseymere
拉绒大衣呢　velours overcoating
粗绒大衣呢　fearnaught; fearnought
立绒大衣呢　stand plush overcoating
毛毯大衣呢　blanket cloth
阿尔斯特长毛大衣呢　ulster cloth; ulster
马基诺双面大衣呢　mackinaw coating
博柏利大衣呢　Burberry
珠皮大衣呢　ratine; chinchilla; petersham
手织大衣呢　homespun

8. 化纤和混纺织物　Chemical Fibre and Blended Fabrics

混纺（交织）化纤织物　blended (union) chemical fibre fabric
变形丝织物　textured yarn fabric
仿毛织物　wool-like fabric; imitation wool fabric
仿棉织物　cotton-like fabric

仿丝织物　silk-like fabric
仿麻织物　linen-like fabric
仿纱型织物　spun-like fabric
黏胶纤维织物　spun rayon fabric
合成纤维织物，合纤布　synthetic fabric
涤纶布料　polyester fabric; dacron fabric; terylene fabric
葆莱绒布料　prolivon fabric
涤纶西服料　polyester suiting
涤纶华达呢　polyester gabardine
涤纶哔叽　polyester serge
涤纶花呢　polyester fancy suiting
涤纶低弹牛津布　polyester texture oxford
涤纶桃皮绒（水蜜桃）　polyester microfibre fabric; polyester peachskin fabric
涤纶苔纹桃皮绒　polyester moss microfibre fabric
涤沦全消光桃皮绒　full dull polyester peachskin fabric
涤纶宽斜纹桃皮绒　big twill polyester peachskin fabric
涤锦复合桃皮绒　polyester/nylon peachskin fabric
涤锦交织桃皮绒　polyester/nylon interwoven peachskin fabric
涤纶花缎　polyester brocade
涤纶缎　polyester satin
涤纶纺，涤纶绸　polyester habotai
涤纶绉　polyester crepe
（涤纶）顺纤绉　polyester crinkle fabric
涤纶塔夫绸　polyester taffeta
涤纶蜂巢（状）塔斯纶　polyester honeycomb taslan
涤纶乔其纱，柔姿纱　polyester georgette
涤纶素色乔其纱　polyester georgette plain dyed
涤纶印花乔其纱　polyester georgette printed
卡丹绒　peach twill
绉绒　peach moss
春亚纺　polyester pongee
春亚纺格布　polyester pongee rip-stop
哑富迪　micro polyester pongee
仿毛华达呢　imitation-wool gabardine
仿毛哔叽　imitation-wool serge
仿毛花呢　wool-like fancy suiting; imitation-wool fancy suiting
的确良　dacron
快巴的确良，涤黏布　polyester/viscose blended fabric
涤纶 PVC 涂层布　polyester/PVC coated fabric
涤/棉织物，棉/涤织物　polyester/cotton fabric; T/C fabric
漂白涤/棉布　bleached trueran fabric
染色涤/棉布　dyed trueran fabric
印花涤/棉布　printed trueran fabric
色织涤/棉布　yarn-dyed trueran fabric
涤/棉线绢　T/C tussores
涤/棉细布，涤细　polyester/cotton cambric; T/C shirting; truran lawn
涤/棉平布，涤平　polyester/cotton plain cloth; truran sheeting
涤/棉府绸，涤府　polyester/cotton poplin; truran poplin
轧光涤府　T/C gloss finished poplin; T/C chintz finished poplin
拷花涤府　T/C embossed poplin
涂层涤府　T/C coated poplin
磨毛涤府　T/C emerized poplin
色织涤府　T/C yarn-dyed poplin
色织涤/棉格布　T/C yarn-dyed check fabric
涤/棉卡其，涤卡　polyester/cotton drill;

trueran drill
涤/棉凉爽绸 polyester/cotton liangsan
色织涤/棉纬长丝 yarn-dyed T/C welt filament cloth
色织涤/棉灯芯布 trueran yarn-dyed pique
T/R（涤黏）混纺布 T/R cloth
CVC（棉涤）混纺布 CVC cloth
麻/涤布 ramie/polyester blended cloth
麻/黏布 ramie/viscose blended cloth
麻/棉布 ramie/cotton blended fabric; linen/cotton blended fabric
麻/棉平布 ramie/cotton blended plains
麻/棉斜纹布 ramie/cotton blended twill
麻/棉色织布 ramie/cotton blended yarn-dyed fabric
麻/棉交织布 ramie/cotton mixed plains
麻/棉牛仔布 ramie/cotton denim
大麻/棉牛仔布 hemp/cotton denim
大麻棉交织布 bataloni
大麻毛混纺粗花呢 hemp/wool tweed
丝麻交织布 silk/ramie mixed cloth
亚麻交织布 union linen
亚麻棉混纺（交织）布 linen/cotton blended(mixed)fabric
亚麻/棉牛仔布 flax/cotton denim
天丝/竹纤维牛仔布 tencel/bamboo fibre denim
天丝亚麻混纺上衣料 tencel/linen coating
天丝锦纶交织布 tencel/nylon mixed fabric
天丝/毛华达呢 tencel/wool gabardeen
天丝/涤纶华达呢 tencel/polyester gabardeen
天丝/涤纶府绸 tencel/polyester poplin
天丝铜氨交织仿真丝绸 tencel/cuprene mixed silk

天丝棉混纺牛仔布 tencel/cotton denim
锦纶织物，锦纶布 nylon fabric
锦纶绸，锦纶纺 nylonpalace; nylon taffeta;ninon〔法〕
锦纶塔夫绸 nylon taffeta
锦纶塔夫泡泡纱 nylon seersucker taffeta
锦纶绉绸 nylon crinkle fabric
锦纶闪光绸 nylon trilobal
锦纶塔斯纶 nylon taslan
锦纶哔叽 nylon serge
锦纶斜纹布 nylon twill
锦纶牛津布 nylon Oxford
锦纶格子布 nylon rip-stop
锦纶交织布 nylon interweave
锦纶桃皮绒 nylon microfibre fabric
锦纶苔纹桃皮绒 nylon moss microfibre fabric
锦纶丝绒 nylon velvet
锦纶雪纺 nylon chiffon
锦纶网眼纱 nylon meshes
锦纶莱卡 nylon lycra
锦纶PVC（PU）涂层布 nylon taffeta with PVC(PU)coating
锦/棉纺 nylon/cotton fabric
腈纶织物，腈纶布 acrylic fabric
德绒面料 dralon fabric
仿真丝绸 imitated silk fabric
丝/腈织物 silk/acrylic blended fabric
丝/麻绸 silk/ramie fabric
丝/棉绸 silk/cotton fabric; silk/cotton chiffon
绨 bengaline;silk-cotton goods;faille
人造丝纺绸 rayon habotai
人造丝罗缎 rayon bengaline
人造丝塔夫绸 rayon taffeta
人造丝乔其纱 rayon georgette

人造丝丝绒 rayon velvet
人造丝双绉 rayon crepe de chine[法]
人造丝缎背绉 rayon satin backed crepe
有光人丝牛仔布 iridescent rayon denim
人丝华达呢 albatross
人丝西服料 rayon suiting
醋酯缎 acetate satin
毛/涤织物 wool/polyester fabric
毛/涤华达呢 wool/polyester gabardine
毛/涤哔叽 wool/polyester serge
毛/涤花呢 wool/polyester fancy suiting
毛/涤派力司 wool/polyester palace
毛/涤凡立丁 wool/polyester valitin
毛/涤巧克丁 wool/polyester tricotine
毛/涤女衣呢 wool/polyester ladies cloth and dress worsted
毛/涤海力蒙 wool/polyester herringbone
毛/涤直贡呢 wool/polyester venetian
毛/涤马裤呢 wool/polyester whipcord
毛/涤啥味呢 wool/polyester cheviot
毛/黏织物 wool/viscose fabric
毛/黏华达呢 wool/viscose gabardine
毛/黏哔叽 wool/viscose serge
毛/黏花呢 wool/viscose fancy suiting
毛/黏马裤呢 wool/viscose whipcord
毛/黏啥味呢 wool/viscose cheviot
毛/黏格子大衣呢 wool/viscose tartan overcoating
毛/黏隐格大衣呢 wool/viscose hidden tartan overcoating
毛黏混纺麦尔登 wool/viscose blended melton
毛黏混纺海军呢 wool/viscose navy cloth
毛黏混纺制服呢 wool/viscose uniform cloth
毛黏混纺女式呢 wool/viscose ladies cloth
毛/黏法兰绒 wool/viscose flannel
丝毛法兰绒 zephyr flannel

花式法兰绒 fancy flannel
绉纹法兰绒 crepe flannel
毛麻混纺呢绒 wool/linen blended fabric
毛锦混纺呢绒 wool/nylon fabric
毛腈混纺呢绒 wool/acrylic fabric
三合一（毛/涤/黏）啥味呢 wool/polyester/viscose cheviot
毛葛，棉毛呢 paramatta
肯塔基棉毛呢 Kentucky jeans
涤/黏哔叽 polyester/viscose serge
涤/黏凡立丁 polyester/viscose valitin
涤/黏花呢 polyester/viscose fancy suiting
黏/锦华达呢 viscose/polyamide gabardine
涤/腈哔叽 polyester/acrylic serge
涤/麻派力斯 polyester/ramie palace
腈纶格布 acrylic gingham
腈纶花呢 acrylic fancy suiting
腈纶大衣呢 acrylic overcoating
腈纶提花大衣呢 acrylic jacquard overcoating
中长织物 midfibre fabric
中长华达呢 midfibre gabardine
中长哔叽 midfibre serge
中长花呢 midfibre fancy suiting
中长膨体大衣呢 midfibre bulked overcoating

9. 毛皮，皮革 fur and leather

海狸毛皮 beaver
狐皮 fox fur
赤狐皮 red fox fur
沙狐皮 kit fox fur
貂皮 marten fur; mink fur; mink
黑貂皮 sable; zibeline; zibelline
水貂皮 ranch mink fur

毛皮，皮革

扫雪貂皮	stone marten fur
雪貂皮	ferret fur
海獭毛皮	sea otter fur
水獭毛皮	land otter fur
旱獭毛皮	marmot fur
狸獭毛皮	nutria fur
羊毛皮	woolfell; woolskin
山羊毛皮	goat fur
小山羊毛皮	kid fur
羔羊毛皮	lambskin
绵羊毛皮	sheep fur
羊剪绒毛皮	sheared sheepskin; shorn sheepskin
兔毛皮	rabbit fur; rabbit
野兔毛皮	hare fur
獭兔毛皮	beaver fur
黄鼠狼毛皮	weasel fur
狼毛皮	wolf fur
虎毛皮	tiger fur
豹毛皮	leopard fur
浣熊毛皮	raccoon
貂熊毛皮	wolverene; wolverine
狗毛皮	dog fur
猫毛皮	cat fur
豹猫毛皮	leopard cat fur
袋鼠毛皮	kangaroo fur
松鼠毛皮	squirrel fur; squirrel
灰鼠毛皮	chinchilla
鼹鼠毛皮	moleskin
臭鼬毛皮	skunk
服装革	leather for garments; clothing leather
手套革	leather for gloves; gloves leather
制帽革	hat leather
制鞋革	shoes leather
鞋面革	leather for shoe cover
鞋里革	leather for shoe lining
制带革	belt leather; buckle leather
箱包革	luggage leather
衬里革	lining leather
全粒面革，头层革	full grain leather
修面革	corrected grain leather
猪皮（革）	pigskin; pig leather; swine leather
猪光面皮	pig nappa
猪绒面皮	pig suede
猪二层皮	pig split
山羊皮	goatskin; goat leather
小山羊皮	kid; kidskin; kid leather
羔羊皮	lambskin
绵羊皮	sheepskin; sheep leather
牛皮	cowhide; cow leather; cattlehide
小牛皮	calf leather
马皮	horse leather
骡皮	mule hide
驴皮	donkey skin
犀牛皮	rhinoceros leather
羚羊皮	antelope leather
麂皮	chamois; chamois leather
油鞣皮，雪米皮	chamois-dressed leather
鹿皮	deerskin; buckskin
海豹皮	sealskin
鳄鱼皮	crocodile skin; crocodile leather; alligator
鲨鱼皮	shark leather; sharkskin
鲸鱼皮	whale leather
海象皮	walrus skin
蛇皮	snake skin
蟒蛇皮	python skin
蜥蜴皮	lizard leather
狗皮革	dog leather
袋鼠皮革	kangaroo leather
鸵鸟皮革	ostrich leather

拷花皮革　embossed leather
染色毛皮　dyed fur
印花毛皮　printed fur
人造毛皮　fake fur; artificial fur; fur cloth; fur fabric; imitation fur; faux fur
腈纶人造毛皮　acrylic boa
素色（印花，拷花）人造毛皮　plain (printed, embossed) fake fur
仿羔皮织物　astrakhan; sherpa
腈纶长毛绒　acrylic plush
仿獐皮（阿尔坎塔拉）　alcantara
人造革　artificial leather; leatheret; imitation leather; simulated leather; faux leather
人造麂皮　suede fabric; micro suede; suede; suedine; suedette
涤/黏仿麂皮　T/R micro suede
弹力仿麂皮　spendex micro suede
仿麂皮针织布复合料　micro suede bounding with knitting fabric
仿麂皮羊羔绒复合布　micro suede bounding with lamb fur
仿麂皮摇粒绒复合布　micro suede bounding with polar fleece
反绒皮，獐皮　suede leather; suede
牛（羊，猪）獐皮　cow(goat, pig) suede
合成革　synthetic leather
PU革　PU leather
巴西PU革　Brazilian PU leather
PVC革　PVC leather
光面革　slick-surfaced leather; smooth leather; nappa leather
绒面革　napped leather; suede leather; suede
磨面革　boarded leather; buffed leather
磨砂革　nubuck leather
粒面革　grain leather

皱纹革　shrink leather; levant leather
烫金属革　metallized leather
透气皮革　breathable leather
水洗皮　washable leather
激光革　laser leather
软皮　soft leather; casting leather
染色皮　dyed leather
双色皮　two-tone leather
扎染皮　tie-dyed leather
漆皮　patent leather; enamel leather
打蜡皮，上光皮　burnished leather
进口皮革　import leather
二层皮革　split leather
旧皮革　used leather
皮革废料　waste leather

10. 针织物及其他布料　knitted fabrics and other materials

针织品　knitwear
棉（麻、丝、毛）针织品　cotton (linen, silk, woolen) knitwear
合纤针织品　synthetic knitwear
混纺针织品　blended fibre knitwear
针织布　knit fabric; knitted fabric; jersey cloth
漂白针织布　bleached knit fabric
染色针织布　dyed knit fabric
单（双）面针织布　single (double) knit fabric; single (double) jersey
双面毛针织布　woolen double jersey
提花针织布　jacquard knit fabric
毛圈针织布　knitted loop cloth; terry knit fabric
罗纹针织布　rib knit fabric
长毛绒针织布　high pile knit fabric
双反面针织布　pearl fabric; purl fabric
经编针织布　warp knitting fabric; tricot

经编针织平布　warp knitting jersey
双（多）梳栉经编针织布　double(multi)-bar warp knit fabric
纬编针织布　weft knitting fabric
弹力针织布　spandex knit fabric
弹力平纹针织布　lycra jersey
珠地平纹针织布　piqué with jersey〔法〕
斯佩西运动装针织布　Spacie
人丝针织布　rayon jersey
大麻/棉针织布　hemp/cotton knitted fabric
木（竹）纤维针织布　wood（bamboo）jersey fabric
有机棉针织布　organic cotton jersey
亚麻针织布　linen jersey cloth
圆筒针织布　circular knit
匹头针织布　jersey piece goods
珠地布　piqué〔法〕
单（双）珠地布　single(double)piqué
弹力珠地布　lycra piqué
提花珠地布　jacquard piqué
缝编布　stitch-bonded fabric
棉毛布　interlock fabric; interlock rib; interlock
色织（印花，本色）棉毛布　yarn-dyed (printed, grey) interlock fabric
卫衣布　fleece
素色卫衣布　solid fleece
单（双）卫衣布　single(double)fleece
鱼鳞卫衣布　rib double fleece
珠地卫衣布　lacoste fleece
汗衫布　undershirt cloth; single jersey
单（双）纱汗布　single(heavy) jersey
彩条汗布　colour-stripes single jersey
增强汗布　impact jersey
复合汗布　fusing jersey
毛巾布　terry cloth

经编呢　warp-knitting cloth
经编灯芯绒　warp knitted corduroy
经编麂皮绒　warp knitted suede
针织弹力麂皮绒　lycra knitted suede
针织压花麂皮绒　embossed knitted suede
拉舍尔经编长毛绒　raschel plush
驼绒　lambsdown; camel hair cloth
针织丝绒　knitted velour
（针织）平绒　plain velvet; panne velvet
针织人造毛皮　fake fur knitted fabric
针织长毛绒　knitted plush; knitted high pile
针织起绒布　knitted fleece; fleeced fabric
摇粒绒　polyester polar fleece
蚂蚁绒　fleece in one side
珊瑚绒　coral fleece; coral velvet
舒棉绒　Shu velveteen
莱卡弹性织物　lycra
泳装衣料　swimsuit fabric
网眼布　mesh fabric; eyelel fabric
毛毡　felt
石棉布　asbestos cloth
竹（木）纤维布　bamboo（wood）fibre cloth
玻璃纤维布　glass fibre woven cloth
油布　oilcloth; oilskin; oil silk; tarpaulin
塑料布　plastic cloth
防雨布　waterproof cloth; raincloth; crevenette; tarpaulin
单面雨衣胶布　PVC/polyester fabric
双面雨衣胶布　PVC/polyester/PVC fabric

（三）辅料　Accessories

1. 里料，衬里　lining fabric; lining

羽纱　camlet; lustre lining

人字羽纱　herringbone camlet
斜纹羽纱　twill camlet
人丝羽纱　rayon lining
蜡线羽纱　lining silk
美丽绸　rayon lining twill; rayon lining silk
蜡线美丽绸　lining silk
里子缎　lining satin
涤纶里子缎　polyester lining satin
人丝里子缎　rayon lining satin
锦纶里子缎　nylon lining satin
意大利缎　Italian cloth
富春纺　fuchun habotai
电力纺　habotai
斜纹绸　silk twill
里子绸　lining silk; lining taffeta
里子薄绸　sarcenet
里子塔夫绸　lining taffeta
锦纶塔夫绸，锦纶绸，锦纶纺　nylon taffeta
涤纶塔夫绸，涤纶绸，涤丝纺　polyester taffeta
醋酸人丝波纹里子绸　acetate moire lining
醋酸斜纹里子绸　acetate twill lining
醋酸提花里子绸　acetate jacquard lining
锦纶醋酸提花里子绸　nylon/acetate jacquard lining
涤/黏提花里子绸　polyester/viscose jacquard lining
涤纶提花里子绸　polyester jacquard lining
吸湿排汗衬里　moisture wicking lining
小提花衬里　dobby lining
细布衬里；麻纱衬里　lining cambric
帆布衬里　lining duck
亚麻布衬里　linen lining
帆布型亚麻衬里　linen roughs
绒布衬里　fleece lining; lining flannelette; flannel
驼绒衬里　lining lambsdown
丝绒衬里　lining velvet
毛绒衬里　lining pile
马海呢衬里　lining mohair cloth
毛毡衬里　lining felt
皮革衬里　lining leather
毛皮衬里　lining fur; furring
剪毛羊皮衬里　lining shearling
条子斜纹衬里　zeck
棉衬里布　cotton lining; sleek
棉府绸　cotton poplin
棉印花里布　printed cotton lining
珊瑚绒　coral fleece; coral velvet
摇粒绒　polyester polar fleece
印花摇粒绒　polyester polar fleece printed
棉针织里布　cotton knitted lining; jersey lining
涤/棉针织里布　T/C knitted lining
网眼布　mesh; tulle; lace
锦纶网布　nylon meshes
涤纶网布　polyester meshes
杜邦（防绒）纸　Dupoint paper
原身衬里　self lining
彼得舍姆硬衬里　petersham

2. 衬料，衬布　interlining; underlining; interfacing cloth; interfacing

轻薄(厚重)衬布　light (heavy)-weight interlining
薄膜衬　film interlining
树脂衬　resin interlining
树脂领衬　resin collar interlining
领角薄膜衬　membrane collar interlining

布衬 cloth interlining
本白布衬，法西衬 interlining grey cloth
漂白布衬 interlining bleached cloth
麻布衬 bast interlining
粗布衬 interlining canvas
绒布衬 interlining flannel
涤/棉衬 trueran interlining
蝉翼纱衬 interlining organdie
玻璃纱衬 interlining organza
保暖衬 heat-sealable interlining
上浆衬 starched interlining
撒粉衬 dusted interlining
水溶衬 water-soluble interlining
热熔衬 hot-melt-adhesive interlining; fusible interlining; iron-on interlining
黏合衬 adhesive-bonded interlining; fusible interlining
双面黏合衬 double-faced fusible interlining
弹力黏合衬 stretch fusible interlining
非黏合衬 non-fusible interlining
非织造布衬 non-woven interlining
非织造黏合衬 fusible non-woven interlining
有纺衬，机织衬 woven interlining
有纺黏合衬 fusible woven interlining
针织衬 knitted interlining
针织黏合衬 knitted fusible interlining
黑炭衬，毛鬃衬 hair interlining
马尾衬 horsehair interlining; ponytail interlining
化纤衬 chemical fibre interlining
聚酯衬 polyester interlining
黏胶衬 viscose interlining
上蜡软衬 waxed soft interlining
石棉衬 asbestos interlining
玻璃丝衬 glass-silk interlining

3. 填料 filler; filling; stuffing; wad; wadding; padding

棉花 cotton
人造棉 staple rayon; spun rayon; artificial cotton
3D 直立棉 3D vertical cotton
生态棉 ecological cotton
喷胶棉 staple rayon; polyfill; filler; polyester padding
涤纶棉 polyester staple fibre; polyester padding; polyfill
腈纶棉 acrylic staple fibre
针刺棉 punched cotton
热熔棉 thermofusible cotton
太空棉 space cotton
竹纤维棉 bamboo fibre cotton
玉米纤维棉 corn fibre cotton
莫代尔纤维棉 modal fibre cotton
蓬松散棉 spongy filler
软硬棉 soft and stiff filler
水洗棉 washing filler
絮片 wadding; wad; batting
喷胶棉絮片 staple rayon wadding
热熔棉絮片 thermofusible cotton wadding
竹纤维棉絮片 bamboo fibre wadding
玉米纤维棉絮片 corn fibre wadding
丝绵 silk floss; silk wadding
仿丝绵 imitation silk floss
丝光木棉 silk cotton; kapok; ceiba fibre
动物毛绒 animal wool
羊毛 wool
羊绒 cashmere
羽绒 down
鸭绒 duck down; eiderdown
鹅绒 goose down
仿羽绒 fake down

海绵 sponge
泡沫塑料 form plastic
胆料 liner
天然毛皮 natural fur
人造毛皮 fake fur
长毛绒 plush
驼绒 lambsdown
混合材料填料 blended material filler
特殊材料填料 special material filler

4. 线，纽扣，拉链 threads, buttons and zippers

① 线 thread

缝纫用线 sewing thread
面（底）线 top(bottom) thread
细（粗）线 thin(thick) thread
缝纫丝线 twist
粗缝用线 tacking thread; basting thread
绷缝用线 covering thread
包缝用线 overlocking thread
暗缝缲边线 blind hemming thread
锁凤眼线 eyelet buttonholing thread
棉线 cotton thread
精梳全棉白线 combed yarn cotton white thread
麻线 flax thread; linen thread
丝线 silk thread
绣花线 embroidery thread
绣花丝线 floss silk; floss
刺绣绒线 crewel
粗丝线 tailor's twist
弹力线 elastic thread; stretch thread
涤纶线 polyester thread
PP 线 spun polyester thread
涤纶高强线 polyester high tenacity thread
涤/棉线 truerun thread; T/C thread

维纶线 vinylon thread
锦纶线 polyamide thread; nylon thread
合纤线 synthetic thread
石棉线 asbestos thread
皮件线 leather article thread
莱尔线 lisle thread
丝光线 mercerized thread
蜡光线 glacé thread〔法〕
装饰线 ornamental thread
花色线 fancy thread
配色线 matching thread; thread tone in tone
金银线 lamé thread; gold and silver thread; metallic thread
特彩线 plyprism thread
夜光线 luminous thread
阻燃线 fire-retarded thread
混合包芯线 cotton wrapped core spun thread
（制鞋用）蜡线 waxed thread
鞋底线 sole thread
宝塔线 cone of thread
木芯线 reel of thread
纸芯线 cop of thread
纸板线 card of thread
木纱团线 spool of thread
线球 ball of thread

② 纽扣 button; bouton〔法〕

两眼扣 two-hole button
四眼扣 four-hole button
鸽眼扣 eyelet button
面扣 face button
底扣 base button
缝线纽扣 sew on button
有柄纽扣 shank button; bar button
钉脚扣 nail button

加固纽扣　reinforcing button
备用扣　space button
包扣　covered button; wrapping button; buttonmold
原身布包扣　self-fabric covered button
透明扣　transparent button
球式纽扣　ball button; spherical button
微型纽扣　tiny button; mini button
特大纽扣　huge button
儿童纽扣　kid's button
卡通扣　cartoon button
卡通动物扣　animation button
文字图案扣　character button
装饰纽扣　decorative button
时款纽扣　stylish button; fashion button
手工艺扣　handicraft button
流行纽扣　popular button
功能纽扣　functional button
免缝纽扣　non-sewing button
珠饰扣　beaded button
圆点扣　dot button
配色扣　matching button
花色扣　fancy button
花结扣，盘花纽　fancy mandarin button; Chinese button; frog
绳结扣　string button
钩编扣　crocheted button
编结带扣　braid button
标识扣　logo button
计算机雕刻扣　computerize engraving button
浮雕扣　relief sculpture button
异形扣　special-shaped button
羊角扣　claw button
牛角扣，套索扣　toggle button; toggle
鸡心扣　heart button
蝴蝶扣　butterfly button

玫瑰花形扣　rosette button
琵琶扣　pi-pa button
葡萄扣　grape button
桃形扣　peach button
棒形扣　stick-shaped button; toggle button; toggle
纺锤形扣　spindle-shaped button
角形扣　angular button
尼龙扣　nylon button
聚酯扣，涤纶扣　polyester button
聚酯印字扣　polyester printing button
水电扣　ABS button
塑料扣　plastic button
树脂扣　resin button
尿素扣　urea button
酪素扣　casein button
电木扣　bakelite button; urea button
有机玻璃扣，珠光扣　plexiglass button; pearl button
曼哈顿（珠光）扣　Manhattan button
珐琅扣　enamel button
组合扣　combination button
天然原料扣　natural material button
环保纽扣　ecological button
骨扣　bone button
角制扣　horn button
贝壳扣　shell button
彩虹贝壳扣　rainbow shell button
珊瑚扣　coral button
白垩扣　chalk button
矿石扣　ore button
宝石扣　gem button
玉石扣　jade button
翡翠扣　jadeite button
水晶扣　crystal button
石扣　stone button
大理石扣　marble button

陶瓷扣　ceramic button
云花扣　natural tone button
沙砾扣　sand grain button
珍珠扣　pearl button
玻璃扣　glass button
果壳扣　nut button
椰壳扣　coconut shell button
象牙果扣　corozo button
植物种子扣　plant seed button
橡胶扣　rubber button
皮扣　leather button
牛皮扣　cattle hide button
皮包扣　leather nub
毛皮包扣　fur-covered button
布扣　linen button;himobutton
木扣　wood button
竹扣　bamboo button
仿骨扣　imitation bone button
仿角制扣　imitation horn button
仿贝壳扣　imitation shell button
再生壳扣　A-MAK button
仿石扣　imitation stone button
仿钻石扣　rhinestone button
仿珍珠扣　imitation pearl button
仿树皮扣　imitation bark button
仿椰壳扣　imitation coconut shell button
仿木扣　imitation wood button
仿皮扣　imitation leather button
仿麂皮包扣　suede-covered button
合成树脂扣　synthtic resin button
金属扣　metal button
铜扣　copper button;brass button
青古铜扣　bronze button
黄铜扣　brass button
铜片扣　brass-plate button
镀镍铜扣　nickelled brass button
环氧扣　epoxy button

仿金属扣　faux metallic button
合金扣　alloy button
双色金属扣　two-colour metal button
金属牛仔扣，工字纽　jeans button
双针工字纽　twin prong jeans button
通心工字纽　open-top jeans button
旋转工字纽　rotatable jeans button
裤头工字纽　top jeans button
四合纽，撳纽，子母扣，唵纽〔粤〕　snap;snap button;fastener;snap-fastener;press button
金属（塑胶）四合纽　metal（plastic）snap
圈面（带帽）四合纽　ring（cap）press button;ring（cap）snap
珠光四合纽　pearl snap
宝石按扣　diamond snap
五爪按扣　prong snap;popper;nail
带扣　buckle;boucle〔法〕
钻石带扣　rhinestone buckle
云花带扣　natural tone buckle
水电带扣　ABS buckle
角制带扣　horn buckle
金属带扣　metallic buckle
合金带扣　alloy buckle
尼龙带扣　nylon buckle
真皮带扣　genuine leather buckle
仿皮带扣　imitation leather buckle
布包带扣　wrapping buckle
有字母的带扣　lettered buckle
方（圆）带扣　square（round）buckle
钩形带扣　hook buckle
旋筒带扣　roll buckle
子母带扣，对扣　pair buckle
胸罩带扣　bra buckle
插扣，卡子　buckle
日字扣，三线扣　buckle

吊带扣，葫芦扣 suspender buckle
D形扣 D-ring
单环扣 O-ring
绳索扣，拉绳扣，弹簧扣 stopper
吊钟 end stopper
尼龙搭扣，魔术贴 nylon fastener tape; Velcro tape; magic tape; self-gripping fastener; self-adhesive tag
搭钩 agraffe
钩眼扣 hook and eye
领钩 neck hook and eye; hook and eye; collar clasp
裤头钩 trousers hook and eye

③拉链 zipper; zip; zip-fastener; slide fastener

金属拉链 metal zipper
黄铜拉链 brass zipper
古铜拉链，青铜拉链 bronze zipper
铝拉链 aluminium zipper
镀铜拉链 copper-plated zipper
锌合金拉链 zinc alloy zipper
镍齿拉链 nickel teeth zipper
尼龙拉链 nylon zipper
塑胶拉链 plastic zipper
树脂拉链 resin zipper
树脂透明（半透明）拉链 resin transparent(translucent) zipper
树脂银（金）齿拉链 resin silver(gold) teeth zipper
涤纶拉链 polyester zipper
环扣（式）拉链 coil zipper
粗齿拉链，大牙拉链 heavy zipper
细齿拉链，小牙拉链 light zipper
（尾部）分开拉链；开尾拉链 split zipper; opened-end zipper
密尾拉链 closed-end zipper

密封拉链 sealed zipper
气密拉链 airtight zipper
磁吸拉链 magnetic zipper
阻燃拉链 anti-flaming zipper
防水拉链 water-proof zipper
隐形拉链，暗缝拉链 hidden zipper; concealed zipper; invisible zipper
加长拉链 long chain zipper
斜拉链 bias zipper
装饰拉链 decorative zipper
双头拉链 zipper with double sliders; two way zipper; moveable open-end zipper
双向拉链 two way zipper
自动头拉链 autolock slider zipper
半自动头拉链 zipper with semi-autolock slider
拉链头 zipper slider; zipper puller
拉链齿 zipper teeth
拉链带 zipper tape

5. 绳带，唛头，徽章，衬垫 ropes, tapes, labels, badges, pads, etc.

①绳带 ropes and tapes; strings and tapes; trimmings

棉绳 cotton string; cotton rope; cotton cord
原色棉绳 cotton rope in nature
漂白棉绳 cotton rope in white
染色棉绳 cotton rope in colour
尼龙绳 nylon string
内置绳带 interior string
拉绳 draw cord; drawstring
帽拉绳 hood drawstring
腰拉绳 waist drawstring
下摆拉绳 hem drawstring
带子 tape; ribbon; belt; band; girdle; tie;

string; strip; strap
窄带　narrow strip
织带　woven tape
编织带　braided tape
装饰带　fashion tape
印字带　printed tape
商标带　brand belt; label cloth
荧光带　fluorescent tape
热胶贴带　hot air sealing tape; sealing tape; adhesive tape
定位带　stay tape; strip for location
腰头丝里带　waistband tape
丝带，缎带　ribbon; riband
网眼丝带　mesh ribbon
透明丝带　sheer ribbon
玻璃纱带　organza ribbon
人丝缎带　rayon ribbon
涤纶缎带　polyester ribbon
提花缎带　jacquard ribbon
人字带　twill tape
橡筋带，松紧带　elastic; elastic ribbon; elastic cord; elastic band; elastic braid; elastic tape; elastic strap
日本"无比耐"松紧带　MOBILON tape
罗纹带　rib band
罗纹　rib
弹力罗纹　elastic rib; lycra rib
抽针罗纹　drop needle rib
法式罗纹　French rib
提花双面罗纹　jacquard double rib
全棉罗纹　cotton rib
锦纶罗纹　nylon rib
腈纶罗纹　acrylic rib
1×1罗纹　one by one rib

②唛头，商标　label
主唛，主商标　main label; brand label

洗水唛　washing label; care label
成分唛　content label
洗水成分唛　care content label
尺码唛　size label
产地唛　country of origin label; origin label
旗唛　flag label
衬唛　trim label
织唛　woven label
计算机织唛　computerized woven label
平面织唛　plain woven label
缎面织唛　satin woven label
印唛　printed label
双面印唛　double face printed label
激光裁切唛头　laser-cut label
拷花唛头　embossed label
充棉（立体）唛头　cotton-filled label
模压唛头　die-cut label
拉链头织唛　zipper pull label

③徽章　badge; insignia; emblem; patch
金属徽章　metal badge
PU 皮徽章　PU badge
真皮徽章　leather badge
绣花徽章　embroidery badge
臂章　brassard
腰牌　waist tag; back tag; joker
袋牌　pocket patch; pocket flash
皮牌　leather label; leather patch
PVC 标牌　PVC label
PU 标牌　PU label
纸徽章　paper badge
金属商标牌　metal label
铜牌　brass plate
橡胶牌（章）　rubber patch
绣花牌（章）　embroidered patch
反光牌　reflective patch

④衬垫，等　pads, waddings, etc.

肩垫　shoulder pads
布包肩垫　covered shoulder pads
海绵肩垫　sponge shoulder pads
泡沫塑料肩垫　form shoulder pads
活动肩垫　detachable shoulder pads
热定型肩垫　heat-setting shoulder pads
针刺肩垫　needle-punching shoulder pads
特型肩垫　special-shaped shoulder pads
手装肩垫　handpull shoulder pads
衬垫料　padding; pad; wadding
衬垫布　padding cloth; patch
蒸汽定型肩垫　steam moulded shoulder pads
袖窿衬垫；弹袖棉　armhole wadding
胸垫，胸罩衬垫　bust pad; bust form
复合胸垫　compound bust pad
领垫，领底呢　under collar pad; under collar felt; collar felt cloth
膝垫　knee pad; knee patch
臀垫　hip pad; hip patch
肘垫　elbow patch
裙撑，鲸骨圈　pannier; panier; hoops
口袋布　pocketing; bag cloth; bag sheeting; sleek
爪钉　nailhead
饰钉，窝钉　stud
撞钉　rivet
袋口钉　pocket rivet
鸡眼　eyelet; grommet; grummet
双面鸡眼　double-face eyelet
网形鸡眼　net eyelet
长方形鸡眼　rectangular eyelet
垫圈　washer, grommet; grummet
领插角片，领口片　collar stay; collar leaf; collar inlay; fall

（四）包装材料　Packing Materials

包装箱　packing box, packing case
纸箱，卡通箱　carton
三层出口纸箱　3 plies export carton
纸盒　card box; cardboard case; bandbox
外箱　outer carton; carton
内箱，内盒　inner box
编织袋　wearing bag
胶袋　polybag; plastic bag
透明胶袋　clear polybag
自封口胶袋　self-sealing polybag
包袋布　pack cloth; packing sheet
包装纸　packing paper; wrapping paper
马粪纸板，黄纸板　strawboard
牛皮纸　kraft paper; kraft
防潮纸　moistureproof paper
衬纸　paper lining; slip sheet; interleaving paper; tissue paper
扎包绳　baling twine; pack thread; packing cord
打包带　pack belt
打包铁皮　baling ties
大头针　pin
吊牌胶针　tag pin; swifttuck
吊牌，挂牌　hang tag; hangtag; swing tag; tag
标签　tag; label
粘贴标签　adhesive label; stick-on label; stick-on
胶贴，贴纸　sticker
纸箱贴纸，箱唛　carton sticker
胶袋贴纸　polybag sticker
吊牌贴纸　hangtag sticker
尺码贴纸　size sticker

价格贴纸　price sticker
条形码贴纸　bar code sticker
封箱纸　box sealing tape
衬衫胶夹　plastic shirt clip
衬衫别针　shirt pin
领口蝴蝶片　collar fly
领条，领托　collar keeper; collar supporter; collar strip
衬衫胶（纸）领条　plastic (paper) shirt band
衬衫纸板　shirt paperboard
（衬衫）纸腰封　paper belt
衣架　clothes hanger; dress hanger; coat hanger
折叠式衣架　folding hanger
带夹衣架　gripper hanger
塑料衣架　plastic hanger
木衣架　wood hanger
金属衣架　metal hanger
金属丝衣架　wire hanger

四、服装工艺操作
Clothing Technological Operation

(一) 技术室打样　Sample Making in Technical Room

服装生产自动化系统　automated system of clothing production
计算机综合成衣系统　computer integrated clothing system
计算机辅助生产　computer aided manufacture (CAM)
计算机综合制造　computer integrated manufacturing (CIM)
计算机辅助车间操作系统　computer aided machineshop operation system (CAMOS)
服装制作（全过程总称）　dress making
操作　operation
工艺学，工艺　technics; technology
服装工艺　clothing technology
技艺，手艺　workmanship; craftsmanship; useful arts
传统工艺　traditional technology; traditional workmanship
工艺要求　technological requirements
工艺单　process sheet; working sheet; artwork sheet
工艺流程　technological process; process flow
工艺变更　technology change
工艺革新　technology innovation
模板　template; form board
服装模板　clothing template
模板设计　template design
模板技术　template technology
模板制作　template making
模板切割　template cutting
工艺模板设计系统　process template design system
样品，样本　sample
样品卡　sample card
样品间　sample room
标准样　standard sample; type sample
客户样　buyer's sample

开发样　empolder sample
初始样，原样　initial sample; original sample; prototype sample; proto-sample
修正样，复样　revised sample
试身样　fitting sample
对等样　counter sample
确认样　approval sample; confirmed sample
留底样　keep sample; duplicate sample
照相样　photo sample
产前样　pre-production sample; bulk sample
放码样　run size sample
齐码样　full size sample; size set sample
生产厂家样　maker's sample
现货样　purchase sample
测试样　test sample
随机查验样　random sample
洗水样　wash sample
仲裁人样　umpire sample
商用样　trade sample
封口标样　sealed sample
推销样　salesman sample; selling sample
展示样　showroom sample
船样　shipping sample; shipment sample; top sample; fore sample
做样，打样　sample making
打样单　sampling form
打初样　making of original sample
打新样　new sample making
打确认样　sample making for approval
打宣传（推销）样　making of photo(salesman)sample
打船样　making of shipping sample
复样　reproduction of sample
单件制作　making-through
量身定制　made-to-measure

试穿　fitting
三维虚拟试穿　3D virtual try-on
三维虚拟试衣系统　3D virtual fitting system
虚拟缝制试穿技术　virtual stitching and fitting technology
合身　fit
贴身　tight fit
很合身　good fit
舒适合身　snug fit
不合身　ill fitting
纸样制作，制板　pattern making; pattern construction; pattern drafting
纸样放缩　pattern grading
（纸样）放缩点　grading point
用量，耗料　consumption
计算用料　calculation of fabrics utilization and consumption
单件耗料　piece yardage
排板描样　positioning marking
排料　marker laying; marking; making the lay; layout arrangement; layout
套排　economic layout
排唛架　marker laying
唛架制作　marker lay making; marker making; marker plotting
唛架复制　reproduction of marker
修改纸样和排料　alternation of paper pattern and marker
计算机辅助排料　computer aided layout (CAL)
计算机放码　computer grading
直接起图放码　draft grading
叠层式放码　stack grading
轨迹式放码　track grading
射线放码　radial grading
计算机排唛架　computer marker planning;

computer marking 计算机放码和排料系统 computer grading and marking system

(二) 原辅料检验 Check of Fabrics and Accessories

验色差 check off colour
查疵点 check flaw
查污渍 check spot
查纬斜 check grain
分幅宽 sort out fabrics
查里料、衬布色泽 check lining and interlining colour and luster
复米，复尺 remeasure length of fabric
查拉链 check zipper
查纽扣 check button and snap
理化试验 physical and chemical test

(三) 裁剪 Cutting

计算机自动裁剪系统 computer automatic cutting system
铺料，拉布 cloth spreading; fabric spreading
智能拉布 intelligent fabric spreading
(弹性布料铺料前) 松布 fabric relaxing
量裁布料 measure and cut fabric
烫料 iron fabric
检查纸样 check pattern
表层划样 draw pattern; cloth marker; marker
复查划样 check pattern
检查唛架 check marker
裁剪，开剪 cutting; scissoring
手工裁剪 hand cutting
单裁 tailor's cutting; cutting by one-piece
多层裁剪 multiply cutting

直裁 straight cutting
横裁 cross cutting
斜裁 bias cutting; diagonal cutting
对条（格）裁剪 cutting as stripes(checks) matching
平面裁剪 plane cutting
立体裁剪 draping cutting; draping; rittai-saidan〔日〕
贴身裁剪 flawless cutting
模塑裁剪 moulding cutting
自由裁剪 freehand cutting
冲切裁剪 die cutting
激光裁剪 laser cutting
程序自动裁剪 programed automatic cutting
查裁片刀口 check cutting edge
钻眼 drilling; punching; perforating
打剪口 notching
打粉印 chalking
编号 numbering
配零料 cut details
钉标签 attach label
验片 check cutted piece
修片 dressing of cut parts; pieces trimming
织补 darning; mending
换片 change bad quality cutted piece
分片（捆扎）arrange cutted piece; identifying and bundling; bundling
冲领角薄膜 punch collar stay

(四) 缝纫，缝制 Sewing; Stitching

1. 一般名称 general terms

缝纫部位 stitching area

缝纫针迹　stitching mark
缝纫速度　sewing speed
缝纫厚度　sewing thickness
缝纫精度　stitching accuracy
缝纫数据　stitching data
缝纫步骤　sewing sequences
缝纫方法　sewing method
缝纫技巧　sewing skill
缝纫工艺要求　sewing technological requirements
手工缝纫　hand sewing
机器缝纫　machine sewing
全自动缝纫　full automatic sewing
程控缝纫　controlled programmed sewing
模板式快速反应缝纫　modular quick response sewing(MQRS)
多功能缝纫　versatility sewing
装饰缝纫　ornamental sewing
加固缝纫　reinforcing sewing; tacking sewing
稳定缝纫　stay stitching
包边缝纫，锁边缝纫　overlock sewing; overedge sewing
卷边缝纫　hem stitching
搭接缝纫　lap sewing
贴底缝纫　under stitching
曲折缝纫　zigzag sewing
单（双、三）针缝纫　single(double, triple) stitching
粗线缝纫　heavy stitching
明线缝纫　top stitching
装饰线缝纫　decorative stitching
单（双）线缝纫　single(double) stitching
对条（格）缝纫　stitching as stripes(checks) matching
顺向缝纫　forward sewing
倒向缝纫　backward sewing
内外缝纫　in-and-out seaming

2. 缝纫工艺　sewing technology

撇片　trim cutted piece
打线丁　make tailor's tack
剪省缝　cut dart
环缝　catch stitch seam
缉省缝　stitch dart
烫省缝　press open dart
推门　stretch front piece
合刀背缝　stitch French dart
烫刀背缝　press open French dart
缉衬省　stitch dart in interlining
缉胸衬　stitch chest interlining
烫胸衬　press chest interlining
敷胸衬　attach chest interlining to front piece
纳驳头，扎驳头　pad stitch lapel
敷驳口牵条　attach tape to lapel roll line; taping lapel roll line
敷止口牵条　attach tape to front edge; taping front edge
敷挂面　attach facing to fly
滚挂面　pipe inside line of facing
合止口　stitch front edge
修剪止口　trim front edge
（用斜针法）扳止口　hemming stitch front edge
（用长针脚）疏缝止口　baste front edge
繰暗门襟　slip stitch facing
绱明门襟　attach band to front piece
合背缝　stitch back centre seam; join back centre seam
归拔后（衣）片　shrink and stretch back piece; blocking back piece
敷袖窿牵条　attach tape to armhole; ta-

ping armhole
敷背衩牵条 attach tape to back vent;taping back vent
封背衩 bartacking back vent end
扣烫过肩 fold and press back yoke
绱过肩 set-in back yoke
合摆缝 stitch side seam;join side seam
分烫摆缝 press open side seam
扣烫底边 fold and press hem
疏缝底边 baste hem
暗缝缲边 blind hemming
倒钩（针法缝扎）袖窿 backstitch armhole
合肩缝 stitch shoulder seam;join shoulder seam
分烫肩缝 press open shoulder seam
叠肩缝（肩缝头与衬扎牢） slip stitch shoulder seam
做肩垫 make shoulder pad
装肩垫,绱肩垫 set-in shoulder pad
（用倒钩针法）倒扎领窝 backstitch neck line
合领衬 close the centre seam of collar interlining
拼领里 piece together under collar
机缉领里 topstitch under collar
归拔领里 shrink and stretch under collar;blocking under collar
合领面 close the centre seam of top collar
归拔领面 shrink and stretch top collar;blocking top collar
覆领面 pin top collar to under collar together
合领子 stitch collars together
翻领子 turn collar
绱领子 set-in collar

分烫上领缝 press open collar seam
分烫领串口 press open gorge line seam
叠领串口 slip stitch gorge line
包领面 turn in top collar edge and stitch it
包底领 turn in band edge and stitch it
做领舌 make collar band tab
领角薄膜定位 attach collar stay
冲领衬 punch collar interlining
热缩领面 iron top collar for preshrinking
热粘领衬、面 fuse interlining to top collar
热压领角定型 press collar point
垫压薄膜衬 fuse fusible interlining
夹翻领 attach collar to band
归拔偏袖 shrink and stretch sleeve inseam;blocking sleeve inseam
合袖缝 stitch sleeve seam;join sleeve seam
分烫袖缝 press open sleeve seam
袖拼角 piece together gore to sleeve
绱袖衩条 sew placket to sleeve
封袖衩 barrack sleeve slit end
缲袖衩 slip stitch sleeve slit
叠袖里缝 sew together sleeve and its lining
翻袖子 turn sleeve
收袖山 shrink sleeve cap
绱袖,绱袖子 set-in sleeve
绲袖窿（毛边） bind armhole
缲袖窿 slip stitch armhole
合袖襻 stitch sleeve tab
翻袖襻 turn sleeve tab
绱袖襻 set-in sleeve tab
冲袖头衬 punch cuff interlining
缝制袖头 make cuff
绱袖头 attach cuff to sleeve

缝纫工艺

缲领钩	attach hook to band
叠暗门襟（暗门襟眼之间用暗针缝牢） slip stitch facing	
坐烫衣里缝	press open lining seam
（用热风机）胶贴线缝	tape seam
合大身面、里	sew together bodice and its lining
翻里子	turn lining
覆里子（缲大身）	attach lining to bodice
缲底边	hemming stitch bottom
缲领下口	slip stitch collar to bodice
合风帽缝	close hood seam
绱帽檐	attach tape to hood brim
合帽里、面	sew together hood and its lining
翻风帽	turn hood
绱风帽	set-in hood
缉袋嵌线	sew pocket welt
开袋口	cut pocket mouth
封袋口	bartack ends of pocket mouth
拼袋盖里	splicing pocket flap lining
做袋盖	make pocket flap
翻袋盖	turn flap
做插笔口	make an opening for pen on the flap
绲袋口（毛边）	pipe pocket mouth
拼耳朵皮	piece together flange
做袋爿	make flap
做贴袋	make patch pocket
绱袋，绱口袋	set-in pocket
缲袋	slip stitch patch pocket
合腰带	stitch waistband
翻腰带	turn waistband
绱明门襟	attach front band
绱橡筋	attach elastic
绱罗纹	attach rib
绱拉链	attach zipper
绱唛头，绱商标	sew on label
绱皮牌	sew on leather label
划绗棉线	draw quilting line
绗棉	quilting
填棉	padding; wadding
刮浆	smear paste
袖下缝剪口	clip underarm seam allowance
绲边	binding
门襟和下摆绲边	binding of placket and bottom
镶边	trim edge
镶花边	insert lace
罗纹镶边	ribbing
毛皮镶边	insert fur
嵌饰皮革	insert leather
嵌补布片	insert patch
嵌松紧带	insert elastic
嵌绳带	insert cord
镶嵌线	cording; trimming
缉明线	topstitching
缝单线	single-stitching
缝双线	double-stitching
缝三线	triple-stitching
缝之字线	zigzag-stitching
锁缝	lockstitching
包缝，锁边	overlocking; overcasting; overedging
三线包缝	overlock with three-thread
五线包缝	overlock with five-thread
拔裆	stretch crotch
覆袋口牵条	attach tape to pocket mouth; taping pocket opening
扣烫膝盖绸	fold and press reinforcement for knees
缲膝盖绸	slip stitch reinforcement for

knees
合侧缝　stitch side seam;join side seam; close side seam
钉裤钩　attach hook to waistband
合腰头　sew together waistband and its lining
翻腰头　turn waistband
翻门襟　turn left fly
翻里襟　turn right fly
绱门襟　attach left fly
绱里襟　attach right fly
合裤带襻　stitch belt loop
翻裤带襻　turn belt loop
缉裤腰裥　stitch waist pleat
绱腰头　attach waistband
绱裤带襻　attach belt loop;looping
绱雨水布　attach trousers curtain
双针合缝　double needle felling seam
合下裆缝　stitch inside seam;join inside seam
缝合裤裆　join crotch
缝合小裆　join front crotch
合前后裆缝　sew together front and back rise
封小裆　bartack front rise
钩后裆缝　backstitch back rise
扣烫裤底　fold and press crotch patch
绱大裤底　attach back crotch stay
花绷十字缝　cross stitch crotch
扣烫贴脚条　fold and press heel stay
绱贴脚条　attach heel stay
叠卷脚口　make French tack at cuff
钉裤钩襻　attach eye to waistband
覆裙腰口牵条　attach tape to waist line; taping waist line
绱裙腰头　attach waistband to skirt
合裙缝　stitch side seam;join side seam

覆裙里　attach lining to skirt
抽碎褶　gathering;shirring
弹性抽褶　elastic shirring
装饰抽褶　decorative shirring
叠顺裥　sew one-way pleat

3. 印花，绣花，等　printing, embroidering, etc.

印花　printing
手工（机器）印花　hand (machine) printing
单面印花　one side printing;blotch printing
双面印花　both-side printing;duplex printing
裁片印花　print on cutted piece
成衣印花　garment printing
直接印花　applied printing
转移印花　transfer printing;decal printing
数字印花　digital printing
三维印花　3D printing
涂料印花　coat printing
拔染印花　discharge printing
烂花印花　burn-out printing
胶浆印花　plastic printing;rubber printing
水浆印花　water based printing
发泡印花　applique printing;puff printing
起绒印花　raised printing
珠光印花　pearl printing
反光印花　reflective printing
钻石印花　diamond printing
金（银）粉印花　golden(silver) powder printing
纸印花　paper printing
丝网印花　silk screen printing
滚筒印花　roller printing

喷射印花 jet printing
防染印花 resist printing
蜡防印花 wax printing
满地印花 all over printing; blotch printing
隐约印花 indistinct printing
（裁片绣花部位）刷花 print design
绣花，刺绣 embroidering
手工绣花 hand embroidering
机器绣花，机绣 machine embroidering
计算机绣花 computerized embroidering
满绣 full embroidering
贴花绣，贴绣 appliqué embroidering; patch embroidery
嵌绣 cord embroidering
珠绣 bead embroidering
网眼绣花 net embroidering; eyelet embroidering
轮廓绣花 out line embroidering
金银线绣花 gold and silver embroidery
转移烫钻 hot fixing transfer rhinestone
做手工 hand work
饰珠片，饰亮片 sequining
珠片贴饰 sequin applique and trimming
钉珠 sequins and beading; beading
钉饰纽，打窝钉 studding
绣片 embroider on cutted piece
补缀 patchwork
抽纱 fagoting
修剪，剪线 trimming
修补 mending; repairing
去污 cleaning
返工 re-work

（五）锁钉，水洗，熨烫，包装 Buttonholing, Buttoning, Washing, Pressing, Packing

1. 锁钉 buttonholing, buttoning

划扣眼位 mark buttonhole position
（剪）开扣眼 cut buttonhole
绲扣眼 pipe buttonhole
缲扣眼 slip stitch buttonhole
锁扣眼 lock stitch buttonhole; buttonhole stitching; buttonholing
平头锁眼，锁直扣眼 straight buttonholing
圆头锁眼，锁凤眼 eyelet buttonholing
缲扣襻 slip stitch button loop
钉扣襻 sew button loop
拉线襻 make French tack
盘花扣 make Chinese frog
点纽位 mark button position
钉扣 sew on button; button sewing; button attaching; button stitching; buttoning
机器钉扣 buttoning by machine
手工钉扣 buttoning by hand
钉四合扣 attach snap
钉领钩襻 sew hook and eye
打撞钉 attach rivet
打鸡眼，打气眼 attach eyelet
打套结，打结 bartack; tacking

2. 水洗 wash, washing

机（手）洗 machine(hand) wash
冷（热，温）水洗 cold(hot, warm) wash
布料水洗 fabric wash
成衣水洗，普洗 garment wash; normal wash

轻（重）普洗　light(heavy)garment wash
翻底洗　underside garment wash
柔洗　soft wash
硅油洗　silicone wash
扎洗　tie wash
漂洗　bleach wash
扎漂洗　tie bleach wash
吊漂洗　hang bleach wash
中漂洗　mid bleach wash
石磨洗　stone wash;stonewash
中磨洗　mid stone wash
石漂洗　bleach stone wash
粉磨洗　powder milling wash
胶球洗　dissolved rubber ball wash
砂洗　sand wash
酵素水洗，酵洗　enzyme wash;ferment wash;bio wash
重酵洗　double enzyme wash
酵石洗　enzyme stone wash
轻（重）酵石洗　light(heavy)enzyme stone wash
酵漂洗　enzyme bleach wash
酵石漂洗　enzyme stone bleach wash
酸洗　acid wash
重酸洗　heavy acid wash
雪花洗　snow wash;PP stone vibration wash
化学洗　chemical wash
化石洗　chemical stone wash
水晶洗　crystal wash
退浆保色洗　rinse wash
碧纹洗　pigment dyed wash;pigment wash
伯克利洗　berkely wash
幻影洗　vision wash
猫须洗　scratch wash
喷色　spray colour wash
喷金（银）粉　spray gold(silver)powder wash

喷马骝　monkey wash;PP spray wash
喷砂　sandblast wash;spray sand wash
喷砂酵漂洗　sandblast enzyme bleach wash
喷砂酵石洗　sandblast enzyme stone wash
手擦　hand brush;hand scrapping
猫须　whisker(WHK);moustaches effect
吊色　tinting
收皱　pinching
收皱洗　pinching and creasing wash
压皱洗　wrinkle effect wash
树脂压皱洗　resin wrinkle effect wash
防皱洗　wrinkle free wash
手擦加漂（清）洗　hand brush and rinse wash
手擦酵石洗　hand brush enzyme stone wash
手擦酵石漂洗　hand brush enzyme stone bleach wash
手擦猫须酵石洗加烂边　hand brush whisker,broken edges and enzyme stone wash
手擦酵素胶球洗　hand brush enzyme rubber ball wash
手擦酵石洗加吊色　hand brush enzyme stone wash and tinting
破坏洗　destruction wash;destroy wash
磨烂洗　grinding wash
烂边洗　broken edge wash
怀旧洗　vintage wash;retro wash;dirty wash;antique wash
染色洗　dyed wash
深（浅）色洗　dark(light)wash
深（浅，中）蓝洗　dark(light,medium)blue wash
黑色洗　black wash
黑染黑　black over black wash
蓝黑洗　blue and black wash
套色洗　overdyed wash

浊色洗　dull wash
退浆洗　destarch wash
固色洗　colour fixed wash
漂（清）洗　rinse; rinsing

3. 熨烫　pressing; ironing

部件熨烫　part pressing
部位熨烫　position pressing
成品熨烫，大烫　product pressing; off pressing
半成品熨烫，中烫　under pressing; pressing in process; intermediate pressing
整理熨烫　finish pressing
高温熨烫，热烫　hot ironing
中温熨烫，温烫　warm ironing
低温熨烫　cool ironing
干熨烫　dry ironing
立体熨烫　formatic pressing; body pressing
定型熨烫　permanent pressing; durable press; top pressing
模压熨烫　die press
滚筒熨烫　roll press
火烙熨烫　pressing by conventional iron
电熨烫　electric pressing
蒸汽熨烫　steam pressing
计算机熨烫　computerize pressing
永久熨烫　ever press
不可熨烫　do not ironing
免烫　no press; non ironing; never press; easy care; wash and wear

4. 包装　packing

出口包装　export packing
外（内）包装　outer(inner) packing
包装方法　packing instruction
单色单码　solid colour/size
混色混码　assorted colour/size
齐色齐码　full colour/size
单色单码包装　solid packing
混色混码包装，搭配包装　assorted packing
颜色（尺码）搭配　colour(size) assortment
装箱配比　ratio
短装（少于订单数）　shortage packing
溢装（多于订单数）　overage packing
折叠包装　fold packing
悬挂包装，挂装　hanger packing; garment on hanger(GOH)
中性包装　neutral packing
打吊牌，上挂卡　attach hangtag
分色分码　assort by colour and size
断码　short in size
断色　short in colour
入胶袋　put clothes into polybag
垫防潮纸　put kraft paper into carton
入纸箱，装箱　put clothes into carton; encasement; cartoning
封箱　seal carton by sealing tape
箱唛　shipping mark
侧唛　side mark
箱号　carton No.
填箱号　numbering on carton
打扎箱带　bale carton by belt
制装箱单　fill in packing list

（六）成品检验　Checking of Finished Products

检验，检查　inspection; check
商品检验，商检　commodity inspection
质量检验，质检　quality inspection
织物检验，验布　fabric inspection

查探断针，验针 needle detection
成品检验 inspection of end products
成品外观 finished appearance; final appearance
抽样查验 sampling inspection
随意抽查 random inspection
初次查验，初查 original inspection
产前查验 pre-production inspection
早期查验 early inspection
中期检验，中查 in-process inspection; inspection of semi-finished products; inter inspection
在线查验 in-line inspection
复查 double check
随机抽查 random sampling
抽样检验 sampling inspection
最后检验，尾查 final inspection; inspection of end products
客户查验，客检 customer inspection; buyer inspection
验货结果 inspection result
验货报告 inspection report
返工品 reworks
缺陷，疵点 defect
成品缺陷 defects of end products
成衣缺陷 garment defects
缝制缺陷 sewing defects
熨烫缺陷 pressing defects
水洗缺陷 washing defects
面料疵点 material defects
布料起球疵 pilling
做工欠缺处 workmanship defects
后整理欠缺处 finish defects
领面松 wrinkles at top collar; top collar too loose
领面紧 top collar appears tight; top collar too tight
领面起泡 crumples at top collar; top collar bubbling
领外口松 collar edge appears loose
领外口紧 collar edge appears tight
底领伸出 collar band is longer than collar; collar band too long
底领缩进 collar band is shorter than collar; collar band too short
底领里起皱 wrinkles at collar band facing
底领外露 collar band lean out of collar
绱领偏斜 collar deviates from front centre line; unmatched collar
绱领不圆顺 collar assembling to body not closed
领圈不圆顺 neckline not smooth
领窝不平 creases below neckline; uneven neckline
后领窝起涌 bunches below back neckline
驳头起皱 wrinkles at top lapel
驳头反翘 top lapel appears tight
驳头外口松 lapel edge appears loose
驳头外口紧 lapel edge appears tight
驳口不直 lapel roll line is uneven; uneven breakline
串口不直 gorge line is uneven
领卡脖 tight neckline
领离脖 collar stand away from neck; loose neckline
领不正 misplaced collar
领尖不对称 unequal collar point
领边不对称 collar edge asymmetric
领座不均匀 uneven collar stand
裂肩（小肩起皱） puckers at shoulders
塌肩（衣胸起绺） wrinkles at shoulder; sloping shoulder

肩缝不顺直　uneven shoulder seam
根窝起绺　creases at underarm
抬根缝起绺　puckers at underarm seam; wrinkles at lower armhole
塌胸（衣胸不丰满）　lack of fullness at chest; hollow chest
省尖起泡　crumples at dart point
门襟起泡　placket bubbling
门襟起拱　wave placket
门襟起皱　wrinkles at top fly
门襟扭曲　twisted placket
门襟边隆起　bump at placket edge
门里襟长短不齐　unmatched front fly
里襟外露　exposed under placket
左右衩长不一　uneven vents length
绱拉链起绺　wrinkles at zip fly
拉链不平伏　wave zipper
拉链不能开合　zipper is not moveable
止口不直　front edge is uneven
止口缩角　front edge is out of square
止口反翘　front edge is upturned
止口反吐　facing leans out of front edge
止口豁（止口上下豁开）　split at front edge; unmeet front edge
止口搅（止口下部搭叠过多）　crossing at front edge; closed front edge
底边起绺　wrinkles at hem; uneven hem; twisted bottom
前身起吊　hiking up at front
后身起吊　hiking up at back
背衩豁（背衩下部豁开）　split at back vent; unmeet vent
背衩搅（背衩搭叠过多）　crossing at back vent; closed vent
里布扭曲　twisted lining
里布太紧　too tight lining
里布过多　too full lining

绗棉起绺　puckers at quilting; uneven quilting
絮棉不匀　padded cotton is uneven
空边（边缘缺棉）　empty hem
绱袖不圆顺　diagonal wrinkles at sleeve cap; uneven armhole
袖窿起皱　armhole puckers
袖山起皱　wrinkles at sleeve cap
袖子偏前　sleeve leans to front
袖子偏后　sleeve leans to back
前袖缝外翻　inseam leans to front
袖口起绺　wrinkles at sleeve opening
袖口不齐　sleeve opening is uneven
袖头扭曲　twisted cuff
袖头高不一　uneven height of cuffs
袖长不一　unmatched sleeves
袖里拧（袖里、面错位）　diagonal wrinkles at sleeve lining
袋盖反翘（袋盖面紧）　top flap appears tight
袋盖反吐（袋盖里外露）　flap lining leans out of edge
袋盖不直　flap edge is uneven
袋盖（左右）高低　high/low flaps
袋口角起皱　creases on two ends of pocket mouth
袋口裂　split at pocket mouth
口袋错位　misplaced pocket
袋形走样　deformed pocket
口袋（布面）丝缕不正　pocket out grain
口袋（左右）不对称　unequal pockets
袋唇宽窄不匀　uneven pocket lip
裤耳大小不匀　uneven belt loops
腰头探出　end of waistband is uneven; waistband extension
腰头扭曲　twisted waistband
腰头高低　high/low waistband ends

腰缝起皱　wrinkles at waistband facing; uneven waist seam
裤门襟起绺　uneven fly-facing
裤门襟止口反吐　unfavoring fly-facing
里襟里起皱　creases at right facing
夹裆（横裆紧）　tight crotch
兜裆（直裆短）　short seat; short crotch
后裆下垂　slack seat; baggy seat
小裆不平（前裆起绺）　wrinkles at front rise
裆底十字缝错位　cross crotch is uneven; unmatched cross crotch
裆缝断线　bursting of crotch seam
裤脚前后（一前一后不齐）　two legs are uneven; unbalanced legs
裤脚不对称　unsymmetrical legs
脚口不齐　leg opening is uneven
吊脚（裤侧缝或内缝起吊）　pulling at outseam or inseam; uneven leg
裤缝扭曲　twisted leg seam
烫迹线外撇　crease line leans to outside
烫迹线内撇　crease line leans to inside
腰缝下口涌　bunches below waistline seam
（裤裙）拼腰大小不匀　uneven back yoke
裙裥豁开　split at lower part of skirt; unmeet skirt pleat
裙摆起吊　skirt hem line rides up
裙浪不匀　skirt flare is uneven
裙摆线不齐　uneven skirt hemline
缉线上、下炕（线路偏离）　stitch seam leans out line; run off stitching; sewing bow; gutter
双轨接线（接线不准）　staggered seam; bad join stitching
线迹歪斜　staggering stitch
线缝皱缩　seam puckering
线缝松脱　seam slippage
线缝断开　seam broken; open seam
线缝扭曲　twisted seam
线缝起褶　pleated seam
线缝起拱　wave seam
缝口不牢　weak seams
缝型错误　wrong type of seam
面线不匀　uneven topstitching
格子不匀　uneven plains
对格不准　mismatched checks
对条不准　mismatched stripes
跳针　skipping stitch; skipping
漏针　missed stitch
断线　broken stitches; thread breakage; breakdown
浮线　loose stitch; floating stitch
缝线起毛　thread flaying
扣位不准　uneven button position
扣眼不正　crooked buttonhole
钉扣不牢　insecure button
钉四合纽不牢　insecure snap
套结误打在袋布上　pocketing caught in bartacking
尺码不符　off size; incorrect size
尺寸超出公差　measurement out of tolerance
缝制不良　irregular top stitches; poor stitching
洗水不良　uneven washing effect; poor washing
熨烫不良　improper pressing; poor ironing
熨烫不匀　uneven pressing
熨烫过度　over pressing
熨烫压痕　ironing mark
亮光　iron-shine; glaze mark
水花，水渍印　water stain
洗水痕迹　washing streak
洗水过度　over washing

锈迹　rust
污渍　spot; stain; soil
油渍　oily stain
划粉印　chalk mark
色差　off colour; off shade; colour deviation
褪色　fading; fugitive colour
线头　thread residue; thread end
未清线头　untrimmed loose thread end
衣里内透线头　thread end left inside
活线头　floating thread ends
破洞　broken hole
针孔　needle hole
起泡　bubbling
毛露　raw edge leans out of seam
绣花露印　embroidery design outline is uncovered
绣花错位　misplaced embroidery
线色不配　mismatched of thread
衣片错配，鸳鸯片　misaligned parts
倒缝唛头　upside down label
错缝唛头　wrong label
漏缝唛头　missing label
漏缝纽扣　missing button
漏缝裤耳　missing belt loop
漏打套结　missing bartack
漏挂吊牌　missing hangtag
错挂吊牌　wrong hangtag
错色　wong colour indicated
错码　wrong size indicated
断码　short in size
装箱搭配差错　wrong packing assortment
装箱数量不符　wrong packing quantity
装箱太紧（太松）　cartoning too tight (loose)
箱唛印错　wrong shipping mark

（七）缝　Seam

缝型，缝式　seam type; seam pattern
结构缝　composite seam
普通缝　normal seam
装饰缝　ornamental seam; decorative seam
连续缝　continuous seam
间断缝　broken seam
长缝　long seam
短缝　short seam
内缝　inseam
外缝　outseam
厚缝　thick seam
粗缝　bulky seam
中缝　centre seam
侧缝，边缝　side seam
斜线缝　bias seam
缭缝　slanting seam
弧线缝　curved seam
分割缝　section seam
公主缝　princess seam
刀背缝　princess seam; French dart
转角缝　corner seam
平缝　flat seam; plain seam
平接缝　flat fell seam; butted seam
搭接缝　lapping seam; overlap seam; overhead seam
双搭接缝　double lapped seam
接合缝　joining seam
拼合缝　abutted seam
双线缝　twin seam
镶色缝　contrast colour seam
嵌花缝　appliance seam
绷缝，覆盖缝　flat lock seam; covered seam; coverseam
来去缝　French seam; bag seam

莱卡缝	lycra seam
伸缩缝	stretch seam; elastic seam
漏落缝	self-bound seam
坐倒缝	plain seam
开缝	opening seam
分缉缝	double topstitched seam
压缉缝	topstitched lap seam
坐缉缝	lapping seam; tucked seam
成形缝	finishing seam; fashion seam
明缝，面缝	outseam; topstitch seam
暗缝	blind seam; invisible seam
暗边缝	hind edge seam
包缝	overseas; closed seam; overedge seam; overcast seam; overlock seam; wrap seam
明包缝	flat fell seam
暗包缝	welt seam
锁式线缝	lockstitch seam
联锁缝	interlock seam
链式线缝	chainstitch seam
环缝	catch stitch seam
嵌缝	cording seam
绗缝	quilting seam
格子缝	lattice seam
交叉缝	cross seam
泡状缝	beaded seam
翼状缝	wing seam
锯齿缝	German seam; zigzag seam
曲折缝	zigzag seam
蝶形缝	butterfly seam
辫形缝	plaited seam
凸形缝	raised seam
带形缝	strap seam
三角缝	gore seam
悬缝	hanging seam
保险缝	flat fell seam
加固缝	tacking seam; fastening seam
终止缝	final seam
止口缝	encased seam; enclosed seam
领缝	collar seam; neck seam
肩缝	shoulder seam
背缝	back seam
背嵌缝	back panel seam
背中缝	centre back seam
袖缝	sleeve seam
前（后）袖缝	front(back) sleeve seam
袖中缝	sleeve centre seam
袖下缝	underarm seam
袖窿缝	armhole seam
肘弯缝	elbow seam
门襟缝	front fly seam; placket seam
门襟止口缝	front edge seam
摆缝，侧缝	side seam
衩缝	vent seam
省缝	dart seam
腰缝	waistline seam
高腰缝	empire seam
腰带缝	waistband seam
裤管缝	leg seam
裆缝	crotch seam
前裆缝	front rise seam
后裆缝	back rise seam; seat seam
下裆缝	inside seam; inseam
折叠缝	folding seam
褶裥缝	tucked seam; pleat seam
对折缝	felled seam
卷边缝	hemming seam
贴边缝	welt seam
绲边缝	bound seam; piping seam
织补缝	darning seam
（热胶）贴带缝	taped seam

(八) 线迹，针法 Stitch

1. 基本线迹 basic stitch

平缝线迹	plain stitch; flat stitch
疏缝线迹，假缝线迹	basting stitch; tacking stitch
绷缝线迹	covering stitch; flat-lock stitch
绗缝线迹	quilted stitch
嵌缝线迹	cord stitch
面缝线迹	top stitch; face stitch overstitch
暗缝线迹	invisible stitch; blind stitch
繰缝线迹	slip stitch; running stitch
折缝线迹	fell stitch
倒缝线迹	reversible stitch; backward stitch
包缝线迹	overlock stitch; overcasting stitch; overedge stitch
锁式线迹	lock stitch; lockstitch
链式线迹	chain stitch
单（多）线链式线迹	single (multi-) chain stitch
包边链式线迹	overedge chain stitch
复盖链式线迹	covering chain stitch
链锁线迹	chainlock stitch
联锁线迹	interlock stitch
人字线迹	herringbone stitch
羽状线迹	feather stitch
珠式线迹	pearl stitch
Z形线迹	catch stitch; zigzag stitch
斜Z形线迹	Byzantine stitch
角形线迹	angle stitch
弓形线迹	arched stitch
锯齿形线迹	picot stitch; zigzag stitch
贝壳形线迹	shell stitch
网眼线迹	basket stitch
蜂窝线迹	honeycomb stitch
直形线迹	straight stitch
波形线迹	wave stitch
双十字线迹	double cross stitch
交叉线迹，十字线迹	cross stitch
圆形线迹	round stitch
廓型线迹	outline stitch
曲折形线迹	zigzag stitch
变形线迹	change stitch
钩编线迹	crochet stitch
织补线迹	darning stitch
刺绣线迹	embroidery stitch; crewel stitch
装饰线迹	ornamental stitch; decorative stitch
花式线迹	fancy stitch
花样线迹	pattern stitch
点画线迹	dot dash stitch
对称线迹	counter stitch
比翼线迹	fly stitch
特殊线迹	special stitch
功能线迹	functional stitch
定位线迹	stay stitch
复合线迹	combination stitch; split stitch
复式线迹	double action stitch
双针线迹	twin needle stitch
双重线迹	twice stitch
三重线迹	triple stitch
缝式线迹	seam stitch
初缝线迹	runstitch
加固线迹	fastening stitch; tacking stitch
打结线迹	knotting stitch
扎缚线迹	padding stitch
抽褶线迹	shirring stitch
伸缩线迹	stretch stitch; elastic stitch
绲边线迹	binding stitch
卷边线迹	hemming stitch
暗卷缝线迹	blind hemming stitch
拼合线迹	abutting stitch

间断线迹　broken stitch
跳针线迹　skipping stitch
安全线迹　safety stitch
每分钟针数　stitch per minute(SPM)
每英寸针数　stitch per inch(SPI)

2. 手缝针法　hand stitch;hand basting

缭针法　slip stitch;running stitch
拱针法　prick stitch
明缭针法　fell stitch
暗缭针法　blind hemming stitch
环针法　catch stitch
叠针法　fastening stitch
擦针法　baste stitch
扎针法　pad stitch
扳针法　diagonal basting
绗针法　quilting stitch
锁针法　lock stitch;buttonhole stitch
倒针法　bartack stitch;back stitch
三角针法　herringbone stitch
杨树花针法　feather stitch
花针法　zigzag stitch
跳针法　skipping stitch

(九) 线缝　Stitching

面缝　top stitching

全缝　through stitching
单（双、三）线缝　single(double, triple)stitching
直线缝　straight stitching
伸缩线缝　stretch stitching;elasticity stitching
装饰线缝　ornamental stitching;decorative stitching
曲折线缝　whip stitching;catch stitching
花式线缝　fancy stitching
羽状线缝　feather stitching
抽褶线缝　shirring stitching
无皱缩线缝　puck-free stitching
拼合线缝　abutted stitching
镶色线缝　contrast stitching
锁式线缝　lock stitching
双锁式线缝　double lock stitching
双搭线缝　double whip stitching
卷边线缝　hem stitching
安全线缝　safety stitching
间隔线缝　space stitching

五、服饰品
Furnishings

(一) 帽 Cap; Hat; Headdress; Headgear Headpiece; Chapeau〔法〕

1. 品种和款式 kinds and styles

职业帽 job cap; job hat
工作帽 working cap; working hat
制服帽 service cap; uniform cap
军帽 military cap; service cap
(旧时) 军团帽 legionary cap; legionnaire cap
筒型军帽 shako
高顶皮军帽 busby
(法国) 有檐平顶军帽 kepi
(波兰) 四角军帽 rogatywka
(澳大利亚士兵用) 宽边丛林帽 bush hat
(宽边、有帽绳的) 奔尼帽 bowler hat; bowler
(士兵) 船型帽 garrison cap
迷彩伪装帽 camouflage cap
大檐帽 service cap; hat
水兵帽 seaman cap; sailcap; sailor hat; gob hat
水兵软帽 sailor barret; sailor beret
海员帽 sailor cap
快艇帽 yachting cap
警察帽，警帽 policeman's cap
矿工帽 miner's hat
安全帽 safety hat; safety helmet
防护帽 protective hat; hard hat; helmet
防静电帽 antistatic cap
防水帽 waterproof hat; southwester; tarpaulin
头盔 helmet
全(半)盔 full(half) face helmet
儿童头盔 kid's helmet
矿工头盔 miner's helmet
消防员头盔 fireman's helmet
防弹头盔 bulletproof helmet
芳纶(防弹)头盔 aramid helmet
防撞头盔 crash helmet
滑雪头盔 ski helmet
飞行员头盔 aviator helmet; pilot helmet
摩托头盔 motorcycle helmet
电动车头盔 electric scooter helmet
自行车头盔 bicycle helmet
沙滩车头盔 ATV helmet
越野车头盔 cross-country helmet
运动头盔 sports helmet
滑板头盔 skateboard helmet
橄榄球头盔 rugby helmet
棒球头盔 baseball helmet
宇航员头盔 space helmet
(19世纪波兰) 方顶头盔 crapka
(19~20世纪德国等国) 尖顶头盔, 钉盔 pickelhaube
普鲁士战盔 Prussian pickelhaube
奥特曼头盔 ultraman helmet
宇航员帽 astronaut's cap; flight deck cap
护士帽 nurse cap
摄影师帽 photographer's cap
厨师帽 chef cap; toque〔法〕

面包小子帽　baker boy hat
学生帽　school cap
童子军帽　scout cap
大学方帽　mortarboard cap; trencher cap; square college cap; college cap; cap
伊顿帽　Eton cap
教士宽边帽　shovel hat; shovel
教士便帽　zucchetto
（天主教）红衣主教法冠　cardinal's cap
（天主教）红衣主教帽　red hat; scarlet hat
（天主教）四角帽　biretta
（基督教）贵格帽　Quaker hat
祭司帽　priest's hat
僧帽　capuche; monk cap
海盗帽　pirate hat
骷髅帽　skull cap
幽灵帽　ghost cap
苦力帽　coolie hat
牛仔帽　cowboy hat; gaucho hat
小丑帽　jester's cap; foolscap
（丑角戴）系铃帽　cap and bells
运动帽　sport cap
棒球帽　baseball cap
板球帽　cricket cap
网球帽　tennis cap
马球帽　polo hat; chukka hat
滑雪帽　ski cap
溜冰帽　skate cap
骑马帽　riding cap
（赛马）骑师帽　jockey cap
登山软帽　alpine
游泳帽　swim cap; bathing cap
礼仪帽　cap of ceremony
礼帽　bowler; derby; topper; hard hat〔英〕; iron hat〔美俚〕
大礼帽　top hat; dress hat; topper; silk hat; tall hat; high hat
小礼帽　billycock; bowler hat
常礼帽，圆顶礼帽　derby; bowler hat; bowler; soft felt hat
（帽顶有黑桃图案的）黑桃礼帽　SPADE fedora hat
高顶礼帽　plug hat; plug; chimney-pot hat; tile
宽缘软礼帽　snap brim hat
中褶帽　Hamburg hat; soft felt hat
乌纱帽　black gauze hat
朝冠　cap of ceremony
教皇法冠　tiara
皇冠　crown
王冠　diadem
便帽　easy cap; cap
军便帽　forage cap
无檐便帽　beanie; beany
自由帽　liberty cap
鸭舌帽　peaked cap; hunting cap; casquette〔法〕
长鸭嘴帽　jockey cap
猎帽　hunting cap
猎鹿帽　deerstalker
钓鱼帽　fishing hat
瓜皮帽　Chinese cap; calotte; skull cap; skullcap
（新疆）维吾尔族小圆帽　Uygur cap
太阳帽　sun hat; topee; topi
阔边太阳帽　sunbonnet
硬壳太阳帽　sun helmet; pith helmet
遮阳帽　shutter cap; sun-shade hat
遮阳帽檐　sun visor; visor; sun-shade hat
凉帽　cool hat; tropical hat
风帽　hood
雨帽　rain cap; raincap; rainhat; souʻwester; tarpaulin

品种和款式

斗笠　bamboo split hat; leaf hat
藤帽　rattan hat; rattan helmet
草帽　straw hat
硬草帽　boater
麻编草帽　gunny hat
平顶宽边草帽　skimmer
意大利草帽　leghorn
睡帽　night cap; nightcap
浴帽　bath cap; shower cap
时尚帽　fashion cap
嘻哈帽　Hip Hop cap
草辫帽　straw braid bonnet
贝壳形帽　shell-like bonnet
甲壳虫帽　beetle cap
工艺帽　craft cap
绣花帽　embroidered cap
广告帽　advertising cap
锡纸帽　tin foil hat
圣诞帽，长袜帽　toboggan cap
头巾帽　turban
长前檐帽　brimmer cap
宽边帽　broad-brimmed hat; broadbrim; bush hat; gypsy hat; lierihattu
宽边软帽　floppy hat; slouch hat
带拉链袋帽　hat with zip-pocket
望远镜型帽　telescope hat
女帽　ladies' hat; millinery
童帽　children's hat
新生婴儿帽　newborn hat
钟型女帽　cloche hat; cloche
阔边女帽　sundown; poke bonnet; poke; capeline
系带女帽　capote
无檐平顶女帽　pillbox
(女、童)无边有带帽　bonnet
(女、童)无檐帽　calot
(女)小圆帽　toque[法]

卷边帽　cocket hat
卷边平顶帽　porkpie hat
无边平顶圆帽　Balmoral cap
扁平软帽　barret
蘑菇型扁帽　mushroom
(可折叠的)软呢帽　squash hat
贝雷帽　beret
尼赫鲁帽　Nehru cap; Nehru hat
苏格兰帽　Scotch cap
苏格兰大黑帽　tam-o-shanter
苏格兰便帽　glengarry
(美国)西部风格帽　Western hat
斗牛士帽　toreador hat
爵士帽　fedora hat; fedora
阿波罗帽　Apollo cap
巴拿马帽　Panama hat
秘鲁帽　Peruvian hat
布列塔帽　Breton
汉堡帽，中褶帽　Hamburg hat
土耳其帽　tarboosh
土耳其毡帽　fez
(犹太人)圆形小便帽　yarmulke; yarmulka; kipa; kippa; kippah
八角帽　octagonal cap
方形帽　square cap
桶形帽　bucket hat
折叠三角形帽　chapeau bras; cocked hat
三(四、五、六、七)瓣帽　3(4,5,6,7) panels cap
拼缝帽　patchwork cap
针织帽　knitted hat; knitted beret
绒线帽　chenille beanie
绒球帽　bobble hat
巴拉克拉瓦盔式(针织)帽　balaclava helmet; balaclava
网眼帽　mesh cap; mesh hat
锦纶网眼帽　nylon mesh cap

斜纹布帽 twill cap
牛仔布帽 denim cap;denim hat
洗水帽 washing cap
冬帽,防寒帽 winter cap
棉帽 cotton-padded cap
毡帽 felt cap;felt hat
软毡帽 trilby hat;trilby;sloppy hat
费多拉软毡帽 fedora hat;fedora
铁若兰毡帽 Tyrolean hat
呢帽 stuff cap;woolen cap
毛华达呢帽 wool gabardine cap
粗花呢帽 tweed cap
羊毛帽 woolen cap
丝绒帽 velour cap;velour hat
摇粒绒帽 polar fleece cap
绒面革防水帽 suede waterproof hat
皮帽 leather cap;leather hat
毛皮帽 fur cap;fur hat
海狸皮帽 beaver
浣熊毛皮帽 Davy Crockett hat
(俄罗斯)护耳毛皮帽 ushanka
(中东)羊皮帽 calpac;calpack
风帽 hood
后领内风帽 concealed hood in collar
折叠式风帽 fold-up hood
可脱卸风帽 removable hood
镶毛皮边风帽 hood with fur trim

2. 帽部件 hat parts

制帽 hat making;hatting
制帽材料 hat materials;hatting
帽型 hat shape
帽坯 hat felt
帽里 hat lining
帽衬 hat interlining
帽圈 hat band;sweatband
帽边 hat brim
帽檐 cap peak;peak;visor
帽顶 top of hat;cap knob;crown
帽护耳 cap tab;earflap;earlap;eartab
帽徽 cap cockade;cap insignia;cockade;hat badge
帽饰绒球 cap pompon
帽饰流苏 cap tassel
帽带 hat ribbon
帽木模 hat block;block;cap piece;block-head
帽撑 cap stretcher
帽架 hat tree;hatrack;hat stand
挂帽钉 hat peg

(二) 鞋 Shoes

1. 品种和款式 kinds and styles

布鞋 cotton shoes;cloth shoes
棉鞋 cotton-padded shoes
帆布鞋 duck shoes;canvas shoes
帆布运动鞋 canvas sports shoes
胶底帆布鞋 gum shoes;gumshoes〔美〕;plimsolls〔英〕;sand shoes
灯芯绒鞋 corduroy shoes
缎面鞋 satin shoes
毡呢鞋 felt shoes
驼绒衬里鞋 fleece-lined shoes
毛皮衬里鞋 fur-lined shoes
塑料鞋 plastic shoes;vinyl shoes
注塑鞋 injection molded shoes
大地鞋 earth shoes
皮鞋 leather shoes
上线皮鞋 stitched leather shoes
无带浅口皮鞋 pumps;skimmer
拷花皮鞋 brogues;embossed leather shoes
玛丽珍鞋 (平跟搭襻浅口女童皮鞋)

Mary Jane shoes	校服鞋 school shoes
克里斯提·鲁布托鞋（女式红底高跟皮鞋）Christian Louboutin shoes	护士鞋 nurse shoes
莫卡辛软皮鞋 moccasin	便鞋 casual shoes; walking shoes; slippers; casuals; walkers
女无后跟拖鞋式皮鞋 mule	软底低跟女便鞋 slipper ballet; ballerina
孟克搭襻带皮鞋 monk strap dress shoes	平底便鞋 loafers; ballerina
束带接头皮鞋 toe cap derby tie dress shoes	翻口便鞋 collar slippers
圆头皮鞋 round-toe leather shoes	皮便鞋 leather casuals
尖头皮鞋 sharp-toe leather shoes	绒面革便鞋 suede casuals
高跟尖头皮鞋 high heel pointed toe pumps	易穿便鞋 step-ins
方头皮鞋 square-toe leather shoes	旅游鞋 traveling shoes; tourist shoes
牛皮鞋 cattle hide shoes	徒步旅行鞋 hiker shoes
猪皮鞋 pigskin shoes	时尚鞋 fashion shoes
羊皮鞋 kid shoes	豹纹鞋 leopard print shoes
鹿皮鞋 buckskins; moccasins	窝钉鞋 studded shoes
人造革皮鞋 artificial leather shoes	情侣鞋 sweethearts shoes
仿麂皮鞋，绒面皮鞋 suede shoes	休闲鞋 leisure shoes; casual shoes; boat shoes
黑漆皮鞋 patent leather shoes	上街鞋 town shoes
软皮鞋 soft leather shoes	社交鞋 social shoes
宇航鞋 astronaut's shoes	正装鞋 dress shoes; dressy shoes
工作鞋 work shoes; duty shoes	礼服用鞋 tuxedo shoes; dress shoes; formal shoes
劳保鞋 labour protective shoes; safety shoes	（女）宫廷礼服鞋 court shoes
防护鞋 protective shoes	晚宴鞋 evening shoes
防静电鞋 antistatic shoes	绅士鞋 dress shoes
绝缘鞋 insulated shoes	舞蹈鞋 dance shoes
无尘鞋 dust-free shoes	芭蕾鞋 ballet shoes; ballerina shoes; toe shoes; toe slippers
防滑鞋 antiskid shoes	踢踏舞鞋 tap shoes
甲板鞋，船鞋 deck shoes	轻便舞鞋 pump shoes
胶鞋 rubber shoes	戏装鞋 theatrical shoes
雨鞋 rain shoes; rubber shoes	牛津鞋 Oxford shoes
保暖防水套鞋 arctics	工艺鞋 craft shoes
（套在鞋上）套鞋 overshoes	印花鞋 printed shoes
橡胶套鞋 gumshoe [美]	绣花鞋 embroidered shoes
防水鞋 waterproof shoes	钩编鞋 crocheted shoes
军鞋，解放鞋 military shoes	

珠饰鞋　beaded shoes
（鞋头有小孔的）装饰鞋　medallion shoes
搭扣鞋　strap shoes; strap; tab shoes
T形带襻鞋　T-bar shoes
无带扣鞋　slip-on; step-ins
无带扣低跟女便鞋　pumps
带扣鞋　buckle shoes; bar shoes
后襻鞋　slingback
踝带鞋　ankle-strap shoes
系带鞋　lace-up shoes; lace-ups
角斗士鞋，罗马鞋　gladiator shoes
角斗士式系带鞋　gladiator style lace-ups
马鞍鞋　saddle shoes
有跟鞋　heeled shoes; heels
平跟鞋　low-heeled shoes
中跟鞋　middle-heeled shoes
高跟鞋　high-heeled shoes
坡跟鞋　wedge shoes; wedgies
罗马风情坡跟鞋　gladiator wedge
带扣坡跟鞋　strappy wedge
厚底坡跟鞋　platform wedge
低坡跟鞋　low wedge shoes
满帮鞋　close shoes
软帮鞋　moccasins
耐磨高帮鞋　walking boots
高帮松紧鞋　gaiter
侧帮松紧鞋　side gore shoes
露趾鞋　peep-toe shoes; open toe shoes
果冻鞋　jelly shoes
松糕鞋　clogs
平底鞋　flats; flatties
打结平底鞋　ankle tie flats
低口鞋　low shoes
增高鞋　elevation shoes
厚底鞋　pantshoes; platform shoes; platforms
圆头鞋　round-toe shoes; pumps

厚底圆头鞋　platform pumps
硬底鞋　galoshes
软底鞋　espadrilles
模压底鞋　mould-soled shoes
木底鞋　clogs
木鞋　wooden shoes; sabot
木屐　clogs; wooden shoes
矫形鞋　orthopedic shoes
僧鞋　priest shoes; monk shoes
雪地鞋　snow shoes
沙滩鞋　beach shoes
防滑鞋　cleated shoes
女装鞋　ladies' shoes
少女鞋　misses' shoes
童装鞋　children's shoes
童虎头鞋　tiger head shoes
童平底鞋　cacks
幼儿学步鞋　prewalkers
婴幼毛线鞋　bootee; bootie
婴幼袜口鞋　sock top slippers

2. 运动鞋　sports shoes

智能运动鞋　intelligent sports shoes
专用运动鞋　special sports shoes
轻便运动鞋　sneakers; sneaks; gum shoes
无扣（带）运动鞋　athletic slip-on sneakers
魔术贴搭襻运动鞋　velcro athletic shoes
吸氧健身运动鞋　aerobic shoes
田径鞋　athlete's shoes; racing shoes; running shoes
赛跑钉鞋　spike shoes; track shoes; racing shoes; running shoes
四（五、七）钉跑鞋　running shoes with 4(5,7) spikes
无钉跑鞋　running shoes without spikes
跑步鞋　jogging shoes; joggers; track shoes;

runner
系带跑步鞋　lace joggers
松紧跑步鞋　slip-on joggers
活动扣襻跑步鞋　self-adjusting joggers
跳远鞋　long jumping shoes
投标枪鞋　javelin shoes
击剑鞋　fencing shoes
射击鞋　shooting shoes
骑自行车鞋　cycling shoes
跳伞鞋　parachute jumping shoes
溜冰鞋　skate shoes; ice skates; skates
速度滑冰鞋　speed skates
花样滑冰鞋　figure skates
冰球鞋　hockey skates
旱冰鞋　roller skates; skating boots; skates
滑雪鞋　ski shoes
滑水鞋　water ski shoes
滑草鞋　grass-sliding shoes
滑板鞋　skateboard shoes
快艇鞋　yacht sneakers
网球鞋　tennis shoes
篮球鞋　basketball shoes; basketball boots
排球鞋　volleyball shoes
足球鞋　football shoes
英式足球鞋　soccer shoes
垒球鞋　softball shoes
羽毛球鞋　badminton shoes
高尔夫球鞋　golf shoes
保龄球鞋　bowling shoes
体操鞋　gym shoes
摔跤鞋　wrestling shoes
拳击鞋　boxing shoes
登山鞋　climbing shoes
爬山钉鞋　crampons shoes; crampons; crampoons
(中国) 功夫鞋　Kung-fu shoes

3. 靴子　boots

长筒靴　high boots; long boots; jackboots
(美国) 西部长靴　Western boots
(英俚) "钓伐"靴　bovver boots
伸缩性长筒靴　stretch boots
过膝长筒靴　over-the-knee boots; jack boots; stocking boots; full length boots
有绑腿的长筒靴　gaiter
齐膝长靴　knee length boots
齐臀长靴　hip boots; thigh boots
长筒女靴　kinky boots
长筒皮靴　jumping boots
长筒软靴　slouch boots
半长筒靴　half boots; mid-calf boots; brogues; buskin; bottine〔法〕
七分靴　3/4 length boots
短筒靴　short boots; ankle boots
短筒女靴　bootee; bootie
时款靴　fashion boots
侧帮松紧靴　twin gusset boots; side gore boots
厚底坡跟靴　platform boots
坡跟靴　wedge boots
高跟靴　high-heeled boots
马靴　riding boots; jodhpur boots; cavalier boots
长筒马靴　top boots
短马靴　jodhpurs
骑师靴　jockey boots; half jack boots
运动靴　sports boots
摩托靴　motorcycle boots
远足靴　hiking boots
猎靴　hunting boots
登山靴　mountaineering boots; climbing boots
滑雪靴　ski boots

雪地靴　snow boots
羊毛皮雪地靴　UGG boots; Ugly boots; Uggs; uggs
保暖靴　winter boots; thermoboots
沙漠靴　desert boots
马球靴　polo boots; chukka boots
威灵顿皮靴　Wellington boots
因纽特人毛皮长靴　mukluk
海盗靴　pirate boots
军靴　military boots; combat boots
伞兵靴　paratroop boots; jump boots
宇航靴　space boots
防护靴　protective boots
劳保靴　safety boots
工作靴　working boots; workboots; brogues; brogan
建筑靴　construction boots
工程靴　engineer's boots
矿工靴　miner's boots
伐木靴　logger's boots
高筒防水靴　waders
钓鱼靴　fishing waders
冬靴，保暖靴　winter boots
晴雨靴　all weather boots
绝热保温靴　insulated boots
绝缘靴　dielectric boots; insulated boots
石棉靴　asbestos boots
钢头靴　steel-toe boots
全粒面革靴　full grain boots
牛二层革靴　cow split boots
牛正绒靴　nubuck boots
牛仔靴　cowboy boots
豹纹窝钉靴　studded leopard print boots
芭蕾靴　ballet boots
反绒皮（獐皮）鞋　suede boots
阿尔坎塔拉（仿獐皮）靴　alcantara boots
饰毛皮靴，毛毛靴　fur-trim boots

仿麂皮靴　suede boots
PVC皮靴　PVC leather boots
雨靴　rain boots; rainboots; water-proof boots
长筒胶靴　gum boots; gumboots; gums; galoshes
彩色胶靴　colour rubber boots
镶花边靴　Balmorals
系带靴　laced boots; lace-ups
拉链靴　zip boots
松紧靴　elastic sided boots
搭襻靴　self-fastening boots
带扣靴　buckle boots
尖头有跟靴　pointed toe heeled boots
（靴身）拼缝靴　patchwork boots
毛（绒）衬里靴　fur(fleece)-lined boots
（充棉）枕头靴　pillow boots
（室内）平底便靴　slipper boots; boot slippers
针织软靴　knit moccasin boots

4. 凉鞋　sandals

希腊式凉鞋　Grecian sandals
格蕾蒂女式凉鞋，角斗士凉鞋　gladiator sandals
布凉鞋　cloth sandals
皮凉鞋　leather sandals
人造革凉鞋　leatheret sandals
合成革凉鞋　synthetic leather sandals
塑料凉鞋　plastic sandals; vinyl sandals
露趾凉鞋　peep-toe sandals; open toe sandals
宽带凉鞋　bold strap sandals
多带式凉鞋　strips sandals
踝带平底凉鞋　espadrille
踝带凉鞋　ankle sandals
网眼凉鞋　mesh sandals

拖鞋

花结凉鞋　sandals with knot
户外活动凉鞋　sports sandals
沙滩（凉）鞋　beach sandals
活动扣襻凉鞋　strap runner sandals
T形搭扣凉鞋　T-strap sandals
（夹带式）人字凉鞋　thong sandals; thongs
坡跟凉鞋　wedge sandals; wedge sliders
厚底（坡跟）凉鞋　platform sandals; platform scuffs
红底高跟踝带扣襻时尚凉鞋　toutenkaboucle(buckle)sandals
软木厚底坡跟凉鞋　cork look platform scuffs
赤脚凉鞋　barefoot sandals
T形带襻凉鞋　T-bar sandals
（平底无扣襻）套带凉鞋　tubular sandals

5. 拖鞋　slippers; sandals; house shoes

土耳其式拖鞋　babouche
日式草履拖鞋　zori
平底拖鞋　scuff slippers; scuffs; slides
（夹带式）人字平底拖鞋　thongs; flipflaps
宽带襻拖鞋　one band slides
活动扣襻拖鞋　adjustable scuffs
边帮拖鞋　sidewall scuffs
花式拖鞋　novelty scuffs
休闲拖鞋　casual scuffs
室内拖鞋　indoor slippers; house slippers
卧室拖鞋　bedroom slippers
室内无后跟女拖鞋　mule
浴鞋　bath slippers
绒毛拖鞋　pile slippers
绣花拖鞋　embroidered slippers
珠绣拖鞋　embroidered slippers with beads
印花拖鞋　printed slippers

海绵拖鞋　sponge rubber slippers
塑料拖鞋　plastic slippers; vinyl slippers
泡沫塑料拖鞋　foamed plastic slippers
草编拖鞋　straw slippers
绒面革拖鞋　suede slippers
皮拖鞋　leather slippers
毛皮拖鞋　fur slippers
冬用拖鞋　winter slippers
绒里拖鞋　fleece-lined slippers
木拖鞋　wooden slippers
纸拖鞋　paper slippers
保健拖鞋　health care slippers

6. 鞋部件　shoe parts

制鞋　shoe making
鞋材　shoe materials
鞋面（里）料　shoe upper(lining) material
鞋面　shoe upper; shoe cover; top of shoe; vamp
鞋里　shoe lining
鞋衬　shoe interlining
鞋头　shoe toe; toe; toe cap
尖鞋头　pointed toe; sharp toe
平鞋头　plain toe
方鞋头　square toe
圆鞋头　round toe
蛋形鞋头　egg toe
翼形鞋头　wing toe
鸭嘴式鞋头　duckbilled toe
花结式鞋头　knot toe
孔饰鞋头　medallion toe
鞋头衬垫　toe puff
鞋口　shoe throat; topline
鞋口绲边　topline binding
鞋领　shoe collar
鞋帮　shoe sides; shoe quarter; upper

鞋前（后）帮 toe(heel) part
鞋全帮 whole vamp
鞋鞘 quarter
鞋底鞘 shank
鞋沿条 welt
鞋用护条 foxing
鞋绲条 binding
鞋补强带 reforced tape
鞋垫弓，铁心 shank
鞋装饰片 overlay
鞋舌 shoe flap; shoe tongue; tongue
鞋底 sole
布（鞋）底 cloth sole
皮（鞋）底 leather sole
胶（鞋）底 rubber sole
塑料（鞋）底 plastic sole
模压底 molded sole
防滑底 cleated sole
鞋外底 outsole
鞋内底 insole
鞋中底 midsole; insole
中底垫皮 sock lining
鞋中插 wedge
软木片 cork sheet
鞋跟 heel
透明鞋跟 clear heel
包皮鞋跟 cover heel
细腰形鞋跟 flared heel
锥形鞋跟 cone heel
勺形鞋跟 scoop heel
楔形鞋跟 wedge heel
高跟 high heel
中跟 middle heel
平跟 low heel
细高跟 stiletto heel; high stacked heel
细低跟 kitted heel
无跟 heelless

鞋跟包皮 heel cover
鞋跟垫片 heel pad
鞋掌 shoe tap; tap
鞋钉 shoe tack; shoe nail
鞋扣 shoe buckle, shoe stud
鞋眼 eyelet
鞋搭扣带 shoe strap
鞋拉襻 shoe tab; shoe tag
鞋带 shoelace; shoe string; shoestring; latchet
鞋拔 shoehorn; shoe lift
鞋垫 shoe cushion; inner sole; insole
除臭鞋垫 odour destroying insole
（装饰用）鞋夹 shoe clip
鞋衬垫 shoe pad
鞋刷 shoe brush
鞋油 shoe polish; shoeshine
鞋粉 shoe powder
鞋楦 shoe last; shoe tree; boot tree
鞋撑 shoe stretcher
鞋跟钳 shoe punch pliers
鞋裁刀 cutting die
鞋锤 shoe hammer
鞋锥 shoe awl
靴楦 boottree; boot last
靴筒 boot leg; bootleg; leg
靴襻 bootstrap
靴带 bootlace

（三）袜 Stockings; Socks; Hose; Hosiery; Leg Wear

长袜 stockings; hose
百慕大长袜 Bermuda hose
过膝长袜 over knee stockings
无跟长袜 hose-tops
中筒袜 half hose; half stockings; knee

袜

stockings
齐膝袜　knee stockings; knee-hi
短袜　socks; anklets; anklet socks; midway socks; socquette[法]
齐踝短袜　ankle socks; ankle hose; low cut socks
膝下长短袜　knee-high socks
高筒短袜　tall socks
针织罗口短袜　crew socks
翻口短袜　anklets; socklets; turnover socks
脚尖透明短袜　sheer toe socks
分趾短袜　toe socks
露趾袜　toeless hosiery; open toe socks
花边紧口袜　lace top stay-ups hosiery
紧口长袜　hold up stockings
智能袜　smart socks
吸湿排汗袜　moisture wicking socks
抗菌防臭袜　antibacterial deodorant socks
礼服袜　dress socks
工作袜　work socks; business socks
运动袜　athletic hose; sports socks; trainer socks
棒球袜　baseball socks
高尔夫球袜　golf hose
滑雪袜　ski socks
滑雪板袜　snowboard socks
骑自行车袜　cycling socks
跑步袜　running socks
远足袜　hiking socks
狩猎袜　hunting socks
休闲袜　casual socks
时装袜　fashion socks
口袋袜　anklet pocket socks
双面穿袜　reversible socks
全成形袜　full-fashioned hose
连续线迹袜　non skip socks
睡袜　bed socks

女连裤袜　panty hose; pantyhose; pantihose; panty stockings; pantistockings
低腰连裤袜　hipster hosiery
高腰连裤袜　high waist pantyhose
高腰塑形连裤袜　allover shaping pantihose; control top pantyhose
网眼连裤袜　lace pantyhose; fishnet pantyhose
透明连裤袜　sheer to waist hosiery
豹纹连裤袜　leopard print pantyhose
绅士袜　men's dress socks
少女袜　girl's socks
童袜　children's socks
婴儿袜，宝宝袜　baby socks; bootie
宝宝莫卡辛袜　baby moccasin socks
布袜　cloth stockings
纱布袜　gauze stockings
纱袜　cotton socks; yarn stockings
线袜　thread socks
莱尔（丝光）线袜　Lisle stockings
丝袜　silk stockings; silk socks
透明丝袜　sheer silk stockings
肉（黑）色丝袜　flesh (black) stockings
连衣丝袜　bodystocking
隐形丝袜，空气丝袜　air stocking
人丝袜　rayon stockings
男用短丝袜　dress socks
丝光袜　mercerized stockings
弹力袜　stretch socks; lycra socks; lycra hose; elastic stockings; stretch hosiery
莱卡袜　lycra hose; lycra socks
无弹力袜　non stretch stockings
锦纶袜　polyamide socks; nylon socks; nylon stockings
毛巾袜　towelling socks
毛袜　woolen stockings; wool socks
毛线袜　worsted socks; knitting wool socks

木（竹）纤维袜 wood（bamboo）fibre socks
间棉袜 quilted stockings
厚袜 service stockings；opaque hose
计算机花袜 computer pattern socks；electronic pattern socks
绣花袜 embroidered socks
花袜 figured stockings；patterned stockings；fancy socks
提花袜 jacquard hose；jacquard socks
网眼袜 mesh stockings；lace stockings；net stockings；fishnet stocking；openwork hose；fishnet hosiery
超细网眼袜 micronet hosiery
（大网眼）围栏网袜 fence net stockings
无缝袜 seamless socks
脚掌套袜 footlets
袜套 leg warmer；ankle socks
袜鞋 room socks；home socks；slipper socks
运动袜鞋 sport shoe socks
家用袜鞋 home socks
吊袜带 stocking suspenders；suspenders
吊袜腰带 garter girdle
绑腿 puttee；puttie；uppers；leg wrappings；gaiter；jambiéres〔法〕
防水绑腿 antigropelos
皮绑腿 leathers
护腿 leggings
暖腿套 leg warmer

（四）手套 Gloves；Mitaines〔法〕

连指手套 mittens；mitts
（女）露指手套 mitts；mittens；mitaine〔法〕
拳击手套 mittens；mufflers；cestus；gloves
拳击练习手套 mitts
棒球手套 baseball gloves；mitts；gloves
板球手套 cricket gloves
冰球手套 ice hockey gloves
赛车手套 racing gloves
骑单车手套 bicycle gloves
滑雪手套 ski gloves；ski mittens；snowboard gloves
潜水手套 diving mittens
时尚手套 fashionable gloves
情侣手套 sweethearts gloves
婚纱手套 bridal gloves
（女用）晚宴手套，晚装手套 ladies dress gloves
无扣手套 slip-on gloves；slip-ons
易戴式手套 slip-ons
长手套 long gloves
观剧（社交）手套 opera length gloves
宽口长手套 gauntlets
宽口半长手套 slip-on gloves
厚手套 mufflers
保健手套 healthful gloves
防护手套 protective mitts；protective gloves
航天手套 space gloves
劳工手套，工作手套 working gloves
电子手套 electronic gloves
石棉手套 asbestos gloves
防接触布手套 anti-contact cloth gloves
绝缘手套 insulated gloves
防 X 射线手套 X-ray proof gloves
焊工手套 welding gloves；welder's gauntlets
锦纶防静电手套 antistatic nylon gloves
远红外手套 far-infrared gloves
隔热手套 heat protective gloves
乳胶手套 emulsion gloves
一次性乳胶手套 latex disposable gloves

橡皮手套　rubber gloves
帆布手套　canvas gloves
（棉）纱手套　cotton gloves
棉毛手套　cotton interlock gloves
绒布手套　fleece gloves
珊瑚绒手套　coral fleece gloves
汗布手套　cotton stockinette gloves
皮手套　leather gloves
羊皮手套　kid gloves; suede gloves; cape gloves
猪绒面皮手套　pig suede gloves
猪二层皮手套　pig split gloves
牛二层皮手套　cow split gloves
仿鹿皮手套　suede gloves
人造革手套　leatherette gloves
PVC荧光手套　fluorescent PVC gloves
毛皮里子手套　fur-lined gloves
毛织手套　knitted gloves
羊毛手套　woolen gloves
兔毛手套　angora gloves
腈纶手套　acrylic yarn gloves
钩编手套　crocheted gloves
网眼手套，通花手套　lace gloves; open-work gloves
提花手套　jacquard gloves
绣花手套　embroidered gloves
印花手套　printed gloves
手笼，暖手筒　muff
护腕　wristband
腕带　wristlet
臂套　arm warmer

（五）领带，围巾　Necktie and Scarf

1. 领带，领结　necktie; tie; neck scarf

颈部饰物，颈饰　neckwear; neckpiece; neckcloth; necklet
领饰　collaret; collarette
欧洲大陆领带　Continental necktie
（美国）西部领带　Western tie
阿斯可领带　ascot tie
德贝领带　derby tie
交叉领带　crossover tie
宽领带　choker
时尚领带　stylish necktie
波希米亚时尚领带　Bohemian tie
便装领带　boater tie
防污领带　teflon tie
流星式领带　bola necktie
拉链领带　zip necktie; necktie with zipper; zipper tie
尖头领带　point-end necktie
方头领带　square-end necktie
棒形领带　bar-shaped necktie
筒形领带　rouleau tie
缎带领带　ribbon tie
丝带领带　string necktie
真丝领带　silk necktie; pure silk tie
涤纶领带　polyester necktie
涤纶提花领带　polyester jacquard woven tie
涤纶手工印花领带　polyester tie hand-printed
针织领带　knitted necktie
毛针织领带　wool-knitted tie
毛料领带　woolen necktie
蛇皮领带　reptile necktie
黑领带，黑领结　black tie
白领带，白领结　white tie
蝴蝶领结　bow tie; bow knot; bow
艾伯特领结　Albert tie
绅士领结　esquire knot
水手领结　sailor's knot

装饰花结　knot
领带结　tie knot
四手结，单结　four-in-hand knot
平结　plain knot
温莎结　Windsor knot
半温莎结，十字结　half Windsor knot
普拉特结　Pratt knot
谢尔比结　Shelby knot
交叉结　cross knot
双交叉结　double cross knot
双环结　double knot
艾伯特（亚伯特或阿尔伯特）王子结　Prince Albert knot
马车夫结，简式结　simple knot
肯特结　Kent knot
巴尔萨斯（巴尔蒂斯）结　Balthus knot
多佛结　Dovorian knot
普拉茨堡结　Plattsburgh knot
圣安德鲁结　Saint Andrew knot
开尔文结　Kelvin knot
尼基结　Nicky knot
东方结　oriental knot
汉诺威结　Hanover knot
卡文迪许（凯文狄许）结　Cavendish knot
维多利亚结　Victoria knot
新古典主义结　neoclassical knot
浪漫结　trend knot
杰尼亚结　Zegna knot
奥纳西斯结　Onasiss knot
梅罗文加结　Merovingian knot
大西洋结　Atlantic knot
对角结　diagonal knot
自由式结　free style knot
领带夹　tie holder; tiepin; necktie clip; stickpin; breast pin
领带别针　tie pin; tiepin; stickpin; necktie clip

领带棒　tie bar
领带饰链　tie chain

2. 围巾　scarf; shawl; kerchief; neckerchief; neck handkerchief; muffler; stole; neck scarf; wrappage

棒针围巾　hand-knitted scarf
钩编围巾　crocheted scarf
头巾式围巾　turban scarf
圆圈形围巾　circular scarf
起皱围巾，皱纹围巾　crinkle scarf
印花围巾　printed scarf
杂色围巾　becher
（女用）三角形披肩围巾　triangle scarf; fichu
皮围巾　neckpiece; necklet
毛皮围巾　fur scarf
海狸毛围巾　beaver scarf
貂皮围巾　mink scarf
（女用）长毛皮围巾　boa
羊毛围巾　woolen scarf; wool shawl
兔毛围巾　angora scarf
腈纶围巾　acrylic scarf
摇粒绒围巾　polar fleece scarf
（棉）纱围巾　cotton scarf
远红外加热围巾　far-infrared heating scarf
红领巾　red scarf
领巾　neckcloth; neckerchief; neck-band; scarf
牛仔领巾　bandana
丝绸围巾，丝巾　silk scarf
手绘丝巾　hand-drawn silk scarf
手印丝巾　hand-printed silk scarf
豹纹丝巾　leopard print silk scarf
雪纺丝巾　chiffon silk scarf
涤纶丝巾　polyester silk scarf
披巾　shawl; scarf; stole

绣花丝绸披巾　silk embroidery shawl
腈纶披巾　acrylic shawl
流苏披巾　fringed scarf
西藏丝毛披巾　Tibet scarf
毛皮披肩　fur shawl
帕西米纳羊毛绒披肩　pashmina shawl; pashmina
头巾　scarf; kerchief; headscarf; headpiece; hood; almuce
穆斯林头巾　turban
印度头巾　pagri; puggaree; puggree; puggry
海盗头巾　pirate scarf
嘻哈头巾　Hip Hop scarf
印花头巾　printed scarf
纱巾　gauze kerchief
面纱　veil; fall
双层面纱　yashmak; yashmac
婚礼面纱　wedding veil; bridal veil
（女帽上）短面纱　nose veil
口罩，面具　mask
卫生口罩　flu mask
纱布口罩　gauze mask
一次性口罩　disposable mask
医用口罩　surgical mask; medical mask
防菌口罩　antibacterial mask
抗病毒口罩　antiviral mask
防毒面具　gas mask
防尘面具　anti-dust mask
氧气面具　oxygen mask
假面具　visor
（古代）面盔　visor
盔甲　armour
颈甲　gorget
护胸甲　breast armour; cuirass; plastron
腿甲　cuish
防寒耳套　earcap; earmuffs; earwarmer

手帕　handkerchief; nose rag
印花大手帕　bandana
胸袋饰巾　pocketchief; pocket handkerchief; pocket square

（六）腰带，手袋　Waistband and Handbag

1. 腰带　waistband; waistbelt; belt; girdle

外腰带　outer belt
内腰带　inside belt; inner belt
（女、童用）腰带　sash
护腰带　kidney belt
保健腰带　health belt
医疗腰带　medical belt
智能腰带　smart belt
柔道腰带　judo belt
举重腰带　weight lifter's belt
猎装腰带　safari belt
僧侣腰带　monk's belt
和服腰带　obi
（英国，丹麦等国）稳定腰带　stable belt
意大利式腰带　Italian belt
印度式宽腰带　cummerband
（非洲）棉腰带　pange
（古希腊）腰带　zoster
（古语）腰带　zone
睡衣腰带　waistband for pyjamas
围巾式腰带　muffler cummerband; scarf belt
时尚腰带　fashion belt; stylish belt
窝钉腰带　studded belt
曲线腰带　curved belt
针织腰带　knitted belt
编结腰带　braided belt

蝴蝶结腰带	bow belt
打结腰带	tie belt; lace-up belt
及臀腰带	hip belt
配衬腰带	match belt
固定腰带	set-in belt
连身腰带	belt attached to coat
活动腰带	detachable belt
松紧腰带	elastic belt; elastic waistband
瘦身腰带	skinny belt
可调节腰带	free size belt
可卸腰带	button-off belt
宽腰带	wide belt
细窄腰带	narrow belt
双条腰带	twin belt
双层腰带	double belt
双扣腰带	twin buckle belt
真皮腰带	genuine leather belt
PVC 皮腰带	PVC belt
帆布腰带	canvas belt
麻织宽腰带	woven hemp sash
布腰带	cloth belt
本布腰带	self-belt
金属腰带	metal belt
金属链式腰带	metallic chain belt
绳索腰带	cord belt
纸腰带	paper belt
腰带扣	buckle; belt buckle; belt clasp; belt clamp
皮带	leather belt; belt
（美国）西部皮带	Western belt
山姆布朗（军官）佩带	Sam Browne belt
子弹带	belt
安全带	life belt
束腰带	girdle
裤带	trouser belt
吊裤带，裤背带	braces; suspenders; shoulder straps; gallowses; galluses; bearers
吊带夹	suspender clips; brace clips
装饰带，饰带	ornamental belt
珠饰带	beaded belt
军服饰带	sash
字母饰带	initialed belt
绶带	belt
肩带，肩章	shoulder strap; shoulder patch; flash
臂章	arm band; armband; armlet; patch; chevron

2. 手袋，手提包 handbag; bag

肩挎式手袋	shoulder handbag
臂夹式手袋	underarm handbag
链带手袋	chain handle handbag; chain bag
双带手袋	twin handle bag
（袋口）束带手袋	drawstring handbag
无带手袋	clutch bag; clutch
无带有盖手袋	envelope clutch
时尚手袋	fashion handbag
缂丝巾手袋	bag with silk scarf
饰流苏手袋	fringed bag
窝钉手袋	studded bag
豹纹手袋	leopard print bag
环保手袋	ecological bag
休闲手袋	casual bag
袖珍手袋	mini handbag
马鞍型手袋	bolide bag; saddle bag
积琪莲手袋	constance bag
姬莉手袋	Kelly bag
晚装手袋	evening handbag; evening bag
化妆手袋	cosmetics bag
钩编手袋	crocheted bag
网眼手袋	mesh bag
珠绣手袋	embroidered bag with pearls
串珠手袋	beaded handbag

真皮手袋 leather handbag
反绒皮（獐皮）手袋 suede bag
阿尔坎塔拉（仿獐皮）手袋 alcantara handbag
PVC 皮手袋 PVC leather handbag
拼皮手袋 patch leather handbag
毛毛（毛皮）手袋 fur bag
木柄手袋 wooden handle bag
帆布手袋 canvas handbag
亚麻手袋 linen handbag
尼龙手袋 nylon handbag
腈纶手袋 acrylic handbag
桃皮绒手袋 microfibre handbag
草编手袋 straw bag
儿童手袋 children bag
拉链袋 zipper bag; zip-up bag
运动袋 sports bag
保龄球袋 bowling bag
旅行袋 travelling bag; luggage bag; hold-all
短途小旅行袋 overnight bag
装有脚轮的旅行袋 trolley bag
装有脚轮的背包 backpack rolling bag
行李袋 travelling bag; duffle bag; duffle
购物袋 shopping bag; tote bag
桶形拷袋 barrel bag
搭盖袋 slouch bag
牛仔袋 denim bag
流浪汉袋 hobo bag
邮差（休闲）包 messenger bag
医生手提包 doctor bag
女用手提包 tote bag; purse
波士顿手提包 Boston bag
箱形手提包 tote
公文包 briefcase
多功能电脑包 versatile computer bag
手机袋 cell phone bag

书包 satchel; school bag
背袋 knapsack; rucksack; backpack
带风帽的背袋 hoodie backpack
腰袋 waist bag; waistpack
小钱包，钱包 purse; wallet; pouch; pocket book〔美〕
零钱包 coin bag

（七）首饰，饰物，花边 Jewelry, Ornament, Lace

1. 首饰，珠宝 jewelry; jewellery

金银首饰 gold and silver jewelries
铜首饰 bronze jewelry; brass jewelry
合金首饰 alloy jewelry
玉首饰 jade jewelry
人造首饰 imitation jewelry
皮首饰 leather jewelry
贝壳首饰 shell jewelry
玻璃首饰 glass jewelry
陶瓷首饰 ceramic jewelry
塑料首饰 plastic jewelry
贴身首饰 body jewelry
流行首饰 fashion jewelry
项饰，颈饰 necklet; necklace
珠宝颈饰 carca-net
项链 necklace; lavaliere; pendant; torque
金项链 gold necklace
彩金项链 colour gold necklace
银项链 silver necklace
水晶项链 crystal pendant
闪石项链 rhinestone necklace
天然石项链 natural stone necklace
珊瑚项链 coral necklace
琥珀项链 amber necklace
木珠项链 wood-beaded necklace

贝壳项链　shell pendant
子弹头项链　bullet necklace
泪滴项链　teardrop necklace
毛衣项链　sweater necklace
波希米亚项链　Bohemian necklace
多层式珍珠项链　multistrand pearl necklace
网状项链　netlike pendant
套索式项链　lariat necklace
碎钻项链　diamond set pendant
水钻项链　rhinestone necklace
（人造）白金蛇形项链　white gold snake chain
短项链　choker
项圈　necklet;chaplet;chain
十字形项饰　cross
项链坠　charm
红宝石项链坠　ruby charm
珊瑚项链坠　coral charm
珍珠项链坠　pearl charm
彩玻项链坠　colour glass charm
垂饰物　pendant;pendent
金龟子垂饰物　scarab pendant
耳环　earrings;ear pendant;hoop
夹式耳环　clip-on earring
扣式耳环，耳扣　button earring
大圈耳环　hoop earrings
球形耳环　euroball earrings
嵌珠金耳环　gold pearl earrings
嵌钻（石）耳环　diamond earrings
嵌闪石耳环　rhinestone earring
水钻耳环　rhinestone earring
宝石耳环　jewel earring
波希米亚耳环　Bohemian earring
木耳环　wooden earring
双（三）色耳环　two(tri)-tone earrings
垂挂式耳环　drop earrings

细条垂挂式耳环　thread earrings
耳圈　hoop earring;hoop
彩金耳圈　colour gold hoop
耳坠子　eardrop
贝壳耳坠　shell eardrop
水晶耳坠　crystal eardrop
蓝宝石耳坠　sapphire eardrop
耳钉　ear nail
鼻钉　nose nail
鼻环　nose ring
手镯，手环　bracelet;wristlet;bangle
金手镯　gold bracelet
彩金手镯　colour gold bracelet
银手镯　silver bracelet
镶钻手镯　diamond inlaid bracelet
翡翠手镯　jadeite bracelet
玉手镯　jade bracelet
珍珠手镯　pearl bracelet
木手镯　wooden bracelet
闪石手镯　rhinestone bracelet
窝钉手镯　studded bracelet
链式手镯　chain bracelet
卷叶形手镯　leaf bangle
雕花手镯　engraved bangle
可伸缩的手镯　expandable bangle
彩色玻璃手镯　colour glass bangle
琼琦（超大）手镯　chunky bracelet
踝镯　ankle bracelet
手表镯　watch bracelet
手环　bracelet;wristlet;bangle
智能手环　smart bracelet
手链　hand chain;bracelet;chain bracelet
挂锁式手链　padlock bracelet
彩色水晶手链　colour crystal bracelet
吊坠手链　charm bracelet
镶嵌手链　studded bracelet
木珠手链　wood-beaded bracelet

闪石手链　rhinestone bracelet
陶瓷手链　ceramic bracelet
踝链　ankle chain; foot chain; anklet
手表，挂表　watch
智能手表　smart watch
航天表　space watch
怀表　pocket watch
手镯表　bracelet watch
手链表　chain watch
礼服表　dress watch
运动表　sports watch
防水表　water resistance watch
品牌表　brand watch
纪念表　souvenir watch
古董表　curio style watch
复古表　retro watch; revival watch
时装表　fashion watch
情侣表　sweethearts watches
儿童卡通表　children's cartoon watch
腕带　wristlet
臂环　armlet; arm ring
戒指　ring; hoop
金戒指　gold ring
铂金戒指　platinum ring
彩金戒指　colour gold ring
银戒指　silver ring
玉戒指　jade ring
翡翠戒指　jadeite ring
水晶戒指　crystal ring
祖母绿戒指　emerald ring
宝石戒指　jewel ring
钻石戒指　diamond ring
蓝（红）宝石钻戒　sapphire (ruby) and diamond ring
水钻戒指　rhinestone ring
木戒指　wooden ring
礼服戒指　dress ring
印章戒指　signet ring
双色戒指　two-tone ring
单身戒指　singelringen
情侣对戒　sweethearts rings
订婚戒指　engagement ring
结婚戒指　wedding ring; bridal ring; wedding band
宝石　precious stone; jewel
人造钻石　rhinestone
水钻　crystal rhinestone; rhinestone

2. 饰物，装饰品　ornaments; decorations

装饰　decoration; ornament
时装配饰品　fashion accessories
女用服饰品　confection
服饰件　trimmings
小（件）饰物　trinket; knickknack
小皮件饰物　small leather goods
水晶饰物　crystal ornament; crystal
骨饰物　bone ornament
木饰物　wooden ornament
瓷器饰物　porcelain accessories
宝石饰物　jewel
头饰　headgear; head tire; headdress
头带，头箍　head band; headband
额饰　frontlet
鼻饰　nose ornament
（御寒）耳套　earcaps; earwarmers; earmuffs
发饰　hair ornament; hair accessories
发型　hair style; hairstyle; hair type
发带　hair band; hair ribbon; snood; bandeau; headband
发夹　hair pin; hair clip; hair slide; bodkin
发针　hair pin; hairpin; bodkin

发簪 hair clasp; hair pin; hairpin
碧玉簪 emerald hairpin
发结 hair bow
顶髻 topknot
缠结 tangle
发网 hair-net; hairlace; caul; snood; filet〔法〕
假发 wig; falsies
女子冠状头饰品 coronet
手扇 hand fan
维多利亚时期古董手扇 antique Victorian hand fan
日式手扇 Japanese style hand fan
中式丝绸折扇 silk Chinese folding hand fan
(17~18世纪贵妇用) 饰颜小圆片 patch
人造指甲 imitation fingernail
胸饰 plastron; jobot
假胸 bust form; bust pads
假乳房 falsies
(服装上的) 花饰, 饰花 flower ornament
女服胸饰花 corsage
花冠 chaplet
绢花 silk flower
绒花 velvet flower
人造花 art flower
丝带花结 ribbon bow
中式花结 Chinese knot
肩饰花结 shoulder knot
胸饰花结 breast knot
扣花 brooch
缀闪石扣花 rhinestone brooch
围巾扣 scarf ring
装饰别针 stickpin
钻石别针 diamond pin

饰针 pin; fibula
胸针 breast pin; breastpin; brooch; ouch
水晶胸针 crystal brooch
钻石胸针 diamond brooch
花饰针 flower brooch
珍珠饰针 pearl brooch
黄铜饰针 brass pin
帽饰针 hat pin
胸饰扣, 盘扣 frog; Chinese frog
袖口饰纽 stud; links; cuff links
金属袖口纽 metal cuff links
饰钉 stud
(牛仔服) 饰钉 rivet
眼镜 glasses; eyewear; eyeglasses; spectacles
智能眼镜 smart glasses
太阳镜 sunglasses; sunshades; sunnies
宽边太阳镜 bold framed sunglasses
无框太阳镜 rimless sunglasses
护目镜 spectacles; goggles
滑雪护目镜 ski goggles
(遮) 风镜 goggles
防护眼镜 protective glasses
变色眼镜 colour-changeable glasses
有色眼镜 goggles
眼罩 eye mask
睡眠眼罩 sleep mask
阳伞 sunshade
雨伞 umbrella
手开伞 hand-open umbrella
自动伞 auto-open umbrella
自动悬浮伞 drone-brella
折叠伞 topless umbrella; folding umbrella
三折伞 trifolding umbrella
花伞 variegated umbrella; printed umbrella

布伞　cloth umbrella
油纸伞　oiled paper umbrella
绸伞　silk umbrella
尼龙伞　nylon umbrella
手杖伞　stick umbrella
手杖　cane
礼仪佩剑　ceremonial sword

3. 花边　lace; motif

装饰用流苏　flange; fringe; purl; tassel
装饰用绒球　pompon
花球　ball flower
饰珠　bead
玻璃珠　glass bead
木珠　wooden bead
管形珠　bugle bead
闪光珠片，亮片　paillette; spangle; sequin
闪光小饰物　glitter
透明硬纱　organza
绢网　tulle
闪石　rhinestone
环状饰物　ring
贴花　appliqué〔法〕; applied decoration
花边领饰　jabot
饰带　fascia
挂带　lanyard
编带　braid
花色编带　fancy braid
辫形编带　plait braid
羽毛饰　feather; crest
边饰　furbelow
缘饰　edging; fringe
饰边带　gimp
珠饰带　beaded trimming
装饰用缎带　ribbon; riband; galloon
装饰用金线　aglet; zari
金银线编带　orris

V形饰布　vestee
褶裥饰边　ruche; ruching
荷叶边　flounce; chiffon; falbala; purl edge
月牙边　scallop; scollop
锯齿牙边　picot
镶边　purfle; braid
珠饰镶边　beaded braid
包边　covered edge
绲边　piped edge; rouleau
花边　lace; motif; passementeric〔法〕
爱尔兰花边　Irish lace
希腊花边　Greek lace
罗马花边　Roman lace
仿古花边　antique lace
刺绣花边　embroidered lace
针绣花边　needle lace; point lace; point
抽绣花边　drawnwork lace
编结花边　braid motif; braid lace
针织花边　knitted lace
机织花边　woven lace
钩织花边　crocheted lace
线织花边　thread lace
网眼花边　hairlace; filet lace; filet〔法〕
金银花边　galloon; passementerie〔法〕; gold and silver lace
金银丝花边　bullion
荷叶边花边　ruffle lace
银耳花边　tremella lace
扭曲花边　torsion lace
镶嵌花边　mosaic lace
褶裥花边　plaited lace
牦牛绒毛花边　yak lace
布尔登花边　bourdon lace
装饰花边　decorative motif
珠饰花边　beading lace; beaded motif
镶饰花边　trimming lace

贴饰花边	applique lace
凸纹花边	guimp lace
隐纹花边	shadow lace
饰带花边	torchon lace; torchon
绳带花边	cording lace
缎带花边	ribbon lace
皮革花边	leather lace
弹性花边	elastic lace
锦纶弹性花边	nylon spandex lace
丝弹性花边	silk spandex lace
人丝花边	rayon lace
涤纶花边	polyester lace
腈纶花边	acrylic lace
婚纱服花边	bridal lace
乳罩花边	bra lace
绣饰，绣花	embroidery
东方刺绣	oriental embroidery
苏（湘、粤、蜀）绣	Suzhou(Hunan, Guangdong, Sichuan) embroidery
日本刺绣	Japanese embroidery
印度刺绣	Indian embroidery
阿拉伯刺绣	Arabian embroidery
巴鲁巴刺绣	Baruba embroidery
英国刺绣	English embroidery
巴黎刺绣	Paris embroidery
丹麦刺绣	Danish embroidery
抽绣，抽纱	drawnwork; punchwork
绒线刺绣	crewelwork; woolwork
凸花刺绣	raised embroidery
金银丝刺绣	zardozi〔印度〕; bullion embroidery
十字绣	cross-stitch embroidery
纹章（刺绣）	emblem
烫画	transfer
移画印花	decalcomania; decal printing; decal
转移烫钻	hot-fixing transfer rhinestones
卡通漫画	cartoon
书法艺术	calligraphy
中国结	Chinese knot
涂鸦	graffito; graffiti〔复〕

六、服装设备和工具
Apparel Equipments and Tools

（一）一般名称 General Terms

服装设备	apparel equipment
放码排板设备	grading and marking equipment
裁剪设备	cutting equipment
（专用）缝纫设备	(special) sewing equipment
（自动）整烫设备	(automatic) pressing equipment
抗皱设备	wrinkle-free equipment
水洗设备	washing equipment
干洗设备	dry cleaning equipment
输送设备	conveying equipment
辅助设备	assisting equipment
机器，机械	machine; machinery
服装机械	garment machinery
机床	machine tool
机件	machine parts

设计打样设备和工具

机器故障　machine problem
机械维修　machinery maintenance
设备维修　equipment maintenance
工具　tool; instrument; implements
服装专用工具　clothing special instruments

(二) 设计打样设备和工具　For Designing and Drafting

绘图机,打样机　drafting machine
高速绘图机　high-speed drafting machine
纸样放缩机　pattern grading machine
纸样复印机　pattern duplicating machine
唛架绘印机　marker plotter
唛架复印机　marker copier
服装模板切割机　clothing template cutter
计算机　computer
计算机放样排板机　computerized pattern grading and marker making machine
计算机辅助设计系统　computer aided design system
计算机配色系统　computer colour-matching system
计算机绣花设计打板系统　computer embroidery design and punching system
绘图板　drafting board; drawing board
曲线板　curve gauge; curve board
圆规　compasses
点线器,擂盘　tracing wheel; tracer; roulette
打眼器,冲头　puncher; perforator
记号剪　notcher
模特　model
人形模型,假人　dummy; body stand; mannequin; manikin; lay figure; model

可调节的人形模型　adjustable mannequin
半身模型架　dress form
衣帽架　clothes tree; clothing stand; hallstand
不锈钢挂衣架　stainless steel hanging rack
尺　rule; ruler
米尺,公制尺　meter ruler
英制尺　inch ruler
米尺（一米长）　meterstick
码尺　yard measure; yardstick〔美〕; yardwand〔英〕
一英尺长的尺　a foot-rule
市尺　Chinese ruler
直尺　ruler
弯尺,曲尺　curve ruler; zigzag ruler
自由曲尺　arbitrarily curved ruler
云尺,弧线尺　cloud-shaped ruler
角尺　square measure; L-square; square
丁字尺　T-square
三棱尺　triangular scale
三角尺,三角板　set square; triangle
L形尺　L-shaped ruler
格子尺　checked ruler
计算尺　slide rule; slipstick
比例尺　scale
放码尺　ruler for pattern grading; marking ruler
尺码放缩仪　multi-grader
折尺　folding rule
软尺　tape measure; tailor's tape; flexible rule; soft ruler
带卷尺　tapemeasure; measuring tape
钢卷尺　steel tapemeasure; steel tape ruler
皮卷尺　linen tape ruler; measuring tape
竹尺　bamboo ruler

塑胶尺 plastic ruler
不锈钢直尺 stainless-steel ruler
铝尺 aluminium ruler
套装尺 ruler set
经纬密度尺 densimeter
英寸量规 sewing gauge
褶边规 hem gauge
绘图笔 drawing pen
鸭嘴笔 ruling pen;drawing pen
钢笔 pen
铅笔 pencil
笔记本 notebook
计算器 calculator
绘图纸 drawing paper
描图纸 tracing paper
复写纸 carbon paper
复印纸 duplicating paper
样板纸 pattern paper;draughting paper
马粪纸板 strawboard
图钉 drawing pin
大头针 pin
划粉 tailor's chalk;marking chalk;French chalk;chalk
蜂蜡 beeswax

（三）裁剪设备和工具 For Cutting

验布机 cloth inspection machine
卷布机 cloth winding machine
卷布验布机 cloth winding inspection machine
拉布机，铺布机 cloth spreading machine
智能拉布机 intelligent fabric spreading machine
量裁机 cloth measuring and cutting machine
纵斜纹切布机 straight and bias cloth cutting machine
电子布块秤 electronic fabric scale
计算机自动裁剪系统 computer automatic cutting system
裁剪机 cutting machine;cutter
全自动/计算机裁剪机 fully automatic/computerized cutting machine
半自动裁剪机 semi-automatic cutting machine
激光裁剪机 laser cutting machine
水射流裁剪机 water jet cutting machine
超声波裁剪机 ultrasonic cutting machine
等离子裁剪机 plasma torch cutting machine
摇臂裁剪机 swing arm vertical cutting machine
电剪 electric cloth cutter;cutter
直刀电剪，直刀裁剪机 straight knife cutter;straight knife cutting machine
圆刀电剪，圆刀裁剪机 circular knife cutter;rotary knife cutter;round knife cutting machine
带刀电剪，带刀裁剪机 band knife cutting machine
微型电剪，迷你裁剪机 miniature electric cutter;mini cutter
电池式（不用电线）裁剪机 cordless cutter
带灯裁剪机 light cutter
立式裁剪机 upright cloth cutter
带式裁剪机 band cloth cutter
冲切裁剪机，模切（裁剪）机 die cutting machine;die cutter;hydraulic cutting presser
油压裁剪机 hydraulic cutting press;hydraulic cutter

缝纫设备和工具

毛皮裁剪机　fur cutting machine
罗纹裁剪机　rib cutting machine
缺口机　notcher machine
钻孔机　drill machine
裁门襟机　placket cutting machine;placket trimming machine
切捆条机　ribbon cutting machine;bias tape cutter
裁带机　tape cutting machine
切边机　edge cutting machine
切刀，裁刀　cutting knife;knife;cutter
条刀　band knife;tape knife
直刀　straight knife
弯刀　bend knife
圆刀　circular blade;round knife
波纹刀　wave knife
裁剪型板　die
裁床，裁剪台　cutting table
计算机自动裁床　computerized automatic cutting table
组合裁床　sectional cutting table;flexible cutting table
气垫/真空裁床　air flotation/vacuum cutting table
直线/十字激光灯（裁布定位用）　straight/cross laser marker
压铁　presser;weights
裁布夹　cloth clip
钻眼器　cloth drill;drill
打眼机　punching machine;perforation machine
打号机　marking machine;numbering machine
剪刀　shears;scissors;trimmers
裁缝用剪　dressmaker's shears;tailoring shears
裁缝用直剪刀　dressmaker's straight trimmers
裁缝用弯剪刀　dressmaker's bent trimmers;bent-handled shears
布剪　cloth scissors
纸样剪　pattern scissors
修剪剪刀　trimming scissors
气动剪刀　automatic air scissors
花齿剪　pinking shears;zigzag scissors
嵌花剪　inlaid trimmers
锥子　awl

（四）缝纫设备和工具　For Sewing

缝纫机　sewing machine;sewer;seaming machine;seamer;stitcher
手摇缝纫机　hand sewing machine
脚踏缝纫机　treadle sewing machine
家用缝纫机　home sewing machine
工业缝纫机　industrial sewing machine
长臂缝纫机　long arm sewing machine
双头缝纫机　double head sewing machine
电动缝纫机　electric sewing machine
电子缝纫机　electronic sewing machine
电子花样缝纫机　electronic controlled pattern sewing machine
计算机程控缝纫机　computer controlled programmed sewing machine;computer sewing machine
微控缝纫机　microprocessor controlled sewing machine
自动缝纫机　automatic welting machine
多功能缝纫机　versatile sewing machine;multi-purpose sewing machine
高速缝纫机　high speed sewing machine
超高速缝纫机　ultra-high speed sewing machine
通用缝纫机　universal sewing machine

特种缝纫机　special sewing machine
全（半）自动模板缝纫机　full (semi)-automatic template sewing machine
平缝机，平车　flat sewing machine; plain sewing machine; lockstitch machine; flat seamer
单针平缝机　single-needle flat sewing machine
双针平缝机　double-needle flat sewing machine
多针平缝机　multi-needle flat sewing machine
高速平缝机　high-speed flat sewing machine
自动剪线平缝机　straight lock stitcher with automatic thread trimmer
多针链式缝纫机　multi-needle chain-stitch machine
折缝机　chainstitch feed-off-arm machine; F.O.A. seam machine; fell seaming machine
绷缝机　flat lock machine; cover seam machine; covering stitch machine; butted seaming machine
多针绷缝机　multi-needle flat lock machine
单（双）面绷缝机　bottom (top and bottom) covering stitch machine
疏缝机，粗缝机　basting machine
扎驳机　pinking machine
绗缝机　quilting machine; quilter
羽绒绗缝机　down quilting machine
暗缝机　blind stitching machine; blind tacker
锁链暗缝机　chain blind-stitch machine
环缝机，链式线缝机　chain-stitch machine

曲折缝缝纫机　zigzag sewing machine; zigzagger
人字机　herringbone machine
装饰线缝机　decorative stitching machine
门襟机　placket machine
止口机　front edge sewer
绱袖机　sleeving machine; sleeve attaching machine
缝袖衩机　sleeve-placket machine
缝裤腰头机　waistband machine
裤脚（卷边）机　bottom hemming machine
省缝机，缝省机　dart sewing machine; dart sewer

（五）包缝，锁钉，绣花，针织等设备和工具　For Overlocking, Buttonholing, Buttoning, Embroidering, Knitting, etc.

包缝机，锁边机　overlock machine; overcasting machine; overedge sewing machine; Merrow sewing machine; selvage-seaming machine; overedger; piper
锁缝机　lockstitch machine; lock machine
联锁机　interlock stitch machine; interlock sewing machine
三线（包缝）机　three-thread overlock machine
四线（包缝）机　four-thread overlock machine
五线（包缝）机　five-thread overlock machine
加固缝纫机　fastening machine
打结机，套结机　tacking machine; bar-tack machine; bar tacker; tacker; lock-

stitch bar tacking machine
锁眼机　buttonhole stitching machine; buttonholing machine; buttonholer
连续锁眼机　continuous buttonholing machine
圆头锁眼机，凤眼机　eyelet buttonhole sewing machine; eyelet stitching machine; eyelet buttonholer
平头锁眼机　straight buttonhole machine; lockstitch buttonholer
钉扣机　button attaching machine; button stitching machine; buttoning machine; button sewing machine; button sewer; fastening machine
钉扣机器人　button robot
揿纽钉扣机　snap attaching machine; snap fastening machine; snap fixing machine; snap attacher; fastening machine
手（电）动揿纽钉扣机　manual(electric) snap fastening machine
气动揿纽钉扣机　pneumatic snap fixing machine
剪线钉扣机　button sewing machine with trimmer
包扣机　machine for covering button with cloth
钉鸡眼机，打气眼机　eyelet fastening machine
打撞钉机　rivet machine
绣花机　embroidery sewing machine; embroidery machine; embroiderer; embroider
电子绣花机　electronic embroidery machine
计算机绣花机　computerized embroidery machine
激光绣花机　laser embroidery machine;

emblaser
多功能绣花机　multi-function embroidery machine
单头/双头绣花机　one/two-head embroidery machine
多头自动绣花机　multi-head automatic embroidery machine
自动（手工）印花机　auto-(manual) printing machine
计算机自动针织机　computerized automatic knitting machine
计算机数控无缝成型针织机　CNC seamless forming knitting machine
针织横机　flat knitting machine
经（纬）编机　warp(weft) knitting machine
圆筒针织机，大圆机　circular knitting machine
圆筒网眼纬编机　circular eyelet knitting machine
圆筒钩针纬编机　circular spring needle machine
拉舍尔经编机　raschel loom
针织缝盘机，袜子缝头机　linking machine
提花横机　jacquard flat knitting machine
毛衣编织机　sweater knitting machine
织补机　darning machine
开袋机，袋唇机　pocket-hole sewing machine; pocket welting machine
全自动开袋机　full automatic pocket-hole sewing machine
上袋机，钉袋机　pocket setting machine
折袋机　pocket creasing machine
折袋盖机　pocket flap creasing machine
自动驳领衬垫机　lapel roll padding automat

缝拉链机　zipper assembler
缝标签机，钉唛头机　labeling machine
毛皮条子缝纫机　fur strap sewing machine
毛皮拼接缝纫机　fur abutting machine
带子机　belt machine
带襻机，裤襻机　belt loop sewing machine; looper sewing machine
橡筋机　elastic sewing machine
打褶机　pleat making machine; ruffling machine
抽褶机　shirring machine
缝锁（边）机　edge seaming and trimming machine
折边机　edge folding machine
卷边机　hemming machine
绲边机　piping machine
镶边机　edge taping machine
珠边机　handstitch machine
波边机　fluted hem machine
饰边机　trimming machine
缲边机　hem-edge machine; edge sewing machine
点衬定位机　spot welding machine
点领机　collar marking machine
绲领机　collar binding machine
切领嘴机　collar point trimming machine
翻领角机　collar point turning machine
修领脚机　collar contour trimmer
翻袖头机　cuff turning machine
装置，器件　device
附属装置，附件　attachment
综合缝纫装置　integrated sewing device
绗缝装置　quilter
平缝装置　feller
送料装置　feeding device; feed unit
倒针装置　backstitch mechanism
针距调节装置　stitch regulating device
自动停针位装置　automatic needle positioning device
自动拨（剪）线装置　automatic thread wiping(trimming) device
切料装置　material trimming device
开孔装置　punching device
润滑装置　lubricating device
吸油装置　oil suction device
安全装置　safety device
卷边器　hemmer
绲边器　binder; piper
折边器　folder
修边器　trimmer
打裥器　gatherer; pleater; tucker
抽褶器　shirring blade
起皱器　ruffler; pucker
缩袋器　pocket setter
（纽扣、扣眼）定位器　spacer
钉揿纽器　snap attacher
剪线器　thread cutter
绣花剪　embroidery scissors
扣眼剪　buttonhole scissors
线头剪　thread clippers; thread clips; U-scissors; snippers
针　needle
缝纫针　sewing needle
手缝针　hand needle; sharps needle
缝纫机针　needle for sewing machine
绣花针　embroidery needle; crewel needle
14号针　needle of size 14
备用针　spare needle
织补针　darning needle; mending needle
钩针　crochet hook; hook needle
顶针，针箍　thimble
插针垫　pin cushion; needle book
裁缝用蜡　tailor's wax
拆线刀　seam ripper; ripper; unpicker

钳子　pinchers；pliers
起子　screwdriver
扳手　spanner
镊子　tweezers
锥子　awl；bodkin
冲模　moulds
糨糊　paste
刮浆刀　scraper；tool for smearing paste
工具箱　tool box
工具袋　tool bag；workbag
针线篮　sewing basket
针线包　sewing kit
针线盒　work box；housewife；hussif；sewing box；needlework box
针线袋　workbag

（六）缝纫机部件　Parts of Sewing Machine

机头　machine head
机板　machine bed；machine table board；sewing table
机架　machine pedestal；machine frame
机座　machine set；machine stand
踏板　pedal；treadle
机头（身）　arm
主轴　arm shaft
底轴　bed shaft
手轮　hand wheel；balance wheel
压脚　presser foot；foot
通用压脚，万能压脚　universal foot
多功能压脚　multi-purpose foot
内（外）压脚　inside(outside) foot
单趾压脚　half foot
滚轮压脚　roller foot
卷边压脚　hemming foot
绲边压脚　piping foot

靠边压脚　edge foot
抽褶压脚　shirring foot
手抬压脚　presser bar lifter
膝抬压脚　knee lift
压脚调节器　presser regulator screw
针板　needle plate
面板，门盖　face plate
针杆　needle bar
针夹　needle clamp
回针杆　reverse bar
送布板　feed plate
送布牙　feed dog
送布调节器　feed regulator
针距调节器　stitch regulating dial；stitch regulator
针距指示牌　stitch indicator plate
运油窗　oil check window
护线器　thread guard
过线器　arm-spool pin pretension guide
导线器　thread guide
夹线器　thread tension device；upper thread tension regulator
挑线杆　thread take-up
绕线器　bobbin winder
线架　thread stand
滑梭　shuttle
梭芯　bobbin
梭匣　bobbin case
照明灯　sewing light
油壶　oilcan
活动箱　moveable sewing box

（七）熨烫设备和工具　For Pressing

1. 熨烫机，整烫机，压烫机　pressing machine；press

欧洲计算机熨烫机　European press with

microprocessor
通用熨烫机，万能熨烫机 utility press
万能风压熨烫机 pneumatic utility press
自（手）动万能熨烫机 automatic(manual)utility press
垂直式多功能熨烫机 vertical multi-finisher
立体熨烫机 form pressing machine; formatic press
滚动式熨烫机 rotary pressing machine
蘑菇式熨烫机 mushroom press
模压熨烫机 die press
大熨烫机 flatwork ironer
辅助熨烫机 supplementary press
蒸汽除皱熨烫机 crepe-smoothing steamer
毛皮整烫机 fur polishing and ironing machine
抗皱衬衫熨烫机 wrinkle free shirt press
抗皱裤子熨烫机 wrinkle free trousers press
裙熨烫机 skirt press
立体整烫机，人像机 body press; body shape press; form finisher
高压人像机 high pressure body shape press
特种人像机 special body shape press
人身机 topper-one
裤型整烫机，裤像机 vertical pants finisher; topper
立体裙整烫机，裙人像机 cubic skirt press
熨烫台板 pressing stand; ironing table
蒸汽烫台 steam pressing stand
组合烫台 combined pressing stand
吹吸烫台 vacuum blowing table
抽汽烫台 pressing stand with an exhauster
抽湿台 vacuum table; vacuum board
强力抽湿台 high power vacuum board
(整烫用) 烘箱 oven
抗皱处理烘箱 wrinkle free batch oven
熨烫部输送带系统 pressing department conveyor system

2. 部位熨烫机 Position Press

滚动式衣身熨烫机 rotary body finisher
前身熨烫机 front press
后身熨烫机 back press
侧身熨烫机 side press
双肩（同时）压烫机 double shoulder press
肩部最终压烫机 shoulder off press
袖窿熨烫机 armhole press
压褶机 tuck press; pleat press
成型机 former
领型机 collar former; collar shaping press
冷热双头领型机 collar creasing press with heat and cool head
压领机 collar presser, collar blocking machine
翻领压领机 collar turning and pressing machine
驳领熨烫机 lapel press
衣袖熨烫机 sleeve press
双袖（同时）压烫机 double sleeve press
袖头压烫机 cuff pressing machine
袖山压烫机 armhole top press
衣边压烫机 edge press
袋盖压烫机 pocket flap forming press
腰围（衬里）压烫机 waist press
裤管成型机 trouser forming press
两侧裤管熨烫机 both side trousers press
裆底压烫机 crotch press
背缝熨开机 back seam opening press
肩缝熨开机 shoulder seam opening press
袖窿缝熨开机 armhole seam opening press
袖缝熨开机 sleeve seam opening press
侧缝熨开机 side seam opening press

熨斗

腰头缝熨开机　waistband seam opening press
省缝熨开机　dart seam opening press

3. 熨　斗　iron; pressing iron; flatiron; smoothing iron

火烙熨斗　conventional flatiron
电熨斗　electric iron
蒸汽熨斗　steam iron
电热式蒸汽熨斗　boiler type steam electric iron
电子恒温熨斗　temptronic steam iron
带微型蒸汽炉熨斗　mini-boiler with iron
带蒸汽台熨斗　steam station iron
吊水熨斗　gravity-feed iron with water tank; iron with water container
盒形熨斗　boxiron
长柄熨斗　gooses
滚筒熨斗　roller iron
泵式熨斗　pump iron
意大利熨斗　Italian iron
大熨斗　sadiron
烫皱褶熨斗　gof(f)er
烫衣板　iron board; press board; pressing plate
袖型烫板，烫凳　sleeve board; pressing bench
铁凳　iron stand
烫天鹅绒垫板　velvet board
拱型烫木　tailor's clapper
熨烫馒头　tailor's ham; dressmaker's ham; pressing bum; egg pad
熨烫垫　press felt; press cloth; pressing mat; ironing padding; ironing sponge
熨烫水布　damp rag; sponge cloth; press cloth
熨斗靴　iron shoe

熨斗盘　iron rest
去污喷枪　cleaning gun; spray gun
去污剂　spot remover
喷水壶　sprayer

4. 黏合机，洗衣机，等　fusing press, washer, etc.

黏合机，压衬机　fusing press machine; heat-bonding machine; fusing press
迷你型黏合机　mini-fusing press
门襟黏合机　placket fusing press
热风机，胶贴机　hot air seam sealing machine
洗衣机，洗水机　washing machine; washer
大货洗水机　bulk washer
工业水洗机　industrial washing machine
石磨水洗机　stone-washing machine
脱水机　extractor; whizzer
水洗脱水机　washer-extractor
工业脱水机　industrial whizzer
震动机　shaking machine
干洗机　dry-clean machine
干衣机　drying tumbler; tumbling machine
锅炉　boiler
电热式蒸汽锅炉　electric steam boiler
蒸汽机　steamer
蒸汽管　steam pipe
蒸汽除皱器　crepe-smoothing steamer
蒸汽定型箱　steam chamber; steam box
毛皮蒸汽清理机　fur processing steamer
毛皮滚毛机　fur brushing machine

(八) 包装设备和工具　For Packing

打包机　press; packer; baling press; pressing gadget
打带机　strapping machine

（包装）折叠机　products folding machine
折衫台　shirt folding table
打包工具　packing tools
吊牌枪，胶枪　tagging gun
胶枪针嘴　tagging needle
胶针　tag pin; plastic string
订书机　stapler
订书钉　staple
裁纸刀　paper cutter
（纸箱）唛头版　stencil
（纸箱）唛头刷　brush for stencilling
（检查断针的）验针器　needle detector

台式验针机　desk top needle detector
气垫式验针机　air-cushioned needle detector
自动带式验针机　automatic conveyer needle detector
手提验针器　portable needle detector
电眼感应清线头机　electric eye sensor thread cleaning machine
吸线头机　thread-thrum sucking machine
真空吸线头机　vacuum board for thread-thrum sucking machine

七、其他
Others

（一）名词和词组　Noun and Word Group

1. 一般名称　general terms

消费品　consumer goods
日用品　daily-use goods
进（出）口商品　import (export) goods
纺织品　textile fabrics; textiles; dry goods〔美〕; soft goods〔英〕
环保纺织品　environmentally friendly textile
智能纺织品　smart textile; intelligent textile
家用纺织品　household textiles
服装纺织品　apparel textiles; clothing textiles; clothtech
数字化纺织　digital textile

功能性纺织材料　functional textiles
保健纺织材料　healthcare textiles
纺织服装技术　textile and apparel technology
服装原理　apparel principle
服装基础理论　garments basic theory
服装高等教育　garments higher education
服装课程　clothing programme
服装图案学　clothing graphics
服装语言学　costume linguistics
服装卫生学　garments hygiene
服装生理学　garments physiology
服装符号学　garments mark theory
服装评论　clothing comments
服饰史　apparel history; fashion history
服饰博物馆　dress museum
服装文物　clothing relics
服装资料　dress data

服装信息　clothing information
服装快讯　apparel news; fashion news
服装特刊　apparel focus
服装杂志　apparel magazine
服装潮流　fashion trends
时装表演，时装秀　fashion show; catwalk show; fashionable dress performance
道德时装秀　ethical fashion show
时装表演 T 台，天桥　catwalk; runway; ramp
模特大赛　model contest
时装博览会　fashion collection
时装周　fashion week
高级定制时装周　haute couture week; couture fashion week
中国国际时装周　China Fashion Week(CFW)
中国国际大学生时装周　China Graduate Fashion Week(CGFW)
时装节　fashion festival
服装节　Apparel Festival; Fashion Week
服装功能　garments function
服装卫生安全性能　clothing hygiene and safety property
服用性能　serviceability; wearability; wear property
服装结构　garments construction
服装标准化　garments standardization
服装机器人　garments robot
自有品牌服饰专业零售商　Speciality Retailer of Private Label Apparel(SPA)

2. 服装团体及组织　organization and group

服装工业　apparel industry; garments industry; clothing industry
服装业　drapery〔英〕; rag trade; couture〔法〕
高级时装业　haute couture〔法〕
服装研究设计中心　garments research and designing centre
服装资讯技术中心　apparel information technology centre
服装工艺示范中心　clothing technology demonstration centre
制衣业培训局　clothing industry training authority
服装研究所　clothing institute
服装大学　garments university
服装学院　garments college
纺织服装学院　college of textile and garments
时装设计学院　fashion design institute
时装工艺学院　fashion institute of technology
工艺美术学院　arts and crafts college
服装职业学校　clothing vocational school
服装协会　garments association; institute of apparel
服装公司　garments corporation; dress company
服装企业　clothing enterprise
服装厂，制衣厂　clothing factory; manufacturer of wearing apparel
被服厂　apparel factory; clothing factory
衬衫厂　shirt factory
羽绒制品厂　down products factory
时装厂　fashion factory
童装厂　children dress factory
针织服装厂　knitwear mill
羊毛衫厂　woolen sweater mill
制鞋厂　shoe factory
制帽厂　hatting factory

制板和样衣部　pattern and sample department
生产部　production department
质检部　quality control department
业务部　business department
跟单部　merchandiser department
财务科　finance section
仓库　warehouse
办公室　office
设计室　design room
技术室　technical room
样品间　sample room
车间　workshop; shop
班组　team; group
裁剪间　cutting shop
缝制间　sewing workshop
整烫间　pressing room
包装间　packing room
服装店　clothing store; clothes shop; toggery
时装店　fashion shop
时装连锁店　fashion chain store
网上时装店　online fashion shop
成衣店，西装店　tailor shop
女服店　dress shop; fashion house; boutique
女装帽店　millinery shop
服装专卖店　apparel speciality store
服装部　clothing department
服装柜　clothing counter
衣帽间　checkroom; cloakroom; vestiary
试衣间　fitting room

3. 服装从业人员　the employed

服装设计师　clothing designer; apparel stylist; costume designer
女装设计师　dress designer; couturier〔法〕
时装设计师　fashion designer; couturier〔法〕
时装女设计师　couturière〔法〕
鞋（帽）样设计师　shoe(hat) designer
时装买手　fashion merchandiser
时装摄影师　fashion photographer
服装样品师　apparel sample hand
时装模特　fashion model; living model; model; mannequin; manikin
女模特　mannequin; manikin
服装制作供应商　costumer
裁缝师，裁缝　tailor; clothier; couturier〔法〕
女（或童）装裁缝　dressmaker; modiste
女装裁缝　draper
男装裁缝　tailleur〔法〕
女裁缝　couturière〔法〕
技师，技术员　technician
时装绘图师　fashion drawer
描图员　sketcher
绘图员　draftsman; drawer
干部　cadre
职员　staff; clerk
经理　manager
科长　section chief
厂长　factory director
车间主任　workshop head
主管　supervisor
生产主管　production supervisor
质量主管　quality supervisor
财务主管　controller
会计　accountant
出纳　cashier
跟单员　documentary handler; order supervisor; merchandiser
单证员　documentary secretary
报关员　custom declarer

业务员	merchandiser
采购员	purchaser
推销员，售货员	salesman
仓管员	warehouseman; storekeeper
收发送料员	bundle handler
检验员	inspector
质检员	quality inspector
质控员（俗称QC）	quality controller
中查员	inspector in process
尾查员	final inspector
指导工，师傅	master
班（组）长	team leader
工人	worker
技术工人	skilled worker
纸样师	patternmaker
排料师，唛架师傅	marker
裁剪师，裁剪工	cutter
立体裁剪师	draper
裁剪试样工	fitter
样品制作工	sample maker; sample hand
缝纫工，车工	sewer; seamer
熨烫工，烫工	presser; ironer
压衬工	fusing operator
剪线工	trimmer
包装工	packer
维修工	repairman; mechanic
电工	electrician
辅助工，副工	helper

4. 文化，艺术与美 culture, art and beauty

文化	culture
物质文化	material culture
精神文化	spiritual culture
高级文化	high culture
大众文化	popular culture
深层文化	deep culture
中华文化	Chinese culture
民族文化	national culture
汉文化	Han culture
主流文化	mainstream culture
传统文化	traditional culture
民俗文化	folk culture
时尚文化	fashion culture
服饰文化	dress culture; clothing culture; costume culture; fashion culture
古代服饰文化	ancient clothing culture
敦煌服饰文化	Dunhuang costume culture
牛仔文化	cowboy culture
企业文化	corporate culture
文化遗产	cultural heritage
文化宝藏	cultural treasures
文化内涵	cultural connotation
文化需求	cultural needs
文化元素	cultural elements
文化视角	cultural insights
文化交流	cultural exchange
文化差异	cultural differences; cultural diversity
文化多元化	cultural diversity
艺术，美术	art
现代（古代）艺术	modern (ancient) art
着装艺术	art of dressing; wearing art
造型艺术	formative art
装饰艺术	decorative art
绘画艺术	painting art
人体艺术	body art
通俗艺术	pop art
视觉艺术	optical art; op art; op
民间艺术	folk art
工艺美术	practical art; industrial art
装饰	decoration; adornment; ornament
服饰研究	research of dress and adorn-

ment
讲究穿着 dressiness
着装规范 dress code
美感 sense of beauty
艺术美 beauty of art
自然美 beauty of nature
奔放美 free beauty
服饰美 beauty of dress
形体美 beauty of body
形式美 formal beauty
造型美 beauty of modelling
线条美 beauty of line
曲线美 beauty of curve
对称美 symmetric beauty
均衡美 balance beauty
韵律美 rhythm beauty
和谐美 harmony beauty
女性美 woman's beauty
整体美 whole of beauty; beauty for ensemble

5. 美容与妆扮 beauty treatment

美容学 cosmetology
美容术 beauty culture; cosmesis; facial beauty
美容店 beauty shop; beauty parlor
美容师 beauty culturist; beautician; cosmetician
全身美容，美体 whole-body beauty
除斑 removal of marks
修眉 eyebrows picking
去（青春）痘 zits picking
去皱纹 emollient
整容 face-lifting
整容术 cosmetology
整容院 cosmetic centre
整容师 cosmetologist

隆鼻 nose lifting
隆胸 breast enlarging
隆臀 hip bulging
拉皮 face lifting
健身 keep fit
减肥 slimming
护肤 skincare
梳妆，化妆 toilet; toilette
新娘化妆 bridal make-up
做发型 cutting and styling
发型师 hair stylist; hairstylist
美发师 trichologist
修指甲，美甲 manicure; nail beauty
修甲师，美甲师 manicurist
修甲工具套装 manicure set
修甲剪 nail scissors
指甲钳 nail clippers
修甲锉 nail file
指甲刷 nail brush
指甲油 nail polish; nail enamel
化妆室 dressing room
化妆台 toilet table; dressing table; dresser
梳妆用品 toilet articles
化妆品 cosmetics; equipage; make-up
粉底 powder foundation
粉底霜 cream base; foundation cream; vanishion cream
清洁霜 cleaning cream
洁面乳 facial cleanser; facial form
护肤霜 body cream; lanoline cream
护肤膏 body butter
润肤露 body lotion
美白润肤露 natural white moisture protection cream
保湿露 moisturising lotion
滋养露 nourishing lotion

抗衰老露 age protection lotion
面膜 face pack; face mask
洁肤面膜 purifying mask
泥浆面膜 mud mask
按摩面膜 massage mask
强力紧致面膜 intensive firming mask
睡眠面膜 sleep mask
面霜 face cream
新生塑颜面霜 regenerist micro-sculpting cream
保湿面霜 facial moisturser
面部磨砂膏 facial scrub
喷雾爽肤水 mist toner
玫瑰果油 rosehip oil
抗皱霜 wrinkle smoothing cream
日（晚）霜 day(night)cream
防（皮肤）老化霜 anti-ageing cream
防晒霜 sunscreen
防晒隔离霜 sun block
护手霜 hand cream
颈霜 neck cream
眼霜 eye cream
眼睑膏，眼影 eyeshadow
双色眼影 eye colour contrast
按摩霜 massage cream
消脂瘦身按摩油 anti-cellulite spa oil
胭脂膏 rouge cream
唇膏 lipstick
香水 perfume; fragrance
眉笔 eyebrow pencil
眼线笔 eye-line pencil
眉刷 eyebrow brush
假睫毛 false eyelashes
卷睫毛器 eyelashes curler
化妆盒 dressing case
化妆袋 dressing bag
首饰盒 jewel case

梳妆镜 toilet mirror; dressing glass
穿衣镜 clothing mirror; wardrobe glass

6. 居家用品 daily utensil

床上用品 bedclothes; bedding; bedlinen
床单 bed sheet; bed spread
床罩 bed cover; bed curtain
被子 quilt
被面 quilt cover
被胎 quilt wadding
被褥 bed filling; bed matte; bed pad; clothes
枕头 pillow
棉枕头 cotton pillow
木棉枕头 kapok pillow
丝绸枕头 silkfloss pillow
鸭绒枕头 eiderdown pillow
保健枕头 pillow for health
枕套 pillow case; pillowcase
绣花枕套 embroidered pillowcase
贴花枕套 applique pillowcase
抽纱枕套 fagoting pillowcase
枕巾 pillow towel
棉毯 cotton blanket
毛巾毯 towel blanket
毛毯 woolen blanket
腈纶毛毯 acrylic blanket
舒棉绒毛毯 Shu Velveteen blanket
拉舍尔毛毯 raschel blanket
电热毯 electric blanket
睡袋 sleeping bag; sleepbag; fleabag
毛巾 towel
浴巾 bath towel
沙滩巾 beach towel
（婴儿）带兜帽保暖毛巾 dry hooded towel
衣柜 clothing cabinet; clothespress; wardrobe

衣箱　trunk;suitcase
衣刷　clothes brush;sweater brush
衣钩　clothes hook
洗烫衣服　laundering
洗涤　wash;washing
干洗　dry cleaning
洗衣店　laundry
换洗衣服袋　clothes bag
换洗衣服篮　clothes basket
洗衣板　washboard;scrubbing board
洗衣刷　laundry brush;wash brush
洗衣粉　washing power
洗涤剂　detergent
晒衣架　clothes horse;airer
晒衣绳　clothesline
晒衣夹　clothes-peg;clothespin

7. 管理，质量，市场 management, quantity, market

管理,经营　management;administration
计算机辅助管理　computer aided management(CAM)
虚拟管理　virtual management
价值管理　value management(VM)
知识管理　knowledge management(KM)
人力资源管理　human resource management(HRM)
客户关系管理　customer relationship management(CRM)
风险管理　risk management(RM)
应变管理　change management(CM)
项目管理　project management(PM)
业务流程管理　business process management(BPM)
物流管理　logistics management(LM)
库存管理　inventory management(IM)
供应链管理　Supply Chain Management(SCM)
制衣业计算机管理信息系统　computer management/ information system for apparel industry
新制造系统　new manufacturing system
智能制造系统　intelligent manufacturing system
智能定制服装系统　intelligent custom-made clothing system
打样系统　sampling system
报价系统　quotation system
订单管理系统　order management system
配额管理系统　quota management system
成本控制系统　cost control system
物流信息系统　logistics information system
布料/辅料存货系统　fabric/accessory inventory system
跟单系统　project file system
生产管理系统　production control system
质量管理系统　quality control system
视觉控制系统　visual control system
船务系统　shipping system
零售系统　retail P.O.S system
财务管理系统　financial management system
人事管理系统　personnel management system
成本，费用　cost
劳工成本　labour cost
生产成本　production cost
成本分析　cost analysis
成本核算　cost accounting
成本控制　cost control
生产，制作　production
大货生产　bulk production
生产计划　production plan

管理，质量，市场	

生产调度　production dispatching
生产排期　production schedule
生产通知单　production order
生产流程　production flow
生产流水线　production line; assembly line; production chain
单件流水　one-piece flow
整包（多件）流水　bunch flow
工序　process
工序安排　process layout
产品，作品　product
成品　finished products
半成品　semifinished products
正（次）品　standard(substandard) products; standard(defective) goods
中等品　fair average quality
报废品　spoiled products
手工艺品　handicraft
手工制作　handmade
技术革新，科技创新　technical innovation
数字化技术　digital technology
三维技术　3D technology
新制造　new manufacturing
智能制造　intelligent manufacturing
质量，品质　quality
外观（内在）质量　appearance(inner) quality
品质标记　quality symbol
品质鉴定　quality evaluation
品质控制　quality control(QC)
现场质控　field quality control
品质保证　quality assurance(QA)
整体质量保证　integral quality assurance(IQA)
质量水平　quality level
质量等级　quality class

质量认证　quality certification
质量证书　certificate of quality
质检证书　inspection certificate of quality
质量可靠　always quality
质量第一　quality first
优质产品　products of quality
全面质量管理　total quality management(TQM); total quality control(TQC)
标准　standard
国家标准　national standard
部颁标准　ministerial standard
质量标准　quality standard
不符合标准　off standard
商标，牌子　brand; trademark
自家商标　own brand
私有商标　private brand
著名商标，名牌，品牌　famous brand; quality brand
品牌化　branding
服装品牌　clothing brand
传统型品牌　traditional brand
成长型品牌　growth brand
电商品牌　E-commerce brand
领先品牌　leading brand
网红品牌　influencer brand
知名品牌　well-known brand
一线品牌　A-line brand
高端品牌　high-end brand
顶级品牌　top brand
奢侈品牌　luxury brand
设计师品牌　designer brand; designer label
自有品牌　private brand; own brand
原创品牌　original brand
创意品牌　creative brand
制造商品牌　manufacturer brand
品牌意识　brand awareness

品牌战略　brand strategy
品牌创建　brand building
品牌定位　brand positioning
品牌管理　brand management
品牌培育　brand cultivation
品牌发展　brand development
品牌创新　brand innovation
品牌优势　brand advantage
品牌价值　brand value
品牌影响力　brand influence
品牌市场份额　brand market share
品牌代言人　brand spokesperson
品牌专卖店（柜）　brand monopolized shop (counter)
注册商标　registered trademark
高档商品　high-grade goods
低档商品　inferior goods
大路货　stape goods
现货　spot goods
库存货　stock goods
服装营销　apparel management and sales; garments marketing
营销策略　sales strategies; marketing strategy
在线营销　online events marketing
直接营销，直销　direct marketing
一对一营销　one to one marketing
体验营销　experiential marketing
网红营销　influencer marketing
促销活动　in-store promotions
虚拟经营　virtual management
市场　market
国际（国内）市场　international (domestic) market
买（卖）方市场　buyer's (seller's) market
服装市场　clothing market
跳蚤市场　flea market
市场经济　market economy
市场供求　market supply and demand
市场调研　market research
市场分析　market analysis
市场预测　market forecasting
市场开发　market development
市场价格　market price
市场信息　market information
资料，数据　data
资料库，数据中心　data bank
电子数据处理　electronic data processing (EDP)
数据存储　data storage
大数据　big data
数据链　data chain
信息链　information chain
产业链　industrial chain
生产链　production chain
物流链　logistics chain
柔性供应链　flexibility supply chain
创新链　innovation chain
品牌价值链　brand vulue Chain
云计算　cloud computing
云会议　cloud conference
在线直播，云直播　online live streaming
线上展览，云会展　online exhibition
电子商务　E-commerce; E-business; business through internet
电商直播　e-commerce livestreaming
网络　web
信息网　information web
销售网　sales web
网址　web address
网页　web side
邮箱　post office box
电子邮箱　e-mail box

邮件 mail
电子邮件 e-mail
传真 facsimile;fax
电话 telephone
手机 mobile phone;cell phone
智能手机 smartphone
贴牌生产，来料（样）加工 original equipment manufacturer(OEM)
贴牌设计生产 original design manufacturer(ODM)
来料加工 process materials supplied by customers
来样加工 processing according to investor's sample
按订单制作 make-to-order(MTO)
按尺寸定做 make-to-measure(MTM)
订单 order
客户订单 buyer's order
采购订单 purchase order
在线订单 online order
试单 trial order
正式订单 formal order
订单号 order number;order No.
订单处理 order processing
合同 contract
销售合同 selling contract
加工合同 manufacturing contract
柔性合同 flexibility contract
签合同 signing of contract
违约 breach of contract
数量 quantity
配额，定额 quota
配额管理 quota management
（报价时）包配额 including quota
（报价时）不包配额 excluding quota
许可证 license
进（出）口许可证 import(export) license

（二）动词和词组 Verb and Word Group

穿衣 attire;clothe;dress
穿戴 wear
打扮 dress up;prink up;attire
穿上，戴上 put on
脱下 take off
系上 tie on
扣住 buckle up
给……脱衣 undress
解开……纽扣 unbutton
解开，松开 untie
拉拉链 zip
设计 design
制图 draft;draw
量身 measure
打样 draft
纸样放缩 grade
铺料 spread
排料 make the lay;lay marker
划样 draw;mark
裁剪 cut
剪掉 cut away
裁下 cut off
裁开 cut open
裁出 cut out
裁得出 cut up
钻孔 drill
打眼 punch;perforate
打粉印 chalk
打号 number
标记（于） mark
缝纫 sew;stitch
缝上 sew on

缝拢	sew up	打褶（于）	crease
缝缉	stitch	折叠	fold
缝，绱	attach; set in	折起	fold up
缝合	sew in; close; join	翻折	turn
拼接	piece together	打褶裥（于）	pleat; gather; ruffle
平缝	fell	抽碎褶（于）	gather
正面缝纫，缉明线	topstitch	抽多行碎褶	shir
粗缝，疏缝，假缝	baste; tack	打横褶（于）	tuck up
绗缝	quilt; stitch out	开省（于）	dart
锁缝	interlock; lockstitch	皱缩	pucker; ruffle; shrink
加固缝	bartack; tack; back tack	卷起	turn up; tuck up
倒缝，回针	backstitch	翻出	turn out
缝制，制作	made	修剪	trim; clip
用……制成	be made of…	剪去	snip off
制成	make up	穿线（于）	thread
把……制成……	make…into	钉扣（于）	button
把（衣服）翻新	make over	开扣眼（于）	buttonhole
拆开线缝	unpick	水洗	wash
给……镶边	edge	熨烫	press; iron
给……绲边	bind; pipe	扣烫	fold and press
给……卷边	hem	分烫	press open
给……贴边	welt	烫平	iron out
给……锁（拷）边	overlock; overcast; overedge; whip	归拔	shrink and stretch
装饰	adorn; decorative; ornament	试穿	fit on; try on
饰珠片，饰亮片	sequin	调节	adjust
钉饰纽，打窝钉	stud	放宽，放松	ease
镶嵌	insert; trim	放长	let down
补缀	patch	放大	let out
贴花	applique〔法〕	改动	alter
绣花	embroider	改小	take in
印花	print	改瘦	take off
绱衬（于）	interline	改短	take up
绱衬里（于）	line	更换	change
铺棉（于），装衬垫（于）	pad; wad	返工	rework
黏合	fuse	翻新	turn; bushel
刮浆	smear paste	检查，检验	check; inspect
		检讫	check off

包装	pack	修补	mend; repair
装箱	box; case; pack	洗涤	wash
编织	knit	洗烫	launder
织补	darn; mend		

# 英汉服装词汇
English-Chinese Clothing Dictionary

A

aba [ˈæbə] 骆驼毛（或山羊毛等的）织物；阿巴原色粗呢；（阿拉伯式）宽大无袖长袍

Abai [əˈbaɪ] 阿巴依外衣（一种土耳其男装，丝绸或薄毛料制，袖上有粗金线装饰）

abbotsford [ˈæbətsfɔːd] 艾博茨福德女式呢（一种有暗格子的斜纹轻薄料）

abdomen [ˈæbdəmən] 腹，腹部
 large *abdomen* 凸腹（体型）
 upper（lower）*abdomen* 上（下）腹

abito [əˈbɪtə] 服装（服装统称，意大利用语）

abrasion [əˈbreɪʒən] 磨损；磨损处
 abrasion proof 耐磨
 abrasion resistance 耐磨牢度，耐磨性

absorption [əbˈsɔːpʃən] （织物的）吸湿性

abstractionism [æbˈstrækʃənɪzəm] 抽象派艺术

abutting [əˈbʌtɪŋ] （缝纫中）拼接，拼合

academicals [ˌækəˈdemɪkəlz] 〔复〕大学礼服，学位服（典礼时所穿戴的学位袍与学位帽）

accent [ˈæksənt] 〔设〕强调

accessory [ækˈsesəri] 辅料，配料；服饰品；附属品，附件
 accessory inventory 辅料储存
 accessories of lady's dress 女装服饰品
 ecological *accessories* 环保辅料
 fashion *accessories* 时装配饰，服饰品
 garment *accessories* 服装辅料，衣服配料
 hair *accessories* 发饰，头饰
 porcelain *accessories* 瓷器服饰品

acetate [ˈæsɪteɪt] 醋酸纤维（合成纤维）

acrylic [əˈkrɪlɪk] 聚丙烯腈纤维，腈纶
 sportswear in *acrylic* 腈纶运动服

activewear [ˈæktɪvweə] 运动服装
 seamless *activewear* 无缝运动装

adaptation [ˌædæpˈteɪʃən] （服装，服饰的）模仿改制

administration [ədˌmɪnɪsˈtreɪʃən] 管理，经营

adornment [əˈdɔːnmənt] 装饰，装饰品
 dress and personal *adornment* 服饰

aesthetics [iːsˈθetɪks] 〔复〕（用作单）美学
 oriental *aesthetics* 东方美学

after dark [ˈɑːftə dɑːk] 晚礼服，晚装

after six [ˈɑːftə sɪks] 晚宴服

agaric [ˈægərɪk] 毛巾布

aglet [ˈæglɪt] （装饰用）金线；（军装）肩带

agraffe [əˈgræf] （盔甲、衣服上的）搭扣，搭钩

airer [ˈeərə] 〔英〕晾衣架，烘衣架
alb [ælb] （牧师、神父穿的）白长袍；麻布僧衣
albatross [ˈælbətrɒs] 海马绒；精纺绉呢；婴儿细绒布；人造丝华达呢
alcantara [ælˈkɒntʌrə] 阿尔坎塔拉（俗称假猄绒布，仿猄皮，涤纶制，天鹅绒般手感）
alligator [ˈælɪgeɪtə] 鳄鱼皮
all-in-one [ˈɔːlɪnwʌn] 上下合一装，连胸紧身女内衣，连身塑形内衣，紧身女胸衣（胸罩、紧腹带、吊袜带合一）
allowance [əˈlaʊəns] （缝纫中）放缝；容许误差
  seam *allowance* 缝份
almuce [ˈælmjuːs] 头巾；风帽
alpaca [ælˈpækə] （南美）羊驼毛；羊驼毛织物
alpine [ˈælpaɪn] 登山软帽
alteration [ˌɔːltəˈreɪʃən] （衣、鞋）修改，改动
  alteration of paper pattern and marker 修改纸样和唛架
alternation [ˌɔːltəːˈneɪʃən] 〔设〕交替
amice [ˈæmɪs] （僧侣用）长方形白麻布围巾；（僧侣）带兜帽的衬皮披肩
angora [æŋˈɡɔːrə] 安哥拉兔毛（或山羊毛）
ankle [ˈæŋkl] 踝；踝节部；（齐踝）裤脚边
  narrow *ankle* 紧窄脚口
anklet [ˈæŋklɪt] 脚镯，踝环，踝链；踝饰；（女、童穿）有脚踝扣带的鞋；略高于踝部的女短袜；翻口袜；罗纹口（做袖口、脚口）

anorak [ˈænəræk] 带风帽短派克大衣；皮猴；滑雪衣，〔粤〕雪褛
  drawstring *anorak* 束带滑雪衣
antigropelos [ˌæntɪˈɡrɒpɪləʊz] （单复同）防水绑腿
Ao Dai [ˈaʊˈdaɪ] 奥黛（又译为袄代，越南传统女套装，两侧高开衩紧身长裙加喇叭脚口长裤）
apparel [əˈpærəl] 衣服，衣着；服装；服饰；外观
  anti-infection *apparel* 防感染服
  astronaut's *apparel* 宇航服
  apparel factory 服装厂，被服厂
  apparel festival 服装节
  apparel focus 服装特刊
  apparel history 服装史
  apparel industry 服装工业，服装行业
  apparel magazine 服装杂志
  apparel news 服装快讯
  apparel principle 服装原理
  apparel sample hand 服装样品师
  apparel stylist 服装设计师
  apparel with shell sets 贝珠衣（高山族首领礼服）
  carbon neutral *apparel* 低碳服装
  career *apparel* 职业服装
  customized *apparel* 定制服装
  eco-friendly *apparel* 生态友好型服装
  intelligent medical *apparel* 智能医疗服装
  men's and women's *apparels* 男女服装
  outdoor(indoor) *apparel* 户外（室内）服装
  protecting *apparel* 防护服（具防

辐射，防化学腐蚀，防火等功能的各类劳保服装）
Speciality Retailer of Private Label Apparel（SPA） 自有品牌服饰专业零售商（一种全新的商业模式）
wearing apparel （总称）服装，衣服

appearance [ə'pɪərəns] 外貌，外观，外表，外形
appearance after laundering（ironing） 洗涤（熨烫）后外观
appearance retention 外观保持性
final appearance 最终外观，成品外观
finished appearance 完工后外观，成品外观

appliqué [æ'pliːkeɪ] 〔法〕缝饰；镶饰；贴绣，贴花，补花，〔粤〕贴布
appliqué and cross-stitch 挑补花
appliqué work 镶嵌；补花缝饰；补花制品

apron ['eɪprən] 围裙；工作裙
embroidered apron dress （童）绣花连衫围裙

arctic ['ɑːktɪk] （常用复）保暖防水套鞋（或罩靴）

argouges ['ɑːɡuːdʒɪs] 漂白亚麻布（法国制）

arm [ɑːm] 臂，上肢；袖子；（缝纫机）机头身
arm cover 袖套
arm length 袖长
arm shaft （缝纫机）主轴
feed-off arm felling 绷袖；〔粤〕埋夹
short under arm 袖下不足

under arm 手臂下；腋下；袖下；〔粤〕夹底
upper（top）arm 上臂

armhole ['ɑːmhəʊl] 袖圈，袖窿，袖孔；挂肩；〔粤〕夹圈
armhole around 袖窿围长
armhole depth 袖窿深，袖深
full armhole 袖窿全长
armhole's length of curves 袖窿弯量
armhole width straight 袖窿直量
uneven armhole 〔检〕绱袖不圆顺

armlet ['ɑːmlɪt] （套在上臂的）臂环，臂章

armo（u）r ['ɑːmə] 盔甲，甲胄；防护服
aramid body armo（u）r 芳纶防弹衣（含超能性芳纶纤维，防弹率极高）
breast armo（u）r 护胸甲
flyer's armo（u）r 防弹飞行服
leather armo（u）r 皮甲
submarine armo（u）r 潜水服

armpit ['ɑːmpɪt] 腋下，腋窝，〔粤〕夹底

array [ə'reɪ] 衣服
fine array 盛装，华服
in fine array 穿着漂亮衣服

art [ɑːt] 艺术；美术
art flower （服饰用）人造花
art of dressing 穿着艺术
art of painting 绘画艺术
art of wearing 着装艺术，服装艺术
body art 人体艺术
decorative art 装饰艺术
folk art 民间艺术

formative art 造型艺术
industrial art 工艺美术，工艺
modern(ancient) art 现代（古代）艺术
optical(op) art 光效应绘画艺术，视幻艺术，欧普艺术
pop art 通俗艺术，流行艺术，波普艺术
practical art 工艺美术，实用美术
useful arts 手艺，工艺
article [ˈɑːtɪkl] 物品，商品，货品
 article number 品号，货号
 article of haberdashery 服饰品
 leather article 皮件
 toilet article 梳妆用品
artwork [ˈɑːtwɜːk] 设计图稿；工艺花样图
assembling [əˈsemblɪŋ] 装配；缝合
assortment [əˈsɔːtmənt] 分类；花色品种；（颜色、尺码）搭配
 packing assortment 包装搭配
 rich(large) assortment 花色齐全
 size(colour) assortment 尺码（颜色）搭配
astrakhan [ˌæstrəˈkæn] 仿羔皮毛织物
asymmetry [ˌeɪˈsɪmətri] 〔设〕不对称
attaching [əˈtætʃɪŋ] （将衣服部件）连接，缝上，钉上

attaching fastener 钉按扣
attaching interlining to collar 〔工〕缝领衬
attaching sleeve 〔工〕绱袖
attaching yoke 〔工〕绱过肩
attaching zipper 〔工〕缝拉链
button attaching 钉扣，〔粤〕打纽
attachment [əˈtætʃmənt] 连接物；附件；附属装置
 attachment for pressing 整烫附属装置
 sewing machine attachments 缝纫机附件
 tacker attachments 打结机附件
attire [əˈtaɪə] 服装，衣着；装束；盛装
 business attire 商务着装；职业服装
 evening attire 晚礼服
 formal attire 礼服
 spare(casual) attire 休闲服装
 splendid attire 华美的装束，盛装
attirement [əˈtaɪəmənt] 衣服；服饰
attitude [ˈætɪtjuːd] 〔设〕态度；意见；看法；观念
awl [ɔːl] 锥子（裁片钻眼用）；鞋钻子
 shoe awl 鞋钻子，鞋锥

B

babouche [bəˈbuːʃ] 土耳其式拖鞋
babywear [ˈbeɪbɪweə] 婴儿服，〔粤〕BB 装
 appliqued babywear 贴花婴儿服

back [bæk] 背，背部；（衣服的）后身；后片；反面
 across back 背宽，〔粤〕背阔
 bare back 露背装

  cross *back* 十字交叉形后身
  half across *back* 半背宽
  hiking up at *back* 〔检〕后身起吊
  racer *back* 丁字形后身
  x-*back* 背宽
background [ˈbækgraʊnd] （布料）底子；底色
backing [ˈbækɪŋ] 衬垫；里衬
backless [ˈbæklɪs] 露背装
backlining [ˈbæklaɪnɪŋ] 反面衬布，反面贴布
backpack [ˈbækpæk] 背包，背袋
  hoodie *backpack* 带风帽的背袋
backpart [ˈbækpɑːt] （裁片）后幅，后片
  pants(coat) *backpart* 裤（衣）后片
backstitch [ˈbækstɪtʃ] （缝纫中）回针，倒缝；扣针（脚）
  backstitch mechanism 倒针装置
  backstitch sewing 回针加固缝纫
badge [bædʒ] 徽章
  children *badge* 儿童徽章
  embroidery *badge* 绣花徽章
  hat *badge* 帽徽
  metal *badge* 金属徽章
  paper *badge* 纸皮牌
  PU and leather *badges* PU 皮和真皮徽章
bag [bæg] 包；袋；手袋；口袋
  backpack rolling *bag* 装有脚轮的背包
  bag-out 袋缝（两块布缝合后将正面翻出）
  bag with silk scarf 缀丝巾手袋
  barrel *bag* 桶状袋，桶袋
  bolide *bag* 马鞍形手袋

Boston *bag* 波士顿手提包（长方形、双把手、宽底）
bowling *bag* 保龄球袋（包）（常设计为时装款女手提包）
casual *bag* 休闲手袋
cell phone *bag* 手机袋
chain *bag* 链带手袋
children *bag* 儿童手袋
clutch *bag* 女用（无带）手握包
coin *bag* （装硬币）小钱包，零钱包
constance *bag* 积琪莲手袋（一种 H 型开关锁扣手袋，以当年肯尼迪夫人携用而出名）
cosmetics *bag* 化妆用手袋
crochet *bag* 钩编手袋
denim *bag* 牛仔袋
doctor *bag* 医用手提包
dressing *bag* 化妆袋
duffel *bag* 行李袋
ecological *bag* 环保手袋
embroidered *bag* with pearls 珠绣手袋
evening *bag* 晚装手袋，晚宴手袋，晚会手袋
fringed *bag* 缘饰手袋；饰流苏手袋
fur *bag* 毛皮袋，毛毛手袋
hobo *bag* 流浪汉袋（现代常表现为一种女式休闲短带挎袋款式）
Kelly *bag* 凯莉手袋（一种以女明星名字命名的马鞍形手袋）
leopard print *bag* 豹纹手袋
luggage *bag* 旅行袋
mesh *bag* 网眼手袋；网袋
messenger *bag* 信使包；邮差包（一种长肩带斜背休闲包）

overnight bag 短途小旅行袋
plastic bag 塑料袋，胶袋
saddle bag 鞍形手袋
shopping bag 购物袋
sleeping bag 睡袋
slouch bag 搭盖袋
sport bag 运动袋
straw bag 草编手袋
studded bag 窝钉手袋
suede bag 反绒皮手袋，猄皮手袋
suit bag 外衣口袋；西装套袋
tote bag （较大的）女用手提包，托特包（音译）；大手提袋；购物袋；行李袋
tool bag 工具袋
travelling bag 旅行袋
trolley bag 装有脚轮的旅行袋
twin handle bag 双挎带手袋
versatile computer bag 多功能电脑包
waist bag 腰袋
weaving bag 编织袋
wooden handle bag 木柄手袋
zipper(zip-up) bag 拉链袋，拉链包

baggies [ˈbæɡɪz] 〔美〕宽松式男子游泳裤；宽翻边男长裤；萝卜裤

bags [bæɡz] 〔复〕〔英口语〕裤子，宽松便裤
    Oxford bags 牛津裤，宽大裤

balaclava [ˌbæləˈklɑːvə] 带巴拉克拉瓦盔式帽的厚（羊毛）大衣；巴拉克拉瓦盔式帽（套头至颈部仅露双眼）

balance [ˈbæləns] 缝份；(衣服穿挂)平衡性；〔设〕均衡，平衡

dynamic balance 动态均衡
formal (informal) balance 对称（不对称）均衡

baling [ˈbeɪlɪŋ] 打包；捆，扎
    baling ties 打包铁皮
    baling twine 打包绳

ball-dress [ˈbɔːldres] （女）舞会礼服

ballerina [ˌbæləˈriːnə] 女式低跟软底便鞋

balmacaan [bælməˈkæn] 巴尔马干大衣（亦译作巴尔玛外套）；连肩袖轻便外套

balmoral [bælˈmɒrəl] 〔B-〕一种条纹衬裙；〔B-〕一种平顶无边圆帽；一种镶花边的靴子（或鞋）

balong [bəˈlɒŋ] 巴龙（音译，菲律宾传统服装）

band [bænd] 带子；带状物；镶边；嵌条
    arm band 臂章
    collar band is longer (shorter) than collar 〔检〕底领伸出（缩进）
    collar band leans out of collar 〔检〕底领外露
    cuff band 假翻袖头
    elastic band 松紧带
    front band 门襟翻边，明门襟
    hair band 发带
    head band 头带；头箍
    hemming band 镶边
    muffler cummer band 围巾腰带（将有装饰的围巾作腰带用）
    plastic (paper) shirt band 衬衫胶（纸）领条
    rib band 罗纹带
    rubber band 胶带；橡筋带
    waist band （裤）腰头；腰带

wedding *band*　结婚戒指

bandana [bænˈdænə]　印花大手帕；鲜艳丝巾；牛仔方巾；印花绸布

bandbox [ˈbændbɒks]　（装帽子、衣领的）硬纸盒

bandeau [ˈbændəʊ]　细带；束发带；紧狭胸罩

bangle [ˈbæŋgl]　手镯，手环，脚镯
　　engraved *bangle*　雕花手镯
　　expandable *bangle*　可伸缩的手镯
　　leaf *bangle*　卷叶形手镯

banian [ˈbænɪən]　宽松的衬衫或上衣（印度）

bar [bɑː]　（铁、木）条，杆，棒；（布料）横档疵
　　needle *bar*　（缝纫机）针杆
　　reverse *bar*　回针杆
　　tie *bar*　领带棒（领带夹的一种）

barege [bəˈdʒ]　一种薄纱（女服和面纱用）

barret [ˈbærɪt]　一种扁平软帽；毛呢雨衣
　　sailor *barret*　水兵软帽

bartack [ˈbɑːtæk]　（缝纫加固）倒回针；打结；加固缝
　　hidden *bartack*　隐形套结，〔粤〕暗枣
　　missing *bartack*　〔检〕漏打套结

basque [bæsk]　（女）紧身短上衣；紧身胸衣

basting [ˈbeɪstɪŋ]　疏缝，假缝，粗缝；〔复〕疏缝针脚
　　diagonal *basting*　扳针法
　　hand *basting*　手缝针法

bataloni [bætəˈlɒnɪ]　麻棉交织布（大麻经，棉纬）

bather [ˈbeɪðə]　浴衣；泳衣

　　2-piece *bather* set　两件式女泳衣
　　racer *bather*　（一件式）赛用女泳衣

bathrobe [ˈbɑːθrəʊb]　浴衣，浴袍；晨衣
　　coral fleece *bathrobe*　珊瑚绒浴袍
　　cotton towelling *bathrobe*　全棉毛巾布浴袍
　　kimono *bathrobe*　和服式浴袍
　　shu velveteen *bathrobe*　舒棉绒浴袍
　　wood（bamboo）fibre *bathrobe*　木（竹）纤维浴袍

batik [bəˈtiːk]　（爪哇）蜡防印花〔法〕；蜡防法印染花布

batiste [bæˈtiːst]　法国上等细亚麻布；细薄棉布；细薄毛织物
　　wool *batiste*　全毛细薄呢

batting [ˈbætɪŋ]　棉胎，棉絮；絮片（片状填料）

beachwear [ˈbiːtʃweə]　〔总称〕海滨服（包括泳服、短裤、浴衣）

bead [biːd]　（装饰用）有孔小珠；珠饰
　　bead-work　珠绣；珠饰细工
　　bugle *bead*　管形珠
　　glass *bead*　玻璃珠
　　wooden *bead*　木珠

beading [ˈbiːdɪŋ]　串珠状边缘饰；珠饰品；钉珠

beanie（=**beany**）[ˈbiːnɪ]　学生戴的小帽；无檐便帽
　　chenille *beanie*　绒线帽

bearcloth [ˈbeəklɔːθ]　粗毛呢

bearer [ˈbeərə]　支座；托架；口袋挂襻；〔复〕背带

beauty [ˈbjuːtɪ]　美，美丽
　　balance *beauty*　均衡美

beauty culture 美容术
beauty culturist 美容师
beauty in colour 色彩美
beauty of art 艺术美
beauty of body 形体美
beauty of curve 曲线美
beauty of dress 服饰美
beauty of line 线条美，曲线美
beauty of modeling 造型美
beauty of nature 自然美
beauty parlor 美容院
beauty shop 美容店
beauty treatment 美容，美容术
facial *beauty* 美容
formal *beauty* 形式美
free *beauty* 奔放美
harmony *beauty* 和谐美
nail *beauty* 美化指甲，美甲
rhythm *beauty* 韵律美，节奏美
science of *beauty* 美学
sense of *beauty* 美感
symmetric *beauty* 对称美
whole body *beauty* 全身美容，美体
whole of *beauty* 整体美
woman's *beauty* 女性美

beaver [ˈbiːvə] 海狸毛皮；海狸呢绒；海狸皮帽

bed [bed] （缝纫机）台板；床
　　bed curtain 床罩
　　bed shaft （缝纫机）底轴
　　bed spread 床单

bedclothes [ˈbedkləʊðz] 床上用品（床单、枕、被等）

bedgown [ˈbedɡaʊn] （女）睡衣

bedizen [bɪˈdaɪzn] 华丽而俗气的衣服（或打扮）

bedlinen [ˈbedlɪnɪn] 床用织物；床上用品

beehive [ˈbiːhaɪv] （把头发向上盘绕成圆锥形）女子蜂窝式发型

beeswax [ˈbiːzwæks] 蜂蜡，黄蜡（旧时作为线的润滑剂，使之更结实、更好用）

beeveedee's [ˈbiːvɪdɪs] 男短裤

behind [bɪˈhaɪnd] 臀部；衣服后背

beige [beɪʒ] 原色哔叽；混色线呢；薄斜纹呢

belcher [ˈbeltʃə] 夹有大白点的蓝色围巾；杂色围巾

bells [belz] 〔复〕喇叭裤
　　elephant *bells* 喇叭裤；大裤脚

belt [belt] 带；腰带；皮带；肩带；绶带；标识带；饰带；武装带；〔英〕吊袜带；紧身胸衣；（包装衬衫用）纸封腰带
　　abdominal *belt* 腹带，肚带，兜肚（妇女、儿童用）
　　beaded *belt* 珠饰腰带
　　belt attached to coat 连身腰带
　　belt keep 腰带襻
　　belt of gown 衣带，袍带
　　belt through 穿引腰带
　　bow *belt* 蝴蝶结腰带
　　braided *belt* 编织腰带
　　button-off *belt* 可解下来的腰带（用纽扣连接的半条腰带）
　　canvas *belt* 帆布腰带
　　cloth *belt* 布腰带
　　cord *belt* 绳索腰带
　　curved *belt* 曲线腰带
　　detachable *belt* 活动腰带
　　double *belt* 双层腰带
　　elastic *belt* 松紧腰带

fashion belt 时尚腰带
free size belt 可调节（长度）腰带
genuine leather belt 真皮腰带
half-back belt 半腰带（缝在上衣后腰节处）；（西装背心）后襻带
health belt 保健腰带
hip belt 及臀腰带
initialed belt 织有字母的腰带
inside(inner) belt 内腰带
Italian belt 意大利式腰带
judo belt 柔道腰带
kidney belt 护腰带
knitted belt 针织线带；针织腰带
lace-up belt 系结腰带
leather belt 皮带
life belt 安全带
match belt 配衬腰带
medical belt 医用腰带
metal belt 金属腰带
metallic chain belt 金属链腰带
monk's belt 僧侣腰带
narrow belt 细窄腰带，〔粤〕幼腰带
ornamental belt 装饰腰带
outer belt 外腰带
paper belt 纸腰带；（包装衬衫用）纸封腰带，纸腰封
plastic belt 塑料（腰）带
PVC belt PVC 皮腰带
safari belt 猎装腰带
Sam Browne belt 山姆布朗佩带（一种军官用武装腰带）
scarf belt 围巾（装饰）腰带
self-belt 本布腰带
set-in belt 固定腰带

shoulder belt 肩（饰）带
skinny belt 紧身腰带，瘦身腰带
smart belt 智能腰带（可健康监测，红外加热，按摩瘦身，防意外摔倒等）
stable belt 稳定腰带（英国，丹麦等国军队佩戴的彩色条纹腰带）
studded belt 窝钉皮带；窝钉腰带
stylish ladies' belt 时尚女装腰带
suspender belt 吊袜带
tie belt 打结腰带
trousers belt 裤带
twin belt 并排双皮带，双条腰带
twin buckle belt 双扣腰带（外套，毛衣，连衣裙等用）
weight lifter's belt 举重宽腰带
Western belt （美国）西部皮带
wide belt 宽腰带

beltline ['beltlaɪn] 裤腰，腰头
bench [bentʃ] 长板凳；工作台
bengaline ['beŋɡəliːn] 罗缎；绨
　　rayon bengaline 人造丝罗缎
benjamin ['bendʒəmɪn] 〔美〕（俚语）男子紧身大衣
beret ['bereɪ] 贝雷帽；无檐软帽
　　knitted beret 针织软帽
　　sailor beret 水兵软帽
bertha(= berthe) ['bɜːθə] （女服）披肩；宽圆花边披领
betweeners [bɪ'twiːnəz] 〔复〕（女）紧身衣
bias ['baɪəs] （成衣的）斜线；斜布条
　　lapel bias （西装上衣）驳领斜度
bib [bɪb] （童）围兜；围裙的上部；背带裤护胸

biceps ['baɪsəps] 臂围；袖宽，袖肥；〔粤〕袖肶围

bikini [bɪ'kiːni] 比基尼（三点式）女泳装；女三角裤，超短内裤
    detachable *bikini* 可卸式比基尼泳装（胸衣可拆卸）
    hipster *bikini* 低腰比基尼裤
    lace front *bikini* 前饰花边比基尼裤
    low rise *bikini* （短直裆）低腰比基尼裤
    string *bikini* 两侧为弹力窄带的比基尼裤

billycock ['bɪlɪkɒk] 〔英〕小礼帽；宽边低顶毡帽

bind [baɪnd] 绲边；叠边
    inner *bind* 隐缝

binder ['baɪndə] 绲边器（又名绲边喇叭）；〔粤〕绲边蝴蝶

binding ['baɪndɪŋ] 绲边；镶边；绲条
    bias *binding* 斜布绲条
    binding of placket and bottom 〔工〕门襟和下摆绲边
    binding operation 绲边操作
    bottom *binding* 裤贴脚条；〔粤〕裤脚增强位
    contrast *binding* 错色镶边；错色绲条
    topline *binding* 鞋口绲边

biretta [bɪ'retə] （天主教徒的）四角帽

blade [bleɪd] 刀身；刀片
    circular *blade* （裁剪用）圆刀；（制鞋用）皮刀
    shirring *blade* 抽褶器

blanket ['blæŋkɪt] 毯子；毛毯；大张毛皮
    acrylic *blanket* 腈纶毛毯
    cotton *blanket* 棉毯
    electric *blanket* 电热毯
    raschel *blanket* 拉舍尔毛毯
    shu velveteen *blanket* 舒棉绒毛毯
    towel *blanket* 毛巾毯
    wool *blanket* 羊毛毯

blazer ['bleɪzə] 运动夹克；女上衣；运动上衣
    boyfriend *blazer* 男式女（西装）上衣
    cruising *blazer* 巡航运动夹克
    regatta *blazer* 赛船运动夹克
    shirt *blazer* 装衬衫领的短上衣

bleach [bliːtʃ] 漂白；漂白剂

bleaching ['bliːtʃɪŋ] 漂白（工艺）

blend [blend] 混纺；混纺制品

block [blɒk] 帽模，帽楦；木制假头；原型；基本样；定型板；裁剪样板
    basic trousers *block* 裤子基本样
    classic shirt *block* 衬衫基本样

blockhead ['blɒkhed] （陈列帽子用的）木制假头；帽模

blocking ['blɒkɪŋ] 用模具使（衣、帽等）成形；熨烫（衣片）成形（归拔）；模烫
    blocking back piece 〔工〕归拔后（衣）片
    blocking sleeve inseam 〔工〕归拔偏袖
    blocking under (top) collar 〔工〕归拔领里（面）

bloomer ['bluːmə] 旧时一种有短裙和灯笼裤的女装；〔复〕女式灯笼（短）裤
    maillot *bloomers* 〔法〕灯笼裤式样的游泳衣

blouse [blaʊz] 女衬衫；罩衫；宽松上衣；宽大短外套；〔美〕军上装；制服上衣
 Balkan *blouse* 巴尔干宽松上衣（长至臀围，束带，宽摆）
 batwing *blouse* 蝙蝠袖衬衫，蝙蝠衫
 beaded *blouse* 珠饰衬衫
 blouse with long (short) sleeve 长（短）袖衬衫
 blousing *blouse* 宽松上衣
 bosom *blouse* 胸饰衬衫
 capelet *blouse* 小披肩领女衬衫
 Chinese *blouse* 中式上衣
 Chinese and Western style *blouse* 中西式上衣
 collarless *blouse* 无领衬衫
 crochet *blouse* 抽纱衫；通花衣；钩编衫
 dark (light)-coloured *blouse* 深（浅）色衬衫
 embroidered *blouse* 绣花衬衫
 farmer *blouse* 农夫罩衫（一种有皱褶的宽松上衣）
 feminine *blouse* 女式上衣；女式衬衫
 girl's *blouse* 女童衬衫
 gypsy *blouse* 吉卜赛女衫
 Hip Hop *blouse* 嘻哈衬衫
 jabot *blouse* 胸饰女衫
 maternity *blouse* 孕妇衫
 middy *blouse* 水手领女罩衫
 midriff *blouse* 露腰女衬衫
 over *blouse* 外上衣；女式长衬衫
 pleated bosom *blouse* 胸褶衬衫
 pull sleeve *blouse* 泡泡袖衬衫
 ruffle *blouse* 皱褶衬衫
 safari *blouse* 猎装短外套
 sailor *blouse* 水手上衣
 sand-washed silk *blouse* 砂洗丝绸衬衫
 sash *blouse* 女饰带上衣（腰围有装饰花结）
 shirt *blouse* 男式女衬衫
 sleeveless *blouse* 无袖衬衫
 sport *blouse* 运动式罩衫
 terry *blouse* 毛巾衫
 tie-bow *blouse* 打结领女衬衫
 tie front *blouse* 前身（下摆处）打结衬衫
 tuck-in *blouse* 内上衣
 tunic *blouse* 束腰衬衫
 under *blouse* 女衬衫
 wrap *blouse* 卷裹衬衫（无纽随意打结裹在身上）

blousette [blaʊset] 女无袖上衣
blousing [ˈblaʊzɪŋ] 上衣料
blouson [bluzɔ̃] 〔法〕夹克式上衣；束摆短上衣；（制服上衣）松紧带束腰女衫；罩衫；宽松外套
 blousing *blouson* 宽松上衣
 blouson long 〔法〕长的宽上衣
 blouson tunic 束腰宽上衣
 pull *blouson* 套头宽上衣，宽松罩衫

boa [ˈbəʊə] （女用）长毛皮（或羽毛）围巾（或）披肩
 acrylic *boa* 腈纶人造毛皮

board [bɔːd] 木板；纸板；烫衣垫板；台板
 bristol *board* 优质绘图纸板
 curve *board* 曲线板
 drafting *board* 制图板，绘图板
 drawing *board* 绘图板

    form *board* 模板
    high power vacuum *board* （熨烫用）强力抽湿台板
    iron *board* 烫衣板
    planning *board* 经济排板
    press *board* （熨烫用）烫板；烫台
    scrubbing *board* 洗衣板
    sleeve *board* 烫凳；袖型烫板
    table *board* of sewing machine 缝纫机台板
    velvet *board* 烫天鹅绒的垫板

boardies [ˈbɔːdɪz] 〔复〕（板状）宽大裤；（冲浪穿）游泳短裤

boardshorts [ˈbɔːdʃɔːts] 〔复〕一种休闲运动短裤（如冲浪用）
    floral *boardshorts* 印花休闲短裤
    knee length *boardshorts* 齐膝长休闲短裤
    quickdry *boardshorts* 快干冲浪短裤

boater [ˈbəʊtə] 〔英〕硬草帽

bobbin [ˈbɒbɪn] （缝纫机）梭芯，〔粤〕梭仔；绕线管
    bobbin case 梭芯套；梭匣
    bobbin case latch 梭门盖；梭匣柄
    bobbin winder 绕线器

bobbinet [ˌbɒbɪˈnet] 珠罗纱，（花边用）六角网眼纱

bobtail [ˈbɒbteɪl] 晚礼服

bodice [ˈbɒdɪs] 宽大背心；衣服大身部分；紧身围腰；紧胸衣
    fitted *bodice* 紧身衣
    front (back) *bodice* 衣服前（后）身
    halter *bodice* 挂脖式紧身胸衣
    under *bodice* 紧身背心

bodkin [ˈbɒdkɪn] 锥子；粗长针；发夹；束发针

body [ˈbɒdi] 身体；布身；上衣的主要部分（除领、袖外），大身，〔粤〕衫身；女紧身衣
    body briefer （胸罩式背心与短裤相连的）女内衣
    body crop 短胸衣
    body hugging （服装）紧身，贴身
    body stocking （用锦纶等制的）女紧身连衣裤
    different parts of the *body* 身体各部分
    human *body* 人体

bodystocking [ˈbɒdistɒkɪŋ] 连衣丝袜，连体袜衣（一种性感的女内衣）
    fishnet *bodystoking* 连体网衣（性感暴露内衣）

bodysuit [ˈbɒdisjuːt] 女紧身连衣裤；婴幼连衫裤装；连体衣服

bodywarmer [ˈbɒdiwɔːmə] 絮棉（或不絮棉）背心

bodywear [ˈbɒdiweə] 紧身衣裤，紧身运动装，健美服

Bohemian [bəʊˈhiːmjən] 波希米亚时尚，波希米亚风（原意指不合习俗和放荡不羁的人，现为流苏、褶皱、大摆裙等流行服饰的象征，并成为自由洒脱、热情奔放的代名词）

boiler [ˈbɔɪlə] 锅炉
    electric steam *boiler* 电热式蒸汽锅炉

boilersuit [ˈbɔɪləsjuːt] 连衫裤工作服

bolero [bəˈleərəʊ] 鲍莱罗女上衣（无扣有袖或无袖的超短女上衣）

bonding [ˈbɒndɪŋ] 黏合，黏衬

bonnet [ˈbɒnɪt] （无边有带）女帽；童

帽
- baby *bonnet* 婴儿软帽
- Gipsy *bonnet* 妇女儿童戴的宽边帽，吉卜赛帽
- poke *bonnet* 朝前撑起的阔边女帽
- shell-like *bonnet* 贝壳形女帽
- straw braid *bonnet* 草辫帽

boot [bu:t] 靴子；长筒靴
- alcantara *boots* （阿尔坎塔拉）仿猄皮靴
- all weather *boots* 晴雨靴
- anklet *boots* 短靴，高帮鞋
- asbestos *boots* 石棉靴
- ballet *boots* 芭蕾靴（一种长靴，穿着者像芭蕾舞者一样走路）
- basketball *boots* 篮球鞋
- bovver *boots* 〔英俚〕"趵伐"靴（斗殴踢人用的有钉的大皮靴）
- buckle *boots* 带扣靴
- casual *boots* 休闲靴，便靴
- cavalier *boots* 马靴
- combat *boots* （中筒）军靴，战靴
- construction *boots* 建筑工地靴
- cowboy *boots* 牛仔靴
- cow split *boots* 牛二层皮靴
- desert *boots* 沙漠长靴
- dielectric *boots* 绝缘靴
- dress *boots* 高雅靴，礼服靴
- elastic sided *boots* 松紧靴（两侧拼缝松紧带）
- engineer's *boots* 工程靴，机工靴
- fashion *boots* 时款靴
- fleece (fur)-lined *boots* 绒（毛）衬里靴
- full grain *boots* 牛皮粒面靴；光面皮靴
- full length *boots* 齐大腿长靴
- fur-trim *boots* 饰毛皮靴，毛毛靴
- gum *boots* 〔美〕高筒橡皮套靴
- half *boots* 半筒靴
- high (long) *boots* 长筒靴
- high-heeled *boots* 高跟靴
- hiking *boots* 徒步旅行靴，远足靴
- hip *boots* （长至臀部的）齐臀长筒靴；捕鱼长筒靴
- hunting *boots* 猎靴
- insulated *boots* 绝缘靴；绝热靴，保温靴
- jack *boots* 过膝长筒靴
- jockey (half jack) *boots* （赛马）骑师靴
- jodhpur *boots* 马靴
- jumping *boots* 长筒皮靴
- kinky *boots* 〔英〕长筒女靴（长及膝或股）
- knee length *boots* 齐膝长筒靴
- knit moccasin *boots* 针织软靴（针织靴身配软底）
- laced *boots* 系带靴
- 3/4 length *boots* 七分靴（略长于中筒靴）
- logger's *boots* 伐木靴
- mid calf *boots* （长及腿肚的）中筒靴
- military *boots* 军靴
- miner's *boots* 矿工靴
- motorcycle *boots* 摩托靴
- mountaineering (climbing) *boots* 登山靴
- nubuck *boots* 牛皮绒面靴；努伯克革靴
- over-the-knee *boots* 过膝长靴
- paratroop (jump) *boots* 伞兵靴
- patchwork *boots* （靴身）拼缝靴

pillow boots 枕头靴（一种柔软充棉便靴）
pirate boots 海盗靴
platform boots 厚底高跟靴
polo(chukka) boots 马球靴
protective boots 防护靴
PVC leather boots PVC 皮靴
rain boots 雨靴
riding boots 马靴
rubber boots 胶靴；长筒套靴；雨靴
safety boots 防护靴，劳保靴
self fastening boots 搭襻（扣）靴
short boots 短筒靴
side gore boots 两侧松紧靴（两侧拼缝宽橡筋带）
skating boots 溜冰鞋，旱冰鞋
ski boots 滑雪靴
slipper boots （室内）平底便靴
slouch boots （长筒）软靴（靴筒软塌）
snow boots 雪地靴，雪地鞋
space boots 太空靴，宇航靴
sport boots 运动靴
steel-toe boots 钢头靴
stocking boots （长袜形）过膝长筒靴
stretch boots 伸缩性长筒靴
studded leopard print boots 豹纹窝钉靴
suede boots 仿麂皮靴；绒面革靴；反绒皮靴；獐皮靴
thigh boots （长至大腿的）齐臀长靴
top boots 长筒马靴
twin gusset boots 两侧松紧短靴（靴舌与后帮以宽松紧带从两侧拼缝）
UGG (Ugly) boots 雪地靴（皮毛一体的平跟套筒羊皮靴，澳大利亚特产）
walking boots 耐磨高帮鞋
water-proof boots 雨靴，套靴
wedge boots 坡跟靴
Wellington boots 威灵顿皮靴（过膝或半长，柔软皮革制）
Western boots （美国）西部牛仔长靴
winter boots 冬靴，保暖靴
working boots 工作靴
zip boots 拉链靴

bootee(= bootie) [ˈbuːtiː] 轻便短筒女靴；幼儿毛线鞋；婴幼软鞋；婴幼短袜

 pointed toe heeled bootee 女式尖头有跟短靴

bootlace [ˈbuːtleɪs] 靴带，鞋带

bootleg [ˈbuːtleg] 靴筒

bootmaker [ˈbuːtmeɪkə] 制靴（鞋）工人

bootstrap [ˈbuːtstræp] 靴襻

boottree [ˈbuːttriː] 靴楦，（鞋）楦

border [ˈbɔːdə] （女服）绲边，镶边

 border lace 镶饰花边，绲边（装饰花边）

bosom [ˈbʊzəm] 胸；（衣服）胸部；胸状物

 starched bosom （礼服衬衫的）硬衬胸，上浆衬胸

bottine [bɒˈtiːn] 〔法〕半长筒靴；女靴；高帮皮鞋

bottom [ˈbɒtəm] 衣（裙、袍）下摆，衣裾；裤脚口；〔粤〕衫脚，脚围；底部；臀部；布底；底色；下半身的

服装（裤、裙）；〔复〕（两件套中的）下装；短睡裤
    angled *bottom* 有角度的裤脚，斜角裤脚
    asymmetric *bottom* 不对称下摆
    bell *bottom* 喇叭脚口
    bell *bottoms* 喇叭裤
    bottom relaxed 下摆平度（松紧下摆不拉开量度）
    bottom shaping 下摆成型
    bottom stretched 下摆拉度（松紧下摆拉开量度）
    bottom with cuff 脚口卷边，裤卷脚，〔粤〕反脚，脚级
    cuffed(cuffless) *bottom* 卷边（无卷边）裤脚
    curve *bottom* 弧形下摆，曲裾
    denim tops and *bottoms* 牛仔服装
    flaring at the *bottom* 放大下摆
    flat *bottom* 平底，平下摆
    foot *bottom* 袜底
    oxford *bottom* 宽松直裤脚
    petal *bottom* 花瓣形下摆，波形下摆
    plain *bottom* 平脚口；平下摆，平裾
    pyjama *bottoms* （睡衣套装中的）睡裤
    rib *bottom* 罗纹下摆
    round *bottom* 圆下摆
    scalloped *bottom* 扇形下摆
    single *bottom* 平脚口（无卷边）
    slant-cut *bottom* 斜下摆
    sleeve *bottom* 袖口
    square cut *bottom* 平下摆，方角下摆
    straight *bottom* 直边下摆，平下摆
    tail *bottom* 燕尾式下摆，圆下摆
    tapered *bottom* 窄裤脚
    turn-up *bottom* 卷裤脚
    twisted *bottom* 〔检〕下摆（或脚口）扭曲
    zipper *bottom* 拉链脚口
bottomless [ˈbɒtəmləs] 下空装
boubou [ˈbuːbuː] （西非男女穿）宽大长袍
boucle [ˈbuːkleɪ] 珠皮呢，仿羊羔皮呢；〔法〕带扣
    bundle *boucle* 一捆裁片
boutique [ˈbuːtiːk] （妇女）时装用品小商店
bouton [butɔ̃] 〔法〕纽扣，扣子
    bouton de chemise 〔法〕衬衫纽扣
boutonnière [butɔnjɛːr] 〔法〕别在驳领纽孔上的花；驳领眼；扣眼
bow [bəʊ] 蝴蝶结；蝶型结结；花结
    hair *bow* 发结，发带
    ribbon *bow* 丝带花结
bowknot [ˈbəʊnɒt] 蝴蝶领结
bowler [ˈbəʊlə] 圆顶硬礼帽，常礼帽 (=bowler hat)；奔尼帽（一种形似礼帽，有宽大圆边和帽绳的软军帽）
bow-tie [ˈbəʊtaɪ] 蝴蝶领结，〔粤〕煲柂
box [bɒks] 箱，盒
    box sealing tape 封箱胶纸
    card *box* 纸盒
    e-mail *box* 电子邮箱
    inner *box* 内箱；内盒
    needlework *box* 针线盒
    packing *box* 包装箱；包装盒
    post office *box* 邮箱
    steam *box* 蒸汽箱
    tool *box* 工具箱

work(sewing) box 针线盒
boxers ['bɒksəz] 〔复〕拳击短裤；(宽松的)平脚短裤
boxiron ['bɒksaɪən] 匣型熨斗
boyleg ['bɔɪleg] 紧身平脚口短裤 (如内裤，泳裤)
boyswear ['bɔɪsweə] 男童服装
bra [brɑː] (口语) 乳罩，胸罩，文胸 (为 brassiere 的略写)
 air bra 空气胸罩，充气胸罩 (杯垫轻薄，可充气，以增大胸围)
 balconette bra 半幅胸罩，半罩杯乳罩 (便于穿低胸装)
 bikini bra 比基尼胸罩，比基尼上装
 booster bra 可提升乳罩，上托乳罩
 Brazillian bra 巴西乳罩
 contour bra (固定造型的) 廓型乳罩
 convertible bra 可调节的乳罩
 embroidered bra 绣花乳罩
 everyday bra 普通乳罩
 form support bra 硬托乳罩
 French bra 法式乳罩 (无罩杯)
 front opening (closure) bra 前扣乳罩
 full figure bra 全杯形乳罩
 glamour bra 魅力胸罩 (时尚性感)
 half(full) cup bra 半(全)杯形乳罩
 instant uplift bra 立即托高胸罩 (无罩杯或肩带，仅靠一条前扣松紧带来托高双胸，后面的宽带类似运动胸罩)
 lightness bra 轻薄乳罩
 limitless bra 无限制乳罩 (两个罩杯，一组肩带，可自由组合不同穿法，号称最有创意胸罩)
 longline bra 长身线形乳罩
 massatherapy bra 可按摩的乳罩
 maternity bra 孕妇乳罩
 men's bra 男性胸罩 (日本内衣商为满足个性需求推出)
 minimal bra 迷你乳罩 (肉色轻薄料制，无撑条和衬垫物)
 moulded bra 模压乳罩
 multiway bra 多功能胸罩，多变乳罩 (可卸肩带，多种穿法)
 nappa bra 羊皮乳罩
 nipple bra 乳头胸罩 (胸罩面上凸现乳点)
 Nude(Nu) bra 隐形乳罩 (由两片硅胶及前扣组成，直接粘贴胸部，性感，无束缚)
 nursing bra 哺乳乳罩
 plunge bra 低胸深 V 形乳罩
 printed fashion bra 时尚印花乳罩
 push up bra 上托乳罩
 shapers flexi bra 弹力塑形乳罩
 silicone bra 硅胶乳罩，隐形乳罩
 smart bra 智能胸罩 (内植智能芯片，可测量人体生理指标，能提早预测乳腺肿瘤等疾病)
 soft cup bra 软杯 (高弹) 乳罩
 sports bra 运动胸罩
 strapless bra 无吊带乳罩
 stretch T-shirt bra 弹力 T 恤乳罩
 superscoop bra 低胸大 U 形乳罩
 teardrop bra 泪珠形迷你乳罩
 teen bra (浅杯) 少女乳罩
 transparent bra 透明乳罩
 triangular bra 三角形乳罩
 tube bra (无吊带) 套筒紧身胸罩

underwire *bra* 钢丝托乳罩
uplift *bra* 托高式乳罩
wireless(wirefree) *bra* 无钢丝托乳罩
wood(bamboo) fibre *bra* 木(竹)纤维乳罩

braces [breɪsɪz] 〔英〕〔复〕西裤的背带，吊带
    a pair of *braces* 一副背带
    elastic *braces* 弹力背带

bracelet [ˈbreɪslət] 手镯，〔粤〕手铐；手环；手链；手腕装饰物
    ankle *bracelet* 踝镯
    ceramic *bracelet* 陶瓷手镯
    chain *bracelet* 链式手镯；手链
    charm *bracelet* 吊坠手链
    chunky *bracelet* 琼琦手镯（超大超重款式）
    colour gold *bracelet* 彩金手镯
    diamond-inlaid *bracelet* 镶钻手镯
    gold *bracelet* 金手镯
    jade *bracelet* 玉手镯
    jadeite *bracelet* 翡翠手镯
    link *bracelet* 链式手镯
    padlock *bracelet* 挂锁式手链
    pearl-beaded *bracelet* 珍珠手镯，珠饰手镯
    rhinestone *bracelet* 闪石手镯；闪石手链
    silver *bracelet* 银手镯
    smart *bracelet* 智能手环（可记录佩戴者锻炼、睡眠、饮食等的实时数据）
    studded *bracelet* 窝钉手镯；镶嵌手链
    watch *bracelet* 表镯
    wood-beaded *bracelet* 木珠手链
    wooden *bracelet* 木手镯

braid [breɪd] 镶边；编带；绳带
    beaded *braid* 珠饰镶边
    elastic *braid* 罗纹口；松紧带
    fancy *braid* 花色编带
    hanger *braid* 挂衣襻
    plait *braid* 辫形编带
    waistband *braid* 腰头饰带

braless [ˈbrɑːlɪs] 不戴胸罩，无胸罩

bralette [ˈbrɑːlɪt] 乳罩，胸罩
    body *bralette* 吊带胸衣
    crossover *bralette* 交叠形胸罩

brand [brænd] 商标；牌子；名牌
    A-line *brand* 一线品牌
    brand advantage 品牌优势
    brand awareness 品牌意识
    brand building 品牌创建
    brand clothes 名牌服装
    brand cultivation 品牌培育
    brand development 品牌发展
    brand image 品牌形象
    brand influence 品牌影响力
    brand innovation 品牌创新
    brand management 品牌管理
    brand market share 品牌市场份额
    brand monopolized shop(counter) 品牌专卖店（柜）
    brand positioning 品牌定位
    brand spokesperson 品牌代言人
    brand strategy 品牌战略
    brand value 品牌价值
    clothing *brand* 服装品牌
    creative *brand* 创意品牌
    designer *brand* 设计师品牌
    E-commerce *brand* 电商品牌
    high-end *brand* 高端品牌
    influencer *brand* 网红品牌

leading brand 领先品牌；领导品牌
luxury brand 奢侈品牌
manufacturer brand 制造商品牌
original brand 原创品牌，自主品牌
own brand 自家商标；自有品牌
private brand 私有商标；（私人）自有品牌
quality(famous) brand 品牌，名牌
top brand 顶级品牌，一流品牌
traditional(growth) brand 传统（成长）型品牌
well-known brand 知名品牌，名牌

branding ['brændɪŋ] 品牌化
brassard ['bræsɑːd] 臂章；袖章
brassiere(= **brassière**〔法〕) ['bræzɪə] 乳罩，胸罩，文胸（略写为 bra）
 a(b, c) -cup brassiere 小（中、大）号乳罩
 band brassiere 束带乳罩
 elastic brassiere 弹性乳罩
 eyelet brassiere 网眼乳罩
 front hook brassiere 前扣乳罩
 garter brassiere 吊带乳罩
 lace brassiere 花边乳罩；网眼乳罩
 off-shoulder brassiere 露肩乳罩；无吊带乳罩
 padded brassiere 加垫（海绵等）乳罩
 strapless brassiere 无吊带乳罩

breadth [bredθ] 宽度；（布料）幅宽
 breadth of back 背宽
 hip breadth 臀宽
 single(double) breadth 单（双）幅

breakdown ['breɪkdaʊn] （缝纫中）断线
breakline ['breɪklaɪn] 驳口线；衍缝线
 uneven breakline 〔检〕驳口不直
breast [brest] 乳房；（衣服）胸部；襟
 breast depth 胸高
 breast enlarging （整形）隆胸
 single(double) breast 单（双）襟
breastline ['brestlaɪn] 胸围
breastpin ['brestpɪn] 领带夹；〔美〕胸针，装饰针
breathability [ˌbriːðə'bɪlətɪ] （织物）透气性
breech [briːtʃ] 臀部
breeches ['briːtʃɪz] (= **breeks** [briːks])〔复〕半长西装裤；马裤；短裤；裤子
 climbing breeches 登山裤
 court breeches 大礼服裤（大腿肥，中腿以下瘦）
 hunting breeches 猎装裤
 knee breeches 齐膝短裤
 riding breeches 马裤
Breton ['bretən] 布列塔帽（宽大前缘向上翻折，法国农民戴）
bridalwear ['braɪdlweə] （新娘穿戴）婚纱服
briefcase ['briːfkeɪs] （扁平）公事皮包
briefs [briːfs]〔复〕三角裤；短内裤；女内裤
 bikini briefs 比基尼内裤
 boxer briefs 拳击短裤式三角裤
 boyleg briefs 紧身平脚口短内裤
 control briefs 吊袜短内裤
 disposable briefs 一次性内裤

fitted briefs 合体短内裤
fly front briefs 前开门襟短内裤
full briefs 宽松短内裤
hi-cut briefs 高脚口短内裤
high waist briefs 高腰短内裤
hipster briefs 低腰短内裤
lace briefs 网眼花边短内裤
maternity briefs 孕妇用短内裤
shapers briefs （紧身）塑形短内裤
sports briefs 运动短内裤
string briefs 两侧为弹力窄带的短内裤
tango briefs 极短的三角裤，丁字裤
wood(bamboo) fibre briefs 木（竹）纤维短内裤
Y-front briefs （男式）前开门襟短内裤

bristle [ˈbrɪsl] （动物）鬃毛
broadbrim [ˈbrɔːdbrɪm] 宽边帽
broadcloth [ˈbrɔːdklɔːθ] 比府绸更细密的平布（又名罗纱）；绒面呢；宽幅布
    wool broadcloth 全毛绒面呢
brocade [brəˈkeɪd] 锦缎；织锦；花缎
    Buyi(nationality) brocade 布依锦
    Chunxiang brocade 春香缎
    crepe satin brocade 花绉缎
    Dai(nationality) brocade 傣锦
    Dong(nationality) brocade 侗锦
    kimono brocade 和服花缎，和服绸
    Miao(nationality) brocade 苗锦
    mixed satin brocade 花软缎
    polyester brocade 涤纶花缎
    satin brocade 织锦缎；花库缎
    Shu(Sichuan) brocade 蜀锦
    Song(dynasty) brocade 宋锦
    Suzhou brocade 古香缎
    Tujia(nationality) brocade 土家锦
    velvet brocade 天鹅绒织锦
    Yun(Nanjing) brocade 云锦
    Zhuang(nationality) brocade 壮锦

brogan [ˈbrəʊɡən] 工作靴
brogue [brəʊɡ] 半筒工作靴；拷花皮鞋；高尔夫鞋
brooch [brəʊtʃ] 胸饰；胸针；领针；饰针；扣花
    crystal brooch 水晶胸针
    diamond brooch 钻石胸针
    flower brooch 花饰针；胸花
    rhinestone brooch 缀闪石扣花
brunchcoat [ˈbrʌntʃkəʊt] 家居餐袍；女家常裙装
    button through brunchcoat 全开襟餐袍
brush [brʌʃ] 刷子
    brush for stenciling 唛头刷
    eyebrow brush 眉刷
    hand brush 手擦，手砂（洗水前，用砂纸在衣服的某些部位有规则地上下摩擦，使表层变色）
    laundry brush 洗衣刷
    nail brush （修甲用）指甲刷
    shoe brush 鞋刷
    sweater brush 毛衣刷
bubbling [ˈbʌblɪŋ] 〔检〕起泡（衬布黏合不牢所致）
    top collar bubbling 〔检〕领面起泡
buckle [ˈbʌkl] 扣，带扣，扣襻；日字扣；卡子；插扣；（服装、鞋用）扣形饰物

ABS buckle 水电带扣
alloy buckle 合金带扣
belt buckle 腰带扣
bra buckle 胸罩带扣
genuine (imitation) leather buckle 真（仿）皮带扣
hook buckle 钩扣
horn buckle 角制带扣
lettered buckle 有字母的带扣
metallic buckle 金属带扣
natural tone buckle 云花带扣
nylon buckle 尼龙带扣
pair buckle 字母带扣，对扣
rhinestone buckle 钻石带扣
roll buckle 旋筒带扣，〔粤〕辘扣
shoe buckle 鞋扣襻，鞋扣
skirt buckle 裙带扣
square (round) buckle 方（圆）形带扣
suspender buckle 吊带扣；葫芦扣
wood buckle 木扣
wrapping buckle 布包扣

buckskin [ˈbʌkskɪn] 鹿皮革；〔复〕鹿皮马裤；鹿皮鞋

build [bɪld] 身材；体型；造型

bullion [ˈbʊljən] 金银丝花边（或饰带）

bunch [bʌntʃ] 起皱，起涌，收缩
  bunches below the back neckline 〔检〕后领窝起涌
  bunches below the waistline seam 〔检〕腰缝下口起涌

bundle [ˈbʌndl] （一）包；（一）捆
  bundle code （裁片）扎号
  bundle system 打包装置；（料片）包扎制度

bundling [ˈbʌndlɪŋ] （裁片料的）捆扎，〔粤〕执扎
  identifying and *bunding* 〔工〕（裁剪后）分片，分扎

buns [bʌnz] 〔复〕臀部

Burberry [ˈbɜːrɪˌberi] （商品名）博柏利雨衣；博柏利风衣；博柏利防水斜纹布；博柏利大衣呢

burka (= burqa = burqua = burkha = bourkha) [ˈbɜːkə] 布卡罩袍装（或译布尔卡、波卡，系穆斯林传统女装，即长袍加头巾、面罩，仅露出眼睛）

burnoose (= burnouse) [bɜːˈnuːs] （阿拉伯）带头巾的斗篷；纱笼

busby [ˈbʌzbi] 〔英〕高顶皮军帽

buskin [ˈbʌskɪn] 半长筒靴

bust [bʌst] （妇女）胸部；胸围
  bust pads(form) 假胸
  bust top 乳围

bustier [ˈbʌstɪə] （无吊带）女胸衣；似胸罩的紧身短上衣

buttocks [ˈbʌtəks] 〔复〕臀部；臀围

button [ˈbʌtn] 纽扣，扣子，〔粤〕钮，（开关）按钮
  ABS *button* 水电扣
  alloy *button* 合金扣
  A-MAK *button* 再生壳扣（仿壳扣）
  angular *button* 角形扣
  animation *button* 卡通动物扣
  automatic *button* feeding 自动喂扣
  bakelite *button* 电木扣
  ball *button* 球式按钮
  bamboo *button* 竹扣
  bar *button* 杆式纽扣
  base *button* 底扣
  beaded *button* 珠饰扣

beared *button* 裤头扣
bone *button* 骨扣
braided *button* 编结（带）扣
brass *button* （黄）铜扣
brass-plate *button* 铜片扣
bronze *button* 青古铜扣
butterfly *button* 蝴蝶扣
button feeder 喂扣器
button mold （布或皮）包扣
button position 钉扣位
button robot 自动钉扣机，钉扣机器人
button stand 里襟；扣位
cartoon *button* 卡通扣
casein *button* 酪素扣
cattle hide *button* 牛皮扣
ceramic *button* 陶瓷扣
chalk *button* 白垩扣，磁扣
character *button* 文字图案扣
Chinese knot *button* 中国绳结扣，花结扣
claw *button* 羊角扣
coconut shell *button* 椰壳扣
collar *button* 领扣
combination *button* 组合扣
computerized engraving *button* 计算机雕刻扣
copper *button* 铜扣
coral *button* 珊瑚扣
cores *button* 象牙果扣
covered *button* 包扣
crocheted *button* 钩编扣
crystal *button* 水晶扣
decorative *button* 装饰扣，花扣
dot *button* 圆点扣
ecological *button* 环保扣
electroplated *button* 电镀扣

enamel *button* 珐琅扣
epoxy *button* 环氧扣
eyelet *button* 鸽眼扣
face *button* 面扣
fancy *button* 花色扣，时尚扣
fancy mandarin *button* 盘花扣
fashion *button* 时装扣；时尚扣
faux metallic *button* 仿金属扣
front *button* 前扣；（牛仔衣裤）门襟纽扣，前部纽扣
functional *button* 功能纽扣（如能闪光、有香味等）
gem *button* 宝石扣
glass *button* 玻璃扣
grape *button* 葡萄扣
handicraft *button* 手工艺扣
heart *button* 鸡心扣
hip pocket *button* 臀袋纽
horn *button* 角制纽扣（如牛角扣）
huge *button* 特大纽扣
imitation wood（stone，shell，pearl，leather，horn，coconut-shell，bone，bark）*button* 仿木（石、贝壳、珍珠、皮、角质、椰壳、骨、树皮）扣
insecure *button* 〔检〕纽扣钉得不牢
jade *button* 玉石扣
jadeite *button* 翡翠扣
jeans *button* 牛仔纽扣，工字纽
kid's *button* 童装纽扣
lacto *button* 合成树脂扣
leather *button* 皮扣
linen *button* 布扣
logo *button* 标识扣（扣面有特定标记）

Manhattan button 曼哈顿扣（热压成形的珠光扣）
marble button 大理石扣
matching button 配色扣
metal button 金属扣
mini button 微型纽扣，迷你扣
missing button 〔检〕漏缝纽扣
mother-of-pearl button 螺钿扣，珍珠母扣
nail button 钉脚扣
natural material button 天然原料扣
natural tone button 云花扣
nickelled brass button 镀镍铜扣
non-sewing button 免缝纽扣（非线缝扣，如四合纽）
nut button 果壳扣
nylon button 尼龙扣
open-top jeans button 通心工字纽
ore button 矿石扣
peach button 桃子扣
pearl button 珠光扣；珠母扣；有机玻璃扣
pi-pa button 琵琶扣
plant-seed button 植物种子纽扣
plastic button 塑胶扣，塑料扣
plexiglass button 有机玻璃扣
polyester button 聚酯扣；仿壳扣
polyester printing button 耐热印字纽
popular button 流行纽扣
press button 揿纽，四合扣
rainbow shell button 彩虹贝壳扣
reinforcing button 加固扣
relief sculpture button 浮雕扣
resin button 树脂扣
rhinestone button 人造钻石扣
ring(cap) press button 圈面（带帽）揿纽
rosette button 玫瑰花形扣
rotatable jeans button 旋转工字纽
rubber button 橡胶扣
sand grain button 沙粒扣
self-fabric covered button 原身布包扣
sew-on button 缝线扣
shank button 有柄纽扣
shell button 贝壳扣
side button 边扣，侧扣
sleeve(cuff) button 袖口扣，袖扣
snap button 揿纽，四合扣，〔粤〕唥纽
space button 备用扣，〔粤〕士啤纽
special-shaped button 异形扣
spherical button 球形纽扣
spindle-shaped button 纺锤形纽扣
stick-shaped button 棒形扣
stone button 石扣
string button 绳结扣
stylish button 时尚纽扣
suede(fur)-covered button 仿麂皮（毛皮）包扣
synthetic resin button 合成树脂扣
three button grouping 三粒扣组合（将三扣靠近为一组去钉扣）
tiny button 微型纽扣
toggle button 套索扣，牛角扣
top jeans button （牛仔裤）裤头工字纽，顶部纽扣
transparent button 透明扣
twin prong jeans button 双针工字纽
two-colored button 双色（金属）扣
two(four)-hole button 两（四）

眼扣
  uneven *button* position 〔检〕扣位不准
  urea *button* 电木纽扣，尿素扣
  waistband *button* 腰头扣
  wood *button* 木纽扣
  wrapping *button* 布包扣
buttoncatch [ˈbʌtnkætʃ] 裤腰头襻钩
buttoner [ˈbʌtnə] 便携式手工钉扣器
buttonhole [ˈbʌtnhəʊl] 纽孔，扣眼，〔粤〕纽门
  buttonhole dimension 扣眼大小
  buttonhole distance 扣眼档（扣眼间距离）
  buttonhole position 扣眼位
  corded *buttonhole* 嵌线（绳）扣眼
  crooked *buttonhole* 〔检〕扣眼不正
  eyelet *buttonhole* 圆头纽孔，凤眼
  fancy *buttonhole* 花式扣眼
  flat *buttonhole* 平头扣眼
  lapel *buttonhole* 驳头眼，〔粤〕襟钮门
  leather *buttonhole* 皮扣眼
  manual *buttonhole* 手工扣眼
  mock(dummy) *buttonhole* 假扣眼，驳头眼
  straight-end *buttonhole* 平头纽孔，平扣眼
  vertical (horizontal) *buttonhole* 直（横）扣眼
  welt(bound) *buttonhole* 绲边扣眼，绲眼（用面料包做的嵌线扣眼）
buttonholer [ˈbʌtnhəʊlə] 锁眼机
  eyelet *buttonholer* 圆头锁眼机，凤眼机
  lockstitch *buttonholer* 平缝锁眼机
buttonholing [ˈbʌtnhəʊlɪŋ] 锁扣眼，〔粤〕打纽门
buttonhook [ˈbʌtnhʊk] 纽扣钩；（皮鞋上的）襻钩
buttoning [ˈbʌtnɪŋ] 钉扣，〔粤〕打纽
BVDS 〔美口〕（一套）内衣（源自一种男子内衣裤的商标名）

C

caban [kɑbā] 〔法〕（水手穿的）厚呢上衣
cabbage [ˈkæbɪdʒ] 〔英〕（裁缝据为己有的）裁剪余料
cack [kæk] 平底童鞋，婴幼软底鞋
caftan [ˈkæftən] （土耳其等国男用）束带长袖袍；宽松上衣
cage [keɪdʒ] （穿在衣裙上的）通花罩衫；薄纱罩裙
calculator [ˈkælkjuleɪtə] 计算器
calendering [ˈkælɪndrɪŋ] （布料）轧光，压光（俗称油光）
calf [kɑːf] 小牛皮；小腿，腿肚；腿肚围
calico [ˈkælɪkəʊ] 〔英〕白棉布，本白布；〔美〕印花布
  chints *calico* 印花棉布
  coarse *calico* 粗棉布
  printed *calico* 印花布
calkin [ˈkælkɪn] 鞋底铁掌

calligraphy [kəˈlɪgrəfi] 书法；书法艺术；书法字

calot [kəˈlɒt] （女、童）无檐帽

calotte [kəˈlɒt] 小无檐帽；瓜皮帽；教士戴小圆帽

calpac(=calpack) [ˈkælpæk] （中东）羊皮帽

cambric [ˈkeɪmbrɪk] 细棉布；细亚麻布

 linen *cambric* 亚麻平布

 truerun yarn-dyed (printed) *cambric* 色织（印花）涤棉细布

cami [ˈkæmi] （非正式）女贴身背心；吊带背心；背心式女内衣

camiknickers [ˌkæmiˈnɪkəz] 〔复〕〔英〕女连裤紧身衣

camisa [kɑːˈmiːsɑː] 衬衫；女胸衣

camisole [ˈkæmɪsəʊl] （有时略作 cami）女吊带背心（或衬衫）；女花边胸衣；女紧身衣；女宽松外套

camlet [ˈkæmlɪt] 羽纱；轻布料

 herringbone *camlet* 人字羽纱

camo [ˈkæməʊ] 伪装迷彩图案；迷彩服

cane [keɪn] 手杖；藤条

canonicals [kəˈnɒnɪkəlz] 〔复〕牧师法衣

canotier [kəˈnɒtɪə] 康康帽（帽身是平的，四周帽檐是直布纹）

canvas(=canvass) [ˈkænvəs] 粗帆布

 breast *canvas* 帆布胸衬

 linen *canvas* 亚麻帆布

 mercerized and bleached *canvas* 丝光漂白帆布

 pigment dyed *canvas* 涂料染色珠帆布

 printed *canvas* 印花帆布

shoe *canvas* 鞋面帆布

suede *canvas* 磨毛珠帆布

cap [kæp] 帽子（无檐边的）；便帽；鸭舌帽；制服帽；军帽；大学方帽；袖山

advertising *cap* 广告帽

alpine *cap* 登山帽

antistatic *cap* 防静电帽

Apollo *cap* 阿波罗帽（毛毡做的鸭舌帽）

astronaut's *cap* 宇航员帽

baseball *cap* 棒球帽

bath *cap* 浴帽

bathing *cap* （女）泳帽

beetle *cap* 甲壳虫帽

black gauze *cap* 乌纱帽

brimmer *cap* 长前檐帽

camouflage *cap* 迷彩帽

cap and bells （宫廷丑角戴的）系铃帽

cap and gown 方帽长袍（大学学位服装）

cap of ceremony 礼仪帽；（旧时）朝冠

cap piece 帽木

cardinal's *cap* （天主教）红衣主教法冠

chef *cap* 厨师帽

Chinese *cap* 瓜皮帽

clothes, shoes and *caps* 服装鞋帽

college *cap* 大学方帽

cotton-padded *cap* （填）棉帽

craft *cap* 工艺帽

cricket *cap* 板球帽

easy *cap* 便帽

Eton *cap* 伊顿帽（英国伊顿公学制服帽）

fashion *cap* 时款帽
felt *cap* 绒帽，毡帽
flight deck *cap* 宇航帽
forage *cap* 军便帽
garrison *cap* 船型帽
ghost *cap* 幽灵帽（万圣节戴）
Hip Hop *cap* 嘻哈帽
hunting *cap* 猎帽，鸭舌帽
jester's *cap* 小丑帽
jockey *cap* 长鸭舌帽；（赛马）骑师帽
leather *cap* 皮帽
legionary(legionnaire) *cap* （旧时）军团帽
liberty *cap* 自由帽（帽檐狭窄）
mesh *cap* 网眼帽
military *cap* 军帽
mortarboard *cap* 大学方帽
Nehru *cap* 尼赫鲁帽（印度人所戴白色小船型帽）
night *cap* 睡帽
nurse *cap* 护士帽
nylon *cap* 尼龙帽
octagonal *cap* 八角帽
3（4，5，6，7）-panel *cap* 三（四、五、六、七）瓣帽
patchwork *cap* 拼缝帽
peaked *cap* 鸭舌帽
photographer's *cap* 摄影师帽
polar fleece *cap* 摇粒绒帽
policeman's *cap* 警察帽，警帽
rain *cap* 雨帽
riding *cap* 骑士帽
sailor *cap* 海员帽
santa *cap* 圣诞帽
school *cap* 学生帽
Scotch *cap* 苏格兰无边帽

scout *cap* 童子军帽
seaman *cap* 水兵帽
service *cap* 制服帽；军帽；大檐帽
shower *cap* 浴帽
shutter *cap* 遮阳帽
skate *cap* 溜冰帽
ski *cap* 滑雪帽
skull *cap* 无檐绒制便帽；（中式）瓜皮帽；骷髅帽；古时的铁盔
sleeve *cap* 袖山（袖子上端与袖窿相连的部位）
sport *cap* 运动帽；便帽
square *cap* 方形帽，四角帽
square college *cap* 学士方帽
stocking *cap* （圆锥形）绒线帽，针织帽
stuff *cap* 呢帽
swim *cap* 游泳帽
tennis *cap* 网球帽
toboggan *cap* 长袜形帽（如圣诞老人帽）
trencher *cap* 大学方帽
twill *cap* 斜纹布帽
uniform *cap* 制服帽
Uygurzu *cap* （新疆）维吾尔族帽
washing *cap* 水洗帽
winter *cap* 冬帽，防寒帽
woolen *cap* 羊毛帽，毛线帽；呢绒帽
working *cap* 工作帽
yachting *cap* 快艇用帽（与警官帽相似）

capa [ˈkæpə] 斗牛士的红色斗篷
caparison [kəˈpærɪsn] 华丽的服装
cape [keɪp] 披肩，斗篷；大氅
    cyclist's *cape* （骑摩托车、自行车用）防雨斗篷

dress cape 礼服斗篷
dress with cape 带披肩的连衣裙
fake fur cape 人造毛皮披肩
front(back) shoulder cape （衣服的）前（后）披肩
rain cape 斗篷雨衣，雨披
capelet [ˈkeɪplɪt] 小披肩；短斗篷
capeline [ˈkeɪplaɪn] 宽边女帽；（古代）武士铁盔
capote [kəˈpəʊt] 带风帽长大衣（或大氅）；〔美〕系带女帽
Capris [kəˈpriːz] 〔美〕卡普里裤（女式紧身长裤）
capuche [kəˈpuːʃ] 僧帽（一种天主教托钵僧戴的尖顶风帽）
capuchin [ˈkæpjʊʃɪn] （女用）带风帽斗篷；女风帽大衣
caraco [ˈkærəkəʊ] (18世纪后期)女式短大衣，卡拉扣上衣(衣后摆包臀，且长于前摆)
carca-net [ˈkɑːkənet] 一串项珠；珠宝颈饰
carcoat [ˈkɑːkəʊt] 跑车外套，短外套，短大衣，〔粤〕太空褛
cardigan [ˈkɑːdɪɡən] 羊毛衫；羊毛背心；（无领）开襟毛衫；（无领）开襟衫

    beaded embroidery cardigan 珠绣开襟衫
    boyfriend cardigan 男式女开襟衫
    crop cardigan （腰节以上）短开襟衫
    draped front cardigan 垂饰前襟毛衫
    half cardigan 半开襟毛衫
    hand-crochet cardigan 手工镂空开襟衫

hooded cardigan 风帽开襟毛衫
letter cardigan 绣字毛衫
one button short sleeve cardigan 单扣短袖开襟衫
snap cardigan 按扣开襟衫
surplice cardigan 和服式开襟毛衫
V-(crew) neck cardigan 尖（圆）领开襟衫
zipper cardigan 拉链开襟衫
cardinal [ˈkɑːdɪnl] 女式短外套
carton [ˈkɑːtən] 纸板箱（盒），卡通箱，外纸箱；纸板

    carton No. 箱号
    outer(inner) carton 外（内）纸箱
    3 plies export carton 三层出口纸箱

cartoning [ˈkɑːtənɪŋ] 装箱，入箱
    cartoning too tight(loose) 〔检〕装箱太紧（太松）
cartoon [kɑːˈtuːn] 草图；漫画；卡通，动画片
case [keɪs] 箱，盒；柜；套
    cardboard case 纸板盒
    dressing case 化妆盒
    embroidered pillow case 绣花枕套
    jewel case 首饰盒
    packing case 包装箱；包装盒
    pin case 针盒
cashmere (= kashmir) [ˈkæʃmɪə] （山）羊绒；开司米；开司米织物（薄毛呢等）
casing [ˈkeɪsɪŋ] 装箱，包装
casquette [kæskt] 〔法〕鸭舌帽
cassimere [ˈkæsɪmɪə] 薄毛呢
cassock [ˈkæsək] （天主教）法衣，袈裟；教士袍（黑色长袍）
casuals [ˈkæʒʊəlz] 〔复〕便鞋

leather *casuals* 皮便鞋
suede *casuals* 绒面革便鞋
casualwear [ˈkæʒuəlweə] 便装
catch [kætʃ] 线缝布（毛）边
  button *catch* 裤里襟
  fly *catch* 裤里襟
catsuit [ˈkætsjuːt] 紧身连衣裤
cattlehide [ˈkætlhaɪd] 牛皮
catwalk [ˈkætwɔːk] 猫步（时装模特台步）；时装表演 T 型台，天桥
caul [kɔːl] 女帽的后部；发网
celanna [ˈselənə] 醋酸里子布
centrality [senˈtræləti] 〔设〕集中，中心性
centre [ˈsentə] 中心；中间；中心线
  apparel information technology *centre* 服装资讯科技中心
  centre front (back) 前（后）中线
  clothing technology demonstration *centre* 服装工艺示范中心
  cosmetic *centre* 化妆美容院，整容院
  garments research and designing *centre* 服装研究设计中心
  top *centre* 明门襟
cerecloth [ˈsɪəklɒθ] 蜡光布；漆布
cest [sest] （女用）腰带
cestus [ˈsestəs] （古罗马拳击用）皮手套
chain [tʃeɪn] 链条；项圈；表链；经纱
  brand value *chain* 品牌价值链
  data *chain* 数据链
  flexibility supply *chain* 柔性供应链
  foot (ankle) *chain* 踝链
  hand *chain* 手链
  industrial *chain* 产业链
  information *chain* 信息链
  innovation *chain* 创新链
  logistics *chain* 物流链
  production *chain* 生产链；流水线
  white gold snake *chain* （人造）白金蛇形项链
chalk [tʃɔːk] 粉笔；划粉片
  French *chalk* （裁缝用）滑石，划粉
  marking *chalk* 划粉片
  tailor's *chalk* （裁缝用）划粉片
chalking [ˈtʃɔːkɪŋ] 〔工〕（裁片间）打粉印
chambray [ˈʃæmbreɪ] 青年布；薄条纹布
  indigo (black) *chambray* 靛蓝（黑色）青年布
chamois [ˈʃæmwɑː] 麂皮；羚羊毛皮
chapeau [ʃæˈpəʊ] 〔法〕帽子
  chapeau bras 〔法〕可折叠的三角帽
chaplet [ˈtʃæplɪt] 花冠；项圈；（较短的）串珠
charm [tʃɑːm] 项链（或手镯等）上的小饰物；项链坠
  colour glass *charm* 彩玻项链坠
  coral *charm* 珊瑚项链坠
  pearl *charm* 珍珠项链坠
  ruby *charm* 红宝石项链坠
charmeuse [ʃɑːˈmuːz] 查米尤司绉缎（柔顺女装缎料）
chasuble [ˈtʃæzjʊbl] 十字褡（神父穿的无袖长袍）；背心装
check [tʃek] 格子花纹；格子织物，格子布，〔粤〕格仔布；检验，核对
  colour *check* 彩格
  double *check* 双重格子；〔检〕复查
  highland *check* 苏格兰彩格

linen check 亚麻格布
mismatched checks 〔检〕对格不准
pin check 小方格,细格子
pyjama check 格子睡衣布
woolen check 格子呢
checker [ˈtʃekə] 质检员,查验者
checking [ˈtʃekɪŋ] 检验,检查,核查
checkroom [ˈtʃekruːm] 〔美〕衣帽间
chemiloon [ʃəmɪˈluːn] 女连裤内衣
chemise [ʃəˈmiːz] 无腰带宽松女上衣;女无袖衬衫;宽松女内衣;女吊带睡衣;衬裙;〔法〕男衬衫
　chemise American 〔法〕美式男用衬衫
　cross back chemise 背部交叉吊带睡衣
　maxi-chemise 特大衬衫,(长过膝的)衬衫装
　pardessus chemise 〔法〕衬衫式外套
　sur chemise 衬衫上衣
chemisette [ʃemɪˈzet] 女无袖胸衣;紧胸衬衣
chemisier [ʃiːmiːzə] 女衬衫上衣
chenille [ʃəˈniːl] 绳绒线,雪尼尔花线;雪尼尔织物;仿绳绒线;仿绳绒线织物
cheongsam (= cheungsam = chongsam) [tʃɒŋˈsæm] (粤语"长衫"的音译)长衫,长袍;旗袍
　satin embroidered cheongsam 软缎绣花旗袍
chest [tʃest] 胸;胸围
　across chest 胸宽
　chest around 胸围
　chest protector (绒布)护胸

chest under armhole 下胸围
flat chest 平胸
hollow chest 塌胸,凹胸;〔检〕(衣服)塌胸(衣胸不丰满)
lack of fullness at chest 〔检〕(衣服)塌胸
pigeon chest 鸡胸
posterior chest width 背宽
chesterfield [ˈtʃestəfiːld] 柴斯特外套,柴斯特礼服大衣(单排扣或双排扣),饿驳领,左胸手巾袋,有盖侧袋
chesty [ˈtʃesti] 〔澳〕针织汗背心
cheviot [tʃeˈvɪət] 粗纺厚呢;精纺缩绒粗呢,啥味呢
　wool/polyester cheviot 毛涤啥味呢
　wool/viscose cheviot 毛黏啥味呢
chevron [ʃevrən] (军服上的)臂章;人字斜纹呢
chic [ʃiːk] 漂亮,别致;别致的款式;时尚风貌
　Bohemian chic 波希米亚风貌
　casual chic 休闲式样;休闲风貌
　elegant chic 典雅别致;典雅风貌
　street chic 街头风貌
　wild chic 狂野风貌
chiffon [ˈʃɪfɒn] 〔法〕绡,雪纺(轻而透明的薄绸);〔复〕女装服饰
　nylon chiffon 锦纶雪纺
　rayon chiffon 人造丝雪纺
　silk/cotton chiffon 丝棉纺,丝棉绸
　wool chiffon (极细薄的)雪纺呢
childrenswear [ˈtʃɪldrənsweə] 儿童服装,童装
chimere [tʃɪˈmɪə] 主教法衣(宽大无袖长袍)
chinchilla [tʃɪnˈtʃɪlə] 灰鼠毛皮;珠

皮呢（羊毛粗大衣呢）

chinlon [ˈtʃɪnlən] 锦纶（聚酰胺纤维，商名，中国制）

chino [ˈtʃɪnə] 一种丝光卡其布（常用于军服）〔复〕丝光卡其布衣裤

 cotton drill *chinos* 全棉丝光卡其布裤

chinoiserie [ʃɪnˈwɑːzəri] 中国艺术风格；中国风；中国风格物品

chintz [tʃɪnts] 印花棉布；轧光布，油光布

chi-pao(= Qipao) [tʃɪˈpaʊ] 旗袍

chlorofibre [ˈklɒrəfaɪbə] 含氯纤维，氯纶

choker [ˈtʃəʊkə] 短项链；颈箍；（硬）高领；宽领带

chrisom [ˈkrɪzəm] 初生婴儿白色洗礼巾（或洗礼服）

cilice [ˈsɪlɪs] 粗毛布；粗毛布衣服（僧人和忏悔者所穿）

circumference [səˈkʌmfərəns] 周长，围长

 ankle *circumference* 踝围
 biceps *circumference* 上臂围；袖肥，袖围
 collar *circumference* 领围，领长
 elbow *circumference* 肘围
 head *circumference* 头围
 hip *circumference* 臀围
 neck *circumference* 颈围
 palm *circumference* 掌围
 seat *circumference* 臀围
 wrist *circumference* 腕围

civies [ˈsɪvɪz] 〔复〕（俚语）便装，便服

clamdiggers [ˈklæmˌdɪɡəz] 〔复〕（长至膝下）半长裤

clasp [klɑːsp] 带扣；钩子；别针；扣环

 belt *clasp* 腰带扣
 collar *clasp* 领钩
 hair *clasp* 发针，发簪

classic [ˈklæsɪk] （俚语）妇女的传统服装

cleaning [ˈkliːnɪŋ] 清洁；清洗

 dry *cleaning* 干洗
 dry *cleaning* shop 干洗店

clericals [ˈklerɪkəlz] 〔复〕牧师服装

clip [klɪp] 夹子；回形针（曲别针）；指甲剪；〔复〕大剪刀

 brace *clips* 吊裤带夹
 cloth *clip* 裁布夹（固定布片）
 hair *clip* 头发夹，发夹
 necktie *clip* 领带夹，领带别针
 plastic shirt *clip* 衬衫胶夹
 shoe *clip* （装饰用）鞋夹
 suspender *clip* 吊带夹
 thread *clips* 纱剪；线头剪
 tie *clip* 领带夹，〔粤〕䊥夹

clippers [ˈklɪpəz] 〔复〕剪刀；钳子

 nail *clippers* 指甲轧剪，指甲钳
 thread *clippers* 纱剪；线头剪

cloak [kləʊk] 斗篷；大氅；大衣；无袖外套

cloakroom [ˈkləʊkrʊm] 〔英〕衣帽间；寄存衣物处

clobber [ˈklɒbə] 〔英俚〕衣服

cloche [kləʊʃ] （圆顶狭边）钟形女帽，吊钟帽

clog [klɒɡ] 木底鞋，木屐；松糕鞋

clogué [klɒɡ] 〔法〕泡泡组织织物；泡泡纱布

 guanle *clogué* 冠乐绉

closing [ˈkləʊzɪŋ] 缝合；闭合；（裤）

腰头门襟；衣物可扣上的部分
 snap closing 按扣门襟
 tab closing 搭襻门襟，扣襻门襟
 tied closing 系带门襟
closure ['kləuʒə] 闭合，封闭；服装闭合物（纽扣、拉链等）；门襟
 toggle closure 套索纽门襟
 velcro closure 尼龙搭扣门襟，魔术贴门襟
cloth [klɒθ] 织物；布；呢绒；衣料
 admiralty cloth 海军呢
 air-conditioning cloth （冬暖夏凉）空调布
 anti-bacterial cloth 抗菌布
 anti-mist cloth 防霉织物
 anti-tinea cloth 抗癣布
 army cloth 军装料；军服呢
 asbestos cloth 石棉布
 bag cloth 口袋布
 basket cloth 板丝呢
 bathrobe cloth 浴衣绒布
 beaver cloth 海狸呢
 blanket cloth 毛毯大衣呢
 bleached cloth 漂白布
 box cloth 缩绒厚呢
 camel hair cloth 驼绒（布）
 cheese cloth 干酪包布（粗棉布）
 cloth of gold(silver) 金（银）线织物
 coated cloth 涂层织物；涂层布料
 colour-changeable cloth （颜色可变化的）变色布
 cotton cloth 棉布
 crease-resisting cloth 防皱织物，防皱布
 crinkle cloth 绉布；泡泡绉
 C. V. C. cloth 以棉为主的混纺布料（含棉量50%以上）
 Down-proof cloth 防羽布，防绒布
 dyed cloth 染色布，色布
 embossed cloth 拷花布
 fancy cloth 花色布，装饰布
 feather cloth 羽毛呢
 figured cloth 提花布
 fine(coarse) cloth 细（粗）布
 fur cloth 仿毛皮织物，人造毛皮
 Gaoshan fancy cloth 高山族花布
 gauffered cloth 轧纹布；凸凹轧花布
 glass fibre woven cloth 玻璃纤维布
 grass cloth 夏布
 grey cloth 坯布，本色布
 habit cloth 英国优质呢；女骑装呢
 hand woven cloth 手织土布
 hypnotic cloth （有催眠作用的）催眠布
 indanthrene cloth 阴丹士林蓝布
 interlining cloth 衬布
 Italian cloth 〔英〕黑色直贡呢；意大利缎（衬里用）
 jersey cloth 针织布
 knitted loop cloth 毛圈针织布
 ladies cloth （粗纺）女式呢
 ladies cloth and dress worsted （精纺）女衣呢
 linen cloth 亚麻布
 linen jersey cloth 亚麻针织布
 lining cloth 衬里布，里子布
 liquid crystal cloth （随温度而变色）液晶布
 luminous cloth （用特殊光学玻璃纤维织成的）发光布
 medium cloth 中厚毛织物，（精纺）中厚呢

melange *cloth* 混色布
military *cloth* 军服呢
mohair *cloth* 马海毛呢（衬里用）
navy *cloth* 海军呢
noil *cloth* 䌷绸；䌷丝织物
odd bits of *cloth* 布头
padding *cloth* 衬垫织物，衬垫布
piece-dyed *cloth* 匹染布
pilot *cloth* 军服呢；海员厚绒呢
plain *cloth* 平布，平纹布；素色布；平纹织物
plastic *cloth* 塑料布
pocket *cloth* 口袋布
press *cloth* 熨烫水布
proofed breathable *cloth* 防水透气布料
rabbit hair *cloth* 兔毛呢
ramie *cloth* 苎麻织物；苎麻布
ramie/polyester(cotton) blended *cloth* 麻涤（棉）混纺织物，麻涤（棉）布
ramie/viscose *cloth* 麻黏混纺布
rayon *cloth* 人造棉织物；人造棉布
reversible *cloth* 双面织物
rim *cloth* 包边布；纽孔布
rubber *cloth* 涂（橡）胶布，雨衣布
sanded *cloth* 磨毛布料
serge *cloth* 哔叽呢
shoe *cloth* 鞋面呢
snowflake *cloth* 雪花呢
sponge *cloth* （熨烫用）湿布
striped *cloth* 条纹布
students'(school)uniform *cloth* 学生呢
terry *cloth* 毛圈织物；毛巾布

thermal insulation *cloth* 保温布
T/R *cloth* T/R 布（涤黏混纺布）
tussah *cloth* 柞丝绸
twine *cloth* 轧光线呢
ulster *cloth* 阿尔斯特长毛大衣呢
undershirt *cloth* 汗衫布
uniform *cloth* 制服呢
union *cloth* 混纺衣料；大众呢
waste *cloth* 零碎布料，布头布尾，〔粤〕布碎
water-proof *cloth* 防水织物；防水布
wax *cloth* 蜡布，油布
weather *cloth* 风雨衣布，晴雨两用布料
wet *cloth* （有淋湿感的）湿光布
wide(narrow) *cloth* 宽（窄）幅布
wood(bamboo) fibre *cloth* 木（竹）纤维布料
woolen *cloth* 呢绒；粗纺毛织物
wrapping *cloth* 包裹布；包装布
yacht *cloth* 游艇呢（浅色薄法兰绒）
yarn-dyed *cloth* 色织布
yarn-dyed T/C weft filament *cloth* 色织涤/棉纬长丝织物
zephyr *cloth* 轻罗绸，薄纱绸

clothes [kləʊðz] 〔复〕衣服，衣裳，〔粤〕衫；服装；（总称）被褥；各种衣物

 adjustable *clothes* （随小孩长大）可调节的衣服
 air-conditioning *clothes* 空调服
 antigas *clothes* 防毒衣
 asbestos *clothes* 石棉衣
 automated system for *clothes* produc-

tion 服装生产自动化系统
ballproof *clothes* 防弹衣
body *clothes* 内衣裤
boy *clothes* 男童服
business *clothes* 工作服,职业服
camouflage *clothes* 伪装迷彩服
civil *clothes* 便服(区别于军警制服)
clinging *clothes* 紧身服装
closed *clothes* 无缝衣服
clothes against theft 防盗服
clothes of ceremony 礼仪服
colour changeable *clothes* 变色服
cotton *clothes* 棉布衣服,全棉服装
custom-made *clothes* 定做的衣服
designer's *clothes* 命名服装;标名服装;名牌服装
eatable *clothes* 可食服
energy storage flash *clothes* 储能闪光服
Eton *clothes* 〔英〕伊顿公学男生制服
evening *clothes* 夜礼服
everyday *clothes* 便服
fatigue *clothes* (军队)劳动装,工作服
flame resistance *clothes* 防火服
floss-padded *clothes* 丝绵袄
fragrant *clothes* 香味服
fur *clothes* 毛皮服装
glad *clothes* (口语)时髦衣服,最好的衣服;晚礼服
gorgeous *clothes* 华丽的衣服
gym *clothes* 运动衣
hat and *clothes* 衣冠
high-tech *clothes* 高科技服装(通过技术增强的服装,以增加其除传统用途外的功能,也可称为智能服装)
inside-out *clothes* 反穿的衣服
investment *clothes* 值得投资的衣服
jade *clothes* sewn with gold thread 金缕玉衣
Kublai Khan *clothes* 忽必烈汗装(东方式立领、宽袖的缎料衣服)
large front *clothes* 大襟衫(中国旧时门襟开在一边的宽大上衣)
late-day *clothes* 午后服
leather *clothes* 皮革服装;皮衣
leopard print *clothes* 豹纹装
long-*clothes* 婴儿衣服
lovers *clothes* 情侣装
luminous *clothes* 发光服
microwave shielding *clothes* 微波防护服,防微波服
moisture wicking *clothes* 吸湿排汗服
monkey *clothes* (美国俚语)礼服;军礼服
mosquito-repellent *clothes* (防蚊叮咬)驱蚊衣
mulberry paper *clothes* (桑皮纸裁制)桑纸服装
night *clothes* 睡衣
one-piece *clothes* 一件式服装
outsize *clothes* 特大号衣服,超大码服装
plain *clothes* 便服,便装
practice *clothes* 练功服
ragged *clothes* 乞丐装,百衲衣
ready-made *clothes* 现成服装,成衣
reflective *clothes* 反光服
rough *clothes* 粗布衣

second hand *clothes* （穿过的）二手服装
sexy *clothes* （透明或暴露的）性感装
show-off *clothes* 展览服，展示装
siamese *clothes* 连体衣
solar-energy *clothes* 光能服
special measurement *clothes* 特体服装
spring and autumn *clothes* 春秋衫
store *clothes* 现成服装，成衣
studded *clothes* 缀窝钉装
sun *clothes* （露肩或露背的）太阳装
sunday(best) *clothes* （口语）最好的服装，节日服装
swaddling *clothes* 婴儿服装
tailored *clothes* 西装型服装
tawdry *clothes* 花哨而俗气的衣服
thermal insulation *clothes* 保暖服
tricky *clothes* 奇巧服装
upper *clothes* 外衣
wadded(padded) *clothes* 填棉服装，棉衣
Western(Chinese)-style *clothes* 西（中）式服装，西（中）装
working *clothes* 工作服
wrinkle-free cotton *clothes* 全棉免烫服装

clothesbag ['kləʊðzbæg] 放换洗衣物的袋子
clothesbrush ['kləʊðzbrʌʃ] 衣刷
clotheshorse ['kləʊðzhɔːs] 晒衣架
clothesline ['kləʊðzlaɪn] 晒衣绳
clothes-peg ['kləʊðzpeg] 晒衣夹
clothespin ['kləʊðzpɪn] 晒衣夹
clothespress ['kləʊðzpres] 衣柜

clothier ['kləʊðɪə] 布料商；服装商；裁缝
clothing ['kləʊðɪŋ] （总称）服装，衣服；服饰；衣裳；被服
adaptive *clothing* 适应性服装（专为残疾人设计制作的服装）
ambisextrous *clothing* 男女不分的服装
ancient *clothing* culture 古代服饰文化
antibacterial deodorant *clothing* 抗菌防臭服
antistatic *clothing* 抗静电服
anti-trauma *clothing* 防创伤服
articles of *clothing* 衣物（衣、帽、手套等）
bedding and *clothing* 被服
brand *clothing* 名牌服装；品牌装
breathable waterproof *clothing* 防水透气服
business *clothing* 职业服，工作服
ceremony *clothing* 庆典服，礼仪服
China International *Clothing* & Accessories Fair (CHIC) 中国国际服装服饰博览会
Chinese and Western *clothing* 中西服饰，华洋服饰
classic *clothing* 经典服装；古典服装
clothing cabinet 衣柜
clothing comfort 服装的舒适性
clothing comments 服装评论
clothing department(counter) 服装部
clothing design 服装设计

clothing designer 服装设计师
clothing enterprise 服装企业
clothing factory 服装厂，被服厂
clothing headgear and footwear 服装鞋帽；衣着
clothing hygiene and safety property 服装卫生安全性能
clothing industry training authority 制衣业训练局，制衣培训班
clothing institute 服装研究所
clothing manufacture 成衣制作
clothing mirror 穿衣镜
clothing programme 服装课程
clothing relics 服装文物
clothing store(shop) 服装商店
coral fleece warm *clothing* 珊瑚绒保暖服装
3D virtual *clothing* 三维虚拟服装
ecological *clothing* （完全使用天然材料的）生态服装，环保服装
fantasy *clothing* 幻想型服装
far-infrared heating *clothing* 远红外发热服装
fashion *clothing* 时装
flame retardant *clothing* 阻燃服装
functional *clothing* 功能服装
Han Chinese *clothing* 汉服
hand-painting silk *clothing* 真丝手绘服装
health care *clothing* 保健服饰
high(mid, low)-grade *clothing* 高（中、低）档服装
high-tech *clothing* 高科技服装
individual *clothing* 独特服装，个性服装
judicial *clothing* 司法服装
latex *clothing* 乳胶服装

licensed *clothing* 特许服装；品牌服装
light *clothing* 轻便服装，轻装
locking *clothing* 锁扣衣（从后面扣上的衣服，防止病人不适当地脱衣）
low carbon *clothing* 低碳服装
minority *clothing* 少数民族服装
newborn *clothing* 新生婴儿服装
nuclear protective *clothing* 防核辐射服，核防护服
occupational *clothing* 职业服
plus *clothing* 加大码衣服
protective *clothing* 防护服
pure silk embroidered *clothing* 真丝绣衣
religious *clothing* 宗教服装
safety *clothing* 安全防护服
see-through *clothing* 透明服装，透视装
signal *clothing* 信号服（缝有反光荧光条，预防车祸等事故）
smart *clothing* 智能服装（通过技术增强的服装，可以增加其传统用途之外的功能，也称为高科技服装）
tie-dyed *clothing* 扎染服装
utility *clothing* 实用型服装
vegan *clothing* 素食服装
weatherproof *clothing* 风雨衣
women's(men's, kids') *clothing* 女（男、童）装
woolen(silk, cotton) knitted *clothing* 毛（丝、棉）针织服装，毛（丝、棉）织服装
wrinkle-free *clothing* 抗皱服装，免熨烫服

clothtech [ˈklɒθtek] 服装纺织品；技术性衣着纺织品

clout [klaʊt] 零布；婴儿服；鞋底铁掌
   swadding *clouts* 婴儿衣服

clutch [klʌtʃ] （女用）无带手袋；女用小钱包
   envelope *clutch* 披盖手袋

coat [kəʊt] 上衣；外套；〔粤〕褛；大衣；女上装；女西装；童大衣；衬裙；（动物的）皮毛
   A-line *coat* A 型外套
   Afghan *coat* 阿富汗外套（毛皮长外套）
   Alaskan *coat* 阿拉斯加外套（风帽檐上装饰毛皮的防寒外套）
   all-weather *coat* 风雨衣，全天候外套，晴雨外套，〔粤〕干湿褛
   all year round *coat* 四季通用外套
   artificial fur *coat* 人造毛皮大衣
   balmacaan *coat* 巴尔马干大衣（亦译为巴尔玛外套，为开关领、暗门襟、连肩袖及斜插袋的男式大衣）
   belted *coat* 束腰带外套
   blazer *coat* 运动西装；便装上衣
   box *coat* 箱形外套；厚外套
   British warm *coat* 英国厚冬大衣（"一战"时英国军用双排扣大衣）
   bush *coat* 丛林外套（原为狩猎用）
   cameraman *coat* 摄影师外套（防水、防寒，胸前有大盖袋）
   caped *coat* 披肩大衣（无袖，呈钟型），披风外套
   car *coat* 跑车外套，短外套，短大衣，卡曲衫

   cashmere *coat* 羊绒大衣
   casual *coat* 休闲外套
   Chinese-style padded *coat* 中式棉袄
   clasp *coat* （女式无扣）抱合式大衣
   classical *coat* 古女装
   clutch *coat* 无纽大衣，围裹式外套
   coachman's *coat* 马车夫外套（大翻领，双排扣，紧腰）
   cocoon *coat* 茧形外套
   coolie *coat* 苦力外套，劳工外套（立领，衣长至腰下）
   cotton-padded *coat* 棉大衣，〔粤〕间棉褛
   cover *coat* 短外套，短大衣
   covert *coat* （骑马、射击穿）短皮外套
   dinner *coat* 小礼服
   donkey *coat* 驴子外套（针织丝瓜领防寒外套）
   down *coat* 羽绒大衣；羽绒服
   dress *coat* 礼服；燕尾服；西装外套
   dual *coat* 两用衫
   duffel *coat* 达夫尔（连帽）外套，（用绳索浮标纽的）风帽粗呢大衣（或上衣）；起绒粗呢大衣
   duster(dust free) *coat* 防尘外套
   Eton *coat* 伊顿外套（英国伊顿学校男生穿）
   evening *coat* 晚礼服
   fashion *coat* 时尚外套
   flared *coat* 宽摆外套
   frock *coat* 礼服大衣，大礼服
   full dress *coat* 大礼服

fur *coat* 毛皮大衣，裘，〔粤〕皮草
fur lining *coat* 毛皮衬里大衣
gentleman *coat* 绅士外套
guardsman *coat* 卫兵外套（双排扣束腰长外套）
half *coat* （比短外套还短的）齐腰短外套
hooded *coat* 带风帽的外套，防寒外套
Hudson Bay *coat* 哈得孙湾外套（厚毛料，双排扣束腰外套）
hunting *coat* 猎装外套
inverness *coat* 羽袖大衣（袖似披肩）
jean *coat* 牛仔（布）外套
jockey *coat* （赛马）骑师外套
King's (Queen's) *coat* 英国军服
knitted *coat* 针织短外套
lab *coat* 实验室工作外套
leather *coat* 皮大衣，〔粤〕皮褛
leather *coat* inserted with fur 镶饰毛皮的皮大衣
Lenin *coat* 列宁装上衣
loden *coat* 洛登缩绒厚呢大衣（布料防水，深橄榄色）
loose *coat* 外套
lumber *coat* 伐木外套（厚格子衣料制）
mackinaw *coat* 马基诺厚呢短外套（格子纹，配腰带，有过肩及有盖口袋）
mandarin *coat* 中式大衣；中式对襟马褂（外套）；西方女晚礼服外套
maxi *coat* 迷喜外套，长外套
midi *coat* 迷地外套，中长外套，

〔粤〕中褛
military *coat* 军大衣；军装式外套；军装式女上装
mink (marten) *coat* 貂皮大衣，貂裘
morning *coat* 早礼服，晨礼服，昼间礼服
Nehru *coat* 尼赫鲁外衣（紧身，高领，前排纽长大衣）
Newbury *coat* （单排扣）钮贝里外套
oilskin *coat* （防水）油布大衣
outer *coat* 外套，外衣，〔粤〕面衫；轻便大衣
padded *coat* 填棉外套，间棉外套
pilot *coat* 飞行员外套
pea *coat* 水手粗呢上衣，水兵外套；双排扣短上衣
pinchback *coat* 背部紧身外套（半腰带，打褶）
plush *coat* 长毛绒大衣
polar fleece *coat* （涤纶）双面绒大衣，摇粒绒大衣
polo *coat* 波鲁外套，马球外套（剑领，双排扣，有盖大贴袋，翻贴袖头，缉明线）
polyfilled *coat* 涤纶棉大衣，喷胶棉大衣
princess *coat* 公主线外套（束腰大摆）
PVC long *coat* PVC 长外套
pyjama *coat* 睡衣
quilted car *coat* 填棉短外套（又称太空服）
raglan *coat* 连肩袖外套
ranch *coat* 牧场外套（牛仔穿毛皮防寒外衣）

reefer *coat*　双排扣水手外套
reversible *coat*　双面大衣
riding *coat*　骑马外套；女骑装上衣
sack *coat*　男便装上衣；婴儿针织上衣
sassard *coat*　沙沙外套（宽西装领，有肩章束腰带，同堑壕外套）
semi-raglan *coat*　半连肩袖外套
shaped *coat*　合身外套
sheepskin *coat*　羊皮大衣，羊皮外套
shooting *coat*　狩猎上衣
short(long) *coat*　短（长）外套，短（长）大衣，〔粤〕短（长）褛
sleep *coat*　睡衣外套；女长睡衣
slim *coat*　细长外套
Spanish *coat*　西班牙外套（以西班牙式编织有垂片的大领子为特征的轻便外套）
sport *coat*　运动型外套；轻便上衣；替换上衣，两用衫
spring *coat*　夹大衣；风衣；春季外套
stadium *coat*　（观赛御寒穿）运动场外套
straight *coat*　直筒型外套
studded leather *coat*　窝钉皮外套
summer *coat*　夏季薄型外套
sur *coat*　外套，上衣
swagger *coat*　阔步大衣（中长尺寸，开摆），潇洒外套
swallow-tailed *coat*　燕尾服，晚礼服
sweater *coat*　毛衫外套

tail *coat*　燕尾服，晚礼服
three-quarter *coat*　三季节（春、秋、冬）用外套
top *coat*　上装，上衣；春秋外套；轻便大衣
tow *coat*　双排扣外套
travel *coat*　旅行外套
trench *coat*　堑壕外套（双排扣，宽腰带）；军用胶布夹雨衣；军装外套式风雨衣，风衣
tunic *coat*　束腰带外套，紧身短外套
tuxedo *coat*　无腰带女外套
ulster *coat*　阿尔斯特长外套（宽松，厚料制）
Vienna *coat*　维也纳外套（燕尾式后摆）
weather-all *coat*　晴雨大衣，风雨衣，〔粤〕干湿褛
wind(dust) *coat*　风衣
winter *coat*　冬季外套
wool *coat*　毛呢大衣
work *coat*　工作外套
wrap *coat*　宽胸式外套，卷裹外套（无纽而有腰带）
young men's *coat*　青年服
Zhongshan *coat*　中山装

coatdress [ˈkəʊtdres]　外套，大衣（开襟，宽腰，直筒形）
coatee [ˈkəʊtiː]　紧身短上衣
coating [ˈkəʊtɪŋ]　上衣料；大衣料；（织物）涂层
　　acrylic *coating*　亚克力胶涂层；水晶胶涂层
　　army *coating*　军服呢
　　breathable *coating*　透气涂层
　　cashmere *coating*　开司米大衣呢

downproof *coating*　防羽绒涂层
flame(fire)-retardant *coating*　阻燃涂层
foam *coating*　泡沫涂层
frock *coating*　植绒涂层
Hypalon *coating*　海帕伦涂层（橡胶涂层之一）
ladies *coating*　女式呢
mackinaw *coating*　马基诺双面大衣呢
milky *coating*　乳白涂层
navy *coating*　海军呢
oily *coating*　含油涂层
PA（polyamide）*coating*　PA（聚酰胺）涂层
PE（polyethylene）*coating*　PE（聚乙烯）涂层
pearly *coating*　珠光涂层
plain *coating*　素呢
PO（polyolefine）*coating*　PO（聚烯烃）涂层（一种环保涂层）
PU *coating*　PU 涂层
PU colour *coating*　PU 彩色涂层
PU two-tone *coating*　PU 双色涂层
PU wax *coating*　PU 蜡光涂层
PVC *coating*　PVC 涂层
rubber *coating*　橡胶涂层
silver *coating*　（面料）涂银，银涂层
student *coating*　学生呢
tencel/linen *coating*　天丝亚麻混纺上衣料
top *coating*　轻薄大衣料
TPE *coating*　TPE 涂层（涂膜具热塑高弹性，环保无毒）
TPU *coating*　TPU 涂层（热塑聚氨酯涂膜，环保且透湿透气）
twill *coating*　啥味呢；斜纹呢
ULY *coating*　ULY 涂层（俗称优丽胶涂层，PU 涂层的一种）
uniform *coating*　制服料；制服呢
waterproof *coating*　防水涂层
wax(paraffin) *coating*　蜡涂层
wool *coating*　呢料

coattail　[ˈkəʊteɪl]　男上衣后摆，燕尾服下翼；〔复〕女长外衣下摆

cockade　[kɒˈkeɪd]　帽章；帽花结
　cap cockade　帽徽

collar　[ˈkɒlə]　衣领，硬领，假领；领饰；领圈；颈圈；项饰
　arched *collar*　弓形领，半圆形领，拱形领
　attached *collar*　活动礼服衬衫领
　Balmacaan(bal)*collar*　巴尔马干大衣领（连肩袖男大衣小方领）
　band *collar*　直领，立领，带形领
　Barrymore *collar*　巴利摩尔（衬衣）领（长尖领）
　Baster Brown *collar*　巴斯特·布朗领（宽大扁平的圆领）
　belt *collar*　带扣领
　bertha *collar*　贝莎披肩领
　bib *collar*　围兜领
　blizzard *collar*　女服防寒高立领
　Bonaparte *collar*　波拿巴领（高立领及宽驳头）
　bottled *collar*　瓶颈领，〔粤〕樽领
　butterfly(bow)*collar*　蝴蝶结领
　button-down *collar*　下扣领（领角有扣），扣贴领
　California *collar*　（长领尖）加利福尼亚衬衫领
　cape *collar*　披肩领；斗篷领
　carp *collar*　鲤鱼领

cascading *collar* 层叠领；波形领
chin *collar* 颚领（高立领）
Chinese *collar* 旗袍领
clerical *collar* （高约一寸的环状领）
closed *collar* 关门领，立领
club *collar* 棒形领；小圆领
coat *collar* 大衣领
coated *collar* 胶领
collar assembling to body not closed 〔检〕绱领不圆顺
collar band 底领，领座，〔粤〕下级领
collar band too short (long) 〔检〕底领缩进（伸出）
collar corner 领角
collar deviates from the front centre line 〔检〕绱领偏斜
collar fall 衣领翻下部分，〔粤〕上级领
collar fly （衬衫）领口蝴蝶插片，〔粤〕胶蝴蝶
collar keeper 领条，领托，支领
collar leaf 领角衬，领口片
collar point spread 领尖距
collar roll (neck) line 领上（下）口
collar slip 礼服背心上的白色添加领
collar stand 领座，领脚；领脚长
collar stand away from neck 〔检〕领离脖（领子不贴脖）
collar stay 领插角片，领口薄膜，〔粤〕插竹
collar step 领脚，领座
collar stop 领止点
collar supporter 领托，领条

collar turner 翻领器，翻领撞针
collar width with stand 后领高
collarless *collar* 无面领的领型（保留底领）
combat *collar* 战斗领（衬衫领）
convertible *collar* 两用领，开关领，换形领
cross muffler *collar* 交叉领巾领（毛衣领）
cross-over *collar* 横文领，交叉领
cross shawl *collar* 交叉围巾领
crumples at the *collar* 〔检〕领面起泡
cup-shape *collar* 杯形领
darted *collar* 短缝领
detachable *collar* 活动领；假领
dog *collar* （神父的）硬白领；项圈形高直领
double *collar* 双折领，二重领
draped *collar* 褶裥领；垂坠领
drawstring *collar* 束带领
Dutch *collar* 荷兰领（套衫圆口小翻领）
Eton *collar* 伊顿领（英国伊顿公学男生制服上的白色硬阔领）
everclean *collar* 保洁领
eyelet *collar* 针孔领（别针固定领端）
false *collar* 假领（衬衫上的可拆装的活动领）
far-away *collar* 远离领
flat *collar* 平领，平翻领
frill *collar* 波褶领
funnel *collar* 漏斗领
fur *collar* 毛皮领
fused *collar* 热熔黏合领
high roll *collar* 高翻领

horizontal collar 水平领

horse-shoe(horse-hoof) collar 马蹄形领；U形领

intarsia collar （针织）嵌花领

Italian collar 意大利领（V形领线上装短低领）

Johnny collar 约翰尼领（女衬衫小方角领）

knitted rib collar 针织罗纹领

lace collar 花边领

lapel collar 驳头领，驳领

low(high) collar 低（高）领

mandarin collar 中式领，马褂领；旗袍领

Mao collar （中国毛式）立领

middy collar 迷蒂领（女学生制服领）；水手领

Milan collar （意大利）米兰领

misplaced collar 〔检〕领子上得不正（领中点不居中）

miter collar 斜角拼接领

Napoleon collar 拿破仑领（高立领宽驳头）

Nehru collar 尼赫鲁领（立领）

notch collar 缺嘴领，叉口领；翻领

oblong collar 长椭圆形领，比翼领

off collar 一字领

officer collar 军官领；中式立领

one-piece collar 单页领，一片领

open collar 开门领

open-wing collar 平翼领

outside(inside) collar 表（里）领，领面（里）

peaked collar 尖形领，尖领

peaked shawl collar V形丝瓜领

peasant collar 农民领（欧洲农民罩衫领）

pedal collar 踏板形领

petal collar 花瓣翻领

Peter Pan collar 彼得·潘领（女童服上的小圆领）

picture collar 象形领

pierrot collar 小丑领（波褶领）

pinhole collar 别针扣领，针孔领（用别针扣定两边）

piping around collar 绲边领

plain collar 普通（衬衫）领

plush collar 长毛绒衣领

pointed collar 尖领

poke collar 活动礼服衬衫领

polo collar 马球衫领（半开襟小翻领）

Prussian collar 普鲁士领

puritanical collar 清教徒领，小斗篷领

purser collar 巴莎领（前面不开衩的披肩领）

reefer collar 帆形叠领

regular collar 普通（衬衫）领

reverse collar 翻领

ribbed collar 罗纹领

ripple(d) collar 细褶波纹领

roll collar 大翻领

round collar 圆领

ruff collar 襞襟，拉夫领（16、17世纪盛行于上流社会的一种白色轮状皱领）

ruffle collar 皱褶领

sailor collar 海员领，水手领

scarf collar 围巾领，方巾领

semi-clover collar 半苜蓿领，西装圆领

semi-cut-away *collar*　八字领
semi-soft *collar*　半硬领
semi-spread *collar*　半展开领
shawl *collar*　青果领，围巾领，新月领，香蕉领，丝瓜领
shearling *collar*　羊毛皮大翻领
shirt *collar*　衬衫领，小方领
shirt square *collar*　（衬衫）小方领
shoe *collar*　鞋领（鞋口外翻部分）
short rounded *collar*　短圆领
sideway *collar*　偏侧领
single *collar*　单折领
spare *collar*　备用领
spread *collar*　八字领，展开领
square *collar*　方领
stand *collar*　立领，〔粤〕企领，学生服领
stand away *collar*　直离立领
stand-fall *collar*　二重领（立领翻折成二层的领）
starched *collar*　上浆领
step *collar*　（阶）梯形领；领缺口为直角的西装领
stick *collar*　棒形领
stiff(soft) *collar*　硬（软）领
stole *collar*　长围巾领
strap *collar*　窄条领
surplice *collar*　葫芦领（教士法衣领）
swallow *collar*　燕子领
tab *collar*　拉襻领
tailored *collar*　西装领
three-way *collar*　三式领，三用领
tie *collar*　领带领；打结领
top *collar*　领面，上领
top *collar* too tight(loose)　〔检〕领面紧（松）

triangle *collar*　三角领
tunnel *collar*　隧道领
turn-down *collar*　翻下领，翻领，〔粤〕上下级领
turn-over *collar*　翻领
turn up *collar*　可翻起的衣领，翻领
turtle *collar*　珺瑠领，瓶颈领
tuxido *collar*　（无尾）小礼服（宽）领
two-piece *collar*　双页领，上下领
ulster *collar*　阿尔斯特大衣领，倒挂领
under(fold)line of *collar*　领下（上）口
under(over) *collar*　内（外）层领
unmatched *collar*　〔检〕绱领偏斜
Vandyck *collar*　范戴克领（方形大尖角领，常有边饰）
V-with *collar*　V形翻领
wide (long, short) pointed *collar*　大（长、短）尖角领
wied spread *collar*　展宽领，大八字领
Windsor *collar*　温莎领（大八字衬衫领）
winged *collar*　翼形领，燕子领
yoke *collar*　抵肩领，约克领
Zhongshan coat *collar*　中山服领
collaret(=collarette)　[ˈkɒləˈret]　女用领饰，围巾，披肩
collarless　[ˈkɒlələs]　无领
collection　[kəˈlekʃən]　（按某一特点将服装分类形成的）系列；集成
　denim garments *collection*　牛仔服装系列

fashion *collection* 时装总汇；时装博览会
full *collection* of garment auxiliaries 服装辅料大全
Paris *collection* 巴黎时装发布会
spring and autumn women's wear *collection* 春秋女装系列

colour(=color) ['kʌlə] 颜色；色彩；彩色；染料；〔复〕（作为所属团体色彩标记的）绶带，徽章，衣帽

accent *colour* 强调色
additional *colour* 点缀色
advancing *colour* 前进色
begin *colour* 本色，天然色
camouflage *colour* 伪装色，保护色，迷彩色
clash of *colour* 色彩不调和
colour card 色卡
colour code 色码
colour combination 色彩的组合，配色
colour coordination 色彩的调和
colour deviation 色差
colour fastness 色牢度
colour guide 色样
colour hue 色调
colour matching 配色
colour mixing 调色
colour No. (=*colour* number) 色号
colour on *colour* 重色配色法
colour scheme 色彩设计，配色；色系
colour shade 色光，色调，色泽
colour tone 色调
complementary *colour* 补色，互补色；对比色
computer *colour* matching system 计算机配色系统
contracting *colour* 收缩色
contrasting (contrast) *colour* 对比色，衬色，〔粤〕撞色
delicate *colour* 娇嫩色，柔和色
design and *colour* 花色
essential *colour* 基本色
expansive *colour* 膨胀色
fashion *colour* 流行色
fast *colour* 不褪色
fugitive *colour* 褪色
full *colour* 彩色，五彩；齐色
ground *colour* 底色
inter *colour* 国际流行色
intermediate *colour* 中间色
metal *colour* 金属色彩
mismatched *colour* 〔检〕不配色
natural *colour* 自然色，本色
neutral *colour* 中间色
off *colour* 色差
plain *colour* 素色
primary(fundamental) *colours* 原色
protective *colour* 保护色
pure *colour* 纯色
receding(receded) *colour* 后退色
rich in *colours* 色彩丰富；浓色
science of *colour* 色彩学
season *colour* 时新色，流行色
secondary *colour* 混合色
similar *colour* 同类色
soft *colour* 柔和色，嫩色
solid (assorted) *colour* 素（混）色，单独（搭配）色
surface *colour* 表面色
traditional Chinese *colour* 中国传统色
transparent *colour* 透明色

  trend *colour* 流行色
  warm(cold) *colour* 暖（冷）色
  wrong *colour* indicated 〔检〕错色
combinaison [kɔ̃bɪnɛzɔ̃] 〔法〕上下相连的套装
  combinaison de femme 〔法〕女连衫衬裙
  combinaison d'homme 〔法〕连衣裤工作服
  combinaison short (= combinshort) 〔法〕短裤型跳跃套装
combinations [ˌkɒmbɪˈneɪʃənz] 〔复〕上下相连的服装；（女）连裤内衫
comfort [ˈkʌmfət] （织物的）舒适性；穿着舒适；〔美〕盖被
comfortability [ˌkʌmfətəˈbɪləti] 舒适性，贴身性
compasses [ˈkʌmpəsɪz] 〔复〕（制图用）圆规
composition [ˌkɒmpəˈzɪʃən] 构成；构图；成分；合成物
computer [kəmˈpjuːtə] 电子计算机
  computer aided manufacture (CAM) 计算机辅助生产
  computer integrated manufacturing (CIM) 计算机综合制造
conception [ˈkənsepʃn] 〔设〕观念；概念；构思；构想
  original in *conception* 构思新颖
confection [kənˈfekʃən] 时髦女装；女用服饰品；成衣；〔法〕服装业；现成服装，一般成衣
construction [kənˈstrʌkʃən] （面料）组织；（使服装）调和
  fabric *construction* 织物组织（指纱支、密度等）
consumption [kənˈsʌmpʃən] 消耗，耗料
  calculation of fabric utilization and *consumption* 计算用料
  fabric *consumption* 估料，耗料
contour [ˈkɒntʊə] 轮廓，外形；轮廓线
contract [ˈkɒntrækt] 合同，合约，契约
  breach of *contract* 违约
  flexibility *contract* 柔性合同，动态合同（具有灵活性，预防风险）
  manufacturing *contract* 加工合同
  selling *contract* 销售合同
  signing of *contract* 签合同
contrast [ˈkɒntræst] 〔设〕对比，对照；反差
controller [kənˈtrəʊlə] 会计长；审计官；主计员；管理员
  quality *controller* 质量管理员，质控员（俗称QC）
coolmax [ˈkuːlmæks] 一种高科技吸湿透气聚酯纤维
coordinates [kəʊˈɔːdɪnəts] 〔复〕（颜色、质料、式样等）协调的衣服；套装
coordination [kəʊˌɔːdɪˈneɪʃən] 〔设〕协调，调和
  shift *coordination* 可自由替换组合着装法
cope [kəʊp] （教士）斗篷式长袍
copy [ˈkɒpi] 副本；复制品，复印件
cord [kɔːd] 灯芯绒类布；绳；带；线；〔复〕灯芯绒裤
  bedford *cord* 厚实凸条布，经条灯芯绒布
  cable *cords* 宽条灯芯绒
  cotton *cord* 棉绳

draw cord （衣服腰部或下摆等处的）拉绳；松紧绳带
embroidery cord 嵌绣
fustian cord 〔英〕灯芯绒
hair cords 麻纱
packing cord 包装绳，打包绳
piping cord 包边绳
welting cord 绲边带
cording [ˈkɔːdɪŋ] 嵌线
corduroy [ˈkɔːdərɔɪ] 灯芯绒；〔复〕灯芯绒裤；工装裤
    bold(wide)-wale corduroy 粗条灯芯绒，宽条灯芯绒
    fine needle(pinwale) corduroy 细条灯芯绒
    medium wale corduroy 中条灯芯绒
    printed(dyed) corduroy 印花（染色）灯芯绒
    sand-washed corduroy 砂洗灯芯绒
    snowflake corduroy 雪花灯芯绒
    warp knitted corduroy 经编灯芯绒
coronet [ˈkɒrənɪt] （王子或显贵戴的）冠；（女用）冠状头饰
corsage [kɔːˈsɑːʒ] 女服的胸部；女胸衣，紧身衣；〔美〕女服胸部或腰部的饰花
corselet(= **corselette**) [ˈkɔːslɪt] 女胸衣
corset [ˈkɔːsɪt] 女紧身胸衣；（硬）围腰；束腹带
    belted corset 束带紧身衣
    girdle corset （女用）束腹带
    laced corset 穿绳紧身胸衣
    posture aid corset 健美紧身衣，整形紧身衣
    sheath corset 紧身胸衣；紧身腹带
    unbelted corset 无带紧身胸衣

cosmesis [kɒzˈmesɪs] 美容术
cosmetic [kɒzˈmetɪk] 化妆品
cosmetician [ˌkɒzməˈtɪʃən] 化妆师，美容师
cosmetologist [ˌkɒzməˈtɒlədʒɪst] 整容专家，美容师
cosmetology [ˌkɒzməˈtɒlədʒi] 整容术，美容学
Cossack [ˈkɒsæk] 哥萨克服装（立领，束带长衬衫）
cost [kɒst] 成本；费用
    cost accounting 成本核算
    cost analysis 成本分析
    cost control 成本控制
    cost price 成本价
    labour cost 劳工成本，工资成本
    production cost 生产成本
costume [ˈkɒstjuːm] （特定国家或历史时期流行的）服装，全套服饰；服装式样；装束；女服，女套装，戏服
    academic costume 学位服
    bathing(swimming) costume （女）游泳衣，泳装
    Chinese-style costume 中式服装；中装，〔粤〕唐装
    classical costume 古典服装
    costume linguistics 服装语言学
    costume of national minorities 少数民族服装
    costume of the Zang(Mongol, Chaoxian, Gaoshan) nationality 藏（蒙古、朝鲜、高山）族服装
    ethnical costume 民族服装，民俗服装
    fanciful costume 奇装异服
    folk costume 民俗服装
    Han Chinese costume 汉服

historical(ancient) costume 古装
minority costume 少数民族服装
national costume 民族服装
oriental costume 东方女服
riding costume 骑装
skating costume 溜冰服
theatrical(stage) costume 舞台服装；戏装
traditional costume 传统服装
velvet costume 丝绒戏服

costumer [ˈkɔstjuːmə] 服装制作人；服饰供应商；[美]（柱式）衣帽架

costumery [ˈkɒstjuːmərɪ] （costume 的变形）服装，装束；戏装

cotton [ˈkɒtn] 棉花；棉线；棉布
    all cotton 全棉，纯棉
    artificial cotton 人造棉
    Asiatic cotton 亚洲棉
    bamboo fibre cotton 竹纤维棉
    chief value of cotton(C. V. C) 以棉为主（50%或以上）的混纺织物
    coloured cotton 有色棉，彩棉
    combed(carded) cotton 精（普）梳棉
    corn fibre cotton 玉米纤维棉
    cotton tweed 仿粗花呢色织棉布
    3D vertical cotton 3D 直立棉（简称 3D 棉，新型环保非织造立体材料）
    ecological cotton 生态棉（一种超微细丙纶纤维，用作填料）；生态棉布
    far-infrared cotton 远红外棉
    green cotton 绿色棉
    long fibre cotton 长绒棉
    mercerized cotton 丝光棉布；丝光棉
    modal fibre cotton 莫代尔纤维棉
    organic cotton 有机棉（种植全程环保）
    Pima cotton 比马棉（美国埃及种棉）；比马棉衣料
    punched cotton 针刺棉
    sewing cotton 缝纫用棉线
    silk cotton 木棉，木丝棉
    space cotton 太空棉
    the padded cotton is uneven [检]（棉衣）铺棉不匀
    thermofusible cotton 热熔棉
    transgenic cotton 转基因棉
    transition cotton 过渡棉

count [kaʊnt] （布料的）纱支

coupe-vent [kupvɑ̃] [法]风衣

couture [kutyːr] [法]（女服）时装（业），服装业，缝纫，定制衣服；针线活
    haute couture [法]高级女子时装（业）；高级定制女时装

couturier [kutyrje] [法]时装女服设计师；时装女服商；裁缝师

couturière [kutyrjɛr] [法]时装女设计师（或女裁缝）

cover [ˈkʌvə] 套子；罩子；（布）面
    bed cover 床罩
    cloth cover 布面
    heel cover 鞋跟包皮
    machine cover 机罩，机套
    pillow cover 枕套
    quilt cover 被面；被套
    shoe cover 鞋面
    velvet cover 绒面

coveralls [ˈkʌvərɔːlz] [复]衣裤相连的工作服；连衣裤服装；婴幼连衣裤

  diver's *coveralls* 潜水员工作服
  labour protective *coveralls* 劳保工作服, 劳保服
coverseam [ˈkʌvəsiːm] 覆盖缝, 绷缝, [粤] 𠝹骨
cover-up [ˈkʌvəʌp] 海滨装（在海滨穿的长袖有头巾的服装）
cowhide [ˈkaʊhaɪd] 牛皮; 牛皮革
cowl [kaʊl] 带头巾僧袍; 僧袍的头巾
cozzie(= **cossie**) [ˈkɒzɪ] 游泳衣
craft [krɑːft] 工艺; 手艺
  arts and *crafts* 工艺美术
craftsmanship [ˈkrɑːftsmənʃɪp] 技艺, 手艺; （服装）做工
crampons [ˈkræmpənz](=**crampoons**) [kræmˈpuːnz] 〔复〕（便于爬山和冰上行走的）钉鞋
crape [kreɪp] 绉纱; 绉布; 绉绸; 绉呢
  crape cloth 绉纱毛织品
crapka [ˈkræpkə] （19世纪）波兰骑兵的方顶头盔
cravat [krəˈvæt] （旧式）领带; 围巾; 领饰
cream [kriːm] 奶油色, 米色; 雪花膏; 膏状物
  anti-ageing *cream* 防（皮肤）老化霜
  base *cream* 底霜
  beauty *cream* 美容霜
  body *cream* 身体霜, 护肤霜
  cleaning *cream* 清洁霜
  cold *cream* 冷霜
  eye *cream* 眼霜
  face *cream* 面霜
  foundation *cream* 粉底霜
  hand *cream* 护手霜
  lanoline *cream* 润肤霜
  massage *cream* 按摩霜
  natural white moisture protection *cream* 美白润肤霜
  neck *cream* 颈霜
  night(day) *cream* 晚（日）霜
  regenerist micro-sculpting *cream* 新生塑颜面霜
  rouge *cream* 胭脂膏
  vanishing *cream* 雪花膏
  wrinkle smoothing *cream* 抗皱霜
crease [kriːs] （衣裤等）折缝, 折痕; 皱折, 皱痕; 褶裥
  centre *crease* 中折
  crease line leans to outside(inside) 〔检〕裤烫折线外（内）撇
  *crease*s at right facing 〔检〕里襟里起皱
  *crease*s at underarm 〔检〕裉窝起绺
  *crease*s below neckline 〔检〕领窝不平
  *crease*s on two ends of pocket mouth 〔检〕袋口角起皱
  *crease*s in trousers 裤线, 裤腿折痕
  false *crease* 假折缝
  front *crease* 裤腿前折缝
  permanent *crease* 永久折痕; 耐久褶裥
creativeness [kriˈeɪtɪvnɪs] 〔设〕创造性
creeper [ˈkriːpə] （婴儿）连衫裤, 爬行服
crepe(= **crêpe**〔法〕) [kreɪp] 绉纱; 绉布; 绉绸; 绉呢

blister crepe 泡泡纱
cotton crepe 布绉
crêpe de chine 〔法〕双绉
crepe georgette 乔其绉
doupion silk crepe 双宫绉
kabe crepe 碧绉（即印度绸）
linen crepe 亚麻绉布
palace crepe 派力司绉
polyester crepe 涤丝绉
rayon back satin crepe 人造丝缎背绉
rayon crepe de chine 人造丝双绉
sheer crepe 透明薄绉
silk crepe 真丝绉
silk crepe satin 真丝绉缎
wool crepe 全毛绉呢

creping [ˈkrepɪŋ] （布料）起绉（工艺）

crepon [ˈkrepɔ̃] 〔法〕重绉纹织物；树皮绉
    Xihu jacquard crepon 西湖呢（小提花丝质春秋冬装料）

creppella [ˈkreɪplə] 绉纹呢（棉织物）

crest [krest] （旧时）头盔上的羽毛饰

crevenette [ˈkrevenet] 克赖文内特防水处理；防雨卡，雨衣料

crewelwork [ˈkruːəlˌwɜːk] 绒线刺绣

crinkle [ˈkrɪŋkl] 折皱；波状
    washed crinkle 洗水皱褶

crinolette [ˈkrɪnəlɪt] （旧时的）箍形裙衬架

crinoline [ˈkrɪnəlɪn] 硬衬布，（旧时）裙衬架；衬架裙

crochet [ˈkrəʊʃeɪ] 钩织；钩针编织品

crocodile [ˈkrɒkədaɪl] 鳄鱼皮革

cross [krɒs] 十字形；十字形项饰（如十字架等）

crotch [krɒtʃ] (= **crutch** [krʌtʃ]) 大腿根处；裤衩，裤裆底，〔粤〕浪底
    above crotch （裤）衩上
    back crotch stay 大裤底（后裆里子）
    bursting of crotch seam 〔检〕裤裆断线
    cross crotch 裤裆底，〔粤〕十字骨
    cross crotch is uneven 〔检〕裆底十字缝错位
    crotch depth （裤）裆深，股上
    front crotch stay 小裤底（小裆里子）
    low cut crotch 低裤裆，垂裆
    reinforcing crotch 加固裆
    short crotch 〔检〕兜裆（直裆短）
    tight crotch 〔检〕夹裆（后裆太紧）
    unmatched cross crotch 〔检〕裆底十字缝错位

crown [kraʊn] 顶部；帽顶；皇冠
    sleeve crown 袖山

crystal [ˈkrɪstl] 水晶；水晶饰物

cuff [kʌf] 袖口，袖头，卡夫（音译），〔粤〕鸡英，介英；衣袖（或裤脚）卷边
    bead cuff 褶裥（泡状）袖头；珠饰袖头
    bracelet cuff 紧袖口
    convertible cuff 两用袖头，可变袖头
    cuff relaxed (extended) （松紧）袖头收缩（拉开）量
    detachable cuff 活动袖头

double *cuff* 双袖头
elastic *cuff* 松紧袖口
fold-back *cuff* 翻折袖头，双袖头
French *cuff* （法式衬衫）双袖头
half(full) *cuff* 半（全）克夫袖口
mock(imitation) *cuff* 假袖头
pleated *cuff* 多褶袖头
rib-knit *cuff* 罗纹袖口
scalloped *cuff* 扇形袖头
single *cuff* 单袖头
slant(round) corner *cuff* 斜（圆）角袖头
sleeve-*cuff* 袖头，袖口
storm *cuff* 防风（松紧）袖头
strapped *cuff* 束带袖口
three points *cuff* 三尖袖头，尖角袖头
turned *cuff* 翻袖
turn-up *cuff* 裤卷脚；翻袖口；双层袖头
twisted *cuff* 〔检〕袖头扭曲
two(one)-button *cuff* 双（单）扣袖头
uneven height of *cuffs* 〔检〕袖头高不匀
windproof *cuff* 防风袖头

cuffless [ˈkʌflɪs] 无袖口；无裤脚卷边；无脚口卷边裤

cufflinks [ˈkʌflɪŋks] 〔复〕袖口链扣，袖头组

cuirass [kwɪˈræs] 胸甲；妇女胸衣

cuish [kwɪʃ] （古代武士）护腿甲，腿甲

culottes [kjʊˈlɒts] 〔复〕裙裤，裤裙
    culottes anglaise 〔法〕紧身短裤
    culottes de cheval 〔法〕马裤
    golf *culottes* 高尔夫裙裤

robe *culottes* 〔法〕裤裙

culture [ˈkʌltʃə] 文化；文明；习俗
    Chinese *culture* 中国文化，中华文化
    corporate *culture* 企业文化
    cowboy *culture* 牛仔文化
    deep *culture* 深层文化
    Dunhuang costume *culture* 敦煌服饰文化
    fashion *culture* 时尚文化；服饰文化
    folk *culture* 民俗文化；民间文化
    Han *culture* 汉文化
    high *culture* 高级文化
    mainstream *culture* 主流文化
    material *culture* 物质文化
    national *culture* 民族文化
    popular *culture* 大众文化
    spiritual *culture* 精神文化
    tradional *culture* 传统文化

cummerbund [ˈkʌməbʌnd] （印度式）宽幅腰带

cummervest [ˈkʌməvest] （宽带绕成的）卡马背心

cup [kʌp] 乳罩罩杯，乳罩窝
    cup form 罩杯托
    cup size 罩杯尺寸

curve [kɜːv] 曲线，弯曲
    curve gauge 曲线板
    curve of beauty 曲线美

cushion [ˈkʊʃən] 垫子；插针垫
    shoe *cushion* 鞋垫

cut [kʌt] 裁，剪，切，（裁剪）式样
    cross *cut* 横纹裁剪，横裁
    diagonal *cut* 斜裁
    front *cut* （衣服的）止口圆角；前摆

front square cut　方角前摆
straight cut　直线裁剪，直裁
cut-away [ˈkʌtəweɪ]　前摆向后斜切的燕尾服；〔美〕晨礼服
cut-offs [ˈkʌtɒfs]　〔复〕剪截式短裤
cutter [ˈkʌtə]　服装裁剪师；裁剪机；切刀；电剪
　　bias tape cutter　切捆条机
　　buttonhole cutter　扣眼切刀
　　circular knife cutter　圆刀电剪；圆刀裁剪机
　　clothing template cutter　服装模板切割机
　　cordless cutter　电池式裁剪机
　　die cutter　冲切裁剪机
　　electric cloth cutter　电剪
　　hydraulic cutter　液压裁剪机
　　light cutter　带灯光的裁剪机
　　miniature electric cutter　微型电剪，迷你裁剪机
　　paper cutter　裁纸刀
　　rotary knife cutter　旋转式圆刀电剪
　　straight knife cutter　直刀电剪，直刀裁剪机
　　tailor's cutter　成衣店的裁剪师
　　thread cutter　剪线器
　　upright (band) cloth cutter　立（带）式裁剪机
cutting [ˈkʌtɪŋ]　裁，剪，裁剪

cross cutting　横裁
cutting and styling　做发型
cutting as stripes (checks) matching　对条（格）裁剪
cutting by one-piece　单件裁剪
cutting loss　裁剪损耗
cutting table　裁剪案板，裁剪台，裁床
diagonal (bias) cutting　斜裁
die cutting　冲切裁剪
draping cutting　立体裁剪
flawless cutting　贴身裁剪
freehand cutting　自由裁剪（无纸型平面裁剪）
garment cutting　服装裁剪
hand cutting　手工裁剪，手裁
laser cutting　激光裁剪
moulding cutting　模塑裁剪（不留缝份，按体型裁剪，如针织内衣）
multiply cutting　多层裁剪
programmed automatic cutting　程序自动裁剪
plane cutting　平面裁剪
straight cutting　直裁
tailor's cuttings　零剪，单裁
template cutting　（服装）模板切割
cymar [sɪˈmɑː]　（女用）宽大轻便的无袖衣；女便袍

D

dacron [ˈdeɪkrən]　涤纶（俗称的确良），达克龙（译音）；涤纶线；涤纶织物
　　cotton dacron　棉涤纶，棉的确良
　　flax dacron　麻涤纶，麻的确良
　　woolen dacron　毛涤纶，毛的确良
dalmatic [dælˈmætɪk]　（罗马天主教）主教法衣；（英国）国王加冕服

damask [ˈdæməsk] 缎子，锦缎，花缎
 damask silk 绫
 linen *damask* 亚麻锦缎；亚麻花布
 Sangbo crepe *damask* 桑波缎
dancewear [ˈdɑːnsweə] 舞蹈服装
darning [ˈdɑːnɪŋ] 织补，缝补；织补物
 darning for woolens 织补毛料衣服
dart [dɑːt] 省（根据体型，在衣片上缝去的部分，俗称省、省道），褶，裥；短缝
 armhole *dart* 袖窿省
 back *dart* 后身省，后省
 back neck *dart* 后领省
 chest(breast) *dart* 胸省，胸褶
 contour *dart* 曲线省，长腰省
 crumples at *dart* point 〔检〕省尖起泡
 dart apex 褶尖
 dart manipulation 省缝处理（修改，移位）
 diagonal *dart* 对角省，斜省
 double *dart* 对称省
 elbow *dart* 肘省，肘褶
 fish *dart* 鱼形褶，肚省（大袋口位的横省）
 French *dart* 法式省（腰节上2.5~5cm的省），刀背缝，曲线省
 front *dart* （衣）前省；胸褶；（裤）前后省
 front(back) open *dart* 前（后）身通省（肩缝至下摆的开刀缝）
 front(back) shoulder *dart* 前（后）肩省
 front(back)waist *dart* 前（后）腰省
 gorge *dart* 领口省
 lapel *dart* 驳头省，〔粤〕襟褶
 neck *dart* 领省
 neckline *dart* 领口省，领孔褶
 open *dart* 开褶
 side *dart* 横省，胁省，胁褶
 side seam *dart* 侧缝省
 stomach *dart* 肚省，腹褶
 underarm *dart* 胁省，袖底省，〔粤〕夹底褶
 waistline(waist) *dart* 腰褶；裤（裙）腰省
dartless [ˈdɑːtlɪs] 无省道，无省
dashiki [dəˈʃiːkiː] 大稀奇装（颜色花哨的非洲短袖宽袍）
data [ˈdeɪtə] 资料，数据
 big *data* 大数据
 data bank 资料库，数据中心
 data storage 数据存储
 dress *data* 服装资料
 electronic *data* processing（EDP）电子数据处理
 somatic *data* 人体数据
 stitching *data* 缝纫数据
decal [dɪˈkæl]（=decalcomania [dɪˌkælkəˈmeɪnɪə]）移画印花法（将绘在纸上的图案移印到服装上）
decky [ˈdekɪ] 礼服衬衫的胸垫（仅有领部和胸部的内衬装）
décolletage [deɪˈkɒltɑːʒ]〔法〕袒胸露肩衣服；（袒胸衣服的）低领
decoration [ˌdekəˈreɪʃən] 装饰，〔复〕装饰品，饰物
 applique *decoration* 贴花，补花
 decoration hole 装饰扣眼，假扣眼
 research of dress and *decoration* 服

饰研究

deerskin [ˈdɪəskɪn] 鹿皮；鹿皮服装

deerstalker [ˈdɪəstɔːkə] （前后翘起，两侧有耳盖的）猎鹿帽

defect [ˈdiːfekt] 缺陷，欠缺；疵点，〔粤〕鸡

 critical *defects* 致命缺陷（如严重色差，断针，环保超标等）
 defects of end products 成品缺陷
 finish *defects* 后整理欠缺处
 garment *defects* 成衣缺陷
 materials(fabric) *defects* 面料疵点
 minor(major) *defects* 小（大）缺陷，〔粤〕细（大）鸡
 sewing(pressing) *defects* 缝制（熨烫）缺陷
 washing *defects* 水洗缺陷
 workmanship *defects* 做工欠缺处

delineator [dɪˈlɪnɪeɪtə] 裁剪图样

denim [ˈdenɪm] 粗斜纹棉布，牛仔布，劳动布，坚固呢；〔复〕（蓝斜纹布）工作服，工装裤；牛仔裤

 black *denim* 黑牛仔布
 bleached *denim* 漂白牛仔布
 burnt-out *denim* 烂花牛仔布
 coating *denim* 涂层牛仔布
 coloured *denim* 彩色牛仔布
 cosmetic *denim* 装饰牛仔布
 dark-coloured *denim* 深色牛仔布
 dobby *denim* 小提花牛仔布
 double-elastic *denim* （经纬）双弹牛仔布
 functional *denim* 功能牛仔布
 filling-wise elastic *denim* 纬弹牛仔布
 flax/cotton *denim* 亚麻/棉牛仔布
 flax *denim* 纯亚麻牛仔布
 flocked *denim* 植绒牛仔布
 hemp/cotton *denim* 大麻/棉牛仔布
 indigo *denim* 印地可牛仔布，靛蓝牛仔布
 iridescent rayon *denim* 有光人丝牛仔布
 jacquard *denim* 提花牛仔布
 knitted *denim* 针织牛仔布
 laminated *denim* 层压牛仔布
 lightweight *denim* 轻身牛仔布，薄牛仔布
 lurex *denim* 金银丝牛仔布
 necked *denim* 竹节牛仔布
 overdyed *denim* 套色牛仔布
 printed *denim* 印花牛仔布
 punching *denim* 冲孔牛仔布
 ramie/cotton *denim* 麻/棉牛仔布
 recycled *denim* 再生牛仔布
 ring spinning *denim* 环锭纺牛仔布
 rotor spinning *denim* 转杯纺牛仔布
 satin *denim* 缎纹牛仔布
 spandex(elastic) *denim* 弹力牛仔布
 specific washed *denim* 特效洗水牛仔布
 stretch(lycra) *denim* 弹力牛仔布
 stretch necked *denim* 竹节弹力牛仔布
 stripe *denim* 条纹牛仔布
 tencel/bamboo fibre *denim* 天丝/竹纤维牛仔布
 tencel/cotton *denim* 天丝/棉混纺牛仔布
 tie-dyed *denim* 扎染牛仔布
 ultra indigo *denim* 超靛蓝牛仔布

ultrathin *denim* 超薄牛仔布
washed *denim* 水洗牛仔布
white-back *denim* 白背牛仔布
yarn-dyed preshrunk *denim* 色织防缩坚固呢（劳动布）

densimeter [denˈsɪmɪtə]（=densitometer [ˌdensɪˈtɒmɪtə]） 密度计；（测布料用的）经纬密度尺

density [ˈdensəti] （经纱、纬纱的）密度

depth [depθ] 深；深度；厚度；（色彩）浓度
    crotch *depth* （裤）裆深
    front(back)neck *depth* 前（后）领深
    hip *depth* 臀高，臀围长
    scye *depth* 袖窿深

derby [ˈdɑːbi] 〔美〕常礼帽，圆顶礼帽

derriere [ˈderɪə] 臀部

déshabillé [dezabije] 〔法〕女便服；女睡衣

design [dɪˈzaɪn] 设计；图案；款式；花样
    abstract *design* 抽象图案
    clothing template *design* 服装模板设计
    colour *design* 色彩设计
    composite *design* 结构设计
    computer aided *design*(CAD) 计算机辅助设计
    conception of *design* 设计构思
    decorative *design* 装饰设计；装饰图案
    design concept 设计理念；设计观念
    design reduction(enlargement) 图案缩小（放大）
    design room 花样间；设计室，打样间
    detail *design* 细部设计；细节设计
    digital apparel *design* 数字化服装设计
    dress(costume)*design* 服装设计
    3D virtual *design* 三维虚拟设计
    elaborate *design* 精心设计
    embroidery *design* 绣花图案设计
    fabric *design* 衣料设计
    fashion *design* 时装设计
    form *design* 外型设计
    graphic *design* 平面造型设计
    Original *Design* Manufacturer(ODM) 原始设计制造商，贴牌设计生产
    package *design* 包装设计
    pattern *design* 花纹图案，纹眼；图案设计；板型设计；排纸样
    process *design* 工艺设计
    stereotype *design* （服装）板型设计
    technological *design* 工艺设计
    textile *design* 纺织品设计
    the latest *design* 最新设计
    women's(men's) *design* 女（男）装设计

designer [dɪˈzaɪnə] 设计员，设计师；制图（打样）人
    costume(apparel) *designer* 服装设计师
    dress *designer* 女装设计师
    fashion *designer* 时装设计师，时尚服饰设计师
    shoe(hat) *designer* 鞋（帽）样设计师

designing [dɪˈzaɪnɪŋ] 设计，构思；绘

制
 costume *designing* 服装设计
detail ['di:teɪl] （服装或资料的）细目、细节；（设计）详图；零（配）件
 big *details* （服装）细部扩大（如随宽领流行而加宽袋盖，加大折边及用较大纽扣）
 buttoning *detail* 扣配件
 flounced *detail* 荷叶边饰
detergent [dɪ'tɜ:dʒənt] 清洁剂；洗涤剂；去垢剂
device [dɪ'vaɪs] （有特定用途的）设备；装置；器件
 automatic needle positioning *device* 自动停针位装置
 automatic thread wiping (trimming) *device* 自动拨（剪）线装置
 feeding *device* 送料装置
 lubricating *device* 润滑装置
 material trimming *device* 切料装置
 oil suction *device* 吸油装置
 punching *device* 开孔装置
 safety *device* 安全装置
 smart wearable *device* 可穿戴智能设备（具有计算处理能力的新型高科技设备，越来越多地应用于服装行业）
 stitch regulating *device* 针距调节装置
 thread tension *device* （缝纫机上的）夹线器
diadem ['daɪədem] 王冠，冕
diaper ['daɪəpə] 菱形花纹图案；菱形花纹织物；尿片
 disposable *diaper* 一次性纸尿片
die [daɪ] 型，模子；裁剪型板；冲裁刀具

 cutting *die* （制鞋等用）裁刀
dimension [dɪ'menʃən] 尺寸，尺码，尺度
discordance [dɪs'kɔ:dəns] 〔设〕不调和
dishabille [ˌdɪsə'bi:l] 便服；穿着便服（或睡衣）
disharmony [dɪs'hɑ:məni] 〔设〕不调和，不协调；不和谐
diversity [daɪ'vɜ:səti] 〔设〕多样化；多元化；差异
 culture *diversity* 文化多元化；文化差异
dobby ['dɒbi] 小提花织物
doeskin ['dəʊskɪn] 驼丝锦（礼服毛料），礼服呢；仿麂皮织物；软山羊皮
dolman ['dɒlmən] 土耳其式长袍；（宽大袖）女外套
domette [də'met] 双面厚绒布；盖肩衬
dominance ['dɒmɪnəns] 〔设〕统一化，支配
doublet ['dʌblɪt] （旧时）男式紧身上衣；紧身背心，马甲
douppioni (= doupioni) [du:pɪ'əʊni] 双宫丝，双宫绸
down [daʊn] 绒毛，羽绒
 duck *down* 鸭绒
 fake *down* 人造羽绒，仿羽绒
 goose *down* 鹅绒
draft [drɑ:ft] 草图，图样
 net *draft* （无放缩的缝纫）净样图
drafting ['drɑ:ftɪŋ] 制图；打样
 computer aided design and *drafting* (CADD) 计算机辅助设计和制图

pattern *drafting* 原型制图；（服装）制板
draftsman [ˈdrɑːftsmən] 制图员；打样人
dralon [ˈdreɪlɒn] 特拉纶，德拉纶，德绒（一种优异的超级腈纶纤维）
drapability [dreɪpəˈbɪləti] （衣服、布料的）悬垂性
drape [dreɪp] 褶皱；褶裥；服装式样；（布料的）悬垂性；（常用复）窗帘
    side *drapes* 边褶，胁褶
    suit with English *drape* 英式套装
draper [ˈdreɪpə] 布料商；女服裁剪师；立体裁剪师
drapery [ˈdreɪpəri] 〔英〕布匹；布业；服装；服装业
draping [ˈdreɪpɪŋ] 立体裁剪
drawer [ˈdrɔːə] 绘图人；〔复〕衬裤；三角裤；长内裤
    bathing *drawers* 游泳裤
    bottom *drawer* 〔英〕女人为结婚而储存的衣物
    cotton *drawers* 男汗裤
    fashion *drawer* 时装绘图师
    under *drawers* 衬裤，内裤
drawing [ˈdrɔːɪŋ] 绘图，图样，图纸
    charcoal *drawing* 素描
    cutting *drawing* 裁剪图
    decoration *drawing* 图案设计
    design *drawing* 设计图
    effect *drawing* （服装设计）效果图
    fashion *drawing* 时装画，时装款式图，效果图
    outline *drawing* 轮廓图，草图
    resolving *drawing* （衣服部件结构）分解图
    scale *drawing* （按比例的）缩尺图
    schematic *drawing* 示意图
    single line *drawing* 单线绘图
    structural *drawing* 结构图
drawnwork [ˈdrɔːnwɜːk] 抽绣，抽纱；抽绣品，抽纱品
drawstring [ˈdrɔːstrɪŋ] （衣、裤的）束带，拉绳
    hem *drawstring* 下摆拉绳
    hood *drawstring* 风帽拉绳
    waist *drawstring* （衣服）腰部拉绳
dreadnaught (= **dreadnought**) [ˈdrednɔːt] 厚呢大衣；厚呢
dress [dres] 女服；连衣裙，〔粤〕衫裙；服装（指外衣）；服饰，衣着，衣冠；装束；礼服；童装
    academical *dress* 大学礼服
    additive *dress* 多层式女服
    afterfive *dress* 夜生活服（下午五时以后的晚会服）
    afternoon *dress* 日间礼服；便宴服
    A-line *dress* A形连衣裙
    anti-radiation maternity *dress* 防辐射孕妇装
    apron *dress* 连衫围裙装；童围裙衫
    asymmetric *dress* 不对称连衣裙
    babydoll *dress* 婴幼服装，娃娃装
    baby *dress* 婴儿装
    balloon *dress* 气球连衣裙装（短而蓬松）
    bare-back *dress* 露背装
    basic *dress* 基本型服装
    bathing *dress* 泳装
    battle *dress* 军制服；卡其布军装，

战地服装
beaded *dress* 珠饰连衣裙
bicolored *dress* 双色连衣裙
bikini *dress* 比基尼连裙装（露肩，超短）
body *dress* 紧身连衣裙
Bohemian *dress* 波希米亚连衣裙（以印花、刺绣、蕾丝、流苏等作装饰，呈伞状或拼接的过踝连衣裙）
bow-tie *dress* 蝴蝶结系带连衣裙
bridal *dress* 新娘礼服
bubble *dress* 泡泡连衣裙（上紧下膨）
camisole *dress* 花边吊带裙（露肩）
casual *dress* 便装
ceremonial *dress* 礼服
Chanel *dress* 香奈儿女服
coat *dress* 开襟明纽女式长服
cocktail *dress* 鸡尾酒会服，晚礼服
collarless short-sleeved *dress* 无领短袖连衣裙
coloured *dress* 彩衣
court *dress* 朝服；大礼服
couple *dress* 情侣装
covert *dress* （狩猎、骑马穿）轻便短外套；避尘衣
craft *dress* 工艺服装
Creole evening *dress* 古法式晚礼服
crossover V-neck *dress* V形叠领口连衣裙
Crystal Millennium *dress* 水晶千禧裙（名师为千禧年创制的露背连衣裙，水晶烟花图案，紫茄色调）

dancing *dress* 舞蹈服
denim smock *dress* 牛仔缩褶连衣裙
dinner *dress* 晚宴服；晚餐服
disposable *dress* 一次性服装
dress code 着装规范，着装要求
dress culture 服饰文化
dress museum 服饰博物馆
dress worsted 精纺女衣呢
embroidered *dress* with pearls 珠绣服装
evening *dress* （女）夜礼服；燕尾服
everyday *dress* 便服
fancy *dress* 奇巧服装，化装服
fashionable *dress* 时装
fashionable *dress* performance 时装表演
fishnet *dress* 连体网衣；网眼连衣裙
flamenco *dress* 吉卜赛连裙装（大裙摆并饰有荷叶边）
flared bottom *dress* 大摆连衣裙
formal *dress* 礼服
fox *dress* 狐皮服装
full *dress* 大礼服；燕尾服
Gigi-*dress* "吉吉"少女连衣裙（长宽袖，丑角领，短至膝）
girl *dress* 女童装
granny *dress* 阿婆服；婆婆裙（长至足踝的宽松连衣裙）；宽松装
gypsy *dress* 吉卜赛装（皱边，缨穗装饰，丝索或绳索腰带）
halter *dress* 挂脖式露背连衣裙
handkerchief *dress* （四方布拼缝的）手绢服装
high rise *dress* 高腰装

house(home) *dress* 居家服,家常服
hunting *dress* 猎装
informal *dress* 便服
irregular *dress* 不规则连衣裙
Japanese *dress* 和服
jersey *dress* 针织服装;针织连衣裙
jumper *dress* 无袖连衣裙,背心裙;套衫连衣裙
knit *dress* 针织服装;针织连衣裙
leopard print *dress* 豹纹连衣裙
lined *dress* 夹衣
little black *dress*(LBD)(香奈儿时代)小黑裙;黑色迷你裙;基本黑色裙装
long(short,half)sleeve *dress* 长(短、中)袖连衣裙
long-long *dress* 特长服装,超长装
low *dress* 女低领衣;袒胸装
low cut back *dress* 露背连衣裙
mandarin *dress* 旗袍
marten(mink) *dress* 貂皮服装
maternity *dress* 孕妇装
maxi *dress* 超长连衣裙
mercenary *dress*(15~16 世纪)雇佣兵服装
modern *dress* 现代服装
morning *dress* 常礼服
night *dress*(女、童)睡衣
nurse *dress* 护士服
off the peg *dress* 成衣
off-the-shoulder *dress* 露肩衣
official *dress* 官服,朝服
one-piece *dress* 连衣裙
one shoulder *dress* 单肩(带)连衣裙

pageant *dress* 庆典礼服
panel *dress* 派内尔连衣裙(缀有长方块饰布布裙)
pants *dress* 带裤裙的连裙装
paper *dress*(一次性的)纸衣服;纸服装
party *dress* 晚会服;宴会服;社交服
peasant *dress* 村姑连衣裙;农妇装
pencil *dress* 铅笔装(细长、直筒轮廓)
Peter Pan collar *dress*(彼得·潘)小圆领连衣裙
Peter Thomson *dress* 彼得·汤姆森连衣裙(20 世纪初风行美国的水手领连衣裙)
pinafore *dress* 连衫围裙装
pleated *dress* 打褶连衣裙
preteens' *dress*(13 岁以下)少年(少女)服
princess line *dress* 公主线(紧身)连衣裙,公主裙
print *dress* 印花布女装
red bridal *dress* 红色婚礼旗袍
rich *dress* 盛装
sack *dress* 女布袋装(宽松短上装);松身衣裙
safari *dress* 猎装式连衣裙
sand-washed silk *dress* 砂洗丝绸服装
school *dress* 女学生(连衣)裙
semi-evening *dress* 简略式夜礼服
separate *dress*(女)套装
sequined *dress* 缀亮片礼服;亮片连衣裙
service *dress*(SD)制服;军便装
sheath *dress* 紧身连衣裙

shift *dress* 可变化的连衣裙；直筒式连衣裙
shirt *dress* 衬衫连衣裙（上身似男衬衫）
shirtwaist *dress* 衬衫腰线连衣裙；仿男式（束腰）衬衫裙
shoestring strap *dress* 吊带连衣裙
skate *dress* 溜冰连衣裙（长袖紧身短喇叭裙）
skin *dress* 贴身装，紧身装
sleeveless slim-fit *dress* 无袖紧身连衣裙
sun *dress* 太阳装（敞肩露背连衣裙装），（吊带）背心裙
swag *dress* 低垂衣服（柔软衣料随身体曲线下垂）
sweater *dress* 针织连衣裙
sweater-top *dress* 毛衫连裙装（一种上身为毛衫的连衣裙）
swim *dress* 泳装式连裙衫
tank *dress* （针织）大圆领口背心裙
tennis *dress* 网球装
tent *dress* 蓬松连衣裙；（塔型）天幕装
toddlers' *dress* 幼儿装
torso *dress* 低腰紧身连衣裙
tricot sailor *dress* 针织水手裙（香奈儿时代流行款式）
tri-tone *dress* 三色连衣裙
tube *dress* 露肩装，筒形装；筒形连衣裙
tulip *dress* 郁金香连衣裙（围裹式紧身下摆裙装）
tunic *dress* 古希腊、罗马人穿的一种宽袍（短袖或无袖，束腰，长至膝部）；束腰连衣裙（袍）

twofer *dress* （错色）双层连衣裙（看似穿两件，实为一件）
two (three)-piece *dress* 两（三）件套装
two-tone *dress* 双色连衣裙
undershirt *dress* 内衫连衣裙
U-neck *dress* U形领口连衣裙
unlined *dress* 单衣
wedding *dress* 结婚礼服
weird (bizarre) *dress* 奇装异服
Western *dress* 西装
woman's *dress* 连衣裙
working *dress* 工作服
wrap *dress* 卷裹连衣裙

dresscoat [ˈdreskəʊt] 礼服
dresser [ˈdresə] 服装讲究者；服装（管理）员；梳妆台
   best *dresser* 十分讲究衣着的人
dressiness [ˈdresnɪs] 讲究穿着，服装时髦
dressing [ˈdresɪŋ] 穿衣；穿着；化妆；装饰，修饰
   dressing mirror 穿衣镜
   dressing of cut parts 裁片修边
   dressing room 化妆室
dressmaker [ˈdresmeɪkə] 女（或童）服裁缝
   dressmaker's ham 熨烫馒头（熨烫衣胸部和裤臀部的烫垫）
drill [drɪl] 斜纹布，〔粤〕斜布，卡其布；钻孔；钻孔器
   cloth *drill* 钻布针，定位针
   cotton down-proof satin *drill* 纯棉防绒缎
   dyed *drill* 染色卡其布
   florentine *drill* 三上一下斜纹布
   ply yarn *drill* 线卡（其）

single yarn *drill* 纱卡（其）
truean *drill* 涤/棉卡其，涤卡
white(bleached) *drill* 漂白卡其布
drilling [ˈdrɪlɪŋ] 〔工〕（在裁片上）钻孔；斜纹布，卡其布
drone-brella [ˈdrəunbrelə] 自动悬浮伞（一种不用手抓，悬在头上方随人移动的伞，伞柄上装有用手机或遥控器控制伞的装置）
drysuit [ˈdraɪsuːt] 潜水服
duck [dʌk] 帆布，粗布；〔复〕帆布衣裤
    canvas *duck* 粗帆布
    combed *duck* 精梳帆布
    linen *duck* 亚麻帆布
    shoe *duck* 制鞋帆布
dud [dʌd] 衣服
duffel [ˈdʌfəl] (= **duffle** [ˈdʌfl]) 粗厚呢料；（旅行者，狩猎人等的）行囊；行李袋
    duffel cloth 起绒粗呢
dummy [ˈdʌmi] 人形模型，假人
    dummy try-on （在假人上穿粗缝衣服）试身
dungaree [ˌdʌŋɡəˈriː] 粗棉布，劳动布；〔复〕粗布工作服；牛仔裤；工装裤；婴幼背带裤
duos [ˈdjuəuz] 〔复〕不同布料组合的调和套装（上衣和裤子）

durability [ˌdjuərəˈbɪləti] （布料等）耐用性
dustcoat [ˈdʌstkəut] 防尘罩衣，罩衫
duster [ˈdʌstə] 〔美〕防尘衣，风衣（= duster coat）
dye [daɪ] 染色；染料
    azo *dye* 偶氮染料
    azo free *dye* 非偶氮染料，环保染料
    basic(acid) *dye* 碱（酸）性染料
    cation *dye* 阳离子染料
    direct *dye* 直接染料
    disperse *dye* 分散染料
    optical *dye* 荧光染料
    pigment *dye* 涂料染色
    reactive *dye* 活性染料
    sulphur *dye* 硫化染料
    vat *dye* 还原染料
dyeing [ˈdaɪɪŋ] 染色（工艺）
    garment *dyeing* 成衣染色
    piece *dyeing* （布）匹染
    random *dyeing* 多色间隔染色
    space *dyeing* 间隔染色（同一条纱线上间断染色）
    tie *dyeing* 扎染，捆染
    uneven *dyeing* （布料）染色不匀
    yarn *dyeing* 纱线染色
dyestuff [ˈdaɪstʌf] 染料；颜料

E

earcap [ˈɪəkæp] （御寒用）耳套
eardrop [ˈɪədrɒp] 耳坠；耳饰
    crystal *eardrop* 水晶耳坠
    sapphire *eardrop* 蓝宝石耳坠
    shell *eardrop* 贝壳耳坠
earflap [ˈɪəflæp] (= **earlap** [ˈɪəlæp] = **eartab** [ˈɪətæb]) （帽上可放下护耳御寒的）耳扇，帽瓣

earmuffs [ˈɪəmʌfs] 〔复〕（御寒用）耳套

earpiece [ˈɪəpiːs] （帽上的）护耳片

earring [ˈɪərɪŋ] 耳环；耳饰

 Bohemian *earring* 波希米亚耳环（异域风情，奇巧时尚）

 clip-on（button）*earring* 夹（扣）式耳环

 diamond *earring* 嵌钻（石）耳环

 drop *earrings* 垂挂式耳环

 euroball *earrings* 球形耳环

 gold pearl *earrings* 嵌珠金耳环

 hoop *earring* 大圈耳环，耳圈

 jewel *earring* 宝石耳环

 rhinestone *earring* 仿钻耳环，水钻耳环，闪石耳环

 thread *earrings* 细条垂挂式耳环

 two-(tri-)tone *earrings* 双（三）色耳环

 wooden *earring* 木耳环

earwarmer [ˈɪəwɔːmə] （御寒用）耳套

ease [iːz] 放松；放宽；松份；容量；容位

easing [ˈiːzɪŋ] （尺寸，衣服）宽松；放宽；容位

easy-care [ˈiːzɪkɛə] （即洗即穿）免烫

easysuit [ˈɪzɪsjuːt] 便套装；（易穿的）婴幼连衫裤装

E-commerce [ɪˈkɒməs] 电子商务

 E-commerce live streaming 电商直播

Eco-Tex [ˌekəʊˈteks] 生态纺织，生态服装

edge [edʒ] （布、衣的）边

 clean *edge* 光边（脚口等部位无卷边，无明线迹）

 collar *edge* 领外口，领边

 collar *edge* asymmetric 〔检〕领边不对称

 crossing at the front *edge* 〔检〕止口搅（止口下部搭叠过多）

 edge covering 包边

 edge margin 止口

 fabric(cloth) *edge* 布边

 facing leans out of front *edge* 〔检〕止口反吐（里料外露）

 fringed *edge* （布）毛边；〔粤〕散口

 front *edge* 门襟止口

 front *edge* is out of square 〔检〕止口缩角

 front *edge* is uneven 〔检〕止口不直

 front *edge* is upturned 〔检〕止口反翘

 laser *edge* 激光（雕花）衣边

 piped *edge* （衣服）绲边

 purl *edge* 流苏边；荷叶边

 raw *edge* 毛边

 raw *edge* leans out of seam 〔检〕毛露（毛边外露）

 split at the front *edge* 〔检〕止口豁（止口上下豁开）

 straight(round) *edge* 直（圆）边

 tack *edge* 假缝边，粗缝边

 top *edge* 上边，上缘

 unmeet(clossed) front *edge* 〔检〕止口豁（搅）

edging [ˈedʒɪŋ] 饰边，绲边；缘饰

effect [ɪˈfekt] 效果；印象；外观；花纹

 moustaches *effect* 猫须（洗水手擦工艺名词）

three-dimensional effect 立体感
visual effect 视觉效果
eiderdown ['aɪdədaʊn] 鸭绒；鸭绒制品
elastic [ɪ'læstɪk] 橡筋带，松紧带，〔粤〕橡根
elastane [ɪ'læsteɪn] 弹性纤维；弹力棉；氨纶
elbow ['elbəʊ] 肘；（衣服）肘部
electrician [ɪlek'trɪʃən] 电工
elegance ['elɪɡəns] (=elegancy ['elɪɡənsi]) （举止，服饰，风格等的）雅致，漂亮；雅致漂亮的服饰等
element ['elɪmənt] 〔设〕元素；要素
 Chinese element 中国元素
 fashion element 时尚元素，流行元素
 oriental element 东方元素
 popular element 流行元素，盛行元素
emblaser [ɪm'bleɪzə] 激光绣花机，激光刺绣机
emblem ['embləm] （装饰于胸袋等处的）纹章，（刺绣）图案；徽章
embroider (=embroiderer) [ɪm'brɔɪdə] 绣花机，绣花器
embroidering [ɪm'brɔɪdərɪŋ] 绣花，刺绣
 appliqué embroidering 贴花绣，贴绣
 beaded embroidering 珠绣
 both-side embroidering 双面绣
 computerized embroidering 电脑绣花
 cord embroidering 嵌线绣花，嵌绣
 embroidering with pearls 珠绣
 full embroidering 满绣
 hand embroidering 手工绣花，手绣
 machine embroidering 机绣，〔粤〕车花
 net (eyelet) embroidering 网眼绣花
 outline embroidering 轮廓绣花
embroidery [ɪm'brɔɪdəri] 绣花，刺绣；刺绣法；绣制品；贴绣
 allover embroidery （服装）全身绣花，满花绣
 Arabian embroidery 阿拉伯刺绣（几何花纹，色彩明亮）
 Baruba embroidery 巴鲁巴刺绣
 bullion embroidery 金银丝绣花
 cross-stitch embroidery 十字刺绣，十字绣
 Danish embroidery 丹麦刺绣（一种白底白花的抽绣）
 darned embroidery 织补式绣花
 embroidery area 绣花部位
 embroidery frame 绣花绷架
 English embroidery 英国（网眼）刺绣
 gold and silver embroidery 金银线绣花
 Indian embroidery 印度刺绣（衲缝贴花等针绣，链式针迹）
 Japanese embroidery 日式刺绣（色线或金银丝线绣）
 misplaced embroidery 〔检〕绣花错位
 oriental embroidery 东方刺绣
 Paris embroidery 巴黎刺绣（用细白线在棱纹底上绣缎纹针迹）
 patch embroidery 贴布绣，贴绣
 raised embroidery 凸花刺绣

renaissance *embroidery* 透孔绣制品

Suzhou (Hunan, Guangdong, Sichuan) *embroidery* 苏(湘、粤、蜀)绣

emerizing [ˈemərarzɪŋ] (织物表面)磨毛,起绒

emphasis [ˈemfəsɪs] 〔设〕强调,强度感

enbifuku [enˈbɪfʊge] 〔日〕燕尾服

encasement [ɪnˈkeɪsmənt] 装箱;包装

ensemble [ɑ̃sɑ̃:bl] 〔法〕整体;总效果;整套服装;三件套(衣、裙、外套)

    beauty for *ensemble* 整体美

    ensemble costume 女套装

    knit *ensemble* 针织套装

epaulet(=epaulette) [ˌepəˈlet] 肩章;肩章形饰物(肩襻)

equipage [ˈekwɪpɪdʒ] (古时的)化妆品;服饰

    dressing *equipage* 全套化妆用品

equipment [ɪˈkwɪpmənt] 设备;器材;装置

    apparel *equipment* 服装设备

    assisting *equipment* 辅助设备

    automatic pressing *equipment* 自动整烫设备

    conveying *equipment* 输送设备

    cutting *equipment* 裁剪设备

    dry cleaning *equipment* 干洗设备

    equipment maintenance 设备维修

    grading and marking *equipment* 放码排唛架设备

    intelligent wearable *equipment* 可穿戴智能设备(具有计算处理能力的新型高科技设备,越来越多地应用于服装行业)

    Original *Equipment* Manufacturer (OEM) 原始设备制造商,(代工)贴牌生产,来(料)样加工

    special sewing *equipment* 专用缝纫设备

    washing *equipment* 水洗设备

    wrinkle-free *equipment* 抗皱设备

espadrille [ˈespədrɪl] 软底鞋;帆布便鞋;登山帆布鞋;(后跟有踝部系带的)平底凉鞋

eveningwear [ˈiːvnɪŋweə] 晚装

expansion [ɪksˈpænʃən] 张开;扩大部分;(衣服)下摆

extractor [ɪksˈtræktə] (洗衣)脱水机

    washer-*extractor* 洗水脱水机

eye [aɪ] 眼(针眼、扣眼等),孔;索眼;圈,环

    button *eye* 扣眼

    hook and *eye* 一副领钩,风纪扣,钩眼扣,〔粤〕乌蝇扣

    needle *eye* 针眼

    thread *eye* 线环,线襻

eyebrow [ˈaɪbraʊ] 眉;眉毛

    eyebrow picking (美容)修眉

eyeglasses [ˈaɪglɑːsɪz] 眼镜(=glasses)

eyelash [ˈaɪlæʃ] 眼睫毛

    eyelashes curler 卷睫毛器

    false *eyelashes* 假睫毛

eyelet [ˈaɪlɪt] (服装、鞋、帽用的)鸡眼,气眼圈(一种金属小圈片,穿绳带或透气用);针眼;小孔扣眼;网眼

    double-face *eyelet* 双面鸡眼

  eyelet end 扣眼收针；扣眼圆孔，凤眼
  net *eyelet* 网形鸡眼
  rectangular *eyelet* 长方形鸡眼
eyewear [ˈaɪweə] 眼镜（总称，包括框架眼镜和隐形眼镜）

F

fabric [ˈfæbrɪk] 织物；衣料；布
  acid-proof *fabric* 耐酸织物
  acrylic *fabric* 腈纶织物；腈纶布
  adhesive-bonded *fabric* 无纺织物，无纺布
  antibacterial deodorant *fabric* 抗菌防臭织物；抗菌防臭布
  anti-flaming *fabric* 阻燃织物；阻燃布料
  anti-radiation *fabric* 防辐射织物；防辐射布
  antistatic *fabric* 抗静电织物；抗静电面料
  anti-trauma *fabric* 防创伤织物；防创伤布
  anti-UV *fabric* 抗紫外线织物；防紫外线面料
  apparel *fabric* 衣料，面料
  basic *fabrics* 基本衣料（平布、华达呢、劳动布、绒布、针织布等）
  bast *fabric* 麻织物；麻布
  braided *fabric* 编织物；编带
  breathable water-proof *fabric* 透气防水织物
  brushed *fabric* 拉绒织物
  casual wear *fabric* 便装料，休闲衣料
  chamois *fabric* 仿麂皮织物
  checked *fabric* 格子布
  chemical fibre *fabric* 化纤织物，化纤布料
  chenille *fabric* 雪尼尔织物，绳绒线布
  coated *fabric* 涂层织物；涂层衣料
  composite(compound) *fabric* 复合布料
  cotton *fabric* 棉织物；棉布
  cotton *fabric* single side moisture transported 单向导湿棉布
  crepe *fabric* 绉织物
  decorative *fabric* 装饰布
  double-faced *fabric* 双面织物
  double mercerized *fabric* 双丝光布（纱、布先后丝光）
  down-proof *fabric* 防羽绒刺出织物；防绒布
  dralon *fabric* 德绒织物；德绒面料，德绒
  ecological *fabric* 环保织物；环保衣料，生态面料
  electric heat *fabric* 电热织物；电热布
  electrostatic induction *fabric* 静电感应织物
  etched(of burnt)-out *fabric* 烂花织物；烂花布
  eyelet *fabric* 网眼织物；网眼布
  fabric absorption 织物吸湿性
  fabric analysis 织物分析
  fabric breathability 布料透气性

fabric comfort 布料穿着舒适性
fabric end 尾子布，零头布
fabric fall 布料悬垂性
fabric flammability 织物可燃性
fabric for suit(dress) 套装（连衣裙）料
fabric relaxing 〔工〕松布（针织或弹性布料裁剪前一天必须逐卷松散出来，以防裁后回缩）
fabric resilience 布料回弹性
fabric spreading 〔工〕（裁剪前）拉布，铺料
fabric weight(composition, count, density, thickness) 布料的重量（成分、纱支、密度、厚度）
fake fur knitted *fabric* 人造毛皮针织布
figured *fabric* 提花织物；提花布
fine *fabric* 精细织物；细薄布料
flameproof(fireproof) *fabric* 防火织物，防火布料
flame(fire)-retardant *fabric* 阻燃布料
flax *fabric* 亚麻织物，亚麻布；麻纱布
fleeced *fabric* 起绒针织布，针织绒布
flocked *fabric* 植绒织物；植绒布
fragrant *fabric* 芳香型织物；有香味的布料
functional *fabric* 功能织物（指具有防火、防水、防静电等功能的织物）
fur *fabric* 仿毛皮衣料，人造毛皮
garment(apparel) *fabric* 服装面料，衣料
green *fabric* （环保的）绿色织物，绿色布料
hemp/cotton *fabric* 大麻/棉混纺织物；大麻/棉布
hemp/cotton knitted *fabric* 大麻/棉针织布
imitation wool *fabric* 仿毛织物
high performance *fabric* 高性能面料
high-pile knit *fabric* 长毛绒针织布
high technology *fabric* 高科技布料
imitated silk *fabric* 仿真丝绸
indigo blue printed *fabric* 靛蓝花布，蓝印花布
intelligent *fabric* 智能衣料（可以感知人和环境变化，并通过反馈做出反应和调整的高科技面料）
interlining *fabric* 衬布
interlock *fabric* 双罗纹针织布；棉毛布
interwoven *fabric* 交织布
iridescent *fabric* 闪光织物，闪色织物
jacquard *fabric* 提花织物；提花布
knitted *fabric* 针织物；针织布
laminated *fabric* 叠层布料，层压布料
light(heavy) *fabric* 轻薄（厚重）衣料（织物）
linen *fabric* 亚麻织物；亚麻布
linen/cotton mixed(blended) *fabric* 亚麻/棉交织（混纺）布
lining *fabric* 里料，里子布
magnetotherapy *fabric* 磁疗织物；磁性布
massotherapy *fabric* 按摩织物
mesh *fabric* 网眼织物；网眼布

micro suede bounding with knitting fabric 仿麂皮针织布复合布料
microfibre fabric 超细纤维织物；超细麦克布；桃皮绒
midfibre fabric 中长织物，中长布
mixed(blended) fabric 交织（混纺）织物；交织（混纺）布
moisture wicking (absorbing) fabric 吸湿排汗布料
moss microfibre fabric 苔（藓）纹桃皮绒
multi(double)-bar warp knitted fabric 多（双）梳栉经编针织布
napped fabric 起绒布料，拉毛布料
narrow fabric 带织物（松紧带、标签带等）
national traditional fabric 民族传统织物
non-woven fabric 非织造织物；非织造布
nylon/cotton fabric 锦/棉纺，锦/棉绸
nylon crinkle fabric 锦纶绉布
nylon fabric 锦纶织物；锦纶布
nylon PVC coated fabric 锦纶PVC涂层布
one(double)-sided fabric 单（双）面织物
organic cotton fabric （环保）有机棉布
peach fabric 细绒面织物；磨毛布
peachskin fabric 桃皮绒
physiotherapy health fabric 理疗保健织物
pile fabric 起绒织物，绒头织物；绒布

pleated fabric 褶裥织物；褶裥布
polyester/cotton fabric 涤棉混纺织物；涤棉布
polyester crinkle fabric 涤纶绉布；顺纡绉
polyester fabric 涤纶织物；涤纶布
polyester PVC coated fabric 涤纶PVC涂层布
polyester textured fabric 涤纶织物；涤纶布
polyester/viscose fabric 涤黏混纺织物；涤黏布（快巴的确良）
prolivon fabric 葆莱织物；葆莱布料；葆莱（新面料）
puckered fabric 绉纹织物；绉纹布
purl(pearl) fabric 双反面针织布
PVC/polyester fabric （做雨衣用）单面胶布
PVC/polyester/PVC fabric （做雨衣用）双面胶布
radioactive fabric 放射性织物
ramie/cotton blended fabric 苎麻棉混纺布
ramie fabric 苎麻织物；苎麻布
rayon fabric 人造丝织物；人造丝绸
rib fabric 罗纹针织物；罗纹布
rib jacquard fabric 罗纹提花针织布
sandwich fabric 夹心织物；衬垫织物
seaweed fabric 海藻纤维织物；海藻布料（保温保健，抑菌防霉的新型面料）
self-fabric 原身布，大身衣料

shell(outer) fabric （做服装的）面料
silk fabric 丝织物；丝绸
silk/acrylic blended fabric 丝腈混纺织物
silk/ramie(cotton) fabric 丝麻（棉）交织绸
slubbed fabric 竹节花式线织物，竹节布
soil-release fabric 防污织物；防污布
sorting out fabrics 〔工〕衣料分类
spandex silk knit fabric 丝弹力针织布
sportswear fabric 运动衣料
spreading fabric 〔工〕（裁剪前）铺料
spun-like fabric 仿纱型织物
spun rayon fabric 人造棉织物；人造棉布
spun silk fabric 绢丝织物；绢丝绸
stitch-bonded fabric 缝编织物
stretch(elastic) fabric 弹性织物；弹力布
suede fabric 人造麂皮织物
swimsuit fabric 泳衣料
synthetic fabric 合成纤维织物；合纤布
T/C fabric 涤/棉织物；T/C 布，棉涤纶
T/C yarn-dyed check fabric 色织涤/棉格布
tencel/nylon mixed fabric 天丝/锦纶交织布
terry knit fabric 毛圈针织布
textured yarn fabric 变形丝织物

thermal fabric 保暖织物
trueran fabric 涤/棉织物；涤/棉布
twill(plain, satin) weave fabric 斜（平、缎）纹织物；斜（平、缎）纹布
union(blended) chemical fibre fabric 交织（混纺）化纤织物
velvet fabric 绒头织物；丝绒，天鹅绒
washed fabric 水洗布
water(fire) proof fabric 防水（火）织物；防水（火）布
wax resist printed fabric 蜡防花布
weft(warp) knitted fabric 纬（经）编织物；纬（经）编布
wood(bamboo) jersey fabric 木（竹）纤维针织布料
wool/acrylic blended fabric 毛腈混纺呢绒
wool fabric 毛织物；呢绒
wool/nylon fabric 羊毛锦纶混纺呢绒
wool(silk, linen, cotton)-like fabric 仿毛（丝、麻、棉）织物
wool/viscose fabric 毛黏混纺呢绒
worsted(woolen) fabric 精（粗）纺毛织物；精（粗）纺毛料
woven fabric 机织物；机织布，有纺布，平织布
wrinkle fabric 起皱织物，折皱布
yarn-dyed jacquard fabric 色织提花布
yarn-dyed(printed, grey) interlock fabric 色织（印花，本色）棉毛布

face [feɪs] （布料的）正面；外观；

（人体的）脸面
  face-lifting　整容；拉皮去皱
  face pack　（美容用）面膜
  single(double) *face*　单（双）面
facing [ˈfeɪsɪŋ]　贴边，〔粤〕贴；镶边；领面；挂面；〔复〕军装上的领章，袖章
  catching *facing*　（衣裤的）里襟，〔粤〕纽子
  fly-*facing*　门襟贴边；裤遮扣布，裤门襟
  fold back *facing*　用原身布折边，〔粤〕原身出贴
  front *facing*　（衣服）挂面
  inside *facing*　内贴边
  lapel *facing*　驳领面，〔粤〕襟贴
  pocket *facing*　袋口贴边，〔粤〕袋贴
  right *facing*　（裤）里襟里子
  sleeve *facing*　大袖衩
  sleeve under *facing*　小袖衩
  uneven fly-*facing*　〔检〕裤门襟起绺
  unfavoring fly-*facing*　〔检〕裤门襟止口反吐
faconné [ˌfæsəˈneɪ]　〔法〕精巧小花纹织物；提花布
facsimile [fækˈsɪməli]　传真（通常缩写为 Fax, fax）
fad [fæd]　时髦装束（短时期内流行之服饰）；新奇的时尚
fading [ˈfeɪdɪŋ]　（布料、服装的）褪色，〔粤〕甩色
fagoting [ˈfæɡətɪŋ]　抽纱绣，抽绣
faille [feɪl]　〔法〕罗缎；线绢；绫纹绸；绨；（丝/棉交织物）
  polyester tissue *faille*　涤纶薄罗缎

falbala [ˈfælbələ]　（衣裙的）荷叶边
fall [fɔːl]　领子的翻下部分；（外衣的）宽下摆；大脚裤；面纱；（旧时）领带；领角衬；（织物）悬垂性
  top collar *fall*　领高
fal-lal [fælˈlæl]　（服装上的）装饰品，花饰
falsies [ˈfɔːlsɪz]　〔复〕（胸罩的）衬垫物；海绵胸罩；假乳房；假发，假胡须
fan [fæn]　扇子；粉丝（狂热爱好者）
  antique Victorian hand *fan*　维多利亚时期的古董手扇
  hand *fan*　手扇
  Japanese style hand *fan*　日式手扇
  Silk Chinese folding hand *fan*　中国丝绸折扇
fancywork [ˈfænsɪwɜːk]　刺绣品；编织品；钩针织物
fashion [ˈfæʃən]　样子；流行；流行式样；时髦，时尚；时装
  anti-*fashion*　反时装（反欧洲传统服装风格的时装潮流）
  avantgarde *fashion*　前卫流行
  basic *fashion*　（在流行中保持）基本式样
  China *Fashion* Week (CFW)　中国国际时装周
  China Graduate *Fashion* Week (CGFW)　中国国际大学生时装周
  classic *fashion*　经典时尚
  display of *fashion* dress　时装展示
  ecological *fashion*　生态服装，环保时装；生态时尚
  ethical *fashion*　道德时装；绿色时尚（强调绿色环保和公平贸易）
  famous brand and high class *fashion*

名牌高档时装

fashion chain store 时装连锁店
fashion cycle 流行周期
fashion element 流行要素
fashion festival 时装节
fashion forecast 流行预测
fashion history 服装史
fashion made of bamboo 竹制时装
fashion made of refuse （利用废料制的）垃圾时装
fashion magazine 时装杂志
fashion merchandiser 时装买手
fashion news 时装快讯
fashion photographer 时装摄影师
fashion plate 时装图片；穿着时髦者
fashion report 流行报道
fashion show 时装表演，时装秀；时装展览
fashion style 流行风格
fashion theme 流行主题
fashion trends 流行趋势；流行潮流；时尚风向标；时装信息
fashion week 时装周，时装节
fast fashion 快速时尚，快时尚（一种在短时间内，以低廉价格推出新潮服饰的流行商业模式）
following of the fashion 赶时髦
forward fashion 超前流行
gentle fashion 文雅式样
green fashion 绿色时尚
haute couture fashion 高级定制时装
high fashion 时髦款式；高档时装
international (China) fashion 国际（中国）时装
kinetic fashion 运动感的流行款式（用荧光涂料、金属等引起光线效果）
knitted fashion 针织时装
mass fashion 大众时装
militant fashion 军服式时装
millennium fashion and accessories 千禧年服饰
modern fashion 摩登式样；时下流行
neat fashion 清爽时装
occupational (professional) fashion 职业时装
out of fashion 过时，不时兴
plus one fashion 加一时装（在原款式上加一件服装，变换效果）
pop fashion 流行时尚；流行风尚
present fashion 时兴；时款
progressive fashion 进步流行
retro fashion 怀旧时装
see-through fashion 透视装式样
sexy fashion （暴露的）性感时装
street fashion 街头时装，大众时装
the latest fashion （服装的）最新式样；时装
top fashion 最新流行
unisex fashion 男女通用服装，中性服装
vegan fashion 素食时尚，纯素时尚
vintage fashion 复古时尚
year round fashion 四季穿的流行女服（几件衣裙组合穿用）

fastener ['fɑːsnə] 扣件；纽扣；揿纽；钩扣
belt fastener 腰带扣
self-gripping fastener 锦纶搭扣，

〔粤〕魔术贴
snap fastener 揿扣，按扣，子母扣，四合纽，〔粤〕唥纽
zip (slide) fastener 拉链
fastfashion [ˈfɑːstfæʃən] 快时尚（价格便宜且时尚，即平价时尚，系国际服装市场上的一种流行潮流）
fatigues [fəˈtiːgz] 〔复〕（士兵穿的）劳动服，工作服
fearnaught (= **fearnought**) [ˈfɪənɔːt] 粗绒大衣呢；粗绒大衣呢外套
feather [ˈfeðə] 羽毛；服装；服饰
features [ˈfiːtʃəz] 〔复〕容貌，相貌；脸型，特征；特色
fedora [fɪˈdɔːrə] 男式浅顶软呢帽；费多拉帽（音译），软毡帽；爵士帽
feed [fiːd] （缝纫时）进给，送布
automatic *feed* unit 自动进料装置
feed dog （缝纫机）送布牙，〔粤〕狗牙
feed plate （缝纫机）送布板
feed regulator 送料（针距）调节器
reversible *feed* 倒顺送料
upper and lower *feed* 上下进料
feel [fiːl] （织物、衣料、服装的）手感
soft *feel* 柔软手感
feeling [ˈfiːlɪŋ] （对艺术的）感受，（艺术品的）情调
aesthetic *feeling* 美感
fell [fel] （衣服的）平缝；折缝；兽皮；羊毛
feller [ˈfelə] （缝纫机的）平缝装置；折缝工人
felt [felt] 毛毡；毡制品
breast *felt* 绒布胸衬

lining *felt* 毛毡衬里，衬毡
press *felt* 熨烫垫布
under collar *felt* 领底绒衬，领底呢
fez [fez] 土耳其男毡帽（倒置桶形，有黑色缨穗的毡帽）
fibre [ˈfaɪbə] (= **fiber**) 纤维；纤维制品
acetate *fibre* 醋酸纤维
acetate staple *fibre* 醋酯纤维
acrylic staple *fibre* 腈纶短纤；腈纶棉
albumen *fibre* 白蛋白纤维
alginate *fibre* 海藻纤维（以海藻为原料，无毒，阻燃，可生物降解的绿色材料）
aloe *fibre* 芦荟纤维（新型功能纤维）
anion *fibre* 负离子纤维
animal *fibre* 动物纤维
antibacterial deodorant *fibre* 抗菌防臭纤维
anti-radiation *fibre* 防辐射纤维
antistatic *fibre* 抗静电纤维
apocynum *fibre* 罗布麻纤维（具有降压、抗菌等保健功能）
aramid *fibre* 芳香族聚酰胺纤维；芳纶纤维（具有高强度、高模量、耐高温、耐酸碱、重量轻等优良性能），芳纶
artificial *fibre* 人造纤维
artificial cellulose *fibre* 人造纤维素纤维
asbestos *fibre* 石棉纤维
bamboo *fibre* 竹纤维
bamboo pulp (carbon) *fibre* 竹浆（碳）纤维

banana *fibre* 香蕉茎纤维
bast *fibre* 韧皮纤维，麻纤维
carbon *fibre* 碳纤维
casein *fibre* 酪素纤维
ceiba *fibre* 木棉
cellulose *fibre* 纤维素纤维
ceramic *fibre* 硅酸盐纤维；陶瓷纤维
chemical *fibre* 化学纤维，化纤
chitin *fibre* 甲壳素纤维（具有保湿护肤、抑菌除臭功能）
chopped *fibre* 短纤维
coconut-charcoal PET staple *fibre* 椰炭纤维（具有吸臭、保温、防湿等功能）
composite *fibre* 复合纤维
coolmax *fibre* （美国）酷美丝纤维（吸湿排汗聚酯合纤）
coolplus *fibre* （中国台湾）酷帛丝纤维（吸湿排汗聚酯合纤）
cooltech *fibre* （中国台湾）新光合纤（吸湿排汗聚酯合纤）
copper ion *fibre* 铜离子纤维（一种抗菌纤维）
corn protein *fibre* 玉米蛋白纤维
cotton *fibre* 棉纤维
cuprene（cupro）*fibre* 铜氨纤维（再生纤维素纤维）
dralon *fibre* 德绒纤维
ecological *fibre* 生态纤维，环保纤维
elastane *fibre* 弹性纤维，弹力纤维
elastic（elastomeric）*fibre* 弹性纤维
far-infrared *fibre* 远红外纤维
fibre density 纤维密度

fibre fineness 纤维细度
fibre length 纤维长度
fibre strength 纤维强度
fine denier *fibre* 细旦纤维
flame-retardant *fibre* 阻燃纤维
fragrant *fibre* 芳香纤维
functional *fibre* 功能性纤维（具有抗静电、阻燃、吸湿排汗、抗菌除臭等功能的纤维）
glass *fibre* 玻璃纤维
green *fibre* 绿色纤维，环保纤维
health *fibre* 健康纤维
heat-generating *fibre* 发热纤维
high performance *fibre* 高性能纤维
high shrinkage *fibre* 高收缩纤维（受热后收缩率特别高的新型功能性纤维）
inorganic *fibre* 无机纤维
long（short）*fibre* 长（短）纤维
lyocell *fibre* 莱赛尔纤维，天丝纤维，天丝
macaroni（hollow）*fibre* 空心纤维，中空纤维
man-made *fibre* 人造纤维，化学纤维，化纤
metallic（metal）*fibre* 金属纤维
micro *fibre* 超细纤维，微纤维
mid（medium staple）*fibre* 中长纤维
milk protein *fibre* 牛奶蛋白纤维
modal *fibre* 莫代尔纤维（一种新型的高强和高湿模量纤维素纤维，吸湿力比棉质高50%）
modified *fibre* 变性纤维
moisture wicking（absorbing）*fibre* 吸湿排汗纤维
natural *fibre* 天然纤维

newcell fibre 纽富纤维（新一代绿色环保的黏胶长丝）
nylon staple fibre 锦纶短纤
original bamboo fibre 竹原纤维
peppermint fibre 薄荷纤维（具有抗菌除臭、清凉清香等功能）
pineapple fibre 菠萝纤维
PLA（polylactide）fibre 聚乳酸纤维（新型环保合成纤维）
polinosic fibre 富强纤维，富纤，虎木棉
polyacrylonitrile fibre 聚丙烯腈纤维，腈纶
polyamide fibre 聚酰胺纤维，锦纶
polyester staple fibre 涤纶短纤，涤纶棉，聚酯棉
polypropylene fibre 聚丙烯纤维，丙纶
polypropylene staple fibre 丙纶短纤
polystyrene fibre 聚苯乙烯纤维
polyurethane fibre 聚氨基甲酸酯纤维，聚氨酯纤维，氨纶
polyvinyl fibre 聚乙烯纤维
polyvinyl alcohol fibre 聚乙烯醇纤维，维纶
polyvinyl chloride fibre 聚氯乙烯纤维，氯纶
profiled fibre 异形纤维
prolivon fibre 葆莱绒纤维（具有中空结构的聚酯纤维，保温及导湿功能强）
protein fibre 蛋白质纤维
PTT（polytrimethylene terephthalate）fibre PTT 纤维（俗称弹性纤维，新型聚酯纤维，21 世纪的热门新材料）
rayon fibre 黏胶纤维

rayon staple fibre 黏胶短纤，人造（棉、毛）短纤
recycled fibre 再循环纤维，环保纤维
regenerated fibre 再生纤维
richcel fibre 丽赛纤维（具有优异综合性能的植物纤维素纤维，业界称之为"植物羊绒"）
seaweed（alginate）fibre 海藻纤维（以海藻为原料，无毒、阻燃、可生物降解，被业界称为"第三种纤维"的绿色材料）
seed fibre 种子纤维
smart（intelligent）fibre 智能纤维（能够感知外界环境或内部状态的变化并做出反应的纤维）
softwarm fibre 一种发热纤维
soybean protein fibre 大豆蛋白纤维
specialty fibre 特种纤维
staple fibre （人造）短纤维
super fine fibre 超细纤维
synthetic fibre 合成纤维
temperature regulating fibre 温度调节纤维，空调纤维
tencell fibre 坦西尔纤维，天丝纤维，天丝
textile fibre 纺织纤维
thermogear fibre （日本）旭化成发热纤维
topcool fibre （中国台湾）远东合纤（吸湿排汗聚酯合纤）
triacetate fibre 三醋酯酸纤维
ultra-high strength fibre 超高强纤维
urethane elastic fibre 氨基甲酸酯弹性纤维，氨纶
UV-proof fibre 防紫外线纤维
vegetable fibre 植物纤维

viscose *fibre*　黏胶纤维
wood pulp *fibre*　木浆纤维
wool *fibre*　羊毛纤维

fibril [ˈfaɪbrɪl]　原纤维；微小纤维
bamboo *fibril*　竹原纤维（天然纤维）

fibula [ˈfɪbjʊlə]　（古希腊、罗马的）扣衣针，饰针

fichu [ˈfiːʃuː]　（女用）披肩式三角薄围巾

fig [fɪg]　服装，穿着
full *fig*　〔口语〕华服，盛装
in full *fig*　穿上盛装；身着礼服

figure [ˈfɪɡə]　体型，身材；图形；图案
cloth with geometrical *figure*　印有几何图案的布料
erect *figure*　腆胸体型
fat *figure*　肥大体型
human *figure*　人体
lay *figure*　（展览用）服装人体模型
plane *figure*　〔设〕展示图，平面图
standard (special) *figure*　标准（特异）体型
stooping *figure*　曲背体型
stout (slender) *figure*　矮胖（苗条）身材

filament [ˈfɪləmənt]　长丝；单纤维
acetate *filament*　醋酸长丝
acrylic *filament*　腈纶长丝
nylon *filament*　锦纶长丝
polyester *filament*　涤纶长丝
polypropylene *filament*　丙纶长丝
viscose *filament*　黏胶长丝

filet [filɛ]　〔法〕网；网袋；发网；网格花边

filler [ˈfɪlə]　填充料（喷胶棉、羽绒等）
imitation silk floss *filler*　仿丝棉
soft and stiff *filler*　软硬棉
special (blended) material *filler*　特殊（混合）材料填料
spongy *filler*　蓬松棉，散棉
washing *filler*　水洗棉

filling [ˈfɪlɪŋ]　填充料；〔美〕纬纱

findings [ˈfaɪndɪŋz]　〔复〕〔美〕零碎的服装附件、工具（针、线、扣、里布等）
shoe *findings*　修鞋工具，附件等

finery [ˈfaɪnəri]　华丽的服饰，装饰，艳服

finette [faɪnt]　棉哔叽料（常用于睡衣料或衬料）

finger [ˈfɪŋɡə]　手指
finger weaving　手工编织

fingernail [ˈfɪŋɡəneɪl]　手指甲
imitation *fingernail*　人造指甲

finish [ˈfɪnɪʃ]　结束，完成；后工序工作；（纺织）后整理；精细加工
anti-bacterial *finish*　抗菌整理
anti-crease *finish*　防皱整理
anti-flaming *finish*　阻燃整理
anti-odour *finish*　防臭整理，防气味整理
anti-pilling *finish*　防起球整理
anti-radiation *finish*　防辐射整理
anti-shrinking *finish*　防缩整理
anti-soiling *finish*　防污整理
anti-static *finish*　抗静电整理
anti-UV *finish*　抗紫外线整理
aromatic *finish*　芳香整理
calender *finish*　轧光整理

clean *finish* 光边，收边，卷边，〔粤〕还口
doeskin *finish* 仿鹿皮整理
down-proof *finish* 防绒整理
durable *finish* 耐久性整理
easy-care *finish* 免烫整理
fabric *finish* 织物整理，布料后整理
flame(fire)-retardant *finish* 阻燃整理
hydrophilic *finish* 亲水性整理
moisture wicking *finish* 吸湿排汗整理
nano far-infrared *finish* 纳米远红外整理
stabilized *finish* （防缩、防皱）定型整理
super soft bio *finish* 生化超柔处理
water-proof *finish* 防水整理
wrinkles *finish* 褶皱整理（使布料有褶皱效果）
yarn *finish* 纱线整理
fit [fɪt] （服装）合身，适宜
good *fit* 很合身
keep *fit* 保持合适身材，健身
snug *fit* 舒适合身
tight-*fit* 紧身，贴身
fitter ['fɪtə] 裁剪试样工
fitting ['fɪtɪŋ] 试穿，试衣；假缝；装配；〔英〕尺寸
fitting room 试衣间
ill *fitting* 不合身
sleeve *fitting* 袖的假缝
flammability [ˌflæməˈbɪləti] （织物）可燃性
flange ['flændʒ] 镶边；装饰流苏；耳朵皮（前身里或挂面的小块拼接布）
flannel ['flænl] 法兰绒；棉法兰绒，绒布，〔复〕法兰绒衣服
cotton *flannel* 棉法兰绒
crepe *flannel* 绉纹法兰绒
elastic *flannel* 针织法兰绒
fancy *flannel* 花式法兰绒
oxford-grey mixture *flannels* 深灰法兰绒
tweed *flannel* 粗花呢；苏格兰呢
twilled *flannel* 斜纹法兰绒
wool/viscose *flannel* 毛黏混纺法兰绒
worsted *flannel* 精纺法兰绒，啥味呢
Yorkshire *flannel* 约克夏法兰绒（全毛优质英国呢绒）
zephyr *flannel* （丝毛混纺）轻薄法兰绒
flannelet(te) [ˌflænəˈlet] 绒布，棉织法兰绒
check *flannelet(te)* 格绒布
cotton yarn-dyed pre-shrunk checked *flannelet(te)* 全棉色织防缩格绒布
flannelet(te) with stripe and check 条格绒布
printed *flannelet(te)* 印花绒布
reversible *flannelet(te)* 双面绒布
single faced *flannelet(te)* 单面绒布
flap [flæp] 口袋盖；帽边
flap edge is uneven 〔检〕袋盖不直
flap lining leans out of edge 〔检〕盖里反吐（袋盖里子外露）
high/low *flaps* 〔检〕袋盖（左右）高低

pocket *flap*　口袋盖，袋盖
　　shoe *flap*　鞋盖，鞋舌
　　storm *flap*　门襟
　　under(top) *flap*　袋盖里（面）
flare [fleə]　（衣裙）喇叭形宽摆；长袜袜口；〔复〕喇叭裤，阔脚裤
　　flare legs　喇叭裤脚
flash [flæʃ]　（服装，外表的）浮华，虚饰；（军用）肩章；徽章
flatiron [ˈflætaɪən]　熨斗
flats [flæts]　〔复〕平底鞋
　　ankle tie *flats*　（踝部）打结平底鞋
flattie(=flatty) [ˈflæti]　平跟（或无跟）鞋（或拖鞋）
flaw [flɔː]　（布料）织疵，疵点；缺点
flax [flæks]　亚麻；亚麻布
fleabag [ˈfliːbæg]　睡袋
fleece [fliːs]　羊毛；绒头织物；起绒布，〔粤〕抓毛布；卫衣布；长毛大衣呢；绒衬里
　　breast *fleece*　绒布胸衬
　　cashmere *fleece*　开司米起绒布
　　coral *fleece*　珊瑚绒（一种用涤纶超细纤维做的新面料）
　　fleece in one side　蚂蚁布，蚂蚁绒
　　fleece with anti-pilling　防起球绒布
　　fusing jersey *fleece*　复合汗衫布
　　knitted *fleece*　针织起绒布
　　lacoste *fleece*　珠地卫衣布
　　micro suede bounding with polar *fleece*　仿麂皮摇粒绒复合布
　　mohair *fleece*　马海毛大衣呢
　　polyester polar *fleece*　涤纶双面绒布（做服装或衬里用，俗称摇粒绒）
　　rib double *fleece*　鱼鳞卫衣布
　　single(double) *fleece*　单（双）卫衣布
　　solid *fleece*　素色卫衣布
　　top making *fleece*　精纺用毛
　　woolen *fleece*　长毛大衣呢
fleuret [ˈfluərɪt]　粗纺斜纹厚呢；小花似的饰品
flexibility [ˌfleksəˈbɪləti]　（布料的）柔韧性
flip-flops [ˈflɪpflɒps]　（夹脚式）人字平底拖鞋
flocking [ˈflɒkɪŋ]　（织物上）植绒
　　embossed *flocking*　拷花植绒
　　jean *flocking*　牛仔布植绒
　　knitted cloth *flocking*　针织布植绒
　　leatherette *flocking*　人造革植绒
　　overall *flocking*　满地植绒
　　printed(plain) *flocking*　印花（素色）植绒
　　PVC *flocking*　PVC 皮植绒
　　rayon cloth *flocking*　人棉布植绒
　　static *flocking*　静电植绒
　　transfer *flocking*　转移植绒
floss [flɒs]　绣花丝线；丝绵
　　floss silk　丝线；绒线
　　silk *floss*　丝绵
flounce [flaʊns]　（衣裙）荷叶边
flow [fləʊ]　流动；（生产）流量
　　bunch *flow*　整包（多件）流水
　　flow chart　流程表
　　flow control　流程调控
　　one-piece *flow*　单件流水
flower [ˈflaʊə]　花；花卉
　　artificial *flower*　人造花
　　ball *flower*　球心花饰，花球
　　flower piece　花卉画

silk *flower*　绢花
velvet *flower*　绒花

fly [flaɪ]　（衣、裤的）纽扣遮布，门襟，门牌，〔粤〕纽牌
 exposed *fly*　（裤）外露门襟，明门襟，明门襟
 false *fly*　假门襟，暗纽牌
 fly opening　（裤）门襟，（裤）门襟
 fly tongue　门襟暗牌，门襟里搭襻
 front *fly*　门襟，门牌
 left *fly*　（衣）门襟；（裤）门襟，（裤）门襟；外襟
 left *fly* edge　（衣）门襟止口；（裤）门襟止口
 left *fly* lining　（衣）门襟里布，（裤）门襟里布
 mock *fly*　假门襟
 right *fly*　裤里襟；里襟；裤底襟
 top *fly*　衣门襟
 under *fly*　（衣、裤的）里襟，〔粤〕纽子
 unmatched front *fly*　（检）门，里襟长短不齐，〔粤〕长短筒，高低脚
 zip (button) *fly*　拉链（纽扣）门襟

flyaway [ˈflaɪəweɪ]　宽大不合身的衣服

fob [fɒb]　（男裤）表口袋；（怀表的）短链及饰物

foil [fɔɪl]　陪衬物；衬箔；衬色

fold [fəʊld]　折叠；褶
 dead *fold*　死褶
 double *fold*　双折边

folder [ˈfəʊdə]　（缝纫机的）折边器，〔粤〕拉筒蝴蝶，拉筒

folding [ˈfəʊldɪŋ]　折叠；折边

foofaraw [ˈfuːfərɔː]　（衣服）褶边；华丽的装饰

fool's cap (= **foolscap**) [ˈfuːlskæp]　（宫廷弄臣或丑角戴）滑稽帽；（处罚学生用）圆锥形纸帽

foot [fʊt]　（复数为 feet [fiːt] 脚；（缝纫机）压脚；〔粤〕靴仔；英尺
 edge *foot*　靠边压脚
 foot lifting amount　压脚提升高度
 half *foot*　单趾压脚
 hemming *foot*　卷边压脚
 length of *foot*　脚长
 multi-purpose *foot*　多功能压脚
 outside (inside) *foot*　外（内）压脚
 piping *foot*　滚边压脚
 presser *foot*　压脚
 roller *foot*　滚轮压脚
 shirring *foot*　抽褶压脚
 universal *foot*　通用压脚，万能压脚

footage [ˈfʊtɪdʒ]　以英尺表示的长度

footgear [ˈfʊtgɪə]　（总称）鞋袜

footies [ˈfʊtɪz]　〔口语〕婴儿连袜衫裤

footlets [ˈfʊtlɪts]　〔复〕脚掌套袜

footwear [ˈfʊtweə]　鞋袜类总称
 baby's *footwear*　婴幼鞋
 beach *footwear*　沙滩鞋

forearm [ˈfɔːrɑːm]　前臂，小臂

forehead [ˈfɒrɪd]　额，前额；前部

forepart [ˈfɔːpɑːt]　（衣服）前襟；前幅，前片
 right (left) *forepart*　右（左）前襟

form [fɔːm]　形状；（人的）体型；（服装）造型；表格
 bodily *form*　体型
 bust *form*　胸垫

dress form 款型，(半身)模型架
external form 外型
form finisher (熨烫用)人型机
sampling form 抽样单
streamline form 流线型
well-proportioned form 匀称体型
formal [ˈfɔːməl] (口语)夜礼服
former [ˈfɔːmə] 模型；成形设备
collar former 领型机，烫领机
collar-cuff-flap former 领、袖口、袋盖熨烫机
foulard [fuːˈlɑːd] 印花薄绸；印花薄绸制品
foundation [faʊnˈdeɪʃən] 基本纸型，基本内衣；女胸衣；塑形紧身内衣；里衬布；底层化妆
boned foundation 骨架式塑形内衣
foundation twill 斜纹里衬布
powder foundation (化妆的)粉底
stretch foundation 弹力塑形内衣；弹力女胸衣
foxing [ˈfɒksɪŋ] 制鞋用护条，围条
fraying [ˈfreɪɪŋ] 散边，毛边
frill [frɪl] 褶边；绉边；饰边
frillery [ˈfrɪlrɪ] 衣褶边
fringe [frɪndʒ] (服装上的)缘饰；流苏；毛边
frock [frɒk] (女)上衣；(童)外衣；连衫裙；男礼服大衣；工装；僧袍
chemise frock 无袖连衣裙；吊带裙
Chinese style frock 旗袍
girl's frock 女童裙衫
house frock (妇女)家居裙衫
priest frock 道袍
smock frock 长罩衫；工作服

frog [frɒg] 盘花纽扣；胸饰扣
Chinese frog 盘(花)扣(中国服装传统工艺)
front [frʌnt] 前胸；(衣服)正面；前身；门襟；(裤裙)前片
across front 上胸
box pleat front 明门襟
button front 里门襟
cardigan front 毛衫式门襟，开门襟
centre front 前身中心线
cross front 前胸宽
fly front 暗门襟
French front 暗门襟(法式衬衫门襟做法)
front cut away 大圆角前摆
front let 额饰
front of garment 衣襟
front round cut 圆角前摆
front stiff (礼服衬衫的)硬衬胸
front waistband (裤)腰头门襟
full front 前身
full open front 全开襟
hiking up at front 〔检〕前身起吊
left front 左前胸；左前身；左前片
open front 开门襟
plain front 暗门襟；(裤、裙)前片无褶裥
right front 右前胸；里门襟
shirt front 衬衫硬前胸
slanting front 偏襟
square front 方角前摆，平下摆
taped front 有挂面的门襟，贴门襟
zip(zip through) front 前装拉链，拉链开襟

fullness ['fʊlnɪs] 丰满度；宽松度；（打褶后形成的）隆起
fur [fɜː] 毛皮，皮子；〔复〕裘皮制品
 beaver rabbit *fur* 獭兔毛皮
 cat *fur* 猫皮
 dog *fur* 狗皮
 faux (fake) *fur* 假毛皮，人造毛皮
 ferret *fur* 雪貂皮
 fox *fur* 狐皮
 fun *fur* 人造毛皮，（针织）化纤花式毛皮
 goat *fur* 山羊毛皮
 hare *fur* 野兔毛皮，草兔毛皮
 imitation (artificial) *fur* 人造毛皮
 kangaroo *fur* 袋鼠毛皮
 kid *fur* 小山羊皮
 leopard cat *fur* 豹猫毛皮
 leopard *fur* 豹毛皮
 marmot *fur* 旱獭毛皮，土拨鼠毛皮
 micro suede bounding with lamb *fur* 仿麂皮羊羔绒复合布
 mink (marten) *fur* 貂皮
 natural *fur* 天然毛皮
 nutria *fur* 海狸鼠毛皮，狸獭毛皮
 plain (printed, embossed) fake *fur* 素色（印花、拷花）人造毛皮
 rabbit *fur* 兔毛皮
 red (kit) fox *fur* 赤（沙）狐毛皮
 sea (land) otter *fur* 海（水）獭毛皮
 sheep *fur* 绵羊皮
 squirrel *fur* 松鼠毛皮
 stonemarten *fur* 扫雪貂皮
 tiger *fur* 虎皮
 weasel *fur* 黄鼠狼皮
 wolf *fur* 狼毛皮
furbelow ['fɜːbɪləʊ] （衣裙）褶饰；边饰
furnishings ['fɜːnɪʃɪŋz] 〔复〕〔美〕服饰品
furring ['fɜːrɪŋ] 毛皮镶边；毛皮衬里
fusing ['fjuːzɪŋ] 熔化，熔合；黏合，黏衬
fustian ['fʌstɪən] 粗斜纹布；纬起毛织物，纬起绒织物

G

gabardeen (= gabardine) ['gæbədiːn] 斜纹织物；华达呢，轧别丁
 cotton *gabardeen* 棉华达呢
 extra-heavy *gabardeen* 加厚华达呢
 imitation wool *gabardeen* 仿毛华达呢
 midfibre *gabardeen* 中长华达呢
 polyester *gabardeen* 涤纶华达呢
 pure wool *gabardeen* 全毛华达呢
 reversible *gabardeen* 双面华达呢
 tencel/polyester *gabardeen* 天丝/涤纶华达呢
 tencel/wool *gabardeen* 天丝/毛华达呢
 viscose/polyamide *gabardeen* 黏/锦华达呢
 wool one-side *gabardeen* 全毛单面华达呢

wool/polyester *gabardeen* 毛/涤华达呢

wool/satin-backed *gabardeen* 全毛缎背华达呢

wool/viscose *gabardeen* 毛/黏华达呢

gabercord ['gæbəkɔːd] 棉华达呢

gaberdine ['gæbədiːn] 华达呢；宽大的粗布衣；工作服

gaiter ['geɪtə] 鞋罩；绑腿；皮脚套；高帮松紧鞋，有绑腿的高筒靴

galloon [gə'luːn] （装饰用）缎带；金银花边（丝带）

gallowses ['gæləuz] (=**galluses** ['gæləsɪz]) 〔复〕裤背带，吊带

galoshes [gə'lɒʃɪz] 〔复〕高筒套鞋；硬底鞋

gandoura [gādʊra] 〔法〕冈多拉装（非洲情调，有袖或无袖的宽长衣服）

garb [gɑːb] 服装；制服；装束

garment ['gɑːmənt] 衣服，外衣；〔复〕服装，衣着

    anti-pollution *garments* 防污染服装

    bast *garments* 麻类服装

    beaded embroidery *garments* 珠绣服装

    business *garments* 商务服装（白领着装）

    chamois *garment* 麂皮衣服

    chemical fibre *garments* 化纤服装

    chemical-biological protective *garments* 防生化服装

    China *garments* 中国服装

    Chinese *garments* 中国服装，中式服装

    Chinese and Western *garments* 中西式服装

    civil *garments* 民用服装

    combined *garments* 组合服装

    cotton *garments* 棉布服装

    cowhide *garments* 牛皮服装

    denim *garments* 牛仔服装

    drawn work *garments* 抽纱服装

    3D virtual *garments* 三维虚拟服装

    ecological cotton *garments* 环保棉布服装

    embroidered *garments* 绣花服装

    environmental *garments* 环保服装

    ethnical(ethnic) *garments* 民族服装，民俗服装

    formed plastics *garments* 泡沫塑料服装

    foundation *garment* 妇女全套紧身内衣（如束腰，胸罩，吊袜带等）

    fur *garments* 裘皮服装

    fur lining *garment* 毛皮衬里外衣

    garments association 服装协会

    garments construction 服装结构

    garments function 服装功能

    garments high education 服装高等教育

    garments hygiene 服装卫生学

    garment of plain silk gauze 素纱禅衣

    garments physiology 服装生理学

    garments robot 服装机器人

    garment setting 衣服成型

    garments standardization 服装标准化

    goatskin *garments* 山羊皮服装

    hand-crocheted *garments* 手工钩编

服装

integral *garments* 整体服装；全套服装

intelligent *garments* 智能服装（已通过技术增强的服装，可以增加其传统用途之外的功能，也称为高科技服装）

knitted *garments* 针织服装

leather *garments* 皮革服装

military *garments* 军用服装

(not) piece knitted *garments* （非）织片针织服装

organic *garments* 有机服装（面、辅料都环保）

outer *garment* 外套，外衣

pig nappa *garments* 光面猪皮服装

pigskin *garments* 猪皮服装

pig split *garments* 猪二层皮服装

pig suede *garments* 猪绒面皮服装

plastic *garments* 塑料服装

professional *garments* 职业服装

ramie/cotton *garments* 麻棉服装

road safety *garments* 道路安全服装（如反光背心）

series *garments* 系列服装

series of *garment* size designation 服装号型系列

sheepskin *garments* 绵羊皮服装

silk *garments* 丝绸服装

special *garments* 特种服装

stylish *garments* 时款服装，时装

towelling *garments* 毛巾服装

varnished cloth *garments* 漆布服装

woolen *garments* 毛呢服装

woven *garments* 机织服装

garniture [ˈɡɑːnɪtʃə] 服装；服饰

garter [ˈɡɑːtə] 吊袜带

pantie *garter* 内裤吊袜带

gather [ˈɡæðə] （衣裙）褶裥，皱褶；死褶；〔复〕碎褶

gatherer [ˈɡæðərə] 打褶器，打褶装置

gathering [ˈɡæðərɪŋ] 〔工〕打褶；抽碎褶

gauchos [ˈɡaʊtʃəʊz] 〔复〕（裤脚肥大，踝部束紧）女式牧裤

gauntlet [ˈɡɔːntlət] 宽口长手套（驾驶、击剑等防护用）

welder's *gauntlets* 焊工长手套

gauze [ɡɔːz] 纱罗，（棉、丝织）薄纱

embroidered *gauze* 绣花纱罗

gambiered Canton *gauze* 莨纱（香云纱）

Hangzhou silk *gauze* 杭罗

jacquard *gauze* 花纱

Lushan *gauze* 庐山纱

plain *gauze* 素罗

summer night *gauze* 夏夜纱

watered *gauze* 香云纱

Xihu *gauze* 西湖纱（桑蚕丝夏装料）

gear [ɡɪə] （年轻人）时髦服饰；衣服

flight *gear* 飞行服

party *gear* 社交服，聚会服

protective *gear* 防护服

sea *gear* 海洋服

georgette [dʒɔːˈdʒet] 乔其纱

chiffon *georgette* 雪纺乔其纱

cotton *georgette* 纯棉乔其纱

crepe *georgette* 顺纡绉

etched-out *georgette* 烂花乔其纱

printed (dyed) polyester *georgette* 印花（素色）涤纶乔其纱（俗

称柔姿纱)
rayon *georgette* 人造丝乔其纱
ghatpot [ˈɡɑːtpɒt] 绫，真丝绫
gigot [ˈdʒɪɡət] （袖隆松，袖口紧）羊腿袖
gilet [ˈʒiːleɪ]〔英〕马甲，背心；[ʒilɛ]〔法〕（男人穿在衬衫外的）坎肩，背心；女用短背心（短至腰线以上）；内衣，衬衣；女羊毛开襟衫
gimp [ɡɪmp] （装饰用）镶边带，嵌线；辫带
gingham [ˈɡɪŋəm] 格子布；柳条布
 acrylic *gingham* 腈纶格布
 print raising *gingham* （仿色织）印格绒布
 yarn-dyed raising *gingham* 色织格绒布
 zephyr *gingham* 轻薄格子布
girdle [ˈɡɜːdl] 带子；腰带；腹带；〔美〕（女用）紧身褡；束腹内裤
 garter *girdle* 吊袜腰带
 panty *girdle* 内裤吊袜带
girth [ɡɜːθ] （身体有关部位的）围长；腰围；大小；尺寸
 abdomen *girth* 腹围
 ankle *girth* 脚踝围
 bust *girth* 胸围；（女）乳胸围
 calf *girth* 腿肚围
 chest *girth* 胸围长（第一胸围）
 elbow *girth* 肘围
 head *girth* 头围
 hip *girth* 臀围
 knee *girth* 膝围
 neck *girth* 颈围；领围，领弯
 thigh *girth* 股围，腿围
 upper arm *girth* 上臂围
 waist *girth* 腰围
 wrist *girth* 腕围
glass [ɡlɑːs] 玻璃；镜子；〔复〕眼镜
 colour-changeable *glasses* 变色眼镜
 dressing *glasses* 穿衣镜
 protective *glasses*es 防护眼镜
 rimless *glasses*es 无框边眼镜
 smart *glasses* 智能眼镜（具有独立操作系统，可安装软件，联网互动，实现信息传输和图像识别等功能）
 sun *glasses* 太阳镜，墨镜
 toilet *glass* 梳妆镜
 wardrobe *glass* 衣柜镜，穿衣镜
 wide rimmed *glasses* 宽框边眼镜
glengarry [ɡlenˈɡæri] 传统的苏格兰便帽（帽后缀有两根短飘带）
glitter [ˈɡlɪtə] （装饰用）小闪光物
glove [ɡlʌv]〔常用复〕（五指分开）手套，〔粤〕手袜；拳击手套；棒球手套
 a pair of *gloves* 一双手套
 acrylic *gloves* 腈纶手套
 angora *gloves* 兔毛手套
 anti-contact cloth *gloves* 防接触布手套
 antistatic nylon *gloves* 锦纶防静电手套
 asbestos *gloves* 石棉手套
 baseball *gloves* 棒球手套
 berlin *gloves* 毛织手套
 bicycle *gloves* 骑单车手套
 boxing *gloves* 拳击手套
 bridal *gloves* 婚纱手套
 canvas *gloves* 帆布手套
 coral fleece *gloves* 珊瑚绒手套
 cotton interlock *gloves* 棉毛手套

cotton stockinette *gloves* 汗布手套
cotton yarn *gloves* （棉）纱手套
cricket *gloves* 板球手套
crocheted *gloves* 钩编手套
driver *gloves* 司机手套
electronic *gloves* 电子手套
emulsion *gloves* 乳胶手套
far-infrared *gloves* 远红外手套（具增温护肤作用）
fashionable *gloves* 时款手套
fleece *gloves* 绒布手套
fluorescent PVC *gloves* 荧光 PVC 手套
fur-lined *gloves* 毛皮里子手套
garden *gloves* 园艺手套
healthful *gloves* 保健手套
heat protective *gloves* 防热手套
ice hockey *gloves* 冰球手套
insulated *gloves* 绝缘手套
jacquard *gloves* 提花手套
kid(cape) *gloves* 羊皮手套
knitted *gloves* 针织手套
lace *gloves* （网眼）通花手套
ladies dress *gloves* 女士晚装手套，晚宴手套
latex disposable *gloves* 一次性乳胶手套
leather *gloves* 皮手套
leatheret *gloves* 人造皮革手套，仿皮手套
long *gloves* 长手套
open work *gloves* 网眼手套
opera length *gloves* （女士在观剧等社交场合戴的）长手套
pig(cow) split *gloves* 猪（牛）皮手套
printed (embroidered) *gloves* 印（绣）花手套
protective *gloves* 防护手套
racing *gloves* 竞赛手套，赛车手套
rubber *gloves* 橡皮手套
ski *gloves* 滑雪手套
slip-on *gloves* 无带扣手套；宽口半长手套
snowboard *gloves* 滑雪（板）手套
space *gloves* 航天手套
suede *gloves* 仿麂皮手套；羊皮手套
sweethearts *gloves* 情侣手套
welding *gloves* 焊工手套
woolen *gloves* 羊毛手套
working *gloves* 工作手套；劳工手套
X-ray proof *gloves* 防 X 射线手套

goatskin [ˈgəʊtskɪn] 山羊皮；山羊皮革；山羊皮袍

godet [gəʊˈdet] （衣裙上拼缝）三角形布片

gof(f)er [ˈgəʊfə] 皱褶，襞；烫皱褶的熨斗

goggles [ˈgɒglz] 〔复〕护目镜，（遮）风镜；泳镜；有色眼镜
ski *goggles* 滑雪护目镜

goods [gʊdz] 〔复〕商品；货物
cashmere knitted *goods* 羊绒针织品
consumer *goods* 消费品
cotton manufactured *goods* 棉织品
cotton yarn and cotton piece *goods* 棉纱棉布
daily-use *goods* 日用品
double-faced woolen *goods* 双面呢绒
dress *goods* （女、童）外衣料

dry *goods* 〔美〕纺织品；现成衣服
fleeced *goods* 单面起绒针织物
high-grade *goods* 高档商品
import(export) *goods* 进口（出口）商品
inferior *goods* 低档商品；劣质商品
jersey piece *goods* 匹头针织布
knit *goods* 针织品
leather *goods* 皮革制品
lycra woolen *goods* 弹力呢绒
piece *goods* （总称）布匹，匹头
printed *goods* 印花布
ready-made *goods* 成衣
silk(knit) *goods* 丝（针）织品
silk-cotton *goods* 绨；丝棉交织物
silk piece *goods* 绸缎
small leather *goods* 小皮件饰物
soft *goods* 〔英〕纺织品；〔美〕不耐用纺织品
spot *goods* 现货
standard(defective) *goods* 正（次）品
staple *goods* 大路产品，大路货
stock *goods* 库存商品，存货
stuff *goods* 毛织品
tricot *goods* 经编针织物；斜纹毛织物
woolen *goods* 呢绒
yard *goods* 按码出售的织物
gooses [ˈɡuːsɪs] 〔复〕长柄熨斗
gore [ɡɔː] 三角形拼布；辅助布
back *gore* （三角形）后裆拼布
underarm *gore* 袖下三角形镶布条
gorge [ɡɔːdʒ] 咽喉；领子；领缺口；领串口，〔粤〕襟领口位

gorge line is uneven 〔检〕领串口不直
gorget [ˈɡɔːdʒɪt] 盔甲的护喉，颈甲；衣领
gown [ɡaʊn] 长袍；长外衣；女睡衣
ball *gown* （女）舞会礼服
bath *gown* 浴衣
beach *gown* 海滨袍（泳后披穿）
blue cotton *gown* 蓝布大褂
bridal *gown* 新娘装，新娘礼服
button through *gown* 纽扣全开襟袍
Chinese *gown* 旗袍；长衫
coral fleece night *gown* 珊瑚绒睡袍
degree(academic) *gown* 学位袍，学位服
double *gown* 两面穿长袍
dressing *gown* 家居便服；化妆衣，晨衣
dust *gown* 罩袍
evening *gown* （裙长及地的）晚礼服
fur-lined *gown* 毛皮袍
home *gown* 便服
hooded fleece *gown* 风帽绒布袍
hostess *gown* 主妇服，女主人长袍
master's(doctor's, bachelor's) *gown* 硕（博、学）士学位袍
maternity nursing *gown* 产妇育婴袍
morning *gown* 晨衣，〔粤〕晨褛
night *gown* 睡袍
patient's *gown* （治疗中的）病人服
printed polar fleece *gown* 印花摇粒

绒袍
 short(long) gown　短（大）褂
 stuff gown　旧时律师袍
 surgical gown　（外科）手术衣
 tea gown　（茶会时穿）宽松女袍
 wedding gown　结婚礼服，新娘礼服，婚纱服
gradation [grəˈdeɪʃən]　〔设〕渐变；（颜色等的）层次，渐层
grader [ˈgreɪdə]　（纸样）放缩机
 multi-grader　尺码放缩仪
grading [ˈgreɪdɪŋ]　（纸样的）放缩，放码，推板，推档
 computer grading　计算机放码
 draft grading　直接起图放码
 pattern grading　纸样放缩，纸样放码
 radial grading　射线放码
 stack grading　叠层式放码
 track grading　轨迹式放码
graffito [grəˈfiːtəʊ]　〔复〕**graffiti** [grəˈfiːti]　涂鸦，涂鸦艺术（户外视觉字体设计艺术）
grain [greɪn]　纹理；布纹；（皮革等）粗糙面
 straight (cross, bias) grain　直（横，斜）纹
graphics [ˈgræfiks]　图像，图样，图形；图案；图案制图学；制图法
 clothing graphics　服装图样；服装图案学
 computer graphics　计算机图像；计算机图形学
graveclothes [ˈgreɪvkləʊðz]　寿衣，尸衣
greatcoat [ˈgreɪtkəʊt]　〔英〕厚外套，厚大衣

grey(=**gray**) [greɪ]　坯布，本色布；灰色，灰色衣服
 grey poplin (drill)　府绸（卡其）坯布；灰色府绸（卡其）
 ramie/cotton grey plain　麻棉平布坯
gripper [ˈgrɪpə]　五爪纽
grommet [ˈgrɒmɪt] (=**grummet** [ˈgrʌmɪt])　扣环；金属孔眼，鸡眼，垫圈
grosgrain [ˈgrəʊsgreɪn]　葛；罗缎
gross [grəʊs]　（一）罗（等于12打）
 a great gross of buttons　12罗纽扣（即1728粒）
grosuit [ˈgrɒsjuːt]　婴幼连衣裤
G-string [ˈdʒiːstrɪŋ]　狭条布三角裤，丁字裤，作三角裤的狭布条（黄色表演穿）
 embroidered mesh G-string　绣花网眼三角裤
G suit [ˈdʒiːsuːt]　抗超重飞行服，宇航服 (=gravity suit)
guide [gaɪd]　比尺（小型助缝工具）；导引物
 arm-spool pin pretension guide　（缝纫机）过线器
 thread guide　导线器，线耳
guimpe [gɪmp]　（穿在背心裙内）女内衣
gumboots [ˈgʌmbuːts]　〔复〕长筒胶靴，防水胶靴
gums [gʌmz]　〔复〕〔美〕长筒橡皮套鞋
gumshoe [ˈgʌmʃuː]　〔美〕橡胶套鞋；〔复〕橡胶底帆布鞋
gun [gʌn]　枪状物；喷雾器
 clean gun　去污枪；（熨烫用）喷

水枪
    spray *gun* 喷枪
    tagging *gun* （装标签用）胶枪，标签枪
gurus [ˈgʊrʊz] 古鲁衫（音译，低腰设计的上装）
gusset [ˈgʌsɪt] 三角形布片（填补、加固、拼放衣裤用）

crutch *gusset* 裤裆底拼布，裤衩布
underarm *gusset* （衣）腋下镶布
gutter [ˈgʌtə] 槽；缝纫偏离滑落，缉线上、下炕，〔粤〕落坑
gymnasterka [gɪmnʌˈsdrɔːkʌ] （19世纪）俄国军队罩衫（立领，双扣，双胸袋的套头衫）
gymwear [ˈdʒɪmwɛə] 运动服装

H

haberdashery [ˌhæbəˈdæʃəri] 〔总称〕男服饰用品；缝纫用品
habiliments [həˈbɪlɪmənts] 〔复〕制服；礼服；衣服
habit [ˈhæbɪt] 表示宗教级别的衣着；妇女骑装；（古语）衣服；〔法〕服装；衣服；男式上装
    habit de chasse 〔法〕猎装
    habit de soirée 〔法〕晚礼服
    *habit*s de travail 〔法〕工作服
    monk's *habit* 僧袍
    riding *habit* 女骑装（紧身上衣配长裙）
habotai(=habutae =habutai) [ˈhæbɒtaɪ] 纺绸，电力纺
    figured *habotai* 提花纺绸
    fuchun *habotai* 富春纺
    huachun *habotai* 华春纺
    interweaved *habotai* 交织纺绸
    polyester *habotai* 涤丝纺
    rayon *habotai* 人造丝纺绸，人造丝电力纺
    spun silk *habotai* 绢丝纺
haik(=haick) [heɪk] 阿拉伯人的白罩袍；阿拉伯妇女的白色绣花面纱

hair [heə] 头发；动物毛
    camel *hair* 驼毛，驼绒
    goat *hair* 山羊毛
    hair-dye 染发剂
    hair-net 发网
    hair oil 发油
    hair slide 发夹
    rabbit *hair* 兔毛
    specialty *hair* fibre 特种动物毛
    yak *hair* 牦牛毛
hairlace [ˈheəleɪs] 网眼花边；发网
hairpin [ˈheəpɪn] 发针，发簪，发夹
    emerald *hairpin* 碧玉簪
hairstyle [ˈheəstaɪl] 发型
hairstylist [ˈheəstaɪlɪst] 发型师
hallstand [ˈhɔːlstænd] 衣帽架
halter [ˈhɔːltə] （露肩、背）女套头背心，三角背心
Hanbok [ˈhʌnbɒk] 传统韩国服装，韩服
hand [hænd] 手；（衣料、皮革的）手感
    back of the *hand* 手背
    good *hand* 手感好
    hand feed （缝制中）手推动作

hand-made 手工制造
hand scrapping 手擦，擦砂（服装洗水工艺之一）
hand wheel （缝纫机头上的）手轮，上轮
handbag [ˈhændbæg] （女用）手提包，手袋
   alcantara *handbag* 假獴绒布（仿獴皮）手袋
   alligator leather *handbag* 鳄鱼皮手袋
   bamboo *handbag* 竹制手袋
   barrel *handbag* 桶形手袋
   beaded *handbag* 珠饰手袋
   canvas *handbag* 帆布手袋
   chain handle *handbag* 链带手袋
   drawstring *handbag* （袋口）束带手袋
   drum *handbag* 鼓形手袋
   evening *handbag* 晚宴手袋，晚会手袋，晚装手袋
   fashion *handbag* 时款手袋
   genuine leather *handbag* 真皮手袋
   jute *handbag* 麻质手袋
   linen *handbag* 亚麻手袋
   microfibre *handbag* 桃皮绒手袋
   mini *handbag* 袖珍手袋，迷你手袋
   nylon *handbag* 尼龙手袋
   patch leather *handbag* 拼缝皮手袋
   patent leather *handbag* 漆皮手袋
   plastic *handbag* 塑料手袋
   PVC leather *handbag* PVC皮手袋
   rattan *handbag* 藤编手袋
   seagrass *handbag* 海草手袋
   shoulder *handbag* 肩挎式手袋
   snakeskin *handbag* 蛇皮手袋
   underarm *handbag* 臂挟式手袋
handfeel [ˈhændfiːl] 手感
   post wash *handfeel* 洗水后手感
handicraft [ˈhændɪkrɑːft] 手艺；手工艺；〔总称〕手工艺品
handkerchief [ˈhæŋkətʃɪf] 手帕，〔粤〕手巾仔；围巾；头巾
   neck *handkerchief* 围巾；头巾
handle [ˈhændl] （衣料）手感
   firm *handle* 手感厚实
   hard *handle* 手感板硬
   harsh *handle* 手感粗糙
   soft *handle* 手感柔软
handling [ˈhændlɪŋ] 操作；管理；处理，〔粤〕执手
handmade [ˈhændmeɪd] 手工制品
handstitch [ˈhændstɪtʃ] 手缝线迹；珠边线迹
Hanfu [ˈhʌnfu] 汉服（又称汉装，为汉民族的传统服饰）
hanger [ˈhæŋə] 挂物架；挂钩；衣架
   clothes *hanger* 衣架，挂衣钩
   folding *hanger* 折叠式衣架
   garment on *hanger* (GOH) 衣服挂装（走货）
   gripper *hanger* 有夹衣架
   hip *hangers* 挂臀裤（低腰，裤带低束）；低腰裙
   trouser(coat) *hanger* 裤（衣）架
   wire *hanger* （金属）丝衣架
   wood (plastic, metal) *hanger* 木（塑料、金属）衣架
hangtag [ˈhæŋtæg] （服装上的）吊牌，挂牌，〔粤〕挂咭；（商品）标签
   plastic (paper, metal) *hangtag* 塑料（纸、金属）吊牌
   wrong(missing) *hangtag* 〔检〕错

（漏）挂吊牌

harmony [ˈhɑːməni] 〔设〕调和，协调，和谐

    dominant *harmony* 主色调和

    harmony and unity 〔设〕调和与统一

    harmony of contrast 对比调和

    harmony of identity 统一调和

    harmony of similarity 类似调和

    tone *harmony* 色调调和

hat [hæt] （有檐边的）帽子，大檐帽

    baker boy *hat* 面包小子帽（一种仿面包师帽的有檐时装帽）

    bamboo split *hat* 斗笠

    black gauze *hat* 乌纱帽（古代官帽）

    bobble *hat* （针织）绒球帽（帽顶饰有毛线绒球）

    bowler *hat* 圆顶礼帽，小礼帽

    bucket *hat* （有边）桶形帽

    bush *hat* 宽边丛林帽（澳大利亚士兵用）

    chateau *hat* 城堡帽（大帽檐麦秆帽）

    chimney-pot *hat* 高顶礼帽

    cloche *hat* 钟形女帽

    cocked *hat* 卷边帽；三角帽；两端尖的帽子

    cool *hat* 凉帽

    coolie *hat* 苦力帽

    cowboy *hat* 牛仔帽

    Davy Crockett *hat* 大卫帽（浣熊毛皮帽，帽后有尾巴）

    denim *hat* 牛仔（布）帽

    derby *hat* 常礼帽

    dress *hat* 大礼帽

    embroidered *hat* 绣花帽

    fedora *hat* 费多拉帽（一种浅顶软毡帽）；爵士帽

    felt *hat* 毡帽，呢帽

    field training *hat* 作（战）训（练）帽

    fishing *hat* 钓鱼帽

    floppy *hat* （宽帽檐）下垂软帽

    fur *hat* 毛皮帽

    gaucho *hat* 加乌乔牧人帽；牛仔帽

    gob *hat* 水兵帽

    gunny *hat* 麻编草帽

    gypsy *hat* （女童）宽边帽

    Hamburg *hat* 汉堡帽，中褶帽

    hard *hat* 防护帽；〔英〕礼帽

    hat and gown 衣冠

    hat band 帽圈

    hat block 帽木模

    hat brim 帽檐，帽边

    hat felt 帽坯

    hat making 帽子制作，制帽

    hat with zip pocket 带拉链小袋的帽子

    high *hat* 高帽；礼帽

    iron *hat* 〔美俚〕礼帽

    job *hat* 职业帽

    knitted *hat* 针织帽

    ladies'（children's）*hat* 女（童）帽

    leaf *hat* 斗笠

    leather *hat* 皮帽

    mesh *hat* 网眼帽

    miner's *hat* 矿工帽

    monk *hat* 僧帽

    Nehru *hat* 尼赫鲁帽

    newborn *hat* 新生婴儿帽

    Panama *hat* 巴拿马帽（麦秆制遮

阳帽)
Peruvian *hat* 秘鲁帽(帽檐宽,帽身浅,条纹布制)
picture *hat* 宽边花帽(羽毛装饰)
pirate *hat* 海盗帽(前面帽檐向上翻折,用缎带绲边的宽檐帽)
plug *hat* 高礼帽
polo(chukka) *hat* 马球帽
porkpie *hat* 卷边低平顶帽
priest's *hat* 祭司帽(平帽缘圆帽身,有黑色装饰带的帽子)
protective *hat* 防护帽
Quaker *hat* 贵格帽(教徒戴的低帽身卷边帽)
rain *hat* 雨帽
rattan *hat* 藤帽
sailor *hat* (童)水手帽;水兵帽
scarlet(red) *hat* 红衣主教帽
shovel *hat* (教士戴)宽边帽
silk *hat* 大礼帽
slouch *hat* 垂边帽,宽边软帽,软毡帽
snap brim *hat* 软宽檐礼帽
soft felt *hat* 礼帽,中褶帽
SPADE fedora *hat* 黑桃礼帽(因帽顶被模压出扑克牌黑桃图案而命名)
squash *hat* (可折叠)软呢帽
straw *hat* 草帽
suede waterproof *hat* 绒面革防水帽
sun *hat* 阔边太阳帽,太阳帽
sun-shade *hat* 遮阳帽;(空顶)遮阳帽檐
tall *hat* 大礼帽
telescope *hat* 望远镜型帽(帽冠似望远镜圆而凹陷的镜头部分

的宽檐帽)
top *hat* (高顶)丝礼帽,大礼帽
tin foil *hat* 锡(箔)纸帽
toreador *hat* 斗牛士帽
trilby *hat* 一种软毡帽
tropical *hat* 凉帽
tweed *hat* 粗花呢帽
Tyrolean *hat* 铁若兰帽(帽顶小,帽檐窄的软毛毡帽)
velour *hat* 丝绒帽
waterproof *hat* 防水帽
Western *hat* (美国)西部风格帽
wide-brimmed *hat* 宽边帽
hatband [ˈhætbænd] 帽圈
hatpeg [ˈhætpeg] (挂帽用)帽钉
hatstand [ˈhætstænd] (可移动的)帽架
hatting [ˈhætɪŋ] 制帽;制帽材料,帽料
head [hed] 头,头部
   sleeve *head* 袖头,袖口
headband [ˈhedbænd] 扎头带;头箍;束发带
headdress [ˈheddres] 帽子;头巾;头饰
headgear [ˈhedgɪə] 帽类,帽子;头饰;安全帽
headpiece [ˈhedpiːs] 帽子;头盔,头巾
headscarf [ˈhedˌskɑːf] 女用头巾
heel [hiːl] 脚后跟;(鞋、袜)后跟;〔复〕有跟鞋
   clear *heel* 透明鞋跟
   cone *heel* 锥形鞋跟
   cover *heel* 包皮鞋跟
   flared *heel* 细腰形鞋跟
   high stacked *heel* 细高跟

hosiery heel　袜跟
kitten heel　细低跟
low(high) heel　平（高）跟
scoop heel　勺形鞋跟
spike heel　（女皮鞋）高后跟
stiletto heel　特细高跟
wedge heel　楔形鞋跟

heelless [ˈhiːlɪs]　（鞋，袜）无后跟
heelpiece [ˈhiːlpiːs]　鞋后跟
heelplate [ˈhiːlpleɪt]　（钉于鞋后跟）金属鞋掌
heifer [ˈhefə]　海虎绒；小母牛皮
height [haɪt]　高度；身高
cap height　袖山高
collar height　领高
cuff height　袖头高
flap height　袋盖高
neck-rib height　罗纹领高
pocket point height　袋尖高
shoulder height　肩高
sleeve cap height　袖山高
slit height　（旗袍）衩高
waistband height　腰头高
yoke height　过肩高；（裙，裤）机头高

helmet [ˈhelmɪt]　头盔；防护帽
aramid helmet　芳纶头盔（含高性能芳纶纤维，防弹率极高的军用头盔）
ATV (all terrain vehicle) helmet　全地形车头盔；沙滩车头盔
balaclava helmet　（套头至颈部仅露双眼的）巴拉克拉盔式帽
baseball helmet　棒球头盔
bicycle helmet　自行车头盔
bulletproof helmet　防弹头盔
crash helmet　（骑摩托车等用）防撞头盔；安全帽
cross-country velmet　越野头盔
electric scooter helmet　电动车头盔
fireman's helmet　消防员头盔
full face helmet　全护式头盔，全盔
half face helmet　半盔（只盖住上半头）
helmet and armour　盔甲
kids helmet　儿童头盔
miner's helmet　矿工头盔
motorcycle helmet　摩托头盔
pilot (aviator) helmet　飞行员头盔
pith helmet　木髓遮阳帽（一种木纤维做的硬壳阔边太阳帽）
rattan helmet　藤盔
rugby helmet　橄榄球头盔
safety helmet　安全帽
ski helmet　滑雪头盔
space helmet　宇航员头盔
sports helmet　运动头盔
stateboard helmet　滑板头盔
sun helmet　硬壳太阳帽
ultraman helmet　奥特曼头盔

hem [hem]　（衣服）折边，〔粤〕摺脚；底边；贴边，吊边；下摆
baby hem　小卷边；细窄的双折边
blind-stitched hem　暗缝缲边下摆
bottom hem　下摆卷边
bouffant hem　蓬松边
cuff hem　袖口折边
decorative hem　装饰贴边
empty hem　〔检〕空边（棉服边部缺棉花）
faced hem　贴边下摆
fluted hem　绉褶波边下摆；波形衣边
French hem　法式衣边（毛边缝后

折至里面再缝一次）
 handstitch at the hem　手工缲边；珠边
 hem around　下摆围，摆围
 hem gauge　褶边规（一种标有各种深度和边褶的测量装置）
 pocket hem　口袋折边
 roll-down hem　下卷边
 selvage hem　布折边，布卷边
 shirttail hem　衬衫圆下摆
 topstitched hem　压明线下摆
 turn-up hem　反折边，反吊边
 uneven hem　〔检〕底边起绺，下摆不匀

hemline [ˈhemlaɪn]　（衣、裙下摆）底边；贴边；裤脚边；下摆线；脚口线；裙摆线
 uneven hemline　〔检〕裙摆线不齐，裙底边不圆顺

hemmer [ˈhemə]　（缝纫机）卷边器
 pocket hemmer　口袋卷边器

hemming [ˈhemɪŋ]　（缝纫中）卷边，〔粤〕还口；缝边；镶边
 blind hemming　暗缝卷边，暗缝缏边，〔粤〕挑脚
 hand hemming　手工缲边，〔粤〕手针挑脚
 hemming bottom　折缝底边（或裤脚口）
 hemming sleeve　绱袖
 hemming stitch front edge　〔工〕扳止口
 hemming stitch hem　〔工〕缝底边
 inside hemming　内贴边（缝纫）

hemp [hemp]　大麻；大麻纤维

hemstitch [ˈhemstɪtʃ]　卷边线迹；花饰线迹；镶边缘饰

herringbone [ˈherɪŋbəʊn]　人字呢；人字斜纹布；海力蒙（音译，精纺呢绒）；人字（鱼骨）花型；人字呢服装
 wool/polyester herringbone　毛涤海力蒙

hide [haɪd]　皮革
 buffalo hide　水牛皮
 cattle hide　牛皮
 mule hide　骡皮
 pig hide　猪皮
 sheep hide　羊皮

himobutton [ˈhɪmɒbʌtn]　直扣；布扣（布条做的纽扣）

hip [hɪp]　臀部；臀围
 hip bulging　（整形）隆臀
 hip-huggers　低腰紧身长裤
 middle(high, low) hip　中（上，下）臀围
 round hip　臀围
 scooped(prominent) hip　垂（翘）臀
 small(large) hip　小（大）臀
 waist to hip　腰臀高（腰围线至臀围线距离）；（裤）直裆

hipline [ˈhɪplaɪn]　臀围线

hipsters [ˈhɪpstəz]　（低及臀部的）低腰裤；低腰内裤

holdall [ˈhəʊldɔːl]　旅行手提包（袋、箱）

hole [həʊl]　孔眼；破洞；缺陷
 broken hole　〔检〕（衣服、布料上的）破洞
 needle hole　〔检〕（返工留下的）针孔，〔粤〕针窿

homespun [ˈhəʊmspʌn]　手织土布；手工纺织呢；手织大衣呢；火姆司本

(音译，粗纺花呢)

hood [hʊd] 风帽，兜帽；头巾
 academic *hood* 学位服领兜帽
 centre back line of *hood* 风帽帽顶线
 concealed *hood* in collar 后领内风帽
 fold-up *hood* 折叠式风帽
 front line of *hood* 风帽前口线；帽嘴线
 hood height 风帽高
 hood width 风帽宽
 hood with fur trim 镶毛皮边风帽
 master's（doctor's，bachelor's）gown and *hood* 硕（博、学）士服
 removable *hood* 可拆卸式风帽
 under line of *hood* 风帽下口线

hoodie [ˈhʊdi] 带风帽的针织上衣，针织风帽衫
 print *hoodie* 印花风帽衫
 pull-on *hoodie* 套头风帽衫
 stripe vee *hoodie* 条纹V领风帽衫
 zip thru *hoodie* 拉链开襟风帽衫

hook [hʊk] 钩，挂钩
 clothes *hook* 挂衣钩
 crochet *hook* 钩（编）针
 inner *hook* 内钩
 leather belt *hook* 皮带钩
 neck *hook* and eye 领钩，风纪扣
 trouser *hook* and eye 裤头钩扣，裤钩

hoop [hu:p] 箍（圈）状物；戒指；耳环，耳圈；〔复〕（旧时）裙箍，鲸骨圈
 colour gold *hoop* 彩金耳圈

hoopskirt [ˈhu:pskɜ:t] （旧时）有裙箍的女裙，箍撑裙

hopsock [ˈhɒpsæk] 板司呢，席纹呢

hose [həʊz] 长筒袜，短袜；（旧时）男紧身裤（从腰至脚）
 ankle *hose* 短袜
 athletic *hose* 运动袜
 Bermuda *hose* 百慕大长袜
 doublet and *hose* 男紧身衣裤
 full-fashioned *hose* 全成形（女）袜
 golf *hose* 高尔夫长袜
 half *hose* 半长筒袜；短筒袜
 jacquard *hose* 提花袜
 lycra *hose* 弹性莱卡袜
 opaque *hose* 厚袜；不透明袜
 openwork *hose* 网眼袜
 panty *hose* 连裤袜
 stretch *hose* 弹力袜

hose-tops [ˈhəʊztɒps] 无跟长筒袜

hosiery [ˈhəʊziəri] 袜子；〔英〕针织品
 fishnet *hosiery* 网眼袜
 hipster *hosiery* 低腰连裤袜
 lace top stay-ups *hosiery* 花边紧口袜
 micronet *hosiery* 超细网眼袜
 sheer to waist *hosiery* 透明连裤袜
 stretch *hosiery* 弹力袜
 toeless *hosiery* 露趾袜
 tubular *hosiery* 圆筒针织品

hotpants [ˈhɒtpænts] 热裤（女子穿的时尚紧身短裤）

housecoat [ˈhaʊskəʊt] （女用）宽大家便服，主妇服

housewife [ˈhaʊswaɪf] 针线盒

housings [ˈhaʊzɪŋz] 〔复〕服饰

hue [hju:] 色彩，色相，色调

hug-me-tight [ˈhʌgmɪtaɪt] （女）紧身短马甲；（女）针织背心或内衣

hussif [ˈhʌzɪf] 针线盒

I

idea [aɪˈdɪə] （设计）构想，想象；意念，观念；意见
 fashion *idea* 时装意识，流行观念
 novel *idea* 新奇设想；构思新颖
 original *idea* 创见，创意
illusion [ɪˈluːʒən] 薄纱，面纱；错觉
 optical *illusion* 〔设〕视错
illustration [ˌɪləsˈtreɪʃən] 插图；图解；裁剪图，结构图
 fashion *illustration* 时装画，服饰图
 cutting *illustration* 裁剪图，结构图
imitation [ˌɪmɪˈteɪʃən] 仿造，模仿
implement [ˈɪmplɪmənt] （常用复）工具；服装
inch [ɪntʃ] 英寸；〔复〕身高；身材
 stitch per *inch* 每英寸针数（车缝针距）
information [ˌɪnfəˈmeɪʃən] 信息；情报；资料
 clothing *information* 服装信息
 market *information* 市场信息
inlay [ɪnˈleɪ] 镶嵌；镶嵌物
 collar *inlay* 领插角片
inleg [ˈɪnleg] 下档；裤内长
inlet [ˈɪnlet] 镶嵌物
innovation [ˌɪnəˈveɪʃən] 革新；创新；改革
 innovation of style 款式翻新
 technical *innovation* 技术革新；科技创新
inseam [ˈɪnsiːm] （衣、袖）内缝；

（裤）内长；股下线，下档缝，〔粤〕内肶骨
 double needle flat felled *inseam* 〔工〕双针合内缝，〔粤〕双针埋夹
 inseam and outseam （裤）内、外缝，（裤）内、外长
 inseam leans to front 〔检〕前袖缝外翻
 sleeve *inseam* 前袖缝
inserting [ɪnˈsɜːtɪŋ] （衣服上）嵌饰、绣饰；（缝纫中）镶嵌，补
 inserting of fur 毛皮镶边
 inserting of leather 嵌饰皮革
 inserting sleeve 绱袖
 inserting zipper 绱拉链
inset [ˈɪnset] 嵌（补）片；镶边
insignia [ɪnˈsɪgnɪə] 徽章
insole [ˈɪnsəʊl] 鞋内底，中底；鞋垫
 odour destroying *insole* 除臭鞋垫
inspection [ɪnsˈpekʃən] 检查，检验
 China Import and Export Commodity *Inspection* Bureau (CIECIB) 中国进出口商品检验局
 commodity *inspection* 商品检验，商检
 customer (buyer) *inspection* 客户查验，客检
 early *inspection* 早期查验
 fabric (cloth) *inspection* 织物检验；布料检查，验布
 final *inspection* 最后查验，尾查
 in-line *inspection* 在线查验
 in-process (inter) *inspection* 中期查

验，中查
inspection of end products 成品检验，尾查
inspection of semi-finished(finished) products 半成品（成品）检验，中（尾）查
inspection report 验货报告
inspection result 检验结果
original inspection 初次查验，初验
pre-production inspection 产前查验
quality inspection 质量检查，质检
random inspection 随意抽查
sampling inspection 抽样查验

inspector [ɪnsˈpektə] 检验员，查货人
final inspector 成品查货人，尾查员
inspector in process 中期查货人，中查员
quality inspector 质量检验员

inspiration [ˌɪnspəˈreɪʃən] （设计）灵感

instep [ˈɪnstep] 脚背；鞋面；袜背

instrument [ˈɪnstrəmənt] 工具，手工工具；仪器
clothing special instruments 服装专用工具

interfacing [ˈɪntəfeɪsɪŋ] 内层衬布（衬于襟缘及领、袖口）

interlining [ˌɪntəˈlaɪnɪŋ] （服装）内层衬布，衬，〔粤〕朴
adhesive-bonded interlining 黏合衬
asbestos interlining 石棉衬
bast interlining 麻布衬
bottom interlining 下摆衬，底边衬
chemical fibre interlining 化纤衬布
chest(breast) interlining 胸衬
cloth interlining 布衬
collar interlining 领衬
cuff interlining 袖头衬，袖口衬
double-faced fusible interlining 双面黏合衬
dusted interlining 撒粉衬
film interlining 薄膜衬
flap interlining 袋盖衬
fly interlining 门襟衬
front interlining 前身衬，前片衬
front facing interlining 挂面衬
fusible interlining 黏合衬，热熔衬，〔粤〕粘朴
fusible non-woven interlining 非织造黏合衬
fusible woven interlining 有纺黏合衬
fusing collar interlining 〔工〕粘翻领（领面与领衬黏合）
glass silk interlining 玻璃丝衬
hair interlining 毛鬃衬，黑炭衬
heat-sealable interlining 保暖衬
hem interlining 底边衬，下摆衬
hot-melt-adhesive interlining 热熔衬
interlining below waistline 下腰节衬
interlining bleached cloth 漂布衬
interlining canvas 粗布衬
interlining domett 盖肩衬
interlining flannel 绒布衬
interlining grey cloth 本白布衬，法西衬
interlining organdie 蝉翼纱衬
interlining organza 玻璃纱衬
iron-on interlining 热熔黏合衬
knitted interlining 针织衬
knitted fusible interlining 针织黏合衬

  lapel *interlining* 驳头细布衬
  left fly *interlining* 裤门襟衬
  light (heavy)-weight *interlining* 轻薄（厚重）衬布
  main *interlining* 主衬，〔粤〕主朴
  membrane collar *interlining* 领角薄膜衬
  necktie *interlining* 领带衬
  non-fusible *interlining* 非黏合衬，〔粤〕生朴
  non-woven *interlining* 非织造衬，〔粤〕纸朴
  pocket mouth *interlining* 袋口衬
  polyester *interlining* 聚酯衬
  ponytail (horsehair) *interlining* 马尾衬
  resin collar *interlining* 树脂领衬
  right fly *interlining* 里襟衬
  shoe (hat) *interlining* 鞋（帽）衬
  shoulder *interlining* 肩衬
  skirt hem *interlining* 裙边衬
  starched *interlining* 上浆衬布
  stretch fusible *interlining* 弹力黏合衬
  truearan *interlining* 涤棉布衬
  viscose *interlining* 黏胶衬
  waistband *interlining* 腰头衬
  water-soluble *interlining* 水溶衬
  waxed soft *interlining* 上蜡软衬
  woven *interlining* 有纺衬，机织衬
interlock [ˌɪntəˈlɒk] （缝纫中）连锁；双罗纹组织；双罗纹针织物，棉毛布
  drop needle *interlock* 抽针双罗纹针织布
  interlock chainstitch 连锁链式线迹，双链线迹
  lycra *interlock* 弹力双罗纹针织布

intertexture [ˌɪntəˈtekstʃə] 交织；交织织物
intimates [ˈɪntɪmeɪts] 〔复〕贴身衣物
inverness [ˌɪnvəˈnes] 长披风，无袖长外套
iron [ˈaɪən] 烙铁；熨斗
  boiler type steam electric *iron* 电热式蒸汽熨斗
  conventional flat *iron* 火烙熨斗
  electric *iron* 电熨斗
  electric soldering *iron* 电烙铁
  flat *iron* 熨斗
  gravity-feed *iron* with water tank 吊水熨斗
  iron press 熨烫
  iron rest （放置熨斗的）熨斗盘
  iron shoe 熨斗套，熨斗鞋（放置熨斗用）
  iron-shine 〔检〕（熨烫不良产生的）亮光，〔粤〕起镜
  iron stand 铁凳（烫衣肩等部位的工具）
  iron with water container 吊水熨斗
  Italian *iron* （圆筒形）意大利熨斗
  mini-boiler with *iron* 带微型锅炉熨斗
  non-*iron* 免熨
  pressing *iron* 熨斗
  pump *iron* 泵式熨斗
  roller *iron* 滚筒熨斗
  smoothing *iron* 熨斗；烙铁
  steam *iron* 蒸汽熨斗
  steam station *iron* 带蒸汽台的熨斗
  temperature steam *iron* 电子恒温熨斗

ironer [ˈaɪənə] 烫衣服的人，熨烫工；熨烫器
    flatwork *ironer* 大熨烫机
ironing [ˈaɪənɪŋ] 熨烫，整烫；（总称）烫过的（要烫的）衣服
    cool *ironing* 低温熨烫
    do not *ironing* 不可熨烫
    dry *ironing* 干熨烫（不喷水雾）
    no-*ironing* 免烫
    poor *ironing* 〔检〕熨烫不良
    safe *ironing* temperature 安全熨烫温度
    warm(hot) *ironing* 温（热）烫
    water stain in *ironing* 〔检〕水花（熨烫产生的水渍印）
item [ˈaɪtəm] （服装的）种类；项目
    item of business 业务项目
    jeans *item* 牛仔装类

J

jabot [ˈʒæbəʊ] （衣服胸、领部位）装饰皱边，皱襞，花边领饰，绣饰
jack [dʒæk] 无袖皮军衣；（口语）夹克
    French sailor *jack* 法国水兵夹克
    hooded *jack* 带风帽的夹克
    midriff *jack* 露腰短夹克
    trench *jack* 堑壕夹克
jackboot [ˈdʒækbuːt] 过膝长筒靴
jacket [ˈdʒækɪt] 短上衣；短外衣；短外套；夹克（音译），〔粤〕机恤
    alpine *jacket* 登山衣
    anorak *jacket* 防寒夹克衫
    army look *jacket* 军款夹克
    baseball *jacket* 棒球夹克
    battle *jacket* 战斗夹克（二重领，束腰，紧身短上衣）
    bed *jacket* 短睡衣；罩在睡衣外的短上衣
    Beer *jacket* 箱型夹克
    belted *jacket* 束腰带上衣
    biker *jacket* 骑车短夹克，骑士风皮夹克
    bolero *jacket* 白来罗短夹克
    bomber *jacket* （轰炸机）飞行员夹克；束袖收摆短夹克
    box *jacket* 箱型夹克；箱型上衣
    boy's *jacket* （男）童夹克
    bulletproof *jacket* 防弹夹克，防弹衣
    bush *jacket* 丛林夹克；猎装夹克
    camo *jacket* 迷彩夹克
    cardigan *jacket* （无领）开襟夹克；开襟毛衫
    casual *jacket* 便装夹克，休闲夹克
    chamois *jacket* 麂皮夹克
    Chanel *jacket* 香奈儿女装夹克
    chimp *jacket* 津布夹克（布面有拉毛感的短夹克）
    Chinese style *jacket* 中式上衣；中山装
    collarless *jacket* 无领夹克
    corduroy(cord) *jacket* 灯芯绒夹克
    costume *jacket* 女式夹克；短外衣
    cotton-padded *jacket* 棉袄，〔粤〕棉衲
    cowboy's *jacket* 牛仔夹克
    cropped(crop) *jacket* （长至腰节

的）截短夹克，短上衣
denim *jacket* 牛仔（布）夹克，劳动布夹克，牛仔（布）外套
detachable *jacket* 可拆卸夹克，活动上衣
detailed *jacket* （拼接）花式夹克
dinner *jacket* 〔英〕夜小礼服，晚宴夹克
donkey *jacket* 女式防风厚上衣
double-breasted six-button *jacket* 双襟式六扣上衣
double-faced *jacket* 双面夹克
doughboy *jacket* 步兵夹克（美军上衣，立领，肩章，四个有盖的袋）
down *jacket* 羽绒夹克
drizzle *jacket* 毛毛雨夹克（防雨上衣）
dumb *jacket* 短皮夹克（20世纪40~50年代美国年轻人穿用）
Edwardian *jacket* 爱德华夹克（紧身，两侧开衩，高折V领和宽翻领）
Eisenhower *jacket* 艾森豪威尔夹克（美军制服上衣）
Eton *jacket* 伊顿夹克（英国伊顿公学制服上衣），伊顿短外套
fancy *jacket* （式样）奇巧夹克
faux leather *jacket* 人造革夹克
flight(aviator's) *jacket* 飞行员皮夹克
fur *jacket* 毛皮夹克
fur collar *jacket* 毛领夹克
fur-lined *jacket* 短皮袄；毛皮衬里夹克
goatskin *jacket* 羊皮夹克
hacking *jacket* 骑装短上衣，骑马夹克（紧腰宽摆）

half(fully)-lined *jacket* 半（全）里夹克
hood *jacket* 带风帽的上衣，风帽夹克
hooded fleecy sweat *jacket* 带帽绒布夹克
house *jacket* 女便装上衣
hunting *jacket* 狩猎夹克（肩有枪垫布，单排扣）
2 in 1(3 in 1) *jacket* 二（三）合一夹克（将内、外两件衣用拉链连在一起）
knitted *jacket* 针织夹克；针织外套
leather *jacket* 皮夹克；皮外套
leatheret *jacket* 人造革夹克
life *jacket* 救生衣
lined *jacket* 夹衣；夹袄
long line *jacket* 长线型上衣
lounge *jacket* 休闲夹克，普通夹克；西装上衣
lumber *jacket* 伐木人夹克（有腰带的轻快夹克）；格绒布间棉短外套
mandarin *jacket* （旧时）马褂
mannish *jacket* 男式短上衣
Mao *jacket* （立领）毛式上衣
mess *jacket* 白色的晚餐礼服
military *jacket* 军装夹克，军便装
monkey *jacket* 水手短上衣
motorcycle(motor) *jacket* 摩托夹克
Nehru *jacket* 尼赫鲁外衣（紧身、高领前排扣长外衣）
nylon/PVC padded *jacket* 锦纶/PVC填棉夹克
occasion *jacket* 应时夹克
odd *jacket* （套装中的）单件夹克，替换夹克
outer *jacket* 罩衣，外衣

over jacket 外褂，罩衣
padded jacket 充棉夹克，填棉夹克，间棉夹克；填棉外套
pea jacket 水手夹克（粗呢上衣），水手外套；双排扣粗呢短外套
pigskin jacket 猪皮夹克
pig suede jacket 猪皮绒面夹克
pilot jacket 海员厚夹克；飞行员夹克
polka jacket 女紧身夹克
polyester(nylon)/PVC coated jacket 涤纶（锦纶）/PVC涂层夹克
puffer jacket 蓬松夹克，填棉夹克，填棉外套
pyjama jacket 夹克式睡衣
quilted jacket 充（填）棉夹克
quilted and embroidered jacket 充棉刺绣夹克
rain jacket 防雨夹克，雨衣
reefing jacket 双排扣水手上衣；女式紧身双排扣上衣
reversible jacket 双面夹克
rider's(riding) jacket 骑士夹克；骑装外套
round jacket 圆型夹克（肩、袖、大身线均呈弧形）
sack jacket 休闲随意夹克
safari jacket 猎装上衣；旅游服
self-heating smart jacket 自加热智能夹克（内置碳纤维加热垫，可自动调整输出热量，以适应穿着者体温）
shell jacket 军用紧身夹克；男式紧身夹克
sherpa-lined jacket 仿羔皮里夹克
shirt jacket 衬衫式夹克
shooting jacket 射击用夹克

short boxy(square) jacket 女式箱型短上衣
silk jacket 丝绸夹克衫
ski jacket 滑雪夹克，滑雪装
sleeping jacket 短睡衣
smoking jacket 吸烟夹克；男用晚宴准礼服
snowboard jacket 滑雪（板）运动夹克（带风帽，防风，防水，透气）
spare jacket 替换夹克
spencer jacket 短外套式夹克
sport jacket 运动衣；男便衫，轻便夹克
spray jacket 防雨夹克
stretch suedette jacket 弹力绒面革夹克
suede jacket 仿麂皮夹克，绒面革夹克
suit jacket 套装上衣；西装上衣
tailored jacket 西装上衣，西装短外套
T/C poplin coated jacket 涤/棉涂层布夹克
T/C washer jacket 涤/棉洗水布夹克
thin jacket 薄（短）外套
thunder jacket （领内带帽）防雨夹克
tight quilted jacket 紧身棉袄
track jacket 田径上衣；运动夹克
twill jacket 斜纹布夹克
two(one)-button single-breasted jacket 单襟式双（单）扣上衣
unlined jacket 无里夹克
Western jacket （美国）西部夹克
wind jacket 防风夹克；（短及胸部

的）胸式夹克
wooled(leather) jacket 呢（皮）夹克
working jacket 工作上衣
yellow jacket （清朝）黄马褂
zipper front jacket 拉链夹克
Zouave jacket 朱阿夫型女短上装（宽的短上衣）

jaconet ['dʒækənɪt] 白色细薄棉布
jacquard ['dʒækɑːd] 提花，提花织物；提花机
  computer jacquard 电脑提花
  double-knitting jacquard 双面提花织物
  polyester jacquard cloth 涤纶提花布料
jak [dʒeɪk] 背心（日本名称）
jambière [ʒɑ̃bjsːr] 〔法〕绑腿
jams [dʒæmz] 〔复〕（口语 pajamas 之缩略）睡衣裤；（冲浪运动用）游泳裤
jazz [dʒæz] 爵士音乐；爵士舞（曲）；爵士音乐风格
jean [dʒiːn] 粗斜纹棉布；劳动布；牛仔布；〔复〕斜纹布服装；牛仔裤；工装裤
  appliqué jeans 贴绣牛仔裤
  black wash jeans 黑洗水牛仔裤
  bleached jeans 漂白牛仔裤；褪色变白牛仔裤
  bootleg(bootscut) jeans （宽裤脚）配靴穿牛仔裤，牛仔靴裤
  boy's jeans （男）童牛仔裤
  burnt-out jeans 烂花牛仔裤
  Capri jeans 卡普里牛仔裤（女式紧身，微喇牛仔裤）
  cigarette jeans 紧身直筒牛仔裤
  coated jeans 涂层牛仔裤
  coloured jeans 彩色牛仔裤
  cord jeans 灯芯绒牛仔裤
  crinkle effect jeans 皱折（洗水）效果牛仔裤
  cropped(crop) jeans 截筒牛仔裤（截短或卷短，一般至腿肚处）
  cuffed jeans 卷边脚口牛仔裤，翻脚牛仔裤
  cut-off jeans 毛边脚口牛仔短裤
  denim jeans 粗斜纹布牛仔裤，牛仔裤
  easyfit jeans 宽松舒适的牛仔裤
  embossed jeans 拷花牛仔裤
  embroidered jeans 绣花牛仔裤
  fade-out jeans 褪色牛仔裤
  fashion jeans 时髦牛仔裤
  fatigue jeans 工作牛仔裤
  five-pocket jeans 五袋款牛仔裤
  flare jeans 喇叭（脚口）牛仔裤
  formal jeans 正装牛仔裤
  glitter denim jeans 闪光牛仔裤
  goffered jeans 压皱牛仔裤，皱纹牛仔裤
  heavy weight jeans 厚重牛仔裤
  jodhpurs-style jeans 马裤式牛仔裤
  Kentucky jean 棉毛牛仔布，肯塔基棉毛呢
  label jeans 标签牛仔裤（以若干不同的标签饰于牛仔裤上）
  laser engraving jeans 激光雕刻牛仔裤
  leopard print jeans 豹纹牛仔裤
  light jeans 轻薄牛仔裤
  little flared-leg jean 小喇叭脚口牛仔裤，微喇牛仔裤
  loose jeans （低裆）松垮牛仔裤

low-cut crotch jeans 低裆牛仔裤
lycra jeans 弹力牛仔裤
non jeans 不像牛仔裤的牛仔裤
ornament stitch of jeans pocket 牛仔裤袋上的装饰线迹
overall jeans 牛仔围兜套裤，吊带牛仔裤
oversize jeans 阔身牛仔裤；宽松牛仔裤
patchwork jeans 贴布牛仔裤，补缀牛仔裤
printed jeans 印花牛仔裤
ramie/cotton jeans 麻棉牛仔裤
recycled jeans 再生牛仔裤（将几条旧牛仔裤拆成裤片，重新组合的裤）
rigid jeans 硬挺牛仔裤
ripped to fashionable shreds jeans 时尚破烂型牛仔裤
sandblasted denim jeans 喷砂（洗水）牛仔裤
satin jean 光洁厚斜纹布；光亮牛仔布
skinny jeans 紧身牛仔裤
soft jeans 软身牛仔裤
stone-washed jeans 石磨洗牛仔裤
straight jeans 直筒牛仔裤
stretch bootleg jeans 弹力牛仔靴裤
stretch skinny jeans 弹力紧身牛仔裤
studded jeans 窝钉牛仔裤
suede jeans 仿麂皮牛仔裤
tab waistband jeans 搭襻腰头牛仔裤
tight(skinny) jeans 紧身牛仔裤
triple pintuck front jeans 三褶饰牛仔裤（在前裤腿上车缝三道褶作装饰）
turn up cuff jeans 卷边脚口牛仔裤，翻脚牛仔裤
two-tone jeans 双色牛仔裤（拼缝牛仔布的正反面亦可形成双色效果）
utility jeans 实用型牛仔裤
vintage jeans 古旧（洗水效果）牛仔裤；古典牛仔裤（早年的名牌牛仔裤）
waistless jeans 无腰头牛仔裤
wide(skinny)-leg jeans 宽大（紧窄）脚口牛仔裤
wide waistband jeans 宽腰头牛仔裤
zipper(button) fly jeans 拉链（纽扣）门襟牛仔裤

jeanette [ˈdʒeɪnet] 细斜纹布
jeaning [ˈdʒiːnɪŋ] 牛仔裤化
jeanswear [ˈdʒiːnzweə] 牛仔服装（总称）；牛仔裤装
jeggings [ˈdʒegɪŋz] （jeans 与 leggings 组合的新词）舒适修身牛仔裤
jerkin [ˈdʒɜːkɪn] （旧时）男紧身皮制无袖短上衣；女用背心
jersey [ˈdʒɜːzi] 乔赛（音译，针织物总称）平针织物；针织平布，针织汗（衫）布；紧身毛衣；卫生衫；运动衫
  baseball jersey 棒球衫
  colour-striped jersey 针织彩条衫；彩条汗布
  cotton interlock jersey 棉毛衫
  football jersey 足球衫
  fusing jersey 复合汗布
  heavy jersey 厚针织布，双纱针织汗布
  impact jersey 增强汗布

jacquard jersey 提花针织布
lycra jersey 弹力平纹针织布
organic cotton jersey 有机棉针织平布
piqué with jersey 珠地针织布
rayon jersey 黏胶纤维针织平布
single(double) jersey 单(双)面针织布，汗布
tweed jersey 粗花呢运动衫
warp knitting jersey 经编平针织物；经编针织平布
woolen double jersey 双面毛针织布

jewel [ˈdʒuːəl] 宝石；宝石饰物
jewelry [ˈdʒuːəlri] 首饰；珠宝
alloy jewelry 合金首饰
body jewelry 贴身首饰（戴在身体上而非服装上）
bronze(brass) jewelry 铜首饰
ceramic jewelry 陶瓷首饰
fashion jewelry 流行首饰
glass jewelry 玻璃首饰
gold and silver jewelry 金银首饰
imitation jewelry 人造首饰；人造珠宝
jade jewelry 玉石首饰
leather jewelry 皮首饰
plastic jewelry 塑料首饰
shell jewelry 贝壳首饰

jibba(=jibbah) [ˈdʒɪbə] (穆斯林男子)长布袍
jitterbug [ˈdʒɪtəbʌg] （美国）摇滚爵士牛仔裤（宽松轮廓）
jodhpur [ˈdʒɒdpə] 短马靴；〔复〕马裤
cowboy jodhpurs 牛仔马裤
joggers [ˈdʒɒgəz] 〔复〕跑步运动鞋
lace joggers 系带跑步鞋
self adjusting joggers 活动扣襻跑步鞋
slip-on joggers （无扣带）松紧跑步鞋

johnny(=johnnie) [ˈdʒɒni] （住院病人穿）短袖无领罩衫
join [dʒɔɪn] 连接；缝合；接合线
cuff join 袖头缝合
join in trousers 裤接缝
joining [ˈdʒɔɪnɪŋ] 连接；缝合
joining waistband to skirt 接裙腰头
joining yoke 绱过肩
joint [dʒɔɪnt] 接头；接缝
joker [ˈdʒəʊkə] 腰牌，挂卡
jubbah [ˈdʒʊbə] 穆斯林男女所穿的开襟长布袍
jumper [ˈdʒʌmpə] 工作服；水兵短上衣；(女式)有袖或无袖套领罩衫；无袖套领背心裙；女针织套头衫；连帽皮外衣；〔美〕童围兜；〔复〕连衫裤童装；〔英〕毛衣，毛衫
canvas jumper 帆布工作服
2 for 1 jumper and shirt 二合一衬衫领套衫（套衫内衬出衬衫领和袖头，看似两件，实为一件）
school jumper 女学生背心裙
shirt jumper 衬衫式短夹克上衣
stadium jumper 运动短外衣
stripe roll neck jumper 翻领条纹套头衫
suspender jumper 背心裙，吊带裙
two-tone jumper 双色短夹克上衣
type jumper 三角形短上衣（领口、袖口、下摆为罗纹）
warm-up jumpers and pants 运动衫裤，运动服
wind jumper 防风上衣

jumpshorts [ˈdʒʌmpʃɔːts] 连身短裤装
jumpsuit [ˈdʒʌmpsjuːt] 连衣裤装，连体装；连体长裤
jupe [ʒyp] 〔法〕女裙
 jupe-culotte 〔法〕裤裙
 jupe en bias 〔法〕斜裙
 jupe fronce 〔法〕皱褶裙
 jupe pantaloons 裙式马裤
 jupe petale 〔法〕花瓣裙
 jupe plissée 〔法〕褶裥裙，细襞裙

jupette [ʒypɛt] 〔法〕短裙
jupon [ˈdʒuːpɒn] 〔英〕(无袖紧身) 铠甲罩衣；衬裙；裤子 [ʒypɔ̃] 〔法〕衬裙
 golf *jupon* 高尔夫西裤；高尔夫裤
 sailor *jupon* (裤脚肥大的) 水兵裤
 ski *jupon* 滑雪裤
justaucorps [ʒystokɔːr] 〔法〕(旧时) 男式齐膝紧身上衣

K

kaftan(= **caftan**) [ˈkæftən] (土耳其式男用) 有腰带的长袖袍；宽松上衣
kamaeri [ˈkɑːməri] (西装或外套的) 平领
kamis [kæˈmɪs] (伊斯兰男用) 长宽衬衫
kapok [ˈkeɪpɒk] 木棉
kapron [ˈkæprən] 卡普纶 (一种锦纶纤维)
kasmir (= **cashmere**) [kæʃˈmɪə] (山) 羊绒；开司米；开司米织物 (薄毛呢)
keneri [ˈkenəri] 剑领 (西装礼服领)
kepi [ˈkeɪpi] (法国的) 平顶有檐军帽
kerchief [ˈkɜːtʃɪf] 方头巾；围巾
 gauze *kerchief* 纱巾
kersey [ˈkɜːzi] 克瑟手织粗呢；克瑟密绒厚呢；棉毛绒 (粗绒布)
kerseymere [ˈkɜːzɪmɪə] 克瑟梅尔短绒大衣呢
khaki [ˈkɑːki] 卡其布；〔复〕卡其布服装 (尤指军装)
 mercerized *khaki* 丝光卡其
 reversible (double-sided) *khaki* 双面卡其
kid [kɪd] 小山羊皮；小山羊革；〔复〕小山羊皮制品 (如手套、皮鞋等)
kidskin [ˈkɪdskɪn] (用于制手套、皮鞋用的) 小山羊皮
kidswear [ˈkɪdsweə] 儿童服装，童装
kilt [kɪlt] (苏格兰式男用) 叠褶短裙 (通常用格子呢做)；(儿童穿) 苏格兰短裙
kimono [kɪˈməʊnəʊ] 〔日〕和服；和服式女晨衣
 bolero *kimono* 〔法〕和服袖短上衣 (高腰无扣)
 pull *kimono* 和服袖套衫
 robe *kimono* 〔法〕和服袖连衣裙
kipa [kɪˈpə](= **kippa** = **kippah**) 犹太小圆帽；无檐便帽
kirtle [ˈkɜːtl] 女长衫；女裙；男外衣
knapsack [ˈnæpsæk] (军用或旅行用,

帆布或皮制的）背袋，背包
knee [niː] 膝盖；（裤子长袜的）膝部；膝围，（裤）中腿；中裆
  knee lift 膝抬压脚装置
  reinforcement for *knees* （裤）膝盖绸
  waist to *knee* 腰线至膝长
knee-hi [ˈniːhaɪ] 齐膝高；齐膝裤（裙）；齐膝袜
knickerbockers [ˈnɪkəbɒkəz] 〔复〕（膝下扎起）灯笼裤
knickers [ˈnɪkəz] 〔复〕（膝下扎起）灯笼裤，尼卡裤（音译）；女用扎口短衬裤
  sanitary *knickers* （女用）月经裤
knickknack [ˈnɪknæk] 小饰物
knife [naɪf] 裁刀；刀片，切刀
  band *knife* 条刀；带式裁剪刀
  bend *knife* 弯刀；万能式裁剪刀
  buttonhole *knife* 扣眼切刀
  cutting *knife* 裁布刀，裁刀
  paper *knife* 裁纸刀
  round *knife* 圆形裁刀，圆刀
  shoe maker *knife* 鞋匠刀
  straight *knife* 直裁刀
  tape *knife* 条形裁刀，条刀
  wave *knife* 波纹形裁刀，波刀
knit [nɪt] 针织，编织；针织物；针织衣服
  circular *knit* 圆筒针织布
  roll neck *knit* 高翻领针织衫
knitter [ˈnɪtə] 针织机，编织机
knitting [ˈnɪtɪŋ] 针（编）织；针（编）织品
  double *knitting* 双面针织品；双面织
  hand *knitting* 手织，棒针编结

  weft (warp) *knitting* 纬（经）编
  woolen yarn for hand *knitting* 棒针毛线
knitwear [ˈnɪtweə] 〔总称〕针织品；编结衣物
  blended fibre *knitwear* 混纺针织品
  cashmere *knitwear* 开司米针织品
  cotton *knitwear* 棉针织品
  ramie *knitwear* 苎麻针织品
  ramie/cotton *knitwear* 麻棉针织品
  silk *knitwear* 丝针织品
  synthetic fibre *knitwear* 合纤针织品
  woolen *knitwear* 毛针织品
knot [nɒt] 结头；（装饰）花结；领带打结处
  Atlantic *knot* 大西洋结（以领带窄端缠绕打结）
  Balthus *knot* 巴尔萨斯结（亦译作巴尔蒂斯结，领带背面先朝外，再开始打结）
  breast *knot* 胸结
  Cavendish *knot* 凯文狄许结（亦译作卡文迪许结，领带双平结）
  Chinese *knot* 中式花结；中国结
  cross *knot* 交叉（领带）结
  diagonal *knot* 对角（线）结（以领带窄端缠绕打结）
  double cross *knot* 双交叉（领带）结
  double *knot* 双环（领带）结
  Dovorian *knot* 多佛结（即普拉茨堡结，领带背面朝外，再开始打结）
  esquire *knot* 绅士领结
  four-in-hand *knot* 四手结（四步骤完成的领带单结，常用领带结之一）

free style *knot* 自由式（领带）结
half Windsor *knot* 半温莎结（又称十字结，常用的领带结之一）
Hanover *knot* 汉诺威结（领结宽度介乎温莎结与半温莎结之间）
Kelvin *knot* 开尔文结（领带背面朝外，再开始打结）
Kent *knot* 肯特结（系法简单，领结小）
Merovingian *knot* 梅罗文加结（以领带窄端缠绕打结）
neoclassical *knot* 新古典主义结
Nicky *knot* 尼基结（领带背面朝外，再开始打结）
Onassis *knot* 奥纳西斯结（在四手结的基础上变化成结）
oriental *knot* 东方结（领带背面朝外再打结，最简单的结）
plain *knot* 平结（系法同四手结）
Plattsburgh *knot* 普拉茨堡结（即多佛结）
Pratt *knot* 普拉特结（又称谢尔比结，领带背面先朝外，再开始打结）
Prince Albert *knot* 艾伯特王子结（亦译为亚伯特或阿尔伯特王子结）
sailors-*knot* 水手领结
Saint Andrew *knot* 圣安德鲁结（领带背面朝外，再开始打结）
Shelby *knot* 谢尔比结（即普拉特结，常用的领带结之一）
shoulder *knot* 肩饰花结；肩垫
simple *knot* 简式结（又称马车夫结，为简单易打的领带结）
tie *knot* 领带结
trend *knot* 浪漫（领带）结
Victoria *knot* 维多利亚结（领带宽端缠绕两次后再成结）
Windsor *knot* 温莎结（最正统的领带结，常用领带结之一）
Zegna *knot* 杰尼亚结（配高档衬衫的时尚领带结）
kraft [krɑːft]（包装用）牛皮纸
kurta [ˈkɜːtə]（印度男子）宽松无领长衬衫（或上衣）；库尔塔衫

L

lab-dip [ˈlæbdɪp]（布厂供客户选择的）小块色样，色卡
label [ˈleɪbl] 标签；标记；商标牌（带）；唛头，徽章牌
    adhesive *label* 粘贴商标（或标签）
    brand *label* 主商标，主唛
    care content *label* 洗水成分唛
    care *label* 洗水唛（标明服装洗水方法）
    case pack *label* 外箱贴纸，箱唛
    certificate *label* 合格标记
    computerized woven *label* 电脑织唛
    content *label* 成分唛（标明面、里料成分）
    cotton-filled *label* 充棉（立体）唛头
    country of origin *label* 产地唛
    designer *label* 设计师品牌
    die-cut *label* 模压唛头

lace-up(s)

double face printed label 双面印唛
embossed label 拷花唛头
environmental label 环保标志
flag label 旗唛
label cloth 标签布，商标带
laser-cut label 激光裁切唛头
leather label by hot-stamping 压印皮牌
main label 主唛，主商标
metal label 金属商标牌
missing label 〔检〕漏缝唛头
origin label 产地唛
printed label 印唛
PVC(PU) label PVC（PU）皮牌
satin(plain) woven label 缎（平）面织唛
size label 尺码唛，〔粤〕烟治，烟子
stick-on label 粘贴商标（或标签）
trim label 衬唛（主要装饰上边缘处）
upside down label 〔检〕倒缝唛头
washing label 洗水唛，洗涤商标，洗涤说明
woven label 织唛
wrong label 〔检〕钉错唛头
zipper pull label 拉链头织唛

lace [leɪs] 花边，蕾丝（音译），网眼花边饰布，〔粤〕厘士（音译）；饰带；网眼织物；鞋带；系带

acrylic lace 腈纶花边
antique lace 仿古花边
appliqué lace 贴饰花边
beading lace 珠饰花边
bourdon lace 布尔登花边（涡形花边）
bra lace 乳罩花边
braid lace 编结花边
bridal lace 婚纱服花边
brocade lace 锦缎花边
cording lace 绳带花边
crochet lace 钩编花边
cuff lace 袖口饰边
drawn work lace 抽绣花边
elastic lace 弹性花边
embroidered lace 刺绣花边
filet lace 网眼花边
Greek lace 希腊花边（镂花）
guimp lace 凸纹花边
Irish lace 爱尔兰花边
leather lace 皮革花边
mosaic lace 镶嵌花边
nylon spandex lace 锦纶弹力花边
plaited lace 褶裥花边
point(needle) lace 针绣花边
polyester lace 涤纶花边
rayon lace 黏胶纤维花边
ribbon lace 缎带花边
Roman lace 罗马花边
ruffle lace 荷叶边花边，皱褶花边
shadow lace 隐纹花边
silver(gold) lace 银（金）花边
stay lace 女用束腹带；胸罩（定型）带；绲边带
thread lace 线织花边
torchon lace 镶边花边，饰带花边
torsion lace 扭转花边
tremella lace 银耳花边
trimming lace 镶饰花边
woven(knitted) lace 机（针）织花边
yak lace 牦牛绒毛花边

lace-up(s) [ˈleɪsʌp(s)] 系带鞋；系

带靴

    gladiator style lace-up(s)　角斗士式系带鞋

lacework [ˈleɪswɜːk]　花边；网眼针织品

lacing [ˈleɪsɪŋ]　（衣服）镶边；鞋带

ladieswear [ˈleɪdɪsweə]　女服，女装

lambsdown [ˈlæmsdaʊn]　驼绒（棉背毛绒面针织布料）

lambskin [ˈlæmskɪn]　羔羊毛皮，羊羔皮革

lambswool [ˈlæmzwʊl]　羔羊毛

lamé [lɑːˈmeɪ]　〔法〕金银线织物

    lamé thread　金银线

lanyard [ˈlænjɑːd]　（用于悬挂哨子、证件牌等）挂绳，挂带；系索

lap [læp]　（衣服）下摆；衣（裙）兜；搭接；衣服搭叠

    under lap　里襟；小袖衩边

lapel [ləˈpel]　（西服上衣）翻领，驳领，驳头，〔粤〕襟领

    bellied lapel　弧线翻领
    chin lapel　耸领
    clover leaf lapel　苜蓿叶形翻领
    fish mouth lapel　鱼嘴领（驳领尖为圆形）
    flower lapel　花式翻领
    fold line for lapel　驳头口，驳口
    kimono lapel　和服领
    lapel roll line is uneven　〔检〕驳口不直
    lapel roll padding automat　自动翻驳领衬垫机
    lapel step　驳领宽，〔粤〕襟宽位
    L-shaped lapel　（西装）L 形翻领
    large(inner) lapel　大（小）襟
    narrow lapel　窄驳头；小翻领
    narrow notch lapel　小方领
    notch lapel　平驳头；V 形翻领，菱领，刻领，缺嘴领
    pad stitch for lapel　〔工〕扎驳头
    peaked lapel　戗驳头；尖领，剑领
    regular notch lapel　方角领
    roll lapel　卷领，驳领
    satin lapel　缎面驳领
    semi-clover lapel　西装圆驳领
    semi-peaked lapel　半尖领，半刻领，半菱领
    single (double)-breasted lapel　单（双）襟驳头
    T-shaped lapel　（西装）T 形翻领
    wide lapel　宽驳领，大翻领

last [lɑːst]　鞋型；鞋楦头

    boot last　靴楦，鞋楦

lasting [ˈlɑːstɪŋ]　厚实斜纹织物（制鞋用）

latchet [ˈlætʃɪt]　鞋带

laundering [ˈlɔːndərɪŋ]　洗烫衣服

laundry [ˈlɔːndri]　洗衣；洗衣店

lavaliere [ˌlævəˈlɪə]　项链（系有垂饰物）

lawn [lɔːn]　上等细布；细麻布，细竹布

    boiled lawns　细亚麻平布
    grass lawn　细麻布
    truean bleached lawn　漂白涤/棉细布

layer [ˈleɪə]　（一）层；（衣服的）隔层

    single(double) layer　单（双）层
    wicking layer　（多层衣服的）吸汗层

layette [leɪˈet]　新生婴儿全套衣物

layout [ˈleɪaʊt]　（穿着）搭配；裁片排

板图；排料
 artwork layout　样稿
 computer aided layout（CAL）　计算机辅助排料
 economic layout　套排（可省料）
 layout arrangement　排料
 layout chart　排料图
leather ['leðə]　皮革；皮革制品；〔复〕皮短裤；皮绑腿
 antelope leather　羚羊革
 artificial leather　人造革
 boarded(buffed) leather　磨面革
 Brazilian PU leather　巴西 PU 革
 breathable leather　透气皮革
 buckle(belt) leather　制带革
 buffed leather　磨面革
 calf leather　小牛皮革
 chamois leather　麂皮革
 chamois-dressed leather　油鞣皮，雪米皮
 clothing leather　服装革
 corrected grain leather　修面革
 cow leather　牛皮革
 crocodile leather　鳄鱼革
 dog leather　狗皮革
 dyed leather　染色皮革
 embossed leather　拷花皮革
 enamel leather　漆皮
 faux leather　仿皮革，人造革
 full grain leather　全粒面革，头层革
 gloves leather　手套革
 goat leather　山羊革
 grain leather　珠面皮，粒面革
 hat leather　制帽革
 horse leather　马皮
 imitation leather　人造革
 import leather　进口皮革
 kangaroo leather　袋鼠皮革
 kid leather　小山羊革
 laser leather　激光革（用激光在皮革上蚀刻图案花纹）
 leather for garment　服装革
 leather for gloves　手套革
 leather for shoe lining(cover) leather　鞋里（面）革
 leather nub　皮革包扣
 lizard leather　蜥蜴革
 luggage leather　箱包革
 metallized leather　烫金属革
 napped leather　绒面革
 nubuck leather　磨砂革
 ostrich leather　鸵鸟皮革
 patent leather　（黑亮）漆皮（做鞋、皮带、手袋用），镜面革
 pig leather　猪皮革
 PU(polyurethane) leather　聚氨基甲酸酯合成革，PU 革
 PVC(polyvinyl chloride) leather　聚氯乙烯合成革，PVC 革
 rhinoceros leather　犀牛皮革
 shark leather　鲨鱼革
 sheep leather　绵羊革
 shoe leather　制鞋皮革
 shrink(levant) leather　皱纹革
 simulated leather　仿革织物，人造革
 slick surfaced leather　光面革
 smooth leather　光面革
 soft(casting) leather　软革
 sole leather　鞋底革
 split leather　二层皮革
 suede leather　起毛皮革，绒面革，反绒皮，猄皮

swine *leather* 猪皮革
synthetic *leather* 合成革
tie-dyed *leather* 扎染皮革
two-tone *leather* 双色（调）皮革
used *leather* 旧皮革
washable *leather* 水洗皮革
waste *leather* 皮革废料
whale *leather* 鲸鱼革

leatheret [ˌleðəˈret] 人造革
leatherwear [ˈleðəweə] 皮革服装
leg [leg] 腿，下肢；裤管；袜筒；靴筒；〔复〕裤脚
  boot *leg* 靴筒
  bottom *leg* 裤脚口
  flared *leg* 喇叭裤管
  front(back)*leg* 裤前（后）片
  inside-*leg* 下落裆，下裆
  "K" *legs* K形裤脚
  leg opening is uneven 〔检〕脚口不齐
  leg wear 袜子
  leg width 裤管宽，裤中裆
  out(in)-*leg* 裤外（内）长
  7/8 pants with wide *leg* 宽裤脚九分裤
  straight *legs* 直筒裤管
  tapered *leg* 锥形裤管，小裤脚，窄裤脚
  trousers *leg* 裤腿，裤管
  twisted *leg* 〔检〕裤腿扭曲，〔粤〕扭肶
  two *legs* are uneven 〔检〕两裤脚不齐（一前一后）
  unbalanced *legs* 〔检〕两裤脚不齐
  uneven *leg* 〔检〕吊脚（裤缝起吊）

  unsymmetrical *legs* 〔检〕两裤脚不对称

leggings [ˈlegɪŋz] 〔复〕（帆布或皮制的）护胫，护腿；（童）护腿套裤；女紧身裤；打底裤（女士防走光穿着的针织紧身裤）
  cotton lycra *leggings* 全棉弹力紧身裤
  denim look *leggings* 牛仔风格紧身裤
  sequined *leggings* 缀亮片紧身裤
  stretch denim *leggings* 弹力牛仔紧身裤
  zip crop *leggings* 拉链脚口截筒紧身裤（约3/4裤长）

leghorn [ˈleghɔːn] 意大利草帽
length [leŋθ] （身体、服装的）长度；一段布
  back *length* 背长
  back side *length* 后胁长
  back yoke *length* （裤、裙）后机头长
  centre back *length* 后中长
  collar *length* 领长，领大，领围
  collar point *length* 领尖长
  crotch *length* （裤）裆长，股上
  elbow *length* 肘长
  flap *length* 袋盖长
  fly *length* 门襟长；裤门襟长
  front *length* （衣）前身长
  full *length* 全长，总长
  full back *length* 后全长
  hip *length* 臀直，臀高，臀围长（腰围到臀的长度）
  knee *length* （服装）齐膝长度
  micro-mini *length* 超超短长度
  net *length* 净长

over knee *length* （服装）超膝长度
over seat *length* （服装）超臀长度
piece-*length* （布）匹长
placket *length* 门襟长
pocket *length* 口袋长
posterior full *length* 身高
remeasure the *length* of fabric 〔工〕复米（复查布料长度）
side (inside) *length* 裤外（内）长
sleeve *length* 袖长
small shoulder *length* 小肩宽（单肩宽度）
standing *length* 身高
stitch *length* 针脚长度，针距
underarm(overarm)*length* 内臂（外臂）长；小（大）袖长
waist *length* 腰节长
waist back *length* 后腰节长，背长
width and *length* （布料）幅宽和长度
yoke *length* 过肩长

leno [ˈliːnəʊ] 罗，纱罗；纱罗织物；通花布
　　Hangzhou *leno* 杭罗
leotard [ˈliːətɑːd] （杂技、舞蹈演员穿的）紧身连衣裤；紧身衣
Le Smoking [ləˈsmɒkɪŋ] 〔法〕吸烟（套）装（以修长西服，铅笔裤，配饰礼帽，领结，丝巾，长筒马靴，高跟鞋等的时尚女套装，呈刚柔相济的中性风格）
Levis [ˈliːvaɪz] 李维斯牌牛仔裤（美国名牌裤商标名）
license [ˈlaɪsəns] 许可证
　　import (export) *license* 进（出）口许可证

lierihattu [lɪrɪˈhʌtu] 宽边帽
ligne [ˈlaɪnɪ] 莱尼（纽扣规格，1莱尼＝0.633毫米）
lilion [ˈlɪlɪən] 力莱（新一代聚酰胺超细纤维，透气、吸湿性能佳）
line [ˈlaɪn] （设计、制图的）线条；线形，外形，轮廓；〔复〕设计，制图；作业线
　　A-*line* A 形线形
　　across shoulders *line* 总肩宽线
　　alternate long and (two) short dashes *line* （双）点画线
　　ample *line* 宽松线形
　　armhole *line* 袖隆线
　　armhole depth *line* 袖深线；袖隆斜线；（中装）抬根线
　　assembly *line* 装配线；生产流水线
　　back neck *line* 后领圈线
　　back rise *line* 落裆线，后裆线
　　back rise width *line* 落裆宽线
　　back seam *line* 背缝线
　　back shoulder *line* 后肩线
　　back waist seam *line* （裤）后腰缝线
　　back width *line* 背宽线
　　balloon *line* 气球线形
　　barrel *line* 桶形线形
　　basic *line* 基本线，基础线
　　big *line* 宽大线形
　　body *line* 身体线条；紧身轮廓线，紧身曲线
　　bottom *line* 下摆线；脚口线，脚口围线
　　bottom width *line* 下摆直线
　　box *line* 箱形线形，盒形线形

button and buttonhole position *line* 扣、眼位线

centre back seam *line* 后中缝线，背缝线

centre front (back) *line* 前（后）身中线，前（后）中线

chalk *line* 划粉线

chest (bust) *line* 胸围线

chest (bust) width *line* 胸宽线

collar band roll (neck) *line* 底领上（下）口线

collar style (stand) *line* 翻领外（上）口线

collar width *line* 翻领前宽斜线

crease *line* 折线；裤烫线

crotch *line* 裤裆线

crutch depth *line* 直裆线

curve *line* 曲线，弧线

curve (bias) *line* of front opening （中装）大襟弧（斜）线

curve *line* of front (back) rise 裤前（后）裆弧线

curve *line* of underarm （中装）裉缝弧线

cutting *line* 裁剪线

Daniel *line* 丹尼尔线形

dart *line* 省道线，褶线

decorative *line* 装饰线

dimension *line* 尺寸线

distance *line* 距离线

dotted *line* 虚线，反面轮廓线

drop waist *line* 低腰线形

elbow *line* 袖肘线

elbow girth *line* 肘围线

elbow seam *line* 后袖缝线，小袖内撇线

empire *line* （女装衣裙）高腰线

end *line* of collar band 底领前宽斜线

equation *line* 等分线

feminine *line* 女性线条

finish *line* 成形线，面缝线

finish *line* of cuff 袖头止口线，袖口翘线

finish *line* of front facing 挂面止口线

fit and flare *line* 上贴（身）下散（开）线形

fold facing *line* 翻折线

fold *line* 折叠线

fold *line* of under (top) sleeve 后（前）偏袖线

front (back) yoke *line* 前（后）过肩线

front cut *line* 撇门线，门襟止口圆角线

front edge *line* 门襟止口线

front finish *line* 撇门线

front overlap *line* 搭门线

front pitch *line* （衣）起翘横线

front point *line* 底边弧线（马甲前摆处尖角斜线）

front rise *line* 前直裆线，前裆内撇线

front rise width *line* 小裆宽线

gorge *line* （领圈）串口线

hem *line* 底边线；下摆线；脚口线

hip *line* 臀围线

horizontal *line* 横线

hourglass *line* 沙漏线形

imaginary *line* 假想线，参考线；上平线（两肩顶连成的虚线），〔粤〕膊平

inner *line* of front facing   挂面里口线
inseam *line*   前袖缝线；裤内缝线
inside seam *line*   下裆线，内缝线
knee *line*   中裆线，中裆围线，膝围线
lapel edge *line*   驳头止口弧线
lapel roll *line*   驳口线
length *line*   （服装）长度线
level *line*   水平线；基准线
line and colour   线与色
line of collar point   领尖点线
line of front cut point   门襟圆角点线
long torso *line*   长身线形
medium *line*   中间线
middle hip *line*   中臀围线
natural *line*   自然线形
neck depth *line*   领深线
neck *line*   领圈线，领口线，领窝线
neck-waist *line*   背长线
neck width *line*   领宽线
notch position *line*   领嘴线
oblique *line*   斜线
original pattern *line*   原型线
outside (edge) *line* of left fly   裤门襟外（止）口线
outside (inside) *line* of collar   翻领外（上）口线
outside (inside) *line* of collar band   底领下（上）口线
outside (inside) *line* of right fly   裤里襟外（里）口线
pleat position *line*   裥位线，褶裥线
pocket position *line*   袋位线

princess *line*   公主线（上身及腰部贴紧，裙摆展宽）
production *line*   生产（流水）线
rectangle *line*   矩形线形
return *line*   折线
roll *line*   驳口（驳头翻折位），驳口线，驳折线，翻折线
seam *line*   完成线；沿线，开缝线
section *line*   分割线
shaped (modeling) *line*   造型线，成型线
shoulder *line*   肩宽线，落肩线，小肩线
shoulder slope *line*   肩斜线（过肩肩部斜度线）
shoulder width *line*   肩头外倾线（过肩袖窿斜线）
side *line*   胁线
side seam *line*   （衣）摆缝线；（裤、裙）侧缝线
silhouette *line*   轮廓线，廓线，外形线
sleeve cap *line*   袖山线，袖山弧线
sleeve centre *line*   袖中线
sleeve girth *line*   袖围线
sleeve length *line*   袖长线
sleeve opening *line*   袖口线
sleeve slit *line*   袖衩线
slim *line*   苗条线形
stomach (abdominal extension) *line*   腹围线
straight *line*   直线
structure *line*   结构线
switch *line*   （服装上、下身）转换线
T-*line*   T形线形，T形造型
tapered *line*   锥形线形

tent *line* 帐篷线形
thigh *line* 横裆线
top *line* 上口线；鞋口
top *line* of cuff 袖头上口线
trapeze *line* 梯形线形
trumpet *line* 喇叭线形
T-shirt *line* T恤线形
tube *line* 筒状线形
tuck *line* 塔克线，褶线
under bust *line* 下胸围线
under sleeve depth *line* 小袖深弧线
under(top) *line* of waistband 腰头下（上）口线
under(top) *line* of yoke 过肩下（上）口线
up *line* of armhole 袖窿翘高线
up *line* of waist 裤后翘线
vent *line* 开衩线，衩位线
vertical *line* 垂直线
waist *line* 腰围线，腰节线
waist girth *line* 掐腰线（衣服的中腰围尺寸线）；腰围线
waistband roll *line* 腰头上口线
X-*line* X形线形
Y-*line* Y形线形
yoke width *line* 过肩宽线

linen [ˈlɪnɪn] 亚麻布（或纱、线）；亚麻织物；似亚麻的制品（如衬衫、内衣等）
 bleached *linen* 漂白亚麻布
 body *linen* 内衣用亚麻布
 brown *linen* 本色亚麻布（法国制）
 dress *linen* 外衣用亚麻布
 fine *linen* 亚麻细布
 grass *linen* 夏布
 linen roughs 帆布型亚麻（衬里）布
 union *linen* 交织亚麻布（棉/亚麻交织）
 wool/*linen* blended fabric 毛麻混纺织物
 yarn-dyed *linen* 色织亚麻布

liner [ˈlaɪnə] 衬里；托布（衬胆的套布）
 liner band 衬带
 camlet *liner* 羽纱衬里

lineup [ˈlaɪnʌp] 〔设〕效果图

lingerie [lɛˈʒrɪ] 〔法〕内衣，衬衣；女内衣
 lingerie for men （仿女式）男士内衣
 sexy *lingerie* 性感内衣，情趣内衣
 style *lingerie* 装饰性内衣（如套裙，睡袍，胸罩，内裤等）

lining [ˈlaɪnɪŋ] （服装的）里料，里子，衬里；装衬里
 acetate jacquard *lining* 醋酸提花里子绸
 acetate twill *lining* 醋酸斜纹里子绸
 collar *lining* 领衬里
 cotton *lining* 棉布里子
 crotch *lining* 裤裆里布
 cuff *lining* 袖头衬里
 detachable *lining* 活动里子
 dobby *lining* 小提花衬里
 flap *lining* 袋盖里
 fly *lining* 门襟里，〔粤〕纽牌里
 fleece *lining* 绒布衬里
 forepart *lining* （衣服）前襟里子
 front(back) *lining* 前（后）身里子

fur *lining* 毛皮里子
half back *lining* 后半身衬里
half(full) *lining* 半（全）衬里
hat *lining* 帽里子
hood *lining* 风帽里
jersey *lining* 针织里布
linen *lining* 亚麻布衬里
lining cambric 细布衬里；麻纱衬里
lining duck 棉帆布衬里
lining felling 绱衬里
lining flannelette 绒布衬里
lining for boarding 定型里
lining lambsdown 驼绒衬里
lining leather 皮革衬里，衬里革
lining pile 毛绒里布
lining silk 里子绸；蜡线羽纱；蜡线美丽绸
lining taffeta 里子塔夫绸
lining velvet 丝绒衬里
lustre *lining* 有光衬里；羽纱
mohair cloth *lining* 马海呢衬里
moisture wicking *lining* 吸湿排汗衬里
nylon/acetate jacquard *lining* 锦纶/醋酸提花里子绸
paper *lining* 衬纸
partial *lining* 部分衬里
pocket *lining* 口袋布
polyester/cotton *lining* 涤/棉里布
polyester jacquard *lining* 涤纶提花里子绸
polyester/viscose jacquard *lining* 涤/黏提花里子绸
printed cotton *lining* 全棉印花里布
rayon *lining* 人造丝羽纱里子

rayon *lining* satin 人造丝里子缎，软缎里子
rayon *lining* twill 美丽绸
self *lining* 原身里布（里布同面布）
shoe *lining* 鞋衬里
shrink-proof *lining* 防缩衬里
sleeve *lining* 衣袖里子
sock *lining* （鞋）中底垫皮，鞋衬垫
splicing pocket flap *lining* 〔工〕拼袋盖里
too tight(full) *lining* 〔检〕里布太紧（过多）
twisted *lining* 〔检〕里布扭曲
under(top) sleeve *lining* 小（大）袖里子
waistband *lining* 腰头里子
zip-out *lining* 拉链式活动里子，拉链脱卸里

links [lɪŋks] 〔复〕（衬衫袖口上的）链扣；袖口纽
  cuff *links* 袖口链扣，袖口纽
  metal cuff *links* 金属袖口纽
  sleeve *links* 袖口纽

lip [lɪp] 嘴唇；唇状物（如袋唇）
  uneven *lip* 〔检〕袋唇宽窄不匀，〔粤〕大小唇

lipstick ['lɪpstɪk] 唇膏，口红；口红色

liquette [lɪkɛt] 〔法〕男用外衬衫（宽敞，前领口低）

liseurse [liˈsɜːs] 〔法〕休闲便衣（女装室内上衣）

list [lɪst] 布边；饰边；目录；一览表
  list work 衣边贴饰
  packing *list* 装箱单；花色码单

price list　价格表
livery [ˈlɪvəri]　（侍从、仆人穿的）特殊制服；号衣；行会会员制服
　　　livery cloth　制服呢
　　　out of livery　穿便服
loafer [ˈləʊfə]　平底便鞋；拖鞋
lock [lɒk]　（缝纫中）锁缝
lockstitch [ˈlɒkstɪtʃ]　双线锁缝，二重缝；锁式线迹；锁针法
　　　lockstitch on buttonhole　〔工〕锁扣眼
　　　two-needle lockstitch　双针锁式线迹
loden [ˈləʊdən]　洛登缩绒（防水）厚呢；洛登缩绒（防水）厚呢大衣
logo [ˈləʊgəʊ]　（衣、帽上面的）标识，标记
long [lɒŋ]　（服装）长尺寸；〔复〕长裤
　　　long-johns　〔美〕长内衣裤
longies [ˈlɒŋiːz]　〔复〕〔美〕长内衣裤
longyi [ˈlɒŋgɪ] (=lungyi =lungee =lungi [ˈlʊŋgiː])　腰布；头巾；笼基（缅甸传统下装，一种围裹式筒裙）
look [lʊk]　装式，款式，式样；风格，风貌；〔复〕容貌
　　　African look　非洲款式
　　　bare look　暴露装；暴露款式（如露肩、臂、腹、背等）
　　　bold look　大胆款式
　　　Bond look　邦德款式（宽领西装，后身两边开衩，优雅衬衫配针织领带）
　　　bulky look　庞大款式
　　　business casual look　商务休闲风格
　　　business formal look　商务正装风格
　　　camouflage look　迷彩款式
　　　casual look　轻便款式，便装装式；休闲风格
　　　Chinese look　中国款式
　　　classic look　经典款式
　　　Jimmy look　吉米款式（短外套加裤的组合）
　　　kinky look　古怪款式，反常款式
　　　lingerie look　内衣装式
　　　military look　军装装式
　　　nautical look　船员装式
　　　new(classic)look　新（传统）装式
　　　New Look　新风貌（早年迪奥 女装设计的高雅品位和全新风格）
　　　op art look　视幻艺术装式，欧普风貌
　　　pirate look　海盗款式
　　　one-piece look　上下装
　　　opulent look　富豪款式
　　　oriental look　东方式样
　　　peasant look　农民装式
　　　punk look　朋克风貌（代表一种反叛的、非主流的流行和观念）
　　　retro look　怀旧款式；怀旧风格
　　　safari look　狩猎装式
　　　sensual look　性感款式
　　　smart casual look　穿着得体的休闲风格，精致休闲风格
　　　sportive look　运动装式；轻便装
　　　suit look　套装装式
　　　thirties look　三十年代款式
　　　total look　整体款式
　　　unisex look　男女通用款式，中性款式
　　　Western look　（美国）西部款式
　　　working wear look　工装款式
　　　worn look　残旧风貌
loom [luːm]　织布机

    automatic *loom*　自动织布机
    jacquard *loom*　提花织布机
    raschel *loom*　拉舍尔经编机
    shuttleless *loom*　无梭织机
    special *loom*　特种织布机
loop [ˈluːp]　线环；(布) 环带, 带襻；裤带襻
    belt *loop*　带襻；裤带襻, 裤耳, 〔粤〕耳仔
    button *loop*　扣襻
    hanger *loop*　挂衣襻
    hook and *loops*　(尼龙) 搭扣
    missing belt *loop*　〔检〕漏钉裤耳
    shoulder *loop*　肩襻
    thread chain *loop*　线链襻
    uneven belt *loops*　〔检〕裤耳大小不匀
loungewear [ˈlaʊndʒweə]　家居服, 休闲服

looping [ˈluːpɪŋ]　上带襻 (如裤耳)
lumberjack [ˈlʌmbədʒæk]　短夹克衫
lurex [ˈljʊəreks]　一种金银丝及其织物；金属丝 (纺织纤维)
lustrine [ˈlʌstrɪn]　〔英〕有光斜纹袖里棉布；光亮绸；全丝光亮塔夫绸
lustring [ˈlʌstrɪŋ]　光亮绸；上光丝带
lutestring [ˈljuːtstrɪŋ]　光亮绸；有光丝带
lycra [ˈlaɪkrə]　莱卡 (弹性纤维商品名)；〔粤〕拉架 (布), 弹性织物
    cotton *lycra*　全棉拉架, 全棉弹力布
    nylon *lycra*　锦纶拉架, 锦纶弹力布
lyocell [ˈlaɪəsel]　莱赛尔纤维 (一种全新环保的人造纤维素纤维, 天丝的英文学名)

M

machine [məˈʃiːn]　机器；机械
    auto-(manual) printing *machine*　自动 (手工) 印花机
    automatic buttonhole stitching *machine*　自动锁眼机
    automatic welting *machine*　自动缝纫机 (计算机控制自动开袋、绱袖、绱领等)
    band knife cutting *machine*　带刀电剪, 带刀裁剪机
    bartack *machine*　套结缝纫机, 打结机, 〔粤〕打枣车
    basting *machine*　疏缝机, 假缝机
    belt *machine*　带子机, 带襻机

    belt-loop attaching *machine*　带子机, 裤襻机
    blind-stitch *machine*　暗缝机, 〔粤〕挑脚车；插边机
    bottom hemming *machine*　裤脚卷边机, 〔粤〕辘脚车
    butted seaming *machine*　绷缝机
    button attaching *machine*　钉扣机
    buttonholing *machine*　锁 (扣) 眼机, 〔粤〕纽门车
    buttoning *machine*　钉扣机, 〔粤〕钉纽车
    button-sewing *machine* with trimmer　剪线钉扣机

button stitching machine 钉扣机
chain blind-stitch machine 链式暗缝机，〔粤〕锁链挑脚车
chain stitch machine 链式缝纫机，锁链机，环缝机
chainstitch feed-off-arm machine （裤内缝或袖下缝）折缝机，拼缝机，〔粤〕埋夹车
circular eyelet knitting machine 圆筒网眼纬编机
circular knitting machine 针织圆筒机，大圆机
circular spring needle machine 圆筒钩针纬编机
cloth folding machine 折布机
cloth inspecting machine 验布机
cloth measuring and cutting machine 量裁（布）机
cloth spreading machine 拉布机，平布机
cloth winding machine 卷布机
CNC（computer numerical control）seamless forming knitting machine 计算机数控无缝成型针织机
collar binding machine 滚领机
collar blocking machine 热压领机，〔粤〕焗领机
collar marking machine 点领机
collar point trimming machine 切领嘴机
collar point turning machine 翻领角机
collar turning and pressing machine 翻领压领机
computer controlled programmed sewing machine 计算机程控缝纫机
computerized embroidery machine 计算机绣花机
（continuous）buttonholing machine （连续）锁眼机
covering stitch machine 绷缝机，〔粤〕拉乓车，乓车，虾苏网车
cover seam machine 绷缝机
cuff pressing machine 压袖头机
cuff turning machine 翻袖头机，〔粤〕反鸡英车
cutting machine 裁布机；裁剪机，裁剪电刀
darning machine 织补机
dart sewing machine 缝省机
decorative stitching machine 装饰线缝机（如珠边机）
die cutting machine 冲切裁剪机，模切裁剪机，模切机，〔粤〕啤机
double head sewing machine 双头缝纫机
down quilting machine 羽绒绗缝机
drafting machine 绘图机；打样机
drill machine （裁剪用）钻孔机
dry-cleaning machine 干洗机
edge cutting machine 切边机
edge folding machine 折边机
edge seaming and trimming machine 缝锁（边）机
edge sewing machine 缲边机
edge taping machine 镶边机
edge trimming machine 修边机
elastic sewing machine 橡筋机
electric eye sensor thread cleaning machine 电眼感应清线头机
electronic controlled pattern sewing machine 电子图样机，电子花样机

electronic (electric) sewing *machine* 电子（电动）缝纫机
embroidery sewing *machine* 绣花缝纫机
eyelet fastening *machine* 钉鸡眼机，打气眼机
eyelet stitching *machine* 圆头锁眼机
fastening *machine* （金属扣、纽扣）钉扣机；加固缝纫机
fell seaming *machine* 折缝机，拼缝机，〔粤〕埋夹车
flat knitting *machine* 〔针织〕横机，针织编机
flat-lock *machine* 平针机，绷缝机
flat (plain) sewing *machine* 平缝机，平机，〔粤〕平车
fluted hem *machine* （绉褶）波边机，〔粤〕推波车
F. O. A (feed-off-arm) seam *machine* 折缝机，拼缝机
form pressing *machine* 立体熨烫机
(full automatic) pocket-hole sewing *machine* （全自动）开袋机
full (semi)-automatic template sewing *machine* 全（半）自动模板缝纫机
fully automatic/computerized cutting *machine* 全自动/计算机裁剪机
fur abutting *machine* 毛皮拼接缝纫机
fur brushing *machine* 毛皮滚毛机
fur cutting *machine* 毛皮裁剪机
fur polishing and ironing *machine* 毛皮整烫机
fur strap sewing *machine* 毛皮条子缝纫机
fusing press *machine* 黏合机，压衬机，〔粤〕粘朴机
grading *machine* （纸样）放缩机
heatbonding *machine* 热压黏合机
hem-edge *machine* 缲边机，撬边机
hemming *machine* 卷边机
herringbone *machine* 人字线迹缝纫机，人字机
(high speed) drafting *machine* （高速）绘图机
(high speed) eyelet buttonhole sewing *machine* （高速）圆头锁眼机，凤眼机
high speed(single-needle) flat sewing *machine* 高速（单针）平缝机
hot air seam sealing *machine* 热风线缝胶贴机，热风机
industrial (home) sewing *machine* 工业（家用）缝纫机
industrial washing *machine* 工业洗衣机
intelligent fabric spreading *machine* 智能拉布机
interlining winding (cutting) *machine* 卷（切）衬机
interlock sewing (stitching) *machine* 联锁缝纫机
jacquard flat knitting *machine* 提花横编针织机
knitting *machine* 针织机
labelling *machine* 缝标签机，钉唛头机
laser cutting *machine* 激光裁剪机
laser embroidery *machine* 激光绣花机
linking *machine* 缝袜头机，捆边

机

lockstitch bar tacking *machine* 锁缝打结机

lockstitching *machine* 锁缝机，平缝机

lockstitch(lock) *machine* 锁缝机

long arm sewing *machine* 长臂缝纫机

loops sewing *machine* 裤襻机，〔粤〕耳仔车

machine cabinet 机柜

machine for covering button with cloth 包扣机

machine head 机头

machine parts 机件（零件、配件）

machine pedestal(frame) 机架

machine problem 机器故障

machine set 机座

machine tool 机床

manual(electric) snap fastening *machine* 手（电）动揿纽钉扣机

Merrow sewing *machine* 包缝机，锁边机

microprocessor controlled sewing *machine* 微机控制缝纫机

multi-function embroidery *machine* 多功能绣花机

multi-head automatic embroidery *machine* 多头自动绣花机

multi-needle chainstitch *machine* 多针链式缝纫机，〔粤〕拉筒车

multi-needle flat lock *machine* 多针绷缝机

multi-purpose sewing *machine* 多功能缝纫机

notcher *machine* （裁剪用）缺口机

numbering(marking) *machine* 打号机

overcasting *machine* 包缝机，锁边机

overedge sewing *machine* 包缝机，锁边机

overlook *machine* 包缝机，锁边机，〔粤〕钑骨车

pattern duplicating *machine* 纸样（图样）复印机

perforation *machine* 打眼机

pinking *machine* 扎驳头机

piping *machine* 绲边机

placket *machine* 门襟机，〔粤〕筒车

placket cutting (trimming) *machine* 裁门襟机，〔粤〕切筒车

plasma torch cutting *machine* 等离子裁剪机（利用高温等离子电弧的热量，以射束的形式通过喷嘴进行裁切）

pleat making *machine* 打褶（裥）机

pneumatic snap fixing *machine* 气动揿纽钉扣机

pocket creasing *machine* 折袋机

pocket flat creasing *machine* 折袋盖机

pocket setting *machine* 绱袋机，装袋机

pocket welting *machine* 开袋机，袋唇机

pressing *machine* 熨烫机，整烫机

printing and dyeing *machine* 印染机

products folding *machine* （包装）成品折叠机

punching machine 打眼机
quilting machine 绗缝机
ribbon cutting machine 切捆条机
rib cutting machine 罗纹裁剪机
rivet machine 打撞钉机
rotary pressing machine 滚动式熨烫机，〔粤〕辘烫机
round knife cutting machine 圆刀电剪，圆刀裁剪机
ruffling machine 打褶机，褶边机
seaming machine 缝纫机
selvage-seaming machine 拷边机，包缝机
semi-automatic cutting machine 半自动裁剪机
sewing machine 缝纫机
shaking machine （洗水用）震动机
shirring machine 抽褶机，缩缝机
single(double, multi)-needle sewing machine 单（双、多）针缝纫机
sleeve attaching machine 绱袖机
sleeve-placket machine 缝袖衩机
sleeving machine 绱袖机
snap fastening (attaching, fixing) machine 钉四合纽机，〔粤〕啤纽车
special sewing machine 专用缝纫机，特种缝纫机，花色机
spinning machine 纺纱机
spot welding machine 点衬定位机
steam machine for processing fur 毛皮蒸汽清理机
stone-washing machine 石磨洗水机
straight buttonhole machine 平头锁眼机
straight knife cutting machine 直刀电剪，直刀裁剪机
strapping machine 打带机
sweater knitting machine 毛衣编织机
swing arm vertical cutting machine 摇臂裁剪机
tacking machine 打结机，加固机；归拔机
tape cutting machine 裁带机
thread-thrum sucking machine 吸线头机
three (four, five) thread overlock machine 三（四、五）线包缝机，锁边机，拷边机
top and bottom (bottom) covering stitch machine 双（单）面绷缝
treadle(hand) sewing machine 脚踏（手摇）缝纫机
trimming machine 修边机；饰边机
tuck machine 压褶机
tumbling machine 干衣机
two(one)-head embroidery machine 双（单）头绣花机
ultra-high-speed sewing machine 超高速缝纫机
ultrasonic cutting machine 超声波裁剪机（直接将超声波能量加载到切刀上进行裁剪）
universal sewing machine 通用缝纫机，万能缝纫机
vacuum wet absorbing machine 真空抽湿机
versatile sewing machine 多功能缝

纫机

waistband machine 绱腰头机，裤头机，〔粤〕拉裤头车

washing machine 水洗机，洗衣机

water jet cutting machine 水射流裁剪机（极细的水射流高速通过喷嘴裁剪布料）

weft（warp）knitting machine 纬（经）编机

zigzag sewing machine 之字针迹缝纫机，曲折缝缝纫机，花针机，〔粤〕人字车

machinery [məˈʃi:nəri]（总称）机器；机械

 garment machinery 服装机械

 machinery maintenance 机械维修

mackinaw [ˈmækɪnɔ:]（音译）马基诺厚呢；马基诺厚呢外套

mackintosh [ˈmækɪntɒʃ] 轻薄防水胶布；〔英〕雨衣

made-to-measure [meɪdtəˈmeʒə] 量身定做

Mae West [meɪˈwest]（俚语）（飞行员穿的）救生衣

magua [mʌˈgwɑ:] 中式对襟马褂

mail [meɪl] 铠甲，锁子甲；邮件

 coat of mail 铠甲装，铠甲

 e-mail 电子邮件

maillot [majo]〔法〕（运动员、演员穿的）紧身衣；（衣连裤）女泳衣

 maillot de corps 〔法〕紧身内衣（汗衫，棉毛衫等）

 maillot de danseur 〔法〕舞蹈演员的紧身衣

 maillot nageur 〔法〕运动员穿的好泳衣

make-up [ˈmeɪkʌp] 化妆；化妆品（尤指妇女的）

 bridal make-up 新娘化妆

 make-up kit 化妆箱

making [ˈmeɪkɪŋ] 做，制作；缝制

 clothing template making 服装模板制作

 dress making 服装制作（全过程总称）；服饰制作

 making French tack 拉线襻

 making hood 做风帽

 making left and right fly 缝制裤门襟、里襟

 making the lay （裁剪间）排料

 making-through （一个人）单件制作，做整件

 marker making 唛架制作

 new sample making 打新样

 pattern making 纸样制作，制板

 sample making for approval 打确认样

management [mænɪdʒmənt] 管理；经营

 apparel management and sales 服装营销

 business process management（BPM）业务流程管理

 change management（CM）变革管理；应变管理

 computer aided management（CAM）计算机辅助管理

 customer relationship management（CRM）客户关系管理

 financial management（FM）财务管理

 human resource management（HRM）人力资源管理

 inventory management（IM）存货管

理
knowledge management (KM) 知识管理
logistics management (LM) 物流管理
order management (OM) 订单管理
personnel management 人事管理
production management 生产管理
project management (PM) 项目管理
quota management 配额管理
risk management (RM) 风险管理
supply chain management (SCM) 供应链管理
total quality management (TQM) 全面质量管理
value management (VM) 价值管理
virtual management 虚拟管理；虚拟经营

manicure ['mænɪkjʊə] 修指甲，美甲

manicurist ['mænɪkjʊərɪst] 修甲师，美甲师

mannequin (= **manikin**) ['mænɪkɪn] 服装人体模型；（表演）时装模特儿，女模特儿
 adjustable mannequin 可调节的人身模型

manteau ['mæntəʊ] （女用）披风，斗篷；女开襟外衣；[mɑ̃to]〔法〕大衣，外套，披风，斗篷
 manteau de demi-saison〔法〕春秋大衣
 manteau de pluie〔法〕雨衣
 manteau domino 多米诺外套（化装舞会用，通常有面罩配合）
 manteau veston〔法〕西装大衣

mantelet ['mæntəlɪt] 小斗篷，披风；短外套

mantilla [mæn'tɪlə] （西班牙等国妇女的）薄头罩；黑丝披巾；晚礼服斗篷

mantle ['mæntl] 披风，斗篷
 aba mantle 阿拉伯披风

manufacturing [ˌmænjʊ'fæktʃərɪŋ] 制造，制作；生产
 apparel manufacturing 成衣生产
 intelligent manufacturing 智能制造
 manufacturing flow chart 生产流程图
 new manufacturing 新制造
 original equipment manufacturing (OEM) 贴牌生产，来料（样）加工
 sample manufacturing 样品制作，打样

marinière [marɪnjɛːr]〔法〕水兵服上衣；套头宽大女衬衫

mark [mɑːk] 商标；唛头；符号；标记
 balance mark 叠合印，剪口（打褶标记）
 carton mark 箱唛
 chalk mark〔检〕（留在衣服上的）划粉印
 dart position mark 开省号
 direct mark 顺向符号
 drilling mark 钻眼符号
 easing mark 放宽符号，容位号
 ellipsis mark 省略号
 gather mark 碎褶号
 glaze mark〔检〕（熨烫不良产生的）亮光，〔粤〕起镜
 inspection mark 检验标记
 ironing mark〔检〕（熨烫不良所致）熨烫痕迹，压痕

negative *mark* 否定号
notch *mark* 领嘴符号，刀口符号
out（cutting）line *mark* 净（毛）样符号
overlapping *mark* 重叠号
piece together *mark* 拼接号
pure wool *mark* 纯羊毛标志
radial *mark* 经向号
removal of *mark* （美容）除斑
rib *mark* 罗纹符号
shipping *mark* 装运标记，箱唛
shrink *mark* 归缩号
side *mark* （纸箱）侧唛
stitching *mark* 缝纫针迹
stretch *mark* 拉伸号
symmetry *mark* 对称符号
tacking *mark* 假缝符号
topstitch *mark* 明线号
trade *mark* 商标
vertical *mark* 直角符号

marker [ˈmɑːkə] 排裁片纸样图，排料图；〔粤〕唛架（音译），（在表层布上）划样；排板师；排料人；唛架师傅
 cloth *marker* 表层布上划样；漏画
 computer *marker* planning 计算机排唛架
 marker copier 唛架复印机
 marker laying 排唛架，排料
 marker lay making 按纸板划样，唛架制作
 marker plotter 唛架绘印机
 marker plotting 绘制唛架
 mini *marker* （估料用）小唛架，迷你唛架
 reproduction of *marker* 复制唛架
 stencil *marker* 漏画纸板
 straight/cross laser *marker* （裁床定位用）直线/十字激光灯

market [ˈmɑːkɪt] 市场；销路；行情
 clothing *market* 服装市场
 flea *market* 跳蚤市场，廉价市场
 international（domestic）*market* 国际（国内）市场
 market analysis 市场分析
 market development 市场开发
 market economy 市场经济
 market forecasting 市场预测
 market price 市场价格，市价
 market research 市场研究
 markets supply and demand 市场供求
 poor（good）*market* 销路不好（销路好）
 seller's（buyer's）*market* 卖（买）方市场
 The *market* rose（fell）行情上涨（下跌）

marketing [ˈmɑːkɪtɪŋ] 市场营销
 direct *marketing* 直效营销，直销
 experiential *marketing* 体验营销
 garments *marketing* 服装市场营销
 influencer *marketing* 影响者营销；网红营销
 marketing strategy 营销策略
 one to one *marketing* 一对一营销
 online events *marketing* 在线营销

marking [ˈmɑːkɪŋ] 划样，排料；描样；标记；打号
 computer *marking* 计算机排料
 positioning *marking* 排板描样

Mary Jane [ˈmeərɪˈdʒeɪn]（= Mary Jane shoes）玛丽珍鞋（平跟搭襻浅口女童皮鞋）

mash-up [ˈmæʃʌp] 〔设〕混搭（一种打破传统，随意搭配的时装潮流，追求融合美感）
mask [mɑːsk] 面具，面罩；防护面具；口罩
    antibacterial *mask* 防菌口罩
    anti-dust *mask* 防尘面具；防尘口罩
    antiviral *mask* 防病毒口罩
    disposable *mask* 一次性口罩
    eye *mask* 眼罩
    face *mask* 面膜（美容用）；口罩
    flu *mask* 卫生口罩
    gas *mask* 防毒面具
    gauze *mask* 纱布口罩
    intensive firming *mask* 强力紧致面膜
    massage *mask* 按摩面膜
    oxygen *mask* 氧气面具
    purifying *mask* 洁肤面膜
    sleep *mask* 睡眠面膜；睡眠眼罩
    surgical *mask* 外科手术口罩；医用口罩
matching [ˈmætʃɪŋ] 相配，相衬；配色
    matching jewelry 相配的首饰
    matching stripes(checks) （裁剪，缝制中）对条（格）
matchy-matchy [ˈmætʃɪˈmætʃɪ] 指服装搭配过度工整协调而失去美感
material [məˈtɪəriəl] 材料，原料；织物；布料，料子
    advanced fiber *materials* 先进纤维材料
    basic *material* 主料
    bill of *material* 物料表
    clothing *material* 衣料
    consumption of *material* 耗料
    double-faced *material* 双面布料
    dress *materials* 衣料，面料
    hat *materials* 制帽材料，帽料
    knitted *material* 针织料
    material purchasing 布料采购
    medium-heavy *material* 中厚料子
    packing *material* 包装材料
    placing *material* （裁剪前）排料
    processing raw *materials* on client's demands 来料加工
    pure silk (wool) *material* 真丝（纯毛）料子
    shoe upper (lining) *material* 鞋面（里）料
    striped(spotted) *material* 带条（点）料子
    textile raw *material* 纺织原料
    tie *material* 领带料；领带绸
    underwear *material* 内衣料
    vintage dress *material* 怀旧衣料
    worsted(woolen) *material* 精（粗）纺毛料
maxi [ˈmæksi] 迷喜装（迷喜为音译，即最大的意思），长女服（长裙，长大衣等）
    maxi-chemise 宽大女衬衫
maxicoat [ˈmæksɪkəʊt] 迷喜大衣，加长外套
maxidress [ˈmæksɪdres] 迷喜女装，加长女服
maxijacket [ˈmæksɪdʒækɪt] 加长夹克衫，加长上衣
maxilength [ˈmæksɪleŋθ] 拖地裙装
maxishorts [ˈmæksɪʃɔːts] 长度短裤（较长的短裤）
maxiskirt [ˈmæksɪskɜːt] （长及脚踝

的）迷喜裙，超长裙

measure [ˈmeʒə] 量度，测量；尺寸；量具
 loose *measure* 宽松尺寸
 make-to-*measure*(MTM) 按尺寸定做
 square *measure* 角尺
 tape *measure* 卷尺，软尺，带尺
 yard *measure* 码尺（直尺或卷尺）

measurement [ˈmeʒəmənt] 量身，量度；（测量得的）尺寸
 basic *measurements* 基本量身（法）
 body *measurement* 人体尺寸；量身
 correct *measurements* 正确量身（法）
 finished (cutting) *measurement* 成品（裁剪）尺寸
 girth *measurement* 围度尺寸；腰围尺寸
 horizontal *measurement* 横向尺寸
 measurement allowance 尺寸允许公差
 measurement discrepancy 尺寸差异
 packing *measurement* 包装尺寸
 taking *measurement* 量身，采寸
 vertical *measurement* 垂直尺寸

mellow [ˈmeləʊ] (= overlocking) 包缝，锁边（Mellow 是美国一家著名的包缝机制造商，故常将 *mellow* 作包缝用）

melton [ˈmeltən] 麦尔登呢（音译，一种粗纺呢绒）
 collar *melton* 领底粗呢
 coloured *melton* 彩色麦尔登
 pure wool *melton* 全毛麦尔登
 wool/viscose *melton* 毛黏（混纺）麦尔登
 worsted *melton* 精纺麦尔登

meltonette [ˈmeltənet] 麦尔登薄呢（多用于女服）

mending [ˈmendɪŋ] 织补，缝补；修补

menswear [ˈmenzweə] 男服，男装

merino [məˈriːnəʊ] 美利奴羊毛；美利奴毛织物

mesh [meʃ] 网眼；网织品，网眼布
 polyester(nylon) *meshes* 涤纶（锦纶）网眼布（做棒球帽等用）

metallasse [metalɑs] 〔法〕金属色泽布料

meter(= metre) [ˈmiːtə] 米，公尺

meterstick [ˈmiːtəˌstɪk] 米尺（一米长有刻度的直尺）

micro [ˈmaɪkrəʊ] 超迷你装；迷哥装，露股装（短至大腿中部或更短）

microfibre [ˈmaɪkrəʊfaɪbə] 超细纤维，微纤维；（超细纤维织的）细绒布
 lilion *microfibre* 力莱超细纤维（欧洲新一代聚酰胺纤维）

micro-mini [ˈmaɪkrəʊmɪnɪ] 超迷你裙，迷哥裙

microskirt [ˈmaɪkrəskɜːt] 超迷你裙，迷哥裙

middle [ˈmɪdl] 中间；身体的中部，腰部
 round the *middle* 腰围

middress [ˈmɪddres] 中空装

middy [ˈmɪdi] 水手领上衣（罩衫）

midi [ˈmɪdi] 迷地装；迷地裙（齐腿肚半长裙）

midriff [ˈmɪdrɪf] 中腹部；女露腰上衣，蜜多夫装（音译）
 bare *midriff* 露脐装，露腰装

midsole [ˈmɪdsəʊl] 鞋中底（夹层底）

millinery [ˈmɪlɪnəri] 〔总称〕女帽；

妇女头饰
millinery shop 女装帽店

mini ['mɪnɪ] 迷你装；迷你裙，超短裙

minibikini [ˌmɪnɪbɪ'kiːnɪ] 超短两截式女泳装

minicoat ['mɪnɪkəʊt] 超短外套

minidress ['mɪnɪdres] 超短裙套衫；迷你连衣裙

minikilt ['mɪnɪkɪlt] 超短（苏格兰）褶裙

minipants ['mɪnɪpænts] 〔复〕超短裤，热裤

minisack ['mɪnɪsæk] （宽松）超短上衣

minishorts ['mɪnɪʃɔːts] 〔复〕超短裤

miniskirt ['mɪnɪskɜːt] 超短裙，迷你裙

  micro *miniskirt* 超短迷你裙

minisuit ['mɪnɪsjuːt] 短西服上衣

mink [mɪŋk] 水貂皮；貂皮外衣

mismatch ['mɪs'mætʃ] 错配；配合不当

  mismatch of stripes(checks) 〔检〕对条（格）不准

mitaine [mɪtɛn] 〔法〕手套；（女用）露指手套

mitt [mɪt] （女用）露指长手套；连指手套（拇指单开同 mitten）〔复〕棒球手套；〔复〕拳击练习手套（击沙袋用）

  protective *mitt* 防护手套

mitten ['mɪtn] 连指手套；（女用）露指长手套（=mitt）；〔复〕拳击手套

  diving *mittens* 潜水连指手套
  ski *mittens* 滑雪连指手套

mobile ['məʊbaɪl] 手机（= mobile phone）

moccasins ['mɒkəsɪnz] 〔复〕莫卡辛鞋（音译）；软帮鞋；软拖鞋；（北美印第安人所穿的）鹿皮鞋；软皮革鞋

modal [mɒ'daɪl] 莫代尔（一种新型的高强和高湿模量纤维素再生纤维）

mode [məʊd] 样式；模式；流行；流派；型

  à la *mode* 〔法〕流行
  out of *mode* 不流行，过时
  Paris *mode* 巴黎时装
  the *mode* 时装；时式

model ['mɒdl] 模型；原型；样式；时装（美术）模特儿；人体假人（人体模型）；胸架

  British *model* 英国西装造型
  fashion *model* 时装模特儿
  living *model* 真人模特
  machine *model* 机型
  model contest 模特比赛
  model number 型号
  model of human body 人体模型
  model plate 型板
  standing *model* 做模特儿
  the latest *model* 最新样式
  unisex *model* 不分性别的模特儿

model(l)ing ['mɒdlɪŋ] 造型；建模；模特儿职业

  model(l)ing of trousers(skirt) 裤（裙）型
  sleeve(collar) *model(l)ing* 袖（领）型
  3D human *model(l)ing* 三维人体建模

modiste [məʊ'diːst] 女裁缝；女装店

mohair [ˈməʊheə] 马海毛,〔粤〕嘛唏；马海毛织物
moire [mwɑː] 云纹绸，波纹绸，波纹布
 acetate *moire* 醋酸人丝波纹绸
 moire antique 云纹绸，波纹绸
moleskin [ˈməʊlskɪn] 鼹鼠（毛）皮；厚斜纹棉绒布；〔复〕厚斜纹棉绒布裤
monofilament [ˌmɒnəˈfɪləmənt] 单纤维丝，单丝
monokini [ˌmɒnəˈkiːni] 超短女三角泳裤；（男子）超短裤
motif [məʊˈtiːf] （图案）基本花纹，色彩；（衣服）花边；（设计）主题
 beaded *motif* 珠饰花边
 braid *motif* 编结花边
 colour *motif* 色彩基调
 decorative *motif* 装饰花边
 motif of design 设计主题
motley [ˈmɒtli] 杂色布；杂色呢；杂色衣服，彩衣
motor [ˈməʊtə] 电动机，马达
mo(u)ld [məʊld] 模子，冲模，模具
mourning [ˈmɔːnɪŋ] 丧服（做丧事穿）
movement [ˈmuːvmənt] 〔设〕运动感

muff [mʌf] （女用）皮手筒，手笼，暖袖
muffler [ˈmʌflə] 围巾；厚手套；拳击手套
mufti [ˈmʌfti] 便服，便衣
mukluk [ˈmʌklʌk] 因纽特人用的毛皮长靴（海豹皮制）
mule [mjuːl] 室内穿的无后跟女式拖鞋；无后跟拖鞋式女装皮鞋
mull [mʌl] 漂白细布；人丝薄绸
multicolour [ˌmʌltiˈkʌlə] 多种色彩，多彩
muscle [ˈmʌsl] 袖臂，上袖
 muscle tanker 无袖圆领衫
mushroom [ˈmʌʃrʊm] 女用蘑菇形扁帽
muslin [ˈmʌzlɪn] 平纹细布；薄纱织物，麦斯林（音译）
 foundation *muslin* （上胶）硬衬里纱布
 muslin delaine 细薄平纹毛织物
 polka dot *muslin* 圆点花细平布
 sheer *muslin* 薄纱
 silk *muslin* 全丝薄纱
 twill *muslin* 薄毛哔叽

N

nail [neɪl] （手、脚）指甲；小钉片；五爪纽
 ear *nail* 耳钉
 nail file 修甲锉
 nail polish(enamel) 指甲油
 nose *nail* 鼻钉
 shoe *nail* 鞋钉
nailhead [ˈneɪlhed] 爪钉；点子花纹细呢
nainsook [ˈneɪnsʊk] 南苏克布（音译，全棉薄平布）
nankeen [nænˈkiːn] [nænˈkɪn](=nankin) 南京棉布（本色棉布），〔复〕本色棉布裤

printed blue *nankeen* 毛蓝印花布
nanofiber = (**nano-fiber**) [ˈnænəufaɪbə]
纳米纤维
nape [neɪp] 后颈；领背
    nape to centre waist 背长；腰直
nappa [ˈnæpə] 纳帕革（软羊皮革）；
光面皮
    pig *nappa* 猪光面皮
navel [ˈneɪvəl] 肚脐，脐部
neck [nek] 颈；颈围；领围；领圈
    back *neck* 后领圈
    button *neck* （针织衣）扣领圈
    front (back) *neck* point 前（后）颈点
    low cut *neck* 低开领圈
    neck apart 领圈宽
    neck around 领围
    neck base 颈根
    neck drop 领深
    neck hole 领孔，领圈
    neck waist 背长
    off turtle *neck* 离脖高套领
    open front *neck* 前开领圈
    polo *neck* 马球领；高圆翻领
    rib *neck* 罗纹领圈
    roll *neck* 可翻折的高领；高翻领衣服
    square *neck* 方翻领
    V-*neck* V形领圈，尖领圈
    Y-*neck* Y形领圈（半开襟，呈Y形）
neckband [ˈnekbænd] （装饰用）领圈；领巾，衬衫领；领子；立领
neckcloth [ˈneklɔːθ] 领饰，颈饰，领巾；(旧式) 领结
neckerchief [ˈnekətʃɪf] 围巾，颈巾，领巾；妇女颈饰

necklace [ˈneklɪs] 颈饰；项链，〔粤〕颈链；项圈
    amber *necklace* 琥珀项链
    Bohemian *necklace* 波希米亚项链（异国风情，多层或单层）
    bullet *necklace* 子弹头项链
    colour gold *necklace* 彩金项链
    coral *necklace* 珊瑚项链
    gold(silver) *necklace* 金（银）项链
    lariat *necklace* 套索式项链
    multistrand pearl *necklace* 多层珍珠项链
    natural stone *necklace* 天然石项链
    rhinestone *necklace* 仿钻项链，水钻项链，闪石项链
    sweater *necklace* 毛衣项链
    teardrop *necklace* 晶莹泪滴项链
    wood-beaded *necklace* 木珠项链
necklet [ˈneklɪt] 项饰；小项圈；皮围巾
neckline [ˈneklaɪn] 领口，领窝；领线，领孔线；领圈；颈圈；开领
    asymmetrical *neckline* 不对称领口
    bateau *neckline* 船形领口
    boat *neckline* 船形领口，一字领口
    bottle *neckline* 瓶颈领领口（高领口）
    bow *neckline* 蝴蝶结形领口
    brick *neckline* 矩形领口
    camisole *neckline* 背心领口；花边胸衫领口
    cardigan *neckline* 开襟毛衫领口，卡帝冈领口（音译）
    Chinese *neckline* 中装领领口
    cowl *neckline* 垂褶领领口，卡尔

领口（音译）
crew neckline 水手领领口；圆领口
cut-away neckline 大圆领口
darted neckline 短缝领口
degage neckline 宽松领口；离脖领口
diamond neckline 钻石形领口
double neckline 二重领领口
draped neckline 垂坠领口，自然皱领口
draped twist neckline 褶皱扭结领口
drawstring neckline 束带领领口，伸缩型领口
Dutch neckline 荷兰领领口（圆形领口）
far-away neckline 远离领领口
frilled neckline 饰边领领口
gown neckline 教士袍领领口
halter neckline 挂脖领口，套索领口，吊带领口
heart-shaped neckline 鸡心领口
Henry neckline 亨利领领口（前面半开襟加纽）
hooded neckline （风帽）头巾形领口
horse-hoof neckline 马蹄形领口
keyboard neckline 键盘形领口
keyhole neckline 匙孔形领口
loose neckline 〔检〕领离脖（领口太松）
low (high) neckline 低（高）领口
neckline not smooth 〔检〕领圈不圆顺
oblique neckline 斜领领口
oblong neckline 长椭圆形领口

off neckline 宽开领口，一字领口，离脖领口
off shoulder neckline 低肩领口
open neckline 开衩领口
oval neckline 椭圆形领口，蛋形领口
placket neckline 半开襟领口
plunging neckline 深插领口（大V形）
polo neckline 马球衫领口
rib neckline 罗纹领口
round neckline 圆形领口
scalloped neckline 扇贝形领口
scarf neckline 围巾形领口
scooped neckline 勺形领口，大圆领口
slashed neckline 长裁领口
slit neckline 开缝领口；一字领口
slot neckline 狭长缝领口
square neckline 方形领口
stand-away neckline 直离领领口
strapless neckline 无吊带领口
strap neckline 条形领口
surplice neckline 法衣领领口，和尚袍领口
sweetheart neckline 爱心形领口，情侣领口，鸡心领口
tab neckline 扣襻领口
tie neckline 打结领口
tight neckline 〔检〕领卡脖（领口太紧）
trapeze neckline 梯形领口
tucked neckline 褶裥形领口
turtle neckline 玳瑁领口，龟颈形领口
U-neckline U形领口
uneven neckline 〔检〕领窝不平

V-*neckline* V形领口
waterfall *neckline* 垂瀑形领口
wide vee *neckline* 宽V形领口
wrap *neckline* 披肩形领口
zigzag *neckline* 锯齿形领口

neckpiece ['nekpiːs] 领饰；皮围巾；领圈

necktie ['nektaɪ] 领带；领结
bar-shaped *necktie* 棒形领带
continental *necktie* 欧洲大陆型领带
knitted *necktie* 针织领带
necktie with zipper 拉链领带
pointed end *necktie* 尖头领带
polyester *necktie* 涤纶领带
real(pure) silk *necktie* 真丝领带
reptile *necktie* 蛇皮领带
square-end *necktie* 方头领带
stylish *necktie* 时款领带
woolen *necktie* 毛呢领带

neckwear ['nekweə] （总称）颈部服饰（领子、领带、围巾等）

needle ['niːdl] 针；缝衣针；编织针
air-cushion *needle* detector 气垫式验针机
automatic conveyer *needle* detector 自动带式验针机
crewel *needle* 长眼绣花针
darning *needle* 织补针
desk top *needle* detector 台式验针机
embroidering *needle* 绣花针
hand *needle* 手缝针
hook *needle* 钩针
knitting *needle* 针织机针；毛衣针，棒针
mending *needle* 织补针

needle and thread 针线
needle book （书形）插针垫
needle breakage 断针
needle clamp 针夹
needle detection 探针，验针（检查隐藏在衣物里的断针）
needle detector 验针器
needle for sewing machine 缝纫机针
needle gauge 针幅
needle groove 针槽
needle of size 14 14号针
needle spacing(pitch) 针距
packing *needle* 缝包针
sewing *needle* 缝纫用针
sharps *needle* 手缝针
spare *needle* 备用针
tagging *needle* （打吊牌用的）胶枪针
triple(single, double) *needle* 三（单、双）针

needlebar ['niːdlbɑː] （缝纫机的）针天心，针杆，针把

needlepoint ['niːdlpɔɪnt] 针尖；针绣花边

needlework ['niːdlwɜːk] 缝纫业；刺绣活

negligee ['neglɪʒeɪ]〔英〕(= neglige ['neglɪʒeɪ]〔法〕)（卧室穿）宽松便服；女长睡衣；女式晨衣

nemuriana [nɪ'mjuːjenə] 睡扣眼（不打鸽眼孔的扣眼）

nightcap ['naɪtkæp] 睡帽

nightclothes ['naɪtkləʊðz]〔复〕睡衣

nightdress ['naɪtdres] （女、童）睡衣；睡袍

nightgown ['naɪtɡaʊn] （女、童）睡

衣；睡袍；（宽松，舒适的）夜间室内衣

nightie [ˈnaɪtɪ] 女睡衣；睡袍

nightshirt [ˈnaɪtʃɜːt] 男用长睡衣；衬衫式睡袍

nightwear [ˈnaɪtweə] （总称）睡衣；夜间家用室内衣服

nighty [ˈnaɪtɪ] （女、童）睡衣

ninon [nɪnɔ̃] 〔法〕尼龙绸；薄绸

noil [nɔɪl] 精梳短毛；精梳落棉；针板落绵

noilcloth [ˈnɔɪlklɒθ] 绵绸；绌丝织物

nose [nəʊz] 鼻子
    nose lifting （整容）隆鼻
    nose rag （俚语）手帕

notch [nɒtʃ] 领嘴，缺嘴，领豁口，衩口，剪口，刀口，刻口，〔粤〕（领）扼位
    lapel *notch* 驳领缺嘴

notcher [ˈnɒtʃə] （剪口用）记号剪

notching [ˈnɒtʃɪŋ] 〔工〕（裁剪时）打剪口，打刀口

notebook [ˈnəʊtbʊk] （制图用）笔记本

notions [ˈnəʊʃənz] 〔复〕〔美〕个人衣物（针线等小件用品）

number [ˈnʌmbə] 数；数字；号码（常略写为 No. 复数略写为 Nos.）
    contract *number* 合同号
    order *number* 订单号
    reference *number* 参考号
    style *number* 款号
    telephone (fax) *number* 电话（传真）号码

numbering [ˈnʌmbərɪŋ] 〔工〕（裁片）打号，编号

nylon [ˈnaɪlən] 耐纶（音译），尼龙，锦纶；尼龙制品
    nylon interweave 锦纶交织布
    nylon rip-stop 锦纶格布
    nylon triloba1 锦纶闪光布料（做运动服装等用）
    woolly *nylon* 仿毛锦纶布料

nylonpalace [ˈnaɪlənpælɪs] 锦纶纺，尼丝纺，锦纶绸
    coated *nylonpalace* 涂层尼丝纺
    printed (dyed) *nylonpalace* 印花（染色）尼丝纺

O

obi [ˈəʊbi] （日本妇女、儿童和服用的）宽大腰带

off-the-peg [ɒfðəˈpeg] 现成服装，成衣

off-the-rack [ɒfðəˈræk] 现成服装，成衣

oil [ɔɪl] 油；〔复〕油布雨衣；油布衣裤
    oil check window （缝纫机头）加油窗
    machine *oil* 机油

oilcan [ˈɔɪlkæn] （缝纫机用）油壶，加油器

oilcloth [ˈɔɪlklɒθ] 油布；漆布

oilskin [ˈɔɪlskɪn] 防水油布；油布雨衣；防水套装

one-piece [ˈwʌnpiːs] 一件头服装；连衣裙；连衣裤

op [ɒp] （具抽象派绘画风格的）光效应绘画艺术，视幻艺术，欧普艺术（也作 op art 或 optical art）
opelon [ˈəʊpəlɒn] 奥佩纶（一种聚氨酯弹性纤维）
opening [ˈəʊpənɪŋ] （衣服的）开襟；开门，门档；端口；（裁片上的）剪口
 back *opening* 后开襟，后开门
 backstitch at *opening* end 剪口末端处倒针
 collar *opening* 领口
 crank *opening* 曲襟
 cuff *opening* 袖口；袖衩
 front *opening* 前开门；（中装）大襟，对襟
 hem *opening* 下摆
 leg *opening* 裤脚口；脚口宽，〔粤〕脚阔
 neck *opening* 领口，领圈；领脚长
 neckline *opening* 领围
 opening end （服装）开口止点
 pocket *opening* 袋口
 short *opening* 短开口
 side *opening* 偏襟，大襟，胁开口
 sleeve *opening* 袖口
openwork [ˈəʊpənwɜːk] 透孔织物；网眼制品
operation [ˌɒpəˈreɪʃn] 操作；工序
 cutting *operation* 裁剪操作
 sewing *operation* 缝纫操作
order [ˈɔːdə] 定购；定制；订单；指令；〔设〕秩序；程序；汇票
 buyer's *order* 客户订单
 cutting *order* 裁剪通知单
 formal *order* 正式订单
 make-to-*order*（MTO） 按订单生产制作
 online *order* 网上定购；在线定单
 order of uniform 定做制服
 order processing 订单处理
 production *order* 生产通知单
 purchase *order* 采购订单
 sample *order* 样品订单，板单
 trial *order* 试单
organdie(= organdy) [ˈɔːgəndɪ] 蝉翼纱（透明极薄棉布）
organza [ɔːˈgænzə] 透明硬纱（丝或人丝制）
orlon [ˈɔːlən] 奥纶（聚丙烯腈纤维，美国商品名）
ornament [ˈɔːnəmənt] 装饰；装饰品，饰物
 bone *ornament* 骨饰品
 braid *ornament* 编结装饰
 clothes and *ornament*s of every dynasty 历代服饰
 crystal *ornament* 水晶饰物
 flower *ornament* 花饰，饰花（服装上的花卉装饰）
 hair *ornament*s 发饰；发饰品
 nose *ornament* 鼻饰（鼻环等）
 stitches *ornament* 线迹装饰
 tapes and buttons for *ornament* 装饰带扣
 technical *ornament* 工艺装饰
 wooden *ornament* 木制饰物
orris [ˈɒrɪs] （装饰用）金银线编带，花边，刺绣
ouch [aʊtʃ] 胸针，饰针（指镶宝石）
outergarment [ˈaʊtəgɑːmənt] 外衣
outerwear [ˈaʊtəweə] （总称）外衣；外套；户外装
 knitted *outerwear* 针织外衣

outfit ['aʊtfɪt] （在某种场合穿的）套装
　　gym *outfit* 运动衫裤，运动套装；体操服
　　hacking *outfit* 骑装
outleg ['aʊtleg] 裤外长
outline ['aʊtlaɪn] 轮廓，外形；（服装的）轮廓线；（制图）实线
　　dress *outline* sketch 服装轮廓素描
　　raglan sleeve *outline* 前后连肩线（连肩袖的轮廓线）
outseam ['aʊtsiːm] 明缝，裤侧缝，裤梐缝，〔粤〕外肐骨；裤外长
outsize ['aʊtsaɪz] 超常尺码，特大尺码的服装（鞋、袜）
outsole ['aʊtsəʊl] （鞋、靴的）外底，大底；皮鞋（皮靴）跟
oven ['ʌvn] （整烫用）烘箱，〔粤〕焗炉
　　wrinkle free batch *oven* 抗皱处理烘箱
overall ['əʊvərɔːl] 〔英〕宽大罩衫（家里穿）；〔复〕工作服，背带工装裤；〔英〕紧身军裤
　　bib *overalls* 带护兜的背带工装裤
　　denim *overalls* 牛仔工作服
　　overalls for medical personnel 医务工作服，医护服
overblouse ['əʊvəblaʊz] 女罩衫
overcasting ['əʊvəkɑːstɪŋ] （片料）锁边，拷边，包缝
overclothes ['əʊvəkləʊðz] 外衣，罩衣
overcoat ['əʊvəkəʊt] 大衣，外套，〔粤〕褛；罩衣
　　fleece-lined *overcoat* 驼绒大衣，绒布里大衣
　　fur *overcoat* 裘皮大衣

　　raglan three-piece sleeved *overcoat* 连肩三片袖外套
　　single(double)-breasted *overcoat* 单（双）排扣大衣
　　spring(light) *overcoat* 夹大衣
overcoating ['əʊvəkəʊtɪŋ] 外套料，大衣呢
　　acrylic *overcoating* 腈纶大衣呢
　　acrylic jacquard *overcoating* 腈纶提花大衣呢
　　angora cashmere *overcoating* 兔羊绒大衣呢
　　camel hair *overcoating* 驼毛大衣呢
　　cashmere *overcoating* 羊绒大衣呢
　　embossed *overcoating* 拷花大衣呢
　　fancy *overcoating* 花式大衣呢
　　midfibre bulked *overcoating* 中长膨体大衣呢
　　pure wool *overcoating* 全毛大衣呢
　　rabbit hair *overcoating* 兔毛大衣呢
　　snowflake *overcoating* 雪花大衣呢
　　velours *overcoating* 拉绒大衣呢
　　vicuna *overcoating* 骆马毛大衣呢
　　wool/viscose *overcoating* 毛黏（混纺）大衣呢
　　wool/viscose tartan *overcoating* 毛黏格子大衣呢
　　yak hair *overcoating* 牦牛绒大衣呢
overdress ['əʊvədres] 外衣，外套
overedge ['əʊvəedʒ] 包缝
overedger ['əʊvəedʒə] 包缝机
overedging ['əʊvəedʒɪŋ] 包边缝纫，包缝
overgarment ['əʊvəgɑːmənt] 外衣；大衣
overgown ['əʊvəgaʊn] 罩袍，外袍
overlap ['əʊvəlæp] （门襟）交搭，重

叠,搭门,〔粤〕搭位;袖衩搭边
  front(back) overlap 衣前(后)搭门
overlay ['əʊvəleɪ] (装饰用)包镶物;(鞋)装饰片;(苏格兰英语)领带
overlocking [ˌəʊvəˈlɒkɪŋ] 包缝,锁(拷)边,〔粤〕钑骨
overpull ['əʊvəpʊl] 套衫
overseam ['əʊvəsiːm] 包缝
overshirt ['əʊvəʃɜːt] (穿在内衣外)衬衫;罩衫
overshoes ['əʊvəʃuːz] 〔复〕套鞋(套在鞋上的鞋,防水或取暖)
oversize ['əʊvəsaɪz] 特大型尺码(服装)

overskirt ['əʊvəskɜːt] 〔美〕(穿于裙上较短的)半裙,上套裙,外裙
  astride overskirt 女骑马裙
oversleeve ['əʊvəsliːv] 袖套
overstitch ['əʊvəstɪtʃ] (装饰用)明线迹
oversuit ['əʊvəsjuːt] 外套装(穿在衣服外面的套装)
oxford ['ɒksfəd] 牛津布(棉布料,做衬衫用);〔复〕牛津衫
  army oxford 军用衬衫布
  nylon oxford 锦纶牛津布
  oxford bags 〔英〕(宽松)袋形裤,牛津袋裤
  polyester textured oxford 涤纶低弹牛津布

P

pack [pæk] (一)包;(一)捆;包装;包装材料
  pack belt 打包带
  pack cloth 包装布
  pack thread 包装用线绳
  some packs of cut parts 几捆(裁)片料
package ['pækɪdʒ] 包;捆;包装;装物品
packer ['pækə] 包装工,打包工;打包机
packing ['pækɪŋ] 包装;装箱,打包;包装法
  compression packing 压缩包装法
  export packing 出口包装
  hanger(fold) packing 挂(折)装
  mixed(assorted) packing 搭配包装,混色混码包装,混装
neutral packing 中性包装
outer(inner) packing 外(内)包装
overage packing 溢装(装箱数多于订单数)
packing instruction 包装方法
packing of exports 出口商品包装
packing sheet 包装布;包装单(纸)
packing specification 装箱清单
shortage packing 短装(装箱数少于订单数)
solid packing 单色单码包装,单一装
wrong packing assortment 〔检〕错误装箱搭配

wrong packing quantity 〔检〕装箱数量不符

pad [pæd] 衬垫，垫塞
 bust(bra) pad 胸罩衬垫，胸垫
 compound bust pad 复合胸垫
 covered shoulder pad 布包肩垫
 detachable shoulder pad 活动肩垫
 egg pad （熨烫衣肩等部位用的）馒头烫垫，熨烫馒头
 elbow pad 肘垫，护肘
 evaporative cooling brow pad 散热降温头盔内垫
 felt pad 毡衬垫
 foam shoulder pad 泡沫塑料肩垫
 handpull shoulder pad 手装肩垫
 heat setting shoulder pad 热定型肩垫
 heel pad 鞋跟垫片
 hip pad 臀垫
 knee pad 膝垫，护膝
 lapel pad 驳领垫
 needle punching shoulder pad 针刺肩垫
 packing pad 包装衬垫
 pad joining 〔工〕上衬垫
 shoe pad 鞋衬垫，鞋垫
 shoulder pad 肩垫，垫肩，〔粤〕膊头棉
 special-shaped shoulder pad 特型肩垫
 sponge shoulder pad 海绵肩垫
 steam moulded shoulder pad 蒸汽定型肩垫
 under collar pad 领垫，领底呢

padding [ˈpædɪŋ] （衣服用）衬垫，垫塞；垫料；填料
 ironing padding 熨烫垫布
 paper padding 纸衬垫
 polyester padding 涤纶棉，喷胶棉

pagne [ˈpeɪɡnɪ] （非洲）棉腰带；腰布卷身而成的裙子（非洲风格）

pagri [ˈpʌɡriː] 印度头巾

paillette [pælˈjet] （装饰女服或织物的）闪光珠片；金属亮片
 embroidered dress with paillettes and beads 珠绣服装

pair [peə] 一对，一双
 a pair of scissors 一把剪刀
 a pair of socks(shoes) 一双袜子（鞋子）
 two pairs of trousers 两条裤子

pajamas (= pyjamas) [pəˈdʒɑːməz] 〔复〕睡衣裤，宽松裤
 check pajamas 格子布睡衣裤
 embroidered silk pajamas 绣花丝绸睡衣裤
 floral pajamas 印花睡衣裤
 notch collar pajamas 翻领睡衣裤
 pajamas with waistband 束带睡衣裤
 palazzo pajamas 宽松女套装
 patchwork pajamas 拼缝睡衣裤
 striped pajamas 条纹布睡衣裤
 towelling pajamas 毛巾睡衣裤

palace [ˈpælɪs] 派力司（音译，一种精纺呢绒）
 polyester/ramie palace 涤/麻派力司
 wool/polyester palace 毛/涤派力司

paletot [palto] 〔法〕（男）礼服大衣；（男、女）宽外套，短大衣
 petit paletot 〔法〕宽短外套（衣

摆展宽）

pallium ['pæliəm] （古希腊、罗马）大披肩；（天主教大主教的）白色羊毛披肩；袈裟

palm [pɑ:m] 手掌，掌心

panel ['pænl] （衣服上的）嵌条；镶拼布块；嵌板
 back *panel* 后幅，后片
 back *panel* seam 背面嵌缝
 button *panel* 暗门襟；裤里襟
 centre *panel* 中心嵌条
 front *panel* 前幅，前片
 rib knit *panel* 罗纹针织嵌布
 right(left) side *panel* 右（左）嵌边
 side *panel* 侧嵌条，侧嵌边；侧幅
 stitch *panel* 花型板

panier(=pannier) ['pæniə] （展宽女裙用）裙撑，鲸骨圈

panne [pæn] 平绒

panoply ['pænəpli] 全副甲胄；礼服

pantacourt [pãtɑku:r] 〔法〕（裤腿较短的）中长裤

pantalets [ˌpæntə'lets] 〔复〕宽松女裤；（骑自行车）灯笼裤

pantalon [pãtalɔ̃] 〔法〕裤子，长裤；（旧时）女衬裤
 pantalon à bas évasé 〔法〕喇叭裤
 pantalon collant 〔法〕紧身裤
 pantalon corsaire 〔法〕齐腿肚的小裤脚裤
 pantalon de dame 〔法〕女式长裤
 pantalon de grandpére 〔法〕祖父型裤（高腰，裤头宽松，裤脚细窄）
 pantalon de maharajah 〔法〕君王裤（似马裤，臀围宽敞，膝下渐窄）
 pantalon indian 〔法〕印度风格裤（臀围宽敞，裤脚变细，呈纺锤形）
 pantalon masculin 〔法〕男式风味裤（皮带束腰，斜装，卷边裤脚）
 pantalon minee 〔法〕细长裤，细窄裤

pantaloon [ˌpæntə'lu:n] 〔常用复〕马裤；〔美〕裤子
 clown *pantaloons* 小丑裤（马戏或喜剧表演穿用）

pantdress ['pæntdres] 连衣裙裤，裙裤套装

pantee ['pænti] 女裤；童裤

pantie(=panty) [pæn'ti:] 〔常用复〕（女、童）紧身短裤；内（衬）裤
 garter *panties* 吊袜内裤
 golf *panties* 灯笼裤
 nylon *panties* 锦纶三角短裤
 short *panties* 女紧身三角裤

pantihose(=panty hose) ['pæntihəuz] 〔单同复〕连裤袜
 allover shaping *pantihose* 高腰塑形连裤袜

panting ['pæntiŋ] 裤料

panti-slip ['pæntislip] 背心热裤装

pantistocking ['pæntistɒkiŋ] 连裤袜

panti-tights ['pæntitaits] 卫生衣裤

pantone ['pæntəun] （美国）潘通色彩体系（以开发和研究色彩而闻名全球）；潘通色卡

pants [pænts] 〔复〕〔美〕裤子；（女、童）紧身短衬裤；〔英〕男短衬裤；紧身长衬裤

above-the-knee *pants*　膝上短裤
anchored *pants*　踩脚裤（针织紧身裤，脚口套踩在脚下）
ankle banded *pants*　束脚口裤
ankle tied *pants*　束脚裤
army *pants*　军裤
athletic *pants*　运动裤
baggy *pants*　袋型裤
bell-bottom *pants*　喇叭裤
bib *pants*　围兜裤，围兜式背带裤
body *pants*　（针织）紧身裤
bootleg(bootscut) *pants*　配靴穿的裤子，（配）靴裤
boss *pants*　老板裤（较宽松休闲裤）
brushed *pants*　起绒裤；磨毛裤
business *pants*　职业装裤，办公裤
camouflage(camo) *pants*　迷彩裤
canvas *pants*　帆布裤，珠帆裤
Capri(capri) *pants*　卡普里裤（锥形七分裤），女紧身长裤
cargo *pants*　货船裤（船员工作裤，两侧有大风琴袋）；大贴袋休闲裤
carpenter *pants*　木工用套裤
casual *pants*　便裤，休闲裤
children's *pants*　童裤
Chinese style *pants*　（旧时）中式裤（大腰头，无前后裆之分）
chop *pants*　截筒裤（裤腿长至脚踝上约15厘米处）
cigarette *pants*　香烟裤（直筒裤）
city *pants*　街市裤
corduroy *pants*　灯芯绒裤
corsair *pants*　海盗裤（齐腿肚的紧身裤）
cotton padded *pants*　棉裤
cowboy *pants*　牛仔裤

cuffed *pants*　卷脚裤（脚口外卷边）
culotte *pants*　裙裤
cut off *pants*　截筒裤
dance *pants*　舞蹈裤
deck *pants*　甲板短裤（长于膝的贴身短裤）
denim *pants*　牛仔裤；劳动布裤
dress *pants*　正装长裤，礼服裤
easy *pants*　（宽松舒适的）便裤
far-infrared cotton *pants*　远红外棉裤（具保暖、保健作用）
fitness *pants*　健身裤
fleece skate *pants*　溜冰绒裤
full *pants*　（宽松的）袋型裤
gaucho *pants*　高乔牧人裤；（南美）牛仔裤；女宽松裤
harem *pants*　后宫裤，闺阁裤（裤身宽大，脚口缩紧）
hot *pants*　热裤（超短裤）
houseboy *pants*　工作裤
interlock *pants*　棉毛裤
jean *pants*　牛仔裤
jersey *pants*　针织（布）裤
jockey *pants*　（赛马）骑师裤
jodhpur *pants*　马裤
knee *pants*　齐膝裤，齐膝短裤
3/4(7/8) length *pants*　四分之三（八分之七）裤，七分（九分）裤
leisure *pants*　休闲裤
low-cut crotch *pants*　低裆裤
low-waist *pants*　低腰裤
massotherapy *pants*　（有按摩作用的）按摩裤
maternity *pants*　孕妇裤
military *pants*　军装裤，军裤
multi-pocket *pants*　多袋裤

nappy pants　尿布裤（裤状尿布）
nylon pants　锦纶裤
open-seat pants　（童）开裆裤
over pants　外裤，罩裤，套裤
painter pants　画家工作裤；油漆工裤
palazzo pants　宫殿女裤（裤长及地，裤脚展宽，类似裙裤）；女宽松阔脚裤
parachute pants　降落伞裤（薄料，宽松，束腰和裤脚）
peg top pants　陀螺形裤，萝卜裤
pipestem pants　直筒裤
pirate pants　海盗裤（细裤筒裤）
pleat(flat) front pants　腰头前打褶裥（无褶裥）裤
polyester micro-fibre pants　涤纶桃皮绒裤
pull on pants　松紧裤
quick-dry pants　快干裤
ramie/cotton pants　麻/棉裤
rock pants　摇滚裤（细窄贴脚裤）
roll-leg pants　卷裤管裤，卷脚裤
running pants　赛跑短裤
sailor pants　水手裤
salopette pants　沙罗佩裤（背带工作裤）
semi-baggy pants　半袋型裤
seven-eighths pants　八分之七裤，九分裤
silk jersey blouse and pants　丝针织衫裤
silk noil pants　丝绸裤，绸绸裤
silk spandex pants　丝弹力裤
skinny pants　紧身裤
ski pants　（紧身）滑雪裤
sleep pants　睡裤
slouchy pants　（宽松随意的）懒散裤
snowboard pants　滑雪（板）运动裤（防风、防水、透气）
stovepipe pants　烟囱裤（无折痕的直筒裤）
suit pants　套装裤
suspender pants　吊带裤；背带裤
sweater pants　毛线裤套装
sweat pants　长运动裤，卫生裤
swim pants　游泳裤
tailored pants　西装裤
tapered pants　锥形裤，窄脚裤
T/C pants　涤/棉裤
tin-pants　防水帆布裤
toreador pants　女式半长紧身运动裤；斗牛裤
track pants　长运动裤
twill pants　斜纹布裤，斜裤
tuned air pants　空调裤（用空调布料制）
vertical pants finisher　垂直式裤子熨烫机，裤型烫机
warm pants　暖裤（比热裤稍长的短裤）
warp pants　卷裹裤（宽腰带打结的裤）
3-way convertible pants　三用可变裤（裤腿拉链自由装卸，可变为长、中、短裤）
Western pants　（美国）西部风格裤（牛仔穿）
wide leg pants　喇叭裤
wide waistband pants　宽腰头裤
yoga pants　瑜伽裤
wrinkle free pants　防皱裤，免熨烫裤

zip-off-leg *pants* 裤腿可脱卸裤（拉链开合，变化裤长）

zip skinnyleg *pants* 拉链窄脚裤

pantshoes [ˈpæntʃuːz] 〔复〕厚底鞋；与喇叭裤匹配的皮鞋

pantskirt [ˈpæntskɜːt] 裙裤，裤裙

pantsuit [ˈpæntsjuːt] 衣裤相配的女套装

panty(= pantie) [ˈpænti] （女、童）紧身短裤；内（衬）裤

  above-the-knee *panty* （短至膝上的）短裤

pantyhose [ˈpæntɪhəʊz] 女连裤袜

  control top *pantyhose* 塑身（腰、臀）连裤袜

  high waist *pantyhose* 高腰连裤袜

  lace(fishnet)*pantyhose* 网眼连裤袜

  leopard print *pantyhose* 豹纹连裤袜

pantywaist [ˈpæntɪweɪst] 衣连裤童装（衣裤于腰部用纽扣扣住）

paper [ˈpeɪpə] 纸；文件；票据

  carbon *paper* 复写纸

  commercial *paper* 商业票据

  draughting *paper* 样板纸

  drawing *paper* 绘图纸

  duplicating *paper* 复印纸

  Dupoint *paper* 杜邦纸，防绒纸（一种可增强羽绒服防绒效果的无纺布）

  interleaving *paper* 衬纸

  kraft *paper* 牛皮纸

  marking *paper* （裁剪）纸样

  moistureproof *paper* 防潮纸

  packing *paper* 包装纸

  pattern *paper* 样板纸；花样设计纸

  tissue *paper* （包装衬衫等用的）衬纸

  wrapping *paper* 包装纸

paperboard [ˈpeɪpəbɔːd] 纸板

paramatta [ˌpærəˈmætə] 毛葛；棉毛呢（棉经毛纬精纺）

parasol [ˌpærəˈsɒl] （女用）阳伞

pardessus [pardəsy] 〔法〕男大衣，男外套

parka [ˈpɑːkə] 风雪大衣，〔粤〕雪褛，派克大衣（带风帽）

  airforce *park* 空军风雪大衣

  yacht *parka* 快艇防寒大衣

part [pɑːt] （衣服、机器的）零件；部件；衣片

  clothes *parts* 衣片

  cut-*parts* 裁片，片料

  misaligned *parts* 〔检〕鸳鸯片（衣片错配）

  shaded *parts* 〔检〕部件色差

  spare *parts* 备用零部件

  toe(heel)*part* 鞋前（后）帮

pashmina [pæʃˈmiːnə] 帕西米纳披肩；羊绒围巾；克什米尔山羊绒

passementerie [pasmɑ̃tri] 〔法〕绦子；花边，金银花边；珠饰；绳饰

paste [peɪst] 糨糊

  chemical *paste* 化学糨糊

  smearing *paste* 〔工〕刮浆（在衣服有关位置均匀涂刮糨糊）

  tool of smearing *paste* 刮浆刀

pasties [ˈpæstɪz] 乳饰；乳（头）贴；粘贴式乳罩

patch [pætʃ] 补丁；补片；补缀布片；（加固用）垫布；臂章；饰章，饰牌；饰颜片（17~18世纪贵族妇女脸上的小圆贴片）

  collar *patch* 领章

  elbow *patch* 肘部垫布，肘部贴布

embroidered patch 绣花牌（章）
half-moon patch （衣后中里）半月形贴布，〔粤〕龟背
inserting patch 嵌补片
knee patch 裤膝加固布片，膝垫
leather(PU) patch 皮（PU 皮）牌
pocket patch 袋牌
reflective patch 反光牌（章）
reinforcement patch 加固布片
rubber patch 橡胶牌（章）
shoulder patch 肩章
woven patch 织章

patchwork [ˈpætʃwɜːk] （将布块）拼缝，补缀；拼缝品

pattern [ˈpætən] 式样；花样；（衣服的）纸型，纸样；（裁剪）样板；样品；样布；图案，图样
  abstract pattern 抽象图案
  basic pattern 基本式样；原型；基本纸样
  block pattern 原型纸样，裁剪样板
  card pattern 裁剪纸样
  carpet pattern 地毯图案
  classic pattern 标准纸样，古典花样
  clothing pattern 衣服纸样；服装板型
  cloth pattern （立体裁剪的）布质板样
  computer aided pattern(CAP) 计算机辅助画样
  decorative pattern 装饰图案
  digitizing pattern 数字化纸样
  drafting pattern 直接起图纸样
  dressmaker's pattern 女（童）服装纸样

  embroidered pattern 绣花图样
  fancy pattern 新奇图案
  flat pattern 平面纸样
  geometrical pattern 几何图案
  hand-drawn pattern 手绘图案
  jacquard pattern 提花图样；提花织纹
  key pattern 主样板
  master pattern 主码纸样，基码样板
  modelling (draping) pattern 立体裁剪纸样
  multi pattern （多种花样组合的）复合花样
  nature pattern （回归）自然花样
  net pattern （裁剪）净样板
  paisley pattern 帕斯力图案（色彩鲜明的涡旋花纹）
  paper pattern 纸样；软质纸样
  pasteboard pattern 硬纸板纸样，硬质纸样
  pattern construction 纸样制作；纸样结构
  pattern copy 图样复制
  production pattern 生产纸样，实样
  seam pattern 缝式，缝型
  shoe pattern 鞋样
  silhouette pattern 轮廓样板
  stitch pattern 线迹花样

patternmaker [ˌpætənˈmeɪkə] 纸型制作者；打（纸）样师

peach [piːtʃ] （织物的）磨毛
  carbon peach 炭素磨毛
  peach moss 绉绒（一种涤纶平纹时尚面料）

peachskin [ˈpiːtʃskɪn] 桃皮绒织物；磨毛布

big twill polyester *peachskin* 宽斜纹涤纶桃皮绒

full dull polyester *peachskin* 全消光涤纶桃皮绒

polyester/nylon *peachskin* 涤/锦复合桃皮绒

polyester/nylon interwoven *peachskin* 涤/锦交织桃皮绒

peacoat ['pi:kəʊt] 水手粗呢上衣；双排扣短上衣

    cropped *peacoat* 双排扣截筒上衣（短至腰部以上）

peak [pi:k] （衣着上的）突出部分；帽檐

    cap *peak* 帽檐

    lapel *peak* 驳领尖

peau de peche [pəʊdepɛʃ] 〔法〕仿丝绒薄绒布；桃皮绒

pedal ['pedl] （缝纫机的）踏板，踏脚

    pedal pushers （骑车用）长及小腿的女裤

peg [peg] （晒衣）夹子；〔复〕腿；裤子

    peg tops 上宽下窄裤子（裙子）

    wooden(plastic) *peg* 木（塑料）衣夹

peignoir ['peɪnwɑː] 〔法〕女用宽大便服（如睡衣、晨衣、浴衣等）

pelerine ['pelərːn] 女式披肩

pelisse [pə'liːs] 女用皮制长外套；皮衬里长外衣；无袖外套；斗篷；〔法〕毛皮大衣，皮袄

peltry ['peltri] 〔总称〕毛皮；皮货

pen [pen] 钢笔

    drawing *pen* 绘图笔，鸭嘴笔

    ruling *pen* 鸭嘴笔

pencil ['pensl] 铅笔；眉笔

    eyebrow *pencil* 眉笔

    eyeline *pencil* 眼线笔

pendant(=**pendent**) ['pendənt] 垂饰物（坠子、项链、耳环、手镯等）

    crystal *pendant* 水晶项链

    diamond set *pendant* 碎钻项链（项链坠上镶有细碎钻石）

    ear *pendant* 耳环

    netlike *pendant* 网状项链

    scarab *pendant* （古埃及）金龟子垂饰

    shell *pendant* 贝壳项链

peplum ['pepləm] （饰于女服腰部以下的）褶襞短裙；波形褶襞

percale [pə'keɪl] 高级密织棉布

perforater ['pɜːfəreɪtə] 打眼器

perforating ['pɜːfəreɪtɪŋ] （在裁片上）打眼

perfume ['pɜːfjuːm] 香味，芳香；香水；香料

Persian ['pɜːʃən] 波斯绸

petersham ['piːtəʃəm] 靛青珠皮大衣；靛青珠皮大衣呢；（加固裤腰，裙腰等用）彼得舍姆硬衬里

petticoat ['petɪkəʊt] 衬裙；(旧时女、童穿)裙子

    hoop *petticoat* 带裙箍的衬裙

    mini *petticoat* （极短的）迷你衬裙

    panty *petticoat* 带内裤的衬裙

pettipants ['petɪpænts] 〔复〕〔美〕女半长内（衬）裤

pettiskirt ['petɪskɜːt] 衬裙

phone [fəʊn](=**telephone** ['telɪfəʊn]) 〔口语〕电话；电话机

    mobile(cell) *phone* 手机

pickelhaube ['pɪkəlhʌb] 尖顶头盔,

钉盔（19~20世纪普鲁士和德国军人警察用）
 Prussian pickelhaube 普鲁士战盔
picot ['pɪkəʊ] 饰边小环；毛圈边；锯齿边
picture ['pɪktʃə] 图画，图片；肖像；照片
piece [piːs] 块；片；件；匹
 a piece of cloth 一块布
 a piece of satin 一匹缎子
 arranging cutted pieces 〔工〕分片（整理裁片）
 body piece 大身衣片
 crotch piece 裤裆片料
 cutted pieces 裁片，片料
 dress pieces 衣服片料
 fur piece 毛皮制品
 neck piece 领饰
 one-piece clothes 一件式服装（连衣裤或连衣裙）
 piece mix 单件衣组合（将单件衣服混合组合）
 pocket piece 口袋布
 satin embroidered piece 软缎绣片
 three-piece set knitted 针织三件套（上衣、背心、裙或帽、围巾、手套）
 woolen piece 呢绒
pigskin ['pɪgskɪn] 猪皮，猪革
pile [paɪl] 绒毛；绒面；起绒
 deep pile 长毛绒
 knitted high-pile 针织长毛绒
 raised pile 丝绒
pillbox ['pɪlbɒks] 无边平顶女帽
pilling ['pɪlɪŋ] 〔检〕布料起球疵
pillow ['pɪləʊ] 枕头
 cotton pillow 棉枕头
 eiderdown pillow 鸭绒枕头
 kapok pillow 木棉枕头
 pillow for health 保健枕头
 pillow towel 枕巾
 silk floss pillow 丝绵枕头
 sponge pillow 海绵枕头
pillowcase ['pɪləʊkeɪs] 枕套
 applique pillowcase 贴花枕套
 embroidered pillowcase 绣花枕套
 fagoting pillowcase 抽纱枕套
pin [pɪn] 针；别针；大头针；饰针
 brass pin 黄铜饰针
 breast pin 胸（饰）针；领带夹
 diamond pin 钻石别针
 drawing pin 图钉
 hair pin 发针，发簪；发夹
 hat pin 帽针
 knitting pin 针织棒针
 pearl pin 珍珠针（头部为色珠的大头针）
 pin cushion 插针垫；针插
 pin head 针头
 pin hole 针孔，针眼
 safety pin 安全别针
 scarf pin 围巾别针；领带夹
 shirt pin 衬衫别针
 tag pin （打吊牌用）胶针
 tie pin 领带别针，领带夹
pinafore ['pɪnəfɔː] （小孩用）围涎布；围裙；围兜背心（或短袖）裙
pinchers ['pɪntʃəz] 〔复〕钳子；（美俚）鞋
pinching ['pɪntʃɪŋ] 收皱，收褶
pinking ['pɪŋkɪŋ] 剪锯齿边；打饰孔
pinny ['pɪni] 围涎布；围裙
pinstripe ['pɪnstraɪp] （布料上的）细条子；细条子衣服

pintuck [ˈpɪntʌk] 细褶
piper [ˈpaɪpə] （缝纫机）绲边（拷边）装置，绲边器，〔粤〕绲边蝴蝶；镶边器；包缝机，拷边机
piping [ˈpaɪpɪŋ] 绲边；包缝，拷边；镶边；（衣服上的）管状细饰条
 contrast *piping* 拼色绲条，错色绲条
 double needle stitched *piping* （双针）压条
 piping trim 捆条
piqué [pik] 〔法〕类似灯芯绒的布；凹凸织物；珠地组织
 jacquard *piqué* 提花珠地布
 lycra *piqué* 弹力珠地布
 single(double) *piqué* 单（双）珠地布
 trueran yarn-dyed *piqué* 色织涤棉灯芯布
placket [ˈplækɪt] 前襟布；门襟，〔粤〕门筒，筒位；门襟翻边；（裙腰）开口，裙衩；（男衬衫）袖衩；裙开口处口袋；（古时）衬裙
 bump at *placket* edge 〔检〕门襟边（扣眼位）隆起
 cover(concealed) *placket* 暗门襟，〔粤〕暗筒
 crossed-over(crossover) *placket* 交叉门襟，叠门襟
 deep *placket* 长开襟；宽前襟；长开衩
 exposed under *placket* 〔检〕里襟外露，〔粤〕露底筒
 false *placket* 假门襟，〔粤〕假门筒
 fly *placket* 裤门襻，裤门襟
 half(full) *placket* 半（全）开襟

 neckline *placket* 半开襟；明门襟
 piped *placket* 绲边袖衩
 placket bubbling 〔检〕门襟起泡（黏合衬未粘牢）
 placket front 半开襟；明门襟
 placket hole （裙腰）开口
 short *placket* 短裙衩
 sleeve *placket* 袖衩
 twisted *placket* 〔检〕门襟扭曲
 under *placket* 里襟，〔粤〕底筒
 wavy *placket* 〔检〕门襟起拱（不平伏），〔粤〕筒起蛇
plaid [plæd] 格子花纹；格子花布；格子花呢；格子花呢服装
 block *plaid* 大方格
 Madras *plaid* （传统的）马德拉斯格子
 uneven *plaid* 〔检〕格子不匀
plains [pleɪnz] 〔复〕平纹布；素色布
 cotton *plains* 纯棉平布
 linen *plains* 亚麻平布
 polyester/cotton *plains* 涤/棉平布
 pure ramie *plains* 纯（苎）麻平布
 ramie/cotton blended yarn-dyed *plains* 麻/棉色织平布
 ramie/cotton mixed *plains* 麻棉交织布
plait [plæt] 褶；裥；辫状物
 box *plait* （裙的）箱型褶
 plait braid 辫形编带
 skirt *plait* 裙褶，裙裥
 top centre *plait* 明门襟
plastics [ˈplæstɪks] 〔复〕塑料；塑料制品
 foamed *plastics* 泡沫塑料
plastic TPO [ˈplæstɪktiːpiːəʊ] (**plastic**

Time, Place, Occasion) 一套服装可在任何时间、地点、场合穿用的方法

plastron ['plæstrən] 胸铠；（皮制）护胸甲；女胸衣；胸饰；上浆衬胸；（男用）带有部分前胸的衬领

plate [pleɪt] （金属）牌子；板
 brass *plate* （饰于服装上的）铜牌
 face *plate* （缝纫机上的）面板，门盖
 metal *plate* （饰于服装上的）金属牌
 needle *plate* （缝纫机）针板

platform [plætfɔːm] 〔常用复〕坡形厚底鞋

playsuit ['pleɪsjuːt] （女、童）运动衫裤；游戏套装

playwear ['pleɪweə] 游戏服；运动服

pleat [pliːt] （衣服、裤、裙的）褶、褶裥，活褶
 accordion *pleats* 手风琴褶；多褶，百褶
 action *pleat* 活动褶（上衣后中线对褶）
 barrel *pleat* 管形褶裥
 beehive *pleat* 蜂窝褶
 bellows *pleat* 风箱式活动褶裥
 blind *pleat* 暗褶
 box *pleat* 盒形褶裥，回形褶，对褶，工字褶
 cartridge *pleat* 子弹带褶，圆角形褶
 cluster knife *pleats* 群刀褶
 curved *pleats* 弧形褶裥
 double *pleats* 叠褶，双褶
 durable *pleat* 耐久褶
 fan *pleats* 扇形褶裥
 front *pleats* 西裤的褶裥（前腰头处），前腰褶
 gathered *pleat* 碎褶
 heat set *pleat* 热定型褶裥
 horizontal *pleat* 横褶，水平褶
 inverted box *pleat* 倒箱（盒）型褶裥，内工字褶
 inverted *pleat* 倒褶；暗（对）褶，阴褶
 kick *pleat* 倒褶（裙底部），跨步褶，助行褶
 knife *pleat* 剑褶裥，狭褶裥，刀形褶，顺褶裥
 one-way *pleat* 顺褶裥，顺风褶
 permanent(ever) *pleat* 永久褶裥
 pleat depth 褶深
 pleat position 打褶线，裥位线
 pressed knife *pleat* 压刀褶
 rear *pleat* 后背褶裥
 reverse *pleat* 倒褶裥
 side *pleat* 同边褶
 single *pleat* 单褶
 soft *pleat* 柔和褶裥（褶裥不明显），活褶
 stayed *pleat* 死褶裥
 stitched *pleat* 压线褶
 sun-burst(sunray) *pleat* 阳光褶襞（呈散射状细褶）
 umbrella *pleat* 伞形褶裥
 unmeet skirt *pleat* 〔检〕裙裥豁开
 unpressed *pleat* 无压褶
 visible *pleat* 明褶裥
 waist(trouser) *pleat* 裤腰褶裥

pleater ['pliːtə] 打褶裥装置，打裥器；打裥人

pleating ['pliːtɪŋ] 打褶，打裥

pleuche ['pluːtʃə] 金丝绒

pliers [ˈplaɪəz] 〔复〕钳子；老虎钳
    shoe punch *pliers* 鞋跟钳
plimsoll [ˈplɪmsəl] 〔常用复〕〔英〕胶底帆布轻便鞋
plug [plʌg] 〔美俚〕男用高礼帽
plus-fours [ˌplʌsˈfɔːz] 〔复〕灯笼裤（过膝下 4 英寸）；宽大运动裤（如高尔夫球裤）
plush [plʌʃ] 长毛绒；〔复〕（差役穿的）毛线裤
    acrylic *plush* 腈纶长毛绒
    blended *plush* 混纺长毛绒
    cotton *plush* 棉长毛绒
    knitted *plush* 针织长毛绒
    mohair *plush* 马海毛长毛绒
    pressed *plush* 拷花长毛绒
    pure wool *plush* 全毛长毛绒
    raschel *plush* 拉舍尔经编长毛绒
    woolen *plush* 长毛绒
poche [pɒʃ] 〔法〕衣袋，口袋
    poche appliquée 〔法〕贴袋
    poche bateau 〔法〕船形口袋
    poche coupée 〔法〕插袋，挖袋
    poche gousset 〔法〕（裤腰内侧或背心的）小口袋
    poche jeans 〔法〕牛仔裤斜袋
    poche revolver 〔法〕裤后袋
pochette [pɒʃɛt] 〔法〕小口袋，小钱袋，小腰袋
pocket [ˈpɒkɪt] 衣（裤、裙）袋，口袋
    accordion *pocket* 风琴袋，立体袋，折叠袋
    angle *pocket* 斜口袋
    back *pocket* 后袋
    baggy patch *pocket* 蓬松贴袋，立体贴袋
    bellows *pocket* 风箱袋，老虎袋
    boat *pocket* 船形袋
    bound *pocket* 夹层口袋
    breast *pocket* 胸袋；（西装）饰巾袋
    card *pocket* 卡片袋，名片袋
    cargo *pocket* 大（风琴褶）贴袋
    cash *pocket* 表口袋
    change *pocket* 零钱袋
    chest *pocket* 胸袋
    coin *pocket* 零钱袋；表口袋
    concealed *pocket* 暗袋
    crescent *pocket* 月牙形袋
    cross *pocket* 横开袋
    curved *pocket* 弧形袋
    cut-away *pocket* （裤、裙前腰两侧的）弧形插袋
    deformed *pocket* 〔检〕袋形走样
    double piped *pocket* 双绳边袋，双唇袋
    double welt inside *pocket* 双嵌里袋
    flap and button down *pocket* 扣纽扣的有盖袋
    flap jetted side *pocket* 有盖贴袋
    flap *pocket* 有盖袋，袋爿袋
    flat *pocket* 平盖袋
    fly *pocket* 女裤袋
    fob *pocket* 男裤表口袋
    forward *pocket* 裤侧斜插袋
    fringed *pocket* 缘饰口袋
    frog *pocket* 裤斜插袋
    glasses *pocket* 眼镜袋
    gusset *pocket* 接裆口袋
    hacking *pocket* （有盖）上衣斜袋
    hexagonal *pocket* 六角袋
    hidden *pocket* 暗袋，内袋
    hip *pocket* 裤后袋，臀袋，后枪袋

imitation pocket　假口袋
insert pocket　挖袋，插口袋
inside breast pocket　内胸袋
inside pocket　内袋
inverted (box) pleated pocket　阴（明）裥袋
jetted pocket　嵌线袋，绲边袋
J-shaped pocket　J形袋
kangaroo pocket　袋鼠袋（运动装前衣身大贴袋）
lap pocket　有盖袋
light curved pocket　微弯袋
lining pocket　里袋
misplaced pocket　〔检〕口袋错位
mobile pocket　手机口袋
mock pocket　假口袋
MP3 player pocket　MP3口袋
muff pocket　暖手袋（上衣左右胸前直口袋）
out pocket　明袋，贴袋
out pleat patch pocket　压爿袋
outline pocket　明贴袋
outseam pocket　外侧袋，〔粤〕侧骨袋
patch pocket　贴袋
patch pocket with flap　有盖贴袋
peach-shaped pocket　桃形袋
pin tuck pocket　针纹褶饰袋
piped pocket　绲边袋；无盖暗袋
piping pocket　绲边袋，嵌线袋
pleated pocket　打褶袋，褶裥袋
pocket bearer　口袋襻，吊袋襻
pocket mouth　袋口
pocket out grain　〔检〕口袋（布面）丝缕不正
pocket placement　口袋装法
pocket position　袋位

pocket setter　绱袋器
pocket square　胸袋饰巾
post pocket　开贴袋
reinforcement patch for pocket　袋角加固布；袋角衬
roomy pocket　大贴袋
rounded (pointed) pocket　圆（尖）角底衣袋
ruler shaped pocket　曲尺袋，矩尺袋
sandwich pocket　内贴袋
scarap pocket　扇形袋（有盖袋）
seam pocket　摆缝袋
set-in pocket　插口袋，挖袋
sham pocket　假口袋
shaped pocket　弧形袋
side pocket　侧直袋；（上衣）横袋；（裤）斜直袋
single (double) jetted pocket　单（双）嵌线袋，单（双）绲边袋，单（双）唇袋
single (double) welt pocket　单（双）嵌线袋，单（双）贴边袋
skirt-pocket　裙袋；（裙）侧袋
slant (horizontal) side pocket　侧缝斜（横）袋
slash pocket　挖袋，长缝袋，插口袋，斜袋
slender welt pocket　细嵌线袋
slit pocket　挖袋，开袋
split at pocket mouth　〔检〕袋口裂
square pocket　方形袋，方角袋
tambour pocket　绲边袋
thigh pocket　裤腿贴袋，〔粤〕肶袋
three points pocket　三尖袋，尖角袋

ticket pocket　票袋
tiny pocket　小口袋
1/2（1/4，1/8）top pocket　二（四、八）分之一斜袋
unequal pockets　〔检〕口袋（左右）不对称
upper(lower) pocket　上（下）口袋
utility pocket　实用口袋，大贴袋
vertical(straight) pocket　直口袋
waistcoat(coat, trouser) pocket　背心（衣、裤）袋
waist welt pocket　（西服背心）下口袋
watch pocket　表袋
welt pocket　嵌线袋，嵌边袋，开缝袋
wide welt pocket　一字嵌袋，单嵌线袋
Zhongshan coat pocket　中山装袋，老虎袋
zigzag inside pocket　锯齿形里袋
zip cargo pocket　带拉链的大风琴贴袋
zipper pocket　拉链口袋
pocketbook ['pɒkɪtbʊk]　小笔记本；〔美〕（女用）手袋；钱包；皮夹子
pocket-chief ['pɒkɪtʃiːf]　（西装）胸袋饰巾（= pocket-handkerchief）
pocketing ['pɒkɪtɪŋ]　口袋布，袋布
　　pocketing caught in bartacking　〔检〕套结误打在袋布上
point [pɔɪnt]　点；尖；要点；论点；细目；尖（端）；尖状物；针绣花边（= point lace）
　　anterior armpit point　腋窝前点
　　back shoulder point　后肩（颈）点
　　bust(breast) point　胸（高）点，乳峰点；乳下线
　　bust point to bust point　胸点间距
　　collar point　领尖；领尖点线
　　elbow point　肘点
　　front cut point　门襟圆角点线
　　grading point　（纸样）放缩点，推档点
　　lapel point　驳领缺嘴
　　neck point　肩顶，领点；颈点（领围线和肩线交点）
　　neck shoulder point　领肩点
　　pocket point　袋尖
　　short(long) point collar　短（长）领尖领
　　shoulder point　肩端，肩尖（点）
　　sleeve point　袖山
　　starting point　（制图）基点
　　underarm point　袖下点
　　unequal collar points　〔检〕领尖（左右）不对称
　　wrist point　腕点
poke [pəʊk]　朝前撑起的阔边女帽
polinosic [ˌpɒlɪˈnɒsɪk]（= polinosic fibre = polinosic rayon）　虎木棉，富强纤维，富纤（一种高性能黏胶纤维，软身牛仔布原料之一）
polka ['pɒlkə]　女式紧身短上衣
　　polka dot　（布料上）圆点花纹
polonaise [ˌpɒləˈneɪz]　（18世纪的）连衫裙
polyamide [ˌpɒlɪˈæmaɪd]　聚酰胺，锦纶；尼龙
polybag ['pɒlɪbæg]　（服装包装用）胶袋
　　clear polybag　透明胶袋
　　self-sealing polybag　自封口胶袋

polyester [ˌpɒlɪˈestə] 聚酯，涤纶（俗称的确良）
    polyester chips 聚酯切片（涤纶原料）
    polyester/cotton liangsan 涤/棉凉爽绸
    polyester fancy suiting 涤纶花呢
polyfill [ˈpɒlɪfɪl] 涤纶棉，喷胶棉
pompadour [ˈpɒmpəduə] 女式紧身胸衣
pompon [ˈpɒmpɒn]（衣帽装饰）绒球，丝球；（军帽的）毛球
    cap *pompon* 帽饰绒球
poncho [ˈpɒntʃəʊ] 斗篷装，披巾装（南美人穿，中间开领口）；雨披（尤指橡胶制的）
    poncho cloth 军用雨披
    PVC *poncho* PVC 雨披
pongee [pɒnˈdʒiː] 茧绸，柞丝绸
    cotton *pongee* 棉茧绸
    micro polyester *pongee* 哑富迪
    polyester *pongee* 涤纶仿柞丝绸，春亚纺
    polyester *pongee* rip-stop 春亚纺格布
    reeled *pongee* 纺绸
    Shandong *pongee* 山东柞丝绸
    spun silk *pongee* 绢丝纺绸
    tussah *pongee* 柞丝绸
    twill *pongee* 斜纹绸
    Yajiang tussah *pongee* 鸭江绸
poorboy [ˈpʊəbɔɪ]（套头式）罗纹紧身毛线衫
pop [pɒp] 流行文化；流行音乐
    pop culture 流行文化，通俗文化
poplin [ˈpɒplɪn] 府绸；毛葛
    combed *poplin* 精梳府绸

cotton *poplin* 棉府绸
dobby-*poplin* 小提花府绸
dyed *poplin* 染色府绸
figured *poplin* 提花府绸
noil *poplin* 绵绸，䌷丝绸
poplin broche 织花府绸
rayon *poplin* 人丝府绸
T/C coated *poplin* 涂层涤/棉府绸
T/C embossed *poplin* 拷花涤/棉府绸
T/C emerized *poplin* 磨毛涤/棉府绸
T/C gloss finished *poplin* 轧光涤/棉府绸
tencel/polyester *poplin* 天丝/涤纶府绸
trueran white (printed) *poplin* 漂白（印花）涤/棉府绸
wool *poplin* 精纺平纹呢
yarn-dyed *poplin* 色织府绸
popper [ˈpɒpə]〔英〕拷纽，大口按扣；五爪纽
poral [ˈpɔːrəl]（精纺）波拉呢
pouch [paʊtʃ] 小袋；钱包；手袋；烟草袋
pre-shrinking [priːˈʃrɪŋkɪŋ]（布料）预缩
press [pres] 压；熨烫；熨烫机；打包机
    armhole *press* 袖窿熨烫机
    armhole top *press* 袖山压熨机
    baling *press* 打包机
    body *press* 立体整烫机
    both-side trousers *press* 两侧裤管熨烫机
    collar creasing *press* with heat and cool head 冷热双头领型机

collar press 压领机

collar shaping press 领型机

crotch press 裤裆底熨烫机

cubic skirt press 女装裙整烫机

die press 模型熨烫法；模压熨烫机

double shoulder press 双肩（同时）熨烫机

double sleeve press 双袖（同时）熨烫机

durable press 定型熨烫，耐久压烫

edge press 衣边压烫机

European press with microprocessor 欧洲型电脑熨烫机

ever press 永久熨烫

formatic press 立体熨烫；立体熨烫机

front(back) press 前（后）身熨烫；前（后）身熨烫机

fusing press （热压）黏合机

high pressure body shape press 高压立体整烫机

hydraulic cutting press 油压裁剪机，〔粤〕啤机

lapel press 驳领压熨机

leg forming press 裤管成型熨烫机

mini fusing press 迷你（简易型）黏合机

manual(automatic) utility press 手（自）动万能熨烫机

mushroom press 蘑菇式熨烫机

never press 无需熨烫，免烫

permanent(durable) press 定型熨烫，耐久熨烫（又称 PP 整理）

placket fusing press 门襟黏合机

pneumatic utility press 万能风压熨烫机

pocket flap forming press 袋盖熨烫机

press board （熨烫用）压板

press felt(cloth) 熨烫垫布

roll press 滚筒熨烫

rotary body finishing press 滚动式衣身熨烫机

shoulder off press 肩部最终熨烫机

side press 侧身熨烫机

skirt press 裙压烫机

sleeve press 衣袖压烫机

special body shape press 特种人体整烫机

steam-(electric) press 蒸汽（电）熨烫

supplementary press 辅助熨烫机

trouser topper(leg) press 裤头（管）熨烫机

tuck(pleat) press 压褶机

utility press 通用熨烫机，万能熨烫机

waistband(sleeve, side, shoulder, dart, back, armhole) seam opening press 腰头（袖、侧、肩、省、背、袖窿）缝分开熨烫机

waist press 腰围（衬里）压烫机

wrinkle free shirt(trousers) press 抗皱衬衫（裤）熨烫机

presser ['presə] 压具；压铁；熨烫工

collar presser 压领机

hydraulic cutting presser 冲压裁剪机，〔粤〕啤机

standard presser foot 标准压脚

pressing ['presɪŋ] 压；熨烫

computerize pressing 计算机熨烫

electric pressing 电熨烫
finish pressing 整理熨烫
formatic(body) pressing 立体熨烫
improper pressing 〔检〕熨烫不良
intermediate pressing 中烫
off pressing （最终的）大烫，成品熨烫
over pressing 〔检〕熨烫过头，烫黄
part pressing 部件熨烫
permanent pressing 定型熨烫
position pressing （衣服）部位熨烫
pressing bench （熨烫用）烫凳
pressing bun 熨烫馒头
pressing by conventional iron 火烙熨烫
pressing gadget 打包机
pressing mat 熨烫垫
pressing plate 烫板，压呢板
pressing stand with an exhauster 抽气烫台
pressing in process 中烫，烫半成品
product pressing 成品熨烫
steam pressing 蒸汽熨烫
top pressing 最后的定型熨烫
under pressing 缝制前（或半成品）熨烫，中烫
uneven pressing 〔检〕熨烫不匀

prêt-à-porter [prɛtɑpɔrte] 〔法〕现成服装，高级成衣（介乎高级时装与一般成衣之间）

prewalkers [priˈwɔːkəz] 〔复〕幼儿学步鞋

print [prɪnt] 印花布；印花布服装；印刷
 cotton print 印花棉布
 floral print 印花
 leopard(animal) print 豹（动物）纹印花，豹（动物）纹

printing [ˈprɪntɪŋ] 印花
all over printing （图案）满印，满地印花
applied printing 直接印花
applique printing （发泡式）立体印花
blotch printing 满地印花；底色印花；单面印花
both-side printing 双面印花
burn-out printing 烂花印花
coat printing 涂料印花
3D printing 三维印花；3D 打印
decal printing 转移印花
diamond printing 钻石印花
digital printing 数字印花，数码印花
discharge printing 拔染印花
duplex printing 双面复合印花
dyestuff printing 染料印花
garment printing 成衣印花
indistinct printing 隐约印花
jet printing 喷射印花
machine(hand) printing 机器（手工）印花
one sided printing 单面印花
paper printing 纸印花
pearl printing 珠光印花
printing and dyeing 印染
printing design 〔工〕（在裁片上）印花，刷花
puff printing 发泡印花
raised printing 起绒印花，立体印花
reflective printing 反光印花

  resist *printing* 防染印花
  roller *printing* 滚筒印花
  rubber(plastic) *printing* 胶浆印花
  silk screen *printing* 丝网印花
  silver(golden) power *printing* 银（金）粉印花
  transfer *printing* 转移印花（烫画）
  water based *printing* 水浆印花
  wax *printing* 蜡防印花
process [ˈprəʊses] 工序；制作法
  process flow 工艺流程
  process layout 工序安排
  technological *process* 工艺流程
product [ˈprɒdʌkt] 产品，（设计）作品
  down *products* 羽绒制品
  finished(end) *products* 成品
  semifinished *products* 半成品
  shoe care *products* 护鞋产品
  substandard(spoiled) *products* 次（废）品
  universal *product* code(UPC) 通用产品代码，条形码
production [prəˈdʌkʃən] 生产；制作
  bulk *production* 大货生产
  production capacity 生产能力
  production control 生产管理
  production dispatching 生产调度
  production flow 生产流程
  production group 生产班组
  production plan 生产计划
  production schedule 生产排期，生产进度
  production status 生产情况
profile [ˈprəʊfaɪl] 侧影；轮廓；外形；形象
prolivon [prɒˈlɪvən] 葆莱绒（新一代高中空涤纶短纤维）
proportion [prəˈpɔːʃən] 〔设〕比例；均衡，匀称；调和
  proportion of human body 人体比例
pucker [ˈpʌkə] 皱纹；皱褶；起绉缩拢；起皱器
  armhole *puckers* 〔检〕袖窿起皱
  pucker-free 无皱缩
  puckers at quilting 〔检〕绗棉起绉
  puckers at the shoulders 〔检〕裂肩（衣服小肩起皱）
  puckers at underarm seam 〔检〕抬裉缝起绉
puckering [ˈpʌkərɪŋ] 折叠；缩皱，起皱
  seam *puckering* 〔检〕线缝皱缩，〔粤〕骨位皱
puff [pʌf] 衣服的蓬松部分；被子
puffer [ˈpʌfə] 蓬松物；间棉衣物
puggaree [ˈpʌgəriː] (= puggree, puggry [ˈpʌgəri]) 印度薄头巾，帽后遮阳布
pull [pʊl] 〔法〕羊毛套衫，套头线衫，套领背心
  pull en V 〔法〕V 形领口的套头上衣
pull-on [ˈpʊlɒn] 套穿的衣物（如套头衫、手套等）
pullover [ˈpʊləʊvə] 套衫，〔粤〕过头笠；套头毛衣
  chenille jacquard *pullover* 绒线提花套头毛衣
  crew neck *pullover* 罗口圆领套衫
  Johnny collar *pullover* 约翰尼领套衫（女衬衫领）
  pullover with zipper 拉链套衫
  rib knitted *pullover* 针织罗纹套衫

    snaker knit *pullover* 厚针织套头衫
    turtle neck *pullover* 高翻领套头毛衣
    V-neck *pullover* 尖领套衫
    woolen sleeveless *pullover* 毛背心
pull-under [ˈpʊlʌndə] 毛线背心
pump [pʌmp] 〔常用复〕（舞会穿）无带浅口皮鞋；圆头鞋；轻软运动鞋；无带（扣）低跟女便鞋
    high heel pointed toe *pumps* 高跟尖头皮鞋
    platform *pumps* 厚底圆头鞋
puncher [ˈpʌntʃə] 打眼器，冲头
punching [ˈpʌntʃɪŋ] （裁片上）锥眼；冲压
    punching collar stay 〔工〕冲领角衬
purfle [ˈpɜːfl] 镶边；花边
purl [pɜːl] 金银绉边，边饰；流苏
purse [pɜːs] 女用手提包；钱包

puttee (= **puttie**) [ˈpʌti] （布或皮）绑腿
pyjamas (= **pajamas**) [pəˈdʒɑːməz] 〔复〕睡衣裤；宽松裤
    blanket *pyjamas* 拉绒布睡衣裤
    character *pyjamas* 印字图的睡衣裤
    coral fleece *pyjamas* 珊瑚绒睡衣裤
    creeper *pyjamas* 婴幼连衫裤睡衣
    flannelette *pyjamas* 绒布睡衣裤
    knitted *pyjamas* 针织睡衣裤
    sexy *pyjamas* （女）性感睡衣裤，情趣睡衣裤
    shu velveteen *pyjamas* 舒棉绒睡衣裤
    tropical *pyjamas* （夏天穿）轻薄料睡衣裤
    two (three)-piece *pyjamas* 两（三）件套睡衣裤
    wood (bamboo) fibre *pyjamas* 木（竹）纤维睡衣裤

Q

Qipao (= **chi-pao**) [tʃɪˈpaʊ] 旗袍
    embroidered *Qipao* with beads and paillettes 珠绣旗袍
quality [ˈkwɒləti] 质量，品质；优质
    always *quality* 质量可靠
    certificate of *quality* 质量证书，（产品）合格证
    fair average *quality* (FAQ) 中等品
    field *quality* control (FQC) 现场质量控制
    inner (appearance) *quality* 内在（外观）质量
    inspection certificate of *quality* 品质检验证书，质检证书
    integral *quality* assurance (IQA) 整体质量保证
    process *quality* 工序质量
    products of *quality* 优质产品
    quality assurance (QA) 品质保证
    quality certification 质量认证
    quality class 质量等级
    quality control (QC) 品质控制，质量管理
    quality evaluation 品质鉴定，质量评估
    quality first 质量第一

quality level 质量水平
quality symbol 品质标记，品质符号
seam *quality* （服装）缝制质量
total *quality* control(TQC) 全面质量管理，全面质量控制
quantity ['kwɒntəti] 数量；大量
quarter ['kwɔːtə] 鞋腰，鞋身
quilling ['kwɪlɪŋ] 网眼纱褶裥边饰
quilt [kwɪlt] 被子，被褥
quilter ['kwɪltə] （缝纫机）绗缝附件；绗缝机
quilting ['kwɪltɪŋ] 绗缝；绗针法；绗棉；绗缝料；（衣服）管状褶裥；填棉刺绣（在布与布间垫呢子或棉花作衬而绣花）
uneven *quilting* 〔检〕绗棉起绺
quota ['kwəutə] 配额；限额；定额
including(excluding) *quota* （报价时）包（不包）配额
production *quota* 生产定额，生产指标
quota for garments 服装配额
quota management 配额管理
quota reservation 配额预定

R

rabbit ['ræbɪt] 兔毛皮
rack [ræk] 搁物架，挂物架
hat *rack* 挂帽钩；帽架
shoe *rack* 鞋架
stainless steel hanging *rack* 不锈钢挂衣架
racoon [rə'kuːn] 浣熊毛皮
rag [ræɡ] 碎布；破布；〔复〕破旧衣服
collecting *rag* 碎布回收
damp *rag* （熨烫用）水布
glad *rags* 华丽衣服；晚礼服
rag trade 服装业
raglan ['ræɡlən] 套袖大衣，连肩袖大衣，插肩袖大衣，斜肩袖大衣（袖缝沿肩直至领根部的大衣）；连肩（袖），插肩（袖）
split *raglan* 分开连肩袖，单边连肩袖，半连肩袖
two piece *raglan* 两片连肩袖

raiment ['reɪmənt] （书面语）衣服，服装
rainboot ['reɪnbuːt] 雨靴
raincap ['reɪnkæp] 雨帽
raincape ['reɪnkeɪp] 雨披，斗篷雨衣
raincloak ['reɪnkləuk] （带帽）防雨斗篷，斗篷雨衣
raincloth ['reɪnklɒθ] 防雨布
raincoat ['reɪnkəut] 雨衣；风衣
coir *raincoat* 蓑衣
labour *raincoat* 劳工雨衣，工作雨衣
nylon *raincoat* 锦纶雨衣
PE disposable *raincoat* PE一次性雨衣
plastic *raincoat* 塑料雨衣
PVC/polyester *raincoat* 单面胶雨衣
PVC/polyester/PVC *raincoat* 双面胶雨衣
reversible *raincoat* 晴雨大衣，风雨

衣
 rubbered raincoat 涂胶雨衣
raindress [ˈreɪndres] 雨装，雨衣
rainhat [ˈreɪnhæt] 雨帽
rainproof [ˈreɪnpruːf] 雨衣；雨披
rainsuit [ˈreɪnsuːt] 雨衣；雨衣套装
 camouflage rainsuit 迷彩雨衣
 coveralls rainsuit 连身雨衣裤
 industrial rainsuit 工业雨衣
 neoprene rainsuit 涂（橡）胶雨衣
 rainsuit for motorcycle rider 摩托雨衣（骑摩托穿）
rainwear [ˈreɪnweə] （总称）雨衣；雨披
raising [ˈreɪzɪŋ] （布料）起绒（工艺），抓绒，拉绒
ramie(=ramee) [ˈræmi] 苎麻
 ramie tops 苎麻球（由苎麻精梳条卷绕成球）
ramp [ræmp] 时装表演舞台，天桥
ratine [ræˈtiːn] 珠皮大衣呢；平纹结子花呢
ratio [ˈreɪʃiəʊ] 比率；（服装按花色、尺码）装箱配比
rayon [ˈreɪɒn] 人造（纤维素）纤维；人造丝；人丝织物
 acetate rayon 醋酯人造丝
 bright rayon 有光人造丝
 cuprammonium rayon 铜氨（人造）丝
 polinosic rayon 富纤，虎木棉
 rayon beihua 蓓花绸
 rayon filament 人造丝
 staple(spun) rayon 人造短纤维，人造棉，喷胶棉；人造毛
 viscose rayon 黏胶人造丝
reach-me-downs [ˈriːtʃmɪdaʊnz] 〔复〕现成（穿旧）的衣服
redingote [ˈrədɛ̃gɔt] 〔法〕（旧时）男礼服；小腰身女式大衣；（双排扣）女骑装式外套，长大衣
reefer [ˈriːfə] 双排扣水手上衣；女式紧身双排扣上衣
regalia [rɪˈgeɪliə] 王权的标志（如王冠、王杖等）；团体制服，徽章或标记
regimentals [ˌredʒəˈmentlz] 〔复〕（军队）团的制服；军装
relation [rɪˈleɪʃən] 〔设〕关联
remnant [ˈremnənt] （裁剩的）零段布，尾子布
Renaissance [rəˈneɪsəns] （欧洲 14～16 世纪的）文艺复兴；文艺复兴时期风格
repair [rɪˈpeə] 修理；修补；维修
 shoe(clothes) repairs 修鞋（补衣）
repetition [ˌrepəˈtɪʃən] 〔设〕重复，反复
resilience [rɪˈzɪliəns] （织物的）回弹性
revers [rəvɛːr]〔法〕(=revere [rɪˈvɪə]) 衣服翻边部分（翻领，翻袖等）
 bottes à revers 〔法〕翻口长靴
 col à revers 〔法〕翻领
 habit à revers 〔法〕驳领上衣
 pantalon à revers 〔法〕卷脚裤
reversibles [rɪˈvɜːsəblz] 〔复〕双面织物；两面穿衣服；（晴雨）两用衣服
reworks [riːˈwɜːks] 返工品
rhinestone [ˈraɪnstəʊn] 仿制金刚钻，人造钻石；水钻；闪石（常缀于服装和服饰品上）

  crystal rhinestone 水钻（水晶玻璃切割而成的服饰品）
  hot-fixing transfer rhinestones 〔工〕转移烫钻（将水钻图案烫贴在衣服上）

rhythm [ˈriðəm] 〔设〕律动，韵律；节奏；调和；匀称

rib [rib] 棱纹，凸条，罗纹
  acrylic rib 腈纶罗纹
  bottom rib 下摆罗纹
  cotton rib 全棉罗纹
  dropneedle rib 抽针罗纹
  elastic rib 弹性罗纹，松紧带
  French rib 法式罗纹
  interlock rib 棉毛布
  jacquard double rib 提花双面罗纹
  lycra rib 弹性罗纹，〔粤〕拉架（罗纹）
  neck(cuff) rib 领（袖）口罗纹
  nylon rib 锦纶罗纹
  one by one rib 1×1 罗纹
  rib knitting 罗纹针织品
  waist rib 腰罗纹

riband [ˈribənd] （装饰用）缎带；丝带

ribbing [ˈribiŋ] （领口、袖口、下摆等处的）罗纹镶边

ribbon [ˈribən] 缎带；丝带
  elastic ribbon 松紧带
  hair ribbon 头发丝带，发带
  hat ribbon 帽带
  jacquard ribbon 提花缎带
  mesh ribbon 网眼丝带
  millinery ribbon 女帽缎带
  organza ribbon （透明）玻璃纱带
  polyester ribbon 涤纶缎带
  rayon ribbon 黏胶纤维缎带
  sheer ribbon 透明缎带
  silk ribbon 丝带
  velvet ribbon 丝绒缎带

richcel [ˈritʃsel] 丽赛纤维（具优异综合性能的植物纤维素纤维，有"植物羊绒"的美称）

rig [rig] （口语）华美奇特的装束

rig-out [ˈrigˈaut] 〔英〕一套服装

rim [rim] （衣、帽）边

rindy [ˈraindi] 宽裤腰袋状牛仔裤

ring [riŋ] 环状配件；环状物；戒指
  arm ring 手环，臂环，臂钏
  colour gold ring 彩金戒指
  crystal ring 水晶戒指
  D-ring D 形环；D 形扣，巴黎扣
  diamond ring 钻石戒指，钻戒
  dress ring 礼服戒指
  ear ring 耳环
  emerald ring 祖母绿（绿宝石）戒指
  engagement ring 订婚戒指
  gold ring 金戒指
  jade ring 玉环；玉戒指
  jadeite ring 翡翠戒指
  jewel ring 宝石戒指，钻石戒指
  nose ring 鼻环
  O-ring 环形扣，单环扣
  platinum ring 铂金戒指
  rhinestone ring 仿钻戒指，水钻戒指
  sapphire(ruby) and diamond ring 蓝（红）宝石钻戒
  scarf ring 围巾扣；领带夹
  signet ring 印章戒指
  silver ring 银戒指
  sweethearts ring 情侣对戒
  toe ring 趾环（脚趾戒指）

robe

two-tone *ring* 双色戒指
wedding(bridal) *ring* 结婚戒指
wooden *ring* 木戒指

rinse [rɪns], **rinsing** [ˈrɪnsɪŋ] （服装洗水时）冲洗，漂洗，漂清

ripper [ˈrɪpə] 拆线缝小刀
seam *ripper* 拆线器（拆线或开扣眼）

rise [raɪz] 裤直裆，立裆，〔粤〕裤浪
back *rise* 裤后直裆，落裆，〔粤〕后浪
body *rise* 上裆，直裆
front *rise* 裤前直裆，小裆，〔粤〕前浪
high *rise* 深直裆，深上裆
low *rise* 短直裆，浅上裆

rittaisaidan [gɪˈdaɪˈsaɪdʌn] 〔日〕立体裁剪

rivet [ˈrɪvɪt] （饰于衣服上的）撞钉，爪钉；铆钉
pocket *rivet* （牛仔裤）袋口撞钉
rivet for jeans 牛仔裤撞钉

robe [rəʊb] 长袍，罩袍，浴衣；罩衣；晨衣；〔复〕礼服，法袍；制服，〔法〕[rɒb]〕长袍；长外衣；连衣裙；道袍；（法官、律师等的）礼服
bath *robe* 浴袍，浴衣
ceremonial *robe* 礼服
Chinese *robe* 中式长衫（民国男装常服）
cotton wadded *robe* 棉袍
dragon *robe* 龙袍
floss-padded *robe* 丝绵袍
fox-fur *robe* 狐裘
fur *robe* 毛皮袍
Imperial *robe* 龙袍（中国古代皇帝着装）

judge's *robe* 法官服，法衣
lawyer's *robe* 律师服，法衣
long *robe* 法官袍；教士袍
official *robe* 官服
over *robe* 罩袍
priest *robe* 僧袍
Republican *robe* 民国男装长衫
robe ailee 〔法〕展翅裙（将裙摆挽在手上，一举手就像鸟儿展翅）
robe ascenseur 〔法〕升降机装，电梯装（有腰带可调节衣长，使上衣身蓬松下垂）
robe biblique 〔法〕圣经时代装（圣经时代风格的丝绸长袍）
robe cachecoeur 〔法〕隐心型装（门、里襟类似和服交叉）
robe-chemisier 〔法〕衬衫裳（衬衫样式的衣衫）
robe corolle 〔法〕花开装（华丽晚装，在胸围处装饰白色丝绸制成的花）
robe decolletee 〔法〕袒胸露肩的女夜礼服
robe de mariée 〔法〕新娘礼服
robe gag 〔法〕非常创新的衣服（将两件衣服组合，形成多种穿法）
robe gitane 〔法〕吉卜赛女装
robe housse 〔法〕裹身连裙装（宽敞型）
robe oiseau 〔法〕鸟形装（希腊风格的束腰宽衣）
robe papillon 〔法〕披肩装，蝶翅装
robe-paysage 〔法〕风景花样装（衣上绘风景）

  robe polo 〔法〕波罗装（马球领装）
  robe pull 〔法〕长至臀围的套头夹克
  the long *robe* 法官服；教士服
  wool lined *robe* 衬绒袍
rogatywka [rɒgʌˈtıfkʌ] 四角军帽（早年的波兰军帽）
roll [rəʊl] （一）卷；（一）匹；（衣服）翻边
  sleeve head *rolls* 袖头棉
roll-neck [ˈrəʊlnek] 可翻折的高领；高翻领套头衫
  striped rib *roll-neck* 间条罗纹高翻领衫
roll-on [ˈrəʊlɒn] 女弹力紧身胸衣
rompers [ˈrɒmpəz] 〔复〕（儿童）连裤外衣（或背心）；连体短裤
  infant's *rompers* 婴幼连衫裤装
  newborn *rompers* 新生婴儿连衣裤
  ski *rompers* 填棉童连衣裤
rope [rəʊp] 绳，索
  cotton *ropes* in white (colour, nature) 漂白（染色、原色）棉绳
rouleau [ruːˈləʊ] （女装）绲边，绲条
roulette [ruˈlet] （在纸样衣片上作标记）点线滚轮，点线器，擂盘
rubber [ˈrʌbə] 橡胶；橡皮；橡胶制品，橡皮擦；〔复〕橡胶套鞋
  sponge(form) *rubber* 海绵橡胶
ruche [ruːʃ] (= **ruching** [ˈruːʃɪŋ]) 褶裥饰边
rucksack [ˈrʌksæk] 背袋；旅行背包，双肩背包
ruff [rʌf] 襞襟，拉夫领（16、17 世纪盛行于上流社会的白色轮状皱领）

ruffle [ˈrʌfl] 褶边；皱褶；荷叶边
  lace *ruffle* 花边褶
ruffler [ˈrʌflə] （缝纫机的）打皱褶装置，打褶器
ruffling [ˈrʌflɪŋ] 打皱褶
rule [ruːl] 尺；刻度尺；规律；规则；条例
  a foot-*rule* 一英尺长的尺
  arbitrarily curved *rule* 自由曲线尺
  bamboo *rule* 竹尺
  cloud-shaped *rule* 云尺（云状弧线尺）
  fashion *rule* 流行规律
  flexible *rule* 软尺，卷尺
  folding *rule* 折尺
  L-shaped *rule* L 形尺
  slide *rule* 计算尺
  steel(linen)tape *rule* 钢（皮）卷尺
  zigzag *rule* 曲尺
ruler [ˈruːlə] 尺；直尺；画线板
  aluminium *ruler* （绘纸用）铝尺
  checked *ruler* 格子尺
  Chinese *ruler* 市尺
  curve *ruler* 弯尺，曲尺
  inch *ruler* 英制尺
  marking *ruler* 放码尺
  meter *ruler* 米尺，公制尺
  plastic *ruler* 塑料尺
  ruler for pattern grading 纸样放码尺
  ruler set 套装尺
  soft *ruler* 软尺
  stainless steel *ruler* 不锈钢尺
  triangular *ruler* 三角尺
runner [ˈrʌnə] 跑步鞋
runstitch [ˈrʌnstɪtʃ] 初缝线迹；撩针

runway ['rʌnweɪ] 时装表演 T 型台，天桥

S

sable ['seɪbl] 黑貂皮；〔复〕黑貂皮外衣（或领）；丧服
sabot ['sæbəʊ]（欧洲农民穿）木鞋；木底皮鞋；有襻鞋
sack [sæk]（一）袋，袋子；睡袋；布袋装；宽松短上衣；（附着于衣肩的）丝绸褶裥长拖纱
 double-breasted sack 双排扣袋型上衣
 eiderdown sack 鸭绒睡袋
 mini sack 迷你夹克（极短宽松上衣）
 over sack 短外套
sacking ['sækɪŋ] 粗平袋布，麻袋布
sacque [sæk] 女宽身长服；女宽松夹克；婴幼短上衣
sadiron ['sædaɪən] 大熨斗（两端尖，柄可以拆卸）
saidan [saɪdʌn]〔日〕裁剪，裁断
sailcap ['seɪlkæp] 水手帽
sailsuit ['seɪlsuːt] 水手服
 boy's sailsuit 童水手套装
salopette [salɒpɛt]〔法〕沙罗佩套裤（背带工作裤）；童背带裤；（猎人穿）罩裤
salwar kameez ['sʌlwɔːkə'miːs] 旁遮普服（印度传统女套装，即齐膝宽松上衣，长裤及披巾）
samfoo (= samfu) ['sʌmfuː] 衫裤（音译，源自粤语，旧时中国妇女套装：短上衣与裤）
sample ['sɑːmpl] 样品，〔粤〕办；样本
 approval sample 确认样，封样，验收样
 bulk sample 货前样，〔粤〕先行办（加工厂为大货试做的样，无需交客户批）
 buyer's sample 买方样，客户样
 colour sample 色样
 confirmed sample 确认样
 counter sample 对等货样；回样（根据原样制作并寄回客户确认）
 empolder sample 开发样
 fitting sample 试身样，批样
 fore sample 船样
 full size sample 齐码样
 initial sample 初样，〔粤〕头办
 keep(duplicate) sample 留底样，存档样
 maker's sample 生产厂家样品
 making of original sample 做初样
 original sample 原样
 photo sample 照相样，宣传样，目录样，〔粤〕影相办
 pre-production sample 产前样，开工样（需经客户确认此样后，才能开货）
 processing according to investor's sample 来样加工
 proto-(prototype) sample 初样
 purchase sample 现货样品
 random sample 随机查验样；随机

抽样
reproduction of sample 复样
revised sample 修正样，复样
run size sample 放码样
sample card 样品卡
sample maker(hand) 样品制作工，打样人
sample of fabrics(buttons) 布料（纽扣）样本
sample room 样品间，〔粤〕办房
sealed sample （投标用）封口标样，封样
selling(salesman) sample 推销样，〔粤〕行街办
shipping(shipment) sample 船样，〔粤〕船头办
showroom sample 展示样，销售样
size set sample 齐码样
standard sample 标准样
test sample 测试样（用来测试色牢度，环保指标等）
top sample 齐色齐码样，船样
trade sample 商用样品
type sample 标准样
umpire sample 仲裁人样品
wash sample 洗水样

sampling [ˈsɑːmplɪŋ] 抽样；取样
random sampling 随机抽样，检查
sampling inspection 抽样检验

sandal [ˈsændl] 凉鞋，拖鞋；浅口套鞋；鞋襻
ankle sandals 踝带凉鞋
barefoot sandals 赤脚凉鞋（赤足当鞋，装饰脚背和踝部）
beach sandals 沙滩（凉）鞋
bold strap sandals 宽带凉鞋
cloth sandals 布凉鞋
gladiator sandals （古罗马）角斗士鞋；格雷蒂女士凉鞋（现代仿此鞋款设计出来的具有罗马风情的时款女装鞋）
Grecian sandals 希腊风格凉鞋
leather sandals 皮凉鞋
leatheret sandals 人造革凉鞋
low(high)-heeled sandals 平（高）跟凉鞋
mesh sandal 网眼凉鞋
peep-toe(open toe) sandals 露趾凉鞋
plastic sandals 塑料凉鞋
platform sandals 厚底（坡跟）凉鞋
sports sandals 户外活动凉鞋
strap runner sandals 活动搭襻凉鞋
strips sandals 多带式凉鞋
synthetic leather sandals 合成革凉鞋
T-bar sandals T形带襻凉鞋
thong sandals （夹带式）人字凉鞋
toutenkaboucle(buckle) sandals 一种多搭襻的踝带高跟露趾皮凉鞋（亮漆红底，超长高跟，法国品牌鞋）
T-strap sandals T形搭扣凉鞋
tubular sandals 套带凉鞋（平底无扣襻）
wedge sandals 坡跟凉鞋

sandblast [ˈsændblɑːst] （牛仔服洗水类）喷砂，打砂；喷砂器
sanding [ˈsændɪŋ] （布料）磨毛，磨绒
sans-couture [sãkuty:r] 〔法〕无制作（与传统制衣工艺完全不同，即采用

无衬布，无里布，无伸缩，不缭缝的非缝制方法）
sarcenet(= sarsenet) [ˈsɑːsnet] 素纺；里子薄绸；里子布
saree(= sarrie, = sari) [ˈsɑːrɪ] 莎丽服，卷布装（印度等南亚国家妇女包头缠身的传统服装，现加入流行元素，成为时尚女装或新娘礼服）
sarong [ˈsərɒŋ] （马来民族穿的）纱笼（音译），围裙；用作纱笼的布料
sarrau [saro] 〔法〕（宽松）罩衫，工作服；套头童罩衫
sash [sæʃ] （女、童）腰带；彩带；（军装）饰带；肩带
 woven hemp *sash* 麻织宽腰带
satchel [ˈsætʃəl] （皮或帆布制）书包，小背包；小提包
sateen [sæˈtiːn] 棉缎，横贡缎
satin [ˈsætɪn] 缎，缎子，〔粤〕色丁（音译）
 broche *satin* 修花缎
 cotton down-proof *satin* 全棉防羽缎
 crepe *satin* 绉背缎，绉缎
 crepe *satin* plain 素绉缎
 figured *satin* 花缎
 Guangdong *satin* 广绫
 heavy duchesse *satin* 厚女公爵缎
 kingtiao *satin* 金雕缎
 lining *satin* 里子缎
 lycra *satin* 弹力缎
 nylon lining *satin* 锦纶里子缎
 paillette *satin* 闪光缎
 palace *satin* 库缎，贡缎
 plain *satin* 素缎
 polyester *satin* 涤纶缎
 rayon *satin* 人丝软缎
 reversible *satin* 双面缎
 Sangbo *satin* 桑波缎
 satin drill 直贡缎，泰西缎
 silk *satin* 真丝缎
 tapestry *satin* 织锦缎
 twilled *satin* 直贡缎，直贡呢
 velvet *satin* 丝绒花缎
 wool *satin* 毛缎
satinet [ˈsætɪnet] (= satinette [ˌsætɪˈnet]) 全丝薄缎；棉毛缎
Saxony [ˈsæksənɪ] 萨克森法兰绒；（加拿大）白绒布；美利奴花呢；光毛呢（大衣呢）
scale [skeɪl] 比例尺；缩尺；天平，秤
 eletronic fabric *scale* 电子布块秤（快速算出每码/米的克重）
 triangular *scale* 三棱尺
scallop(= scollop) [ˈskɒləp] 扇形饰边，月牙边
scanties [ˈskæntɪz] 〔复〕（口语）女短内衬裤
scapular [ˈskæpjʊlə] （僧侣披肩）无袖外衣
scarf [skɑːf] 围巾，〔粤〕颈巾；披巾；头巾；领巾；领带
 acrylic *scarf* 腈纶围巾
 angora *scarf* 兔毛围巾
 beaver *scarf* 海狸毛围巾
 circular *scarf* 圆圈形围巾
 cotton *scarf* （棉）纱围巾
 crinkle *scarf* 起皱围巾，皱纹围巾
 crocheted *scarf* 钩编围巾
 far-infrared heating *scarf* 远红外加热围巾
 fringed *scarf* 缘饰围巾；流苏披巾
 fur *scarf* 毛皮围巾
 hand-drawn silk *scarf* 手绘丝巾

hand knitted *scarf* 棒针围巾
　　hand-printed silk *scarf* 手工印花丝巾
　　Hip Hop *scarf* 嘻哈头巾
　　leopard print silk *scarf* 豹纹丝巾
　　mink *scarf* 貂皮围巾
　　neck *scarf* 颈巾；领巾；围巾；领带；领饰
　　pirate *scarf* 海盗头巾
　　polar fleece *scarf* 摇粒绒围巾
　　polyester (chiffon) silk *scarf* 涤纶（雪纺）丝巾
　　printed *scarf* 印花围巾；印花头巾
　　red *scarf* 红领巾
　　silk *scarf* 丝绸围巾，丝巾
　　Tibet *scarf* 西藏丝毛披巾
　　triangle *scarf* 三角围巾
　　turban *scarf* 头巾式围巾
　　woolen *scarf* 羊毛围巾
scarlet [ˈskɑːlɪt] 红色制服，红衣；红布
schoolbag [ˈskuːlbæg] 书包
schoolwear [ˈskuːlwɛə] 学生服，校服
scissoring [ˈsɪzərɪŋ] 裁剪；〔复〕剪下的布条
scissors [ˈsɪzəz] 〔复〕剪刀
　　automatic air *scissors* （自动）气动剪刀
　　buttonhole *scissors* 开扣眼剪
　　cloth *scissors* 布剪
　　embroidery *scissors* 刺绣剪刀
　　nail *scissors* （美甲用）修甲剪
　　pattern *scissors* 纸样剪
　　trimming *scissors* 修边剪
　　u-*scissors* （U形）纱剪；线头剪
　　zigzag *scissors* 花齿剪（刀口为锯齿形的剪刀）

scraper [ˈskreɪpə] （涂糨糊的）刮子
screwdriver [ˈskruːdraɪvə] （螺丝）起子
scuff [skʌf] 平底拖鞋；平底凉鞋
　　adjustable *scuffs* 活动扣襻拖鞋
　　casual *scuffs* 休闲拖鞋
　　cork look platform *scuffs* 软木厚底坡跟凉鞋
　　novelty *scuffs* 新奇拖鞋，花式拖鞋
　　sidewall scuffs 边帮拖鞋（无后帮或后帮低）
scye [saɪ] 袖窿，袖孔
sealskin [ˈsiːlskɪn] 海豹毛皮，海虎绒；海豹皮服装
seam [siːm] 缝，线缝，接缝，缝口
　　abutted *seam* 拼合缝
　　appliance *seam* 嵌花缝
　　armhole *seam* 袖窿缝
　　back *seam* 背缝
　　back panel *seam* 背嵌缝
　　beaded *seam* 泡状缝
　　bias *seam* 斜线缝
　　bound *seam* 绲边缝，包边缝
　　broken *seam* 间断缝
　　bulky *seam* 粗缝，松散缝
　　butted *seam* 平接缝，拼接缝
　　butterfly *seam* 蝶形缝
　　catch stitched *seam* 〔工〕环缝
　　centre *seam* 中缝
　　centre back *seam* 背中缝，〔粤〕后中骨
　　chain stitch *seam* 链式线缝
　　closed *seam* 包缝，暗缝，封头缝
　　collar *seam* 领缝
　　composite *seam* 结构缝（各式线缝统称）

continuous *seam* 连续缝
contrast colour *seam* 镶（拼）色缝
cording *seam* 嵌缝，凸纹缝
corner *seam* 转角缝
covered *seam* 覆盖缝，绷缝，〔粤〕冚骨
cross *seam* 交叉缝，十字缝
crotch *seam* 裆缝
curved *seam* 曲线缝，弧形缝
darning *seam* 织补缝
dart *seam* 省缝，褶缝
double lapped *seam* 双搭缝
double needle felling *seam* 〔工〕双针合缝
double top stitched *seam* 分缉缝，双排明缝
elbow *seam* 肘弯缝，后袖缝
empire *seam* 高腰线缝，〔粤〕胸底骨
enclosed (encased) *seam* 封边缝，止口缝
fastening *seam* 加固缝
felled *seam* 对折缝，双包边缝
final *seam* 终止缝，末缝
finishing (fashion) *seam* 成形缝，面缝
flat-fell *seam* 平接缝，明包缝，保险缝
flat lock *seam* 覆盖缝，绷缝
flat *seam* 扁平缝
folding *seam* 折叠缝
French (bag) *seam* 来去缝，袋缝
front (back) rise *seam* 前（后）裆缝，小（落）裆缝
front (back) sleeve *seam* 前（后）袖缝

front edge *seam* 门襟止口缝
front fly *seam* 门襟缝
German *seam* 锯齿缝
gore *seam* 三角缝
hanging *seam* 悬缝，挂缝
hem *seam* 边缝
hemming *seam* 卷边缝
hind edge *seam* 暗边缝
inside *seam* 下裆缝，内侧缝
interlock *seam* 联锁缝
invisible (blind) *seam* 暗缝
joining *seam* 接合缝，缝合缝
lapping *seam* 搭接缝，坐缉缝
lattice *seam* 格子缝
leg *seam* 裤管缝；袜筒缝
lockstitch *seam* 锁式缝迹缝，包缝
lycra *seam* 莱卡缝
neck *seam* 领缝
normal *seam* 普通缝
open *seam* 〔检〕线缝散开，〔粤〕爆口
opening *seam* 开缝，分缝
ornamental (decorative) *seam* 装饰缝
outlet *seam* 缝份，毛边
overcast *seam* 包边缝，锁边缝
overedge *seam* 包边缝
overlap (overhead) *seam* 搭缝
overlock *seam* 锁缝，包缝
piping *seam* 绲边缝
placket *seam* 门襟缝
plain *seam* 平缝，坐倒缝
plaited *seam* 辫形缝
princess *seam* 公主缝，刀背缝，〔粤〕公主骨
pulling at inside or outside *seam* 〔检〕吊脚（裤下裆缝或裤侧缝起吊）

quilting seam 绗缝
raised seam 凸形缝
seam allowance 缝份
seam broken 〔检〕线缝断开，〔粤〕爆口
seam distance(depth) 接缝长（深）度
seam end 缝止点，缝端
seam opening 开缝；线缝断开
seam puckering 线缝皱缩，〔粤〕骨位皱
seam slippage 线缝松脱
seat seam 后裆缝，〔粤〕后浪中骨
section seam 分割缝
self-bound seam 漏落缝，灌缝
short(long) seam 短（长）缝
shoulder seam 肩缝，〔粤〕膊头骨
side seam 边缝，侧缝，摆缝，〔粤〕侧骨
slanting seam 缭缝，斜缝
sleeve centre seam 袖中缝
sleeve seam 袖缝
staggered seam 〔检〕双轨接线（绱线交错，接线不准）
stitch seam leans out the line 〔检〕绱线上下炕（线绱得或上或下）
strap seam 带形缝
stretch(elastic) seam 伸缩缝，弹性缝
tacking seam 加固缝
taped seam 热风胶贴缝，贴带缝
thick seam 厚缝，密缝
top-stitched lapped seam 压绱缝，明搭接缝
top stitched seam 明缝，面缝

tucked seam 褶裥缝，打裥缝，坐缉缝
twin seam 双行缝，双线缝
twisted seam 〔检〕线缝扭曲，〔粤〕扭骨
underarm seam 袖下缝
uneven waist seam 〔检〕腰缝起皱
vent seam 衩缝
waistband seam 腰带缝
waistline seam 腰缝
wave seam 〔检〕线缝起拱，〔粤〕骨位起蛇
weak seam 〔检〕缝口不牢
welt seam 贴边缝，折边缝，暗包缝
wing seam 翼状缝
wrap seam 包缝，〔粤〕包骨
wrong type of seam 〔检〕缝型错误
zigzag seam 曲折缝，之字缝
seamer [ˈsiːmə] 缝纫机；缝纫工
flat seamer 平缝机
seaming [ˈsiːmɪŋ] 缝纫，缝合，缝接
in-and-out seaming 内外缝纫
lap seaming 下摆缝合
seaming armhole 绱袖
seaming front edge 缝止口
seaming security 缝接牢度
seaming tools 缝纫工具
seat [siːt] （人或裤的）臀部；裤裆，上裆；臀围，〔粤〕坐围
seat circumference 臀围
short seat 〔检〕兜裆（上裆吊紧）
slack(baggy) seat 〔检〕后裆下垂（裤臀部下沉）
section [ˈsekʃən] 切断，分割；零件
golden section 〔设〕黄金分割

seersucker [ˈsɪəˌsʌkə] 绉条纹薄织物；泡泡纱
    nylon *seersucker* taffeta 锦纶塔夫泡泡纱
    seersucker gingham 格子泡泡纱
see-through [ˈsiːθruː] （用透明薄料做）透视装，透明装
selvage [ˈselvɪdʒ] 布边，织边
    broken *selvage* 破边
    fast *selvage* 光边，加固边
    rough *selvage* 毛边
    selvage guide （缝纫机）导布器
sense [sens] 感官；感觉；意识
    colour *sense* 色感
separates [ˈsepərɪts] 〔复〕上下装；（妇女）不配套服装；自由组合套装
sequin [ˈsiːkwɪn] （装饰衣服用的）金属小圆片，珠片，闪光饰片
    sequin appliqué and trimming 珠片贴饰
sequining [ˈsiːkwɪnɪŋ] 饰珠片，饰亮片
serge [sɜːdʒ] 哔叽
    all wool *serge* 全毛哔叽
    cotton *serge* 棉哔叽
    imitation-wool *serge* 仿毛哔叽
    imperial *serge* 细密防水哔叽
    indigo *serge* 藏青哔叽
    melange *serge* 混色哔叽
    midfibre *serge* 中长哔叽
    nylon *serge* 锦纶哔叽
    polyester/acrylic *serge* 涤/腈哔叽
    polyester *serge* 涤纶哔叽
    polyester/viscose *serge* 涤/黏哔叽
    satin *serge* 缎纹哔叽
    serge canvas 小方块纹哔叽
    service *serge* 军服哔叽
    wide wale *serge* 粗纹哔叽
    wool *serge* 全毛哔叽，毛哔叽
    wool/viscose *serge* 毛/黏哔叽
serging [ˈsɜːdʒɪŋ] 包缝（边），粗缝（边）
serviceability [ˌsɜːvɪsəˈbɪləti] 服用性能（服装的实用性）
set [set] （一）套；身材；（衣服穿在身上的）样子
    baby's padded *set* 婴儿填棉套装
    bandeau *set* 抹胸套装（狭胸带及短内裤）
    cabana *set* 男海滩套装（短袖上衣和短裤）
    camisole and tap pant *set* 女睡衣套装
    complete *set* of equipments 整套设备
    co-ordinated *set* （颜色、款式等）协调的套装
    crop *set* 短上衣套装（如短胸衣和短内裤）
    crop top and brief *set* 吊带胸衣和短内裤套装
    girl's(boy's) *set* 女（男）童套装
    hat, scarf and gloves *set* 帽子、围巾、手套三件套
    jogging *set* 慢跑运动套装
    knitwear *set* 针织套装
    manicure *set* 修指甲工具套装
    pram *set* 婴儿全套用品
    safari *set* 猎装
    shirt and tee *set* 衬衫（圆领）T恤套装
    sleep *set* 睡衣套装，睡衣裤
    three-piece skirt *set* 裙三件套
    toilet *sets* 化妆用品

top and bottom *set* 上下套装
twin *set* （女用）两件套（上衣与套头衫）
two-piece *set* 两件套装
sewer [ˈsəʊə] 缝纫工具；缝纫机；缝纫工，〔粤〕车工
    button *sewer* 钉扣机
    dart *sewer* 缝省机，省缝机，短缝机
    front edge *sewer* 止口缝纫机
sewing [ˈsəʊɪŋ] 缝纫，〔粤〕车缝；缝制物
    buttonhole *sewing* 锁扣眼
    button *sewing* 钉扣
    controlled programmed *sewing* 程控缝纫
    cutting and *sewing* 裁剪和缝纫，〔粤〕切驳
    forward(backward) *sewing* 顺（倒）向缝纫
    full automatic *sewing* 全自动缝纫
    hand *sewing* 手工缝纫，手缝
    home *sewing* 家庭缝纫
    integrated *sewing* device 综合缝纫装置
    lap *sewing* 搭接缝纫
    machine *sewing* 机器缝纫，〔粤〕车缝
    materials for *sewing* 缝纫用料
    modular quick response *sewing* (MQRS) 模板式快速反应缝纫
    moveable *sewing* box 活动缝纫箱
    ornamental *sewing* 装饰缝纫
    overlock(overedge) *sewing* 包边缝纫，锁边缝纫
    sample *sewing* 样品缝制
    sewing area 缝纫部位
    sewing basket 针线篮
    sewing bow 〔检〕车缝偏离；缉线上、下炕
    sewing gauge 缝距；英寸量规（带有可移动的指示器）
    sewing kit 针线包
    sewing light （缝纫机上的）缝纫照明灯
    sewing method 缝纫方法
    sewing sequences 缝纫工序，缝制步骤
    sewing skill 缝纫技巧
    sewing speed 缝纫速度
    sewing technological requirements 缝纫工艺要求
    sewing thickness 缝纫厚度
    tacking(reinforcing) *sewing* 加固缝纫
    versatility *sewing* 多功能缝纫
    zigzag *sewing* 曲折缝纫
shade [ʃeɪd] 色调，色度，颜色的深浅
    deep *shade* 深色，浓色
    dominant *shade* 主色调
    off *shade* 色差
    relaed *shade* 相关色调
shading [ˈʃeɪdɪŋ] 调整色光；色差（染色或色光差异）
shako [ˈʃækəʊ] 筒形军帽；毛皮军帽
shalwar [ˈʃʌlwɑː] （南亚妇女穿）宽松长裤
shameuse [ˈʃæmuːs] 留香绉
shank [ʃæŋk] 小腿；袜筒；鞋底腰部（插在鞋底腰的）垫子，铁心；（纽扣柄
shanking [ˈʃæŋkɪŋ] 钉扣打结
Shantung [ˈʃænˈtʌŋ] 山东绸（柞蚕丝

绸）
shape [ʃeɪp] 形状；款式，样式；模型
    collar shape 领型
    hat shape 帽型
    smart shape 时髦款式
shaper [ˈʃeɪpə] 合体服装；塑形装，紧身装
shapewear [ˈʃeɪpweə] 塑形服装，紧身装
sharkskin [ˈʃɑːkskɪn] 鲨鱼皮；鲨皮革；鲨皮布；（精纺）雪克斯金细呢
shawl [ʃɔːl] 围巾，披肩
    acrylic shawl 腈纶披巾
    fur shawl 皮披肩
    pashmina shawl 帕西米纳羊绒披肩
    silk embroidery shawl 丝绸绣花披巾
    wool shawl 羊毛围巾
shearling [ˈʃɪəlɪŋ] 剪取的羊毛；剪毛羊皮
    lining shearling 羊毛皮衬里
shears [ʃɪəz] 〔复〕大剪刀（多用于裁剪多层或厚重布料）
    bent-handled shears 弯柄剪
    dressmaker's shears 裁缝用剪
    pinking shears 花齿剪
    tailoring shears 裁缝剪刀
sheath [ʃiːθ] （女）紧身服装
    sheath corset 紧身腹带
sheepskin [ˈʃiːpskɪn] 绵羊皮；带毛绵羊革；羊皮袄
    shorn(sheared) sheepskin 羊剪绒毛皮（剪了一刀羊绒后余下的带毛羊皮）
sheer [ʃɪə] 透明薄纱；透明薄织物及制品

pure ramie sheer （纯苎麻）爽丽纱，爽丽绸
woolen sheer 轻薄型毛料（夏服用）
sheet [ʃiːt] 被单；纸张；表格；成幅的薄片
    art-work sheet 工艺图纸；工艺单
    bed sheet 床单
    cork sheet （制鞋用）软木片
    pattern sheet 衣服纸样
    process(working) sheet 工艺单
    specification sheet 规格表
sheeting [ˈʃiːtɪŋ] 阔幅平布，细布；被单布
    bag sheeting 口袋布
    bleached sheeting 漂白布
    dyed sheeting 染色细布
    grey sheeting 细布坯，本白平布，白市布
    printed sheeting 印花细布
    pure ramie white sheeting 纯麻漂白细布
    trueran T/C sheeting 涤/棉平布
shell [ʃel] 〔美〕贝壳衫（无领无袖女罩衫）；外表，表面
sherpa [ˈʃɜːpə] 一种粒粒绒仿毛皮，仿羊羔毛皮
Shetland [ˈʃetlænd] 雪特兰呢（苏格兰呢）
shield [ʃiːld] （衣袖腋部）吸汗垫布；遮护物
    fly shield 裤里襟
shift [ʃɪft] （方言）衬衫，女用衫
shimmy [ˈʃɪmi] 女衬衫
shirring [ˈʃɜːrɪŋ] 抽褶（多层抽皱缝纫法），缩缝，多层碎褶
    decorative shirring 装饰抽褶

elastic *shirring* 弹性抽褶

shirt [ʃɜːt] 男衬衫,〔粤〕恤衫;(仿男式)女衬衫;内衣,汗衫

    active T-*shirt* (圆领短袖)运动T恤衫

    advertising T-*shirt* 广告衫

    aloha *shirt* 阿罗哈衬衫(夏威夷开领衬衫)

    Arabian *shirt* 阿拉伯衬衫

    army *shirt* 军式衬衫

    ascot *shirt* 阿斯可衬衫(传统衬衫)

    athletic *shirt* 运动汗衫

    aviator *shirt* 飞行员衬衫

    baggy *shirt* 袋形衬衫(浮离身体、宽松衬衫)

    bedford cord *shirt* 经条灯芯绒衬衫

    body *shirt* 女贴身衬衫(或背心)

    boiled *shirt* 硬胸衬衫

    business *shirt* 商务衬衫、办公衬衫、工作衬衫

    Canadian *shirt* 加拿大衬衫

    casual *shirt* 便装衬衫、休闲衬衫

    ceremonial *shirt* 礼仪衬衫

    chambray *shirt* 青年布衬衫

    Chinese style *shirt* (中式)小褂

    cleric *shirt* 牧师衬衫(领、袖为白色的彩色衬衫)

    coat *shirt* 外套衬衫(无纽长衬衫)

    collar attached *shirt* 装领衬衫(指不结领带衬衫或极正统衬衫)

    college *shirt* 大学生(礼服)衬衫

    coloured spots *shirt* 花点衬衫

    corduroy *shirt* 灯芯绒衬衫

    cotton *shirt* 全棉衬衫

    cowboy *shirt* 牛仔衬衫(美国西部牛仔穿衬衫)

    C. P. O (Chief Petty Officer) *shirt* 美国士官衬衫(左右有盖胸袋、格子料)

    cricket *shirt* 板球衫;运动衫

    denim *shirt* 牛仔(布)衬衫

    dobby *shirt* 小提花衬衫

    double mercerized T-*shirt* 双丝光T恤衫(双丝光布制)

    dress *shirt* 礼服(白)衬衫

    fitted *shirt* 合体衬衫;紧身衬衫

    flannel *shirt* 法兰绒衬衫;绒布衬衫

    flexible *shirt* 两用衬衫

    flannelette(brushed) check *shirt* 格绒衬衫

    flocking *shirt* 植绒布衬衫

    formal *shirt* 正规衬衫、礼仪衬衫、礼服衬衫

    French *shirt* 法式衬衫(一种双叠袖头、双扣高领、暗门襟、无胸袋的高档礼服男衬衫)

    gaucho *shirt* 高乔牧人衬衫(阔领针织衫)

    golf *shirt* 高尔夫球衫

    Hawaii(Aloha) *shirt* 夏威夷衬衫、香港衫

    Henley *shirt* (半开襟、无领短袖)亨利衬衫

    intelligent T-*shirt* 智能T恤衫(将碳纳米管纤维织进T恤,可监控心率、追踪定位等)

    interlock T-*shirt* 棉毛短袖圆领衫

    Ivy *shirt* 常春藤风格衬衫(美国足球联盟服式)

    jacket *shirt* 夹克式衬衫

    kangaroo *shirt* 袋鼠衬衫(有大贴

袋)
knitted *shirt* 针织衬衫
kurtah *shirt* 印度男子无领长衬衫
leisure *shirt* 休闲衬衫
linen *shirt* 亚麻衬衫
longline *shirt* 长身线形衬衫
lumber *shirt* 伐木衬衫(填棉格绒衬衫)
lurex pinstripe *shirt* 饰金银丝的细条衬衫
lycra *shirt* 弹力衬衫
maternity *shirt* 孕妇衬衫
memory *shirt* 免烫衬衫
Menni *shirt* 纽扣门襟针织衬衫
mesh *shirt* 网眼衬衫
Mexican wedding *shirt* 墨西哥婚礼男衬衫(白色,绣花)
neckband *shirt* (配西装用)饰狭窄领带的无领衬衫
open-necked *shirt* 开领衫,翻领衫
oxford *shirt* 牛津布衬衫
paper *shirt* 纸衬衫(一次性穿用)
patterned *shirt* 花样衬衫(有花样图案的衬衫)
pilot *shirt* 飞行员衬衫(有肩章,大口袋)
piqué polo *shirt* 珠地马球衫
pirate *shirt* 海盗衬衫(宽大,蓬松袖)
pleated bosom *shirt* 前胸打细褶的衬衫
polo *shirt* 马球衬衫(开领,短袖);翻领衫
printed *shirt* 印花衬衫,花衬衫
pullover *shirt* 套头衬衫;套头毛衫
quilted lumber jack *shirt* 绗棉衬衫
ramie/cotton *shirt* 麻棉衬衫

rayon *shirt* 人棉衬衫
rayon *shirt* printed 人棉印花衬衫
roll-neck *shirt* (可翻领)高领衫
rugby *shirt* 橄榄球运动衫
running *shirt* 低领无袖运动衫(汗衫)
sailor's striped *shirt* 海魂衫
school *shirt* 学生(制服)衬衫
seersucker *shirt* 泡泡纱衬衫
semi-fitted *shirt* 半紧身衬衫
shirt on *shirt* 双重穿衬衫方法(叠穿在领、袖等部位显露不同效果)
shirt on *shirt* on *shirt* 三重穿衬衫方法
shirt with ruffle in front 前胸饰褶衬衫
short(long)sleeved *shirt* 短(长)袖衬衫
silk(georgette)*shirt* 丝绸(乔其纱)衬衫
skivvy *shirt* 男圆领汗衫
sleep *shirt* 衬衫式睡袍
3/4 sleeve *shirt* 七分袖女衬衫
sleeveless(collarless)*shirt* 无袖(无领)衬衫
smart *shirt* 智能衬衫(可监测体温、呼吸、心率等,有的还具有连网操作功能)
sports *shirt* 运动衫,两用衫;休闲衬衫,便装衬衫
stretch *shirt* 紧身弹力衬衫
stripe(check)*shirt* 条(格)子衬衫
suedette *shirt* 绒面革衬衫
surf T-*shirt* (圆领短袖)冲浪 T 恤衫
swallowtail *shirt* 燕尾衬衫

sweat *shirt* 圆领长袖运动衫；卫生衣
T-(tee)*shirt* 圆领短袖汗衫,〔粤〕T恤,文化衫
tail(straight)bottomed *shirt* 圆（平）下摆衬衫
tapered *shirt* 细长衬衫
towelling *shirt* 毛巾衫
trueran *shirt* 涤棉衬衫,的确良衬衫
tuxedo *shirt* 礼服衬衫
twofer *shirt* （错色）双层衬衫（看似穿两件,实为一件）
uniform *shirt* 制服衬衫,军衬衫
Western *shirt* 西式衬衫；（美国）西部衬衫
work *shirt* 工作衫
zoot *shirt* 阻特衬衫（音译,配套阻特装穿的艳色衬衫）

shirtband [ˈʃɜːtbænd] （衬衫领子或袖头的）衬布

shirtcoat [ˈʃɜːtkəʊt] 衬衫式外套

shirtdress [ˈʃɜːtdres] 衬衫式连衣裙,衬衫裙

shirting [ˈʃɜːtɪŋ] 衬衫料；平布,细布
brushed check *shirting* 格绒衬衫布
calico *shirting* 棉衬衫布
dyed *shirting* 染色细纺
Oxford *shirting* 牛津衬衫料
ramie *shirting* 麻衬衫料,麻细平布
shirting of polyester georgette printed 涤纶印花乔其纱衬衫料
zephyr *shirting* 轻薄衬衫料

shirttail [ˈʃɜːtteɪl] 衬衫下摆（指背部）

curved *shirttail* 圆口衬衫后摆

shirtwaist [ˈʃɜːtweɪst] （仿男式）女衬衫

shirtwaister [ˈʃɜːtweɪstə] 女式衬衫裙,衬衫连衣裙

shoe [ʃuː]〔常用复〕鞋
acrylic crocheted *shoes* 腈纶钩编鞋
aerobic *shoes* 吸氧健身运动鞋
ankle-strap *shoes* 踝带鞋
antiskid *shoes* 防滑鞋
antistatic *shoes* 防静电鞋
artificial leather *shoes* 人造革皮鞋
astronaut's *shoes* 宇航鞋
athlete's *shoes* 田径鞋
badminton *shoes* 羽毛球鞋
ballet(ballerina) *shoes* 芭蕾鞋
bar *shoes* 带扣鞋,条扣鞋
basketball *shoes* 篮球鞋
beach *shoes* 沙滩鞋,海滨鞋
beaded *shoes* 珠饰鞋
boat *shoes* 船形鞋；休闲鞋
bowling *shoes* 保龄球鞋
boxing *shoes* 拳击鞋
brocade *shoes* （提花）锦缎鞋
buckle *shoes* 带扣鞋
canvas sports *shoes* 帆布运动鞋
casual *shoes* 便鞋,休闲鞋
children's *shoes* 童装鞋
Christian Louboutin *shoes* 克里斯提·鲁布托高跟皮鞋（亮漆红底）
cleated *shoes* （鞋底有防滑钉的）防滑鞋
close *shoes* 满帮鞋
corduroy *shoes* 灯芯绒鞋
cotton(cloth) *shoes* 布鞋
cotton-padded *shoes* 棉鞋
court *shoes* （女式）宫廷礼服鞋

（一种浅帮船形鞋）
crampons shoes （便于爬山和冰山行走的）钉鞋
cycling shoes 骑自行车运动鞋
dance shoes 舞蹈鞋
deck shoes 甲板鞋，船鞋
dress shoes 礼服用鞋；正装鞋；绅士鞋
dressy shoes 考究的鞋；正装鞋
duck(canvas) shoes 帆布鞋
dust-free shoes 无尘鞋
earth shoes 大地鞋
elevation shoes 增高鞋
embossed leather shoes 拷花皮鞋
embroidered shoes 绣花鞋
evening shoes 晚宴用鞋
fashion shoes 时尚鞋，时款鞋
felt shoes 毡（呢）鞋
fencing shoes 击剑鞋
football shoes 足球鞋
formal shoes 礼服用鞋
fur(fleece)-lined shoes 毛皮（驼绒）衬里鞋
gladiator shoes 角斗士鞋，罗马鞋（绑绳，系踝，露趾）
glass-sliding shoes 滑草鞋
golf shoes 高尔夫球鞋
gum shoes 〔美〕橡胶套鞋；橡胶底帆布鞋；轻便运动鞋
gym shoes 体操鞋
heeled shoes 有跟鞋
hiker shoes 徒步旅行鞋
house shoes 拖鞋，便鞋
injection moulded shoes 注塑鞋
insulate shoes 绝缘鞋
intelligent sports shoes 智能运动鞋（内置感应器，可通过称重统计穿着者的运动量和体重变化，施行健康管理）
javelin shoes 投标枪鞋
jelly shoes 果冻鞋（浅口平底，果冻色塑胶女装鞋）
jogging shoes 跑步鞋，慢跑鞋
kung-fu shoes （中国）功夫鞋
labour protective shoes 劳保鞋
lace-up shoes 系带鞋
ladies' shoes 女装鞋，淑女鞋
leather shoes 皮鞋
leisure shoes 休闲鞋，便鞋
leopard print shoes 豹纹鞋
long jumping shoes with 6 spikes 六钉跳远鞋
low shoes 低口鞋，半筒鞋
low(middle, high)-heeled shoes 平（中、高）跟鞋
low wedge shoes 低坡跟鞋
Mary Jane shoes 玛丽珍鞋（平跟搭襻浅口女童皮鞋）
misses' shoes 少女鞋
monk shoes 孟克鞋（音译）；僧侣（束带）鞋
monk strap dress shoes 孟克搭襻带皮鞋
moulded sole shoes 模压底鞋
nurse shoes 护士鞋
orthopedic shoes 矫形鞋
Oxford shoes 牛津鞋（浅口便鞋）
parachute jumping shoes 跳伞鞋
patent leather shoes 漆皮革鞋
peep(open)-toe shoes 露趾鞋
plastic shoes 塑料鞋
platform shoes 厚底鞋
priest shoes 僧鞋
printed shoes 印花鞋

protective shoes 防护鞋
pump shoes 轻便舞鞋
rain shoes 雨鞋
round (plain) toe shoes 圆（平）头鞋
rubber shoes 胶鞋，套鞋
running(racing) shoes 田径鞋，赛跑鞋，钉鞋
running shoes with 4 (5, 7) spikes 四（五、七）钉跑鞋
running shoes without spikes 无钉跑鞋
saddle shoes 马鞍鞋
safety shoes 防护鞋，劳保鞋
sand shoes 胶底帆布鞋（沙地穿用）
satin shoes 缎面鞋
school shoes 校服鞋，校鞋
shoe hammer 鞋锤
shoe lift 鞋拔
shoe making 制鞋
shoe materials 制鞋材料，鞋材
shoe polish 鞋油
shoe powder 鞋粉
shoe quarter 鞋后帮
shoe rest 鞋砧
shoe sides 鞋帮
shoe stockings （新式）连袜鞋
shoe stretcher 鞋撑
shoe throat 鞋口
shoe tongue 鞋舌
shoe tree 鞋楦
shoe upper(lining) 鞋面（里）
shooting shoes 射击运动鞋
side gore shoes 两侧拼缝松紧带鞋，侧帮松紧鞋
skate(skating) shoes 溜冰鞋

skateboard shoes 滑板鞋
ski shoes 滑雪鞋
soccer shoes 英式足球鞋
social shoes 社交鞋
softball shoes 垒球鞋
soft leather shoes 软皮皮鞋
special sports shoes 专用运动鞋
spike shoes （赛跑用）钉鞋
sports shoes 运动鞋，球鞋，〔粤〕波鞋
strap shoes 搭扣鞋
studded shoes 窝钉鞋
suede shoes 绒面皮鞋；仿鹿皮鞋；羊皮皮鞋
sweethearts shoes 情侣鞋
tab shoes 拉襻鞋，搭襻鞋
tap shoes 踢踏舞鞋
T-bar shoes T形带襻鞋
tennis shoes 网球鞋
theatrical shoes 戏装鞋
tiger head shoes （小童穿）虎头鞋
toe cap derby tie dress shoes 束带接头皮鞋
town shoes 上街用鞋，街市鞋
track shoes （田径）钉鞋；跑步鞋
travelling(tourist) shoes 旅游鞋
tuxedo shoes 礼服鞋
velcro athletic shoes 魔术贴搭襻运动鞋
vinyl shoes 塑胶鞋
volleyball shoes 排球鞋
walking shoes 散步鞋；便鞋
waterproof shoes 防水鞋
water ski shoes 滑水运动鞋
wedge shoes 坡跟鞋，〔粤〕船踭鞋
wooden shoes 木质鞋；木屐

work(duty) shoes 工作鞋
wrestling shoes 摔跤鞋
shoehorn [ˈʃuːhɔːn] 鞋拔
shoelace [ˈʃuːleɪs] 鞋带
shoemaker [ˈʃuːmeɪkə] 制鞋工，鞋匠
shoeshine [ˈʃuːʃaɪn] 鞋油
shoestring [ˈʃuːstrɪŋ] 鞋带
shortalls [ˈʃɔːtɔːlz] （带围兜的）童装连身短裤
shorts [ʃɔːts] 〔复〕短裤，短西裤
  basketball shorts 篮球短裤
  beach shorts 沙滩短裤
  Bermuda shorts （齐膝）百慕达短裤
  bib shorts （有护胸的）背带短裤
  bike shorts 骑自行车短裤
  bloomer shorts 灯笼短裤
  board shorts 冲浪（或游泳）短裤；沙滩裤
  boxer shorts 松紧腰短内裤；拳击短裤
  burnt-out denim shorts 烂花牛仔短裤
  cargo(cargo pocket) shorts 大贴袋短裤
  casual shorts 便装短裤，休闲短裤
  city shorts 都市短裤，热裤（与运动上衣配套）
  climbing shorts 登山短裤（多口袋短裤）
  cotton twill shorts 全棉斜纹短裤
  cuffed shorts 卷脚口短裤，摺脚短裤
  cycling bib shorts 护裆骑车短裤（裆部有防护衬垫）
  denim shorts 牛仔短裤
  drawstring shorts 束带短裤

  flared shorts 宽脚短裤，喇叭短裤
  gym shorts 运动短裤
  high-rise shorts 高腰短裤
  Jamaica shorts 牙买加短裤，田径短裤
  leisure shorts 休闲短裤
  loose cut shorts 宽松型短裤
  maxi shorts 长短裤
  nylon shorts 锦纶短裤
  pigment dyed canvas shorts 涂料染色帆布短裤
  pull-on shorts 松紧腰头短裤
  ramie/cotton shorts 麻棉短裤
  ripped cuffed shorts 破烂型卷脚（牛仔）短裤（裤身上有意剪破几处）
  rugger shorts 橄榄球短裤
  sandblasted shorts 喷砂（洗水）短裤
  short(mini) shorts 超短裤
  shorts with adjustable waist 腰头可调节的短裤
  silk shorts 丝绸短裤
  skate shorts 溜冰短裤
  sleep shorts 男式短睡裤
  sports shorts 运动短裤；球裤，〔粤〕波裤
  surfer's shorts 冲浪短裤
  swim shorts 游泳外短裤（穿在泳裤外）
  tailored shorts 西装短裤，西短裤
  tennis shorts 网球短裤
  track shorts 田径短裤；运动短裤
  walking shorts 散步短裤（长至膝盖，适于散步用）
  Western shorts （美国）西部短裤
  work shorts 工作短裤

shorty(=shortie) [ˈʃɔːti] 特别短的衣服；超短裤
shoulder [ˈʃəʊldə] 肩，〔粤〕膊头，膊；肩型；肩宽
   across back *shoulders* 总肩宽，横肩宽
   broad *shoulder* 宽肩型
   built-up *shoulder* 起肩型
   classic straight *shoulder* 直肩
   concaved *shoulder* 凹肩型
   cross *shoulder* 肩宽
   drop *shoulder* 落肩，垂肩
   high *shoulder* 耸肩型，高肩
   low *shoulder* 溜肩，塌肩
   off the *shoulder* 露肩
   regular(natural) *shoulder* 正常肩型
   saddle *shoulder* 马鞍形肩
   shoulder broad 肩带，肩襻，肩牌
   shoulder length 肩宽
   shoulder slope 肩斜（度）
   shoulder to bust point 肩至胸点距离
   shoulder to waist line 肩至腰线距离
   sloping (slanting) *shoulder* 斜肩型，垂肩；〔检〕塌肩
   small round *shoulder* 小圆肩型
   square *shoulder* 平肩，方型肩
   standard *shoulder* 标准肩型
   tailored *shoulder* 西装型肩线
   uneven *shoulder* seam 〔检〕肩缝不顺直
   wide(narrow)*shoulder* 宽（窄）肩
shovel [ˈʃʌvl] （教士戴的）宽边帽（= shovel hat）
show [ʃəʊ] 展览（会）；表演
   catwalk *show* 时装表演
   ethical fashion *show* 道德时装秀
   show breeches 马裤
shrink [ʃrɪŋk] （布料）缩水
   shrink-proof 防缩，不缩水
shrinkage [ˈʃrɪŋkɪdʒ] 缩水；缩水率
   fabric *shrinkage* 布料缩水率
   testing of *shrinkage* 试缩水率
shrinking [ˈʃrɪŋkɪŋ] （将布料）缩水
shrug [ʃrʌg] （单扣）女式短上衣
shuttle [ˈʃʌtl] （缝纫机）滑梭，摆梭
   shuttle bobbin （缝纫机）梭芯
   shuttle case （缝纫机）梭芯套
side [saɪd] （布料）面；（身体或服装的）侧边，胁
   back *side* 背面；后胁
   side seaming 缝边缝
   side slippage （缝料）侧向滑移
   wrong (right) *side* of cloth 布料的反（正）面
sidekick [ˈsaɪdkɪk] （裤两边）插袋
sideseam [ˈsaɪdsiːm] 侧缝，边缝，摆缝，胁缝，栋缝；外长
sidesplit [ˈsaɪdsplɪt] （衣服）衩，衩口；边衩
silhouette [ˌsɪluˈet] 剪影；（人体）轮廓；轮廓线，（服装）外形线，廓型
   barrel *silhouette* 桶形轮廓
   basque *silhouette* 巴斯克轮廓（紧身廓型）
   clothing *silhouette* 服装廓型
   dirndl *silhouette* 旦多尔轮廓（紧身连衣裙廓型）
   empire *silhouette* 帝国式服装轮廓（法国拿破仑时代流行的低领口，高腰节裙装廓型）
   hobble *silhouette* 跛行轮廓（窄摆裙

廓型)
hourglass *silhouette* 沙漏型轮廓
long torso *silhouette* 长驱轮廓(低腰廓型)
Mae West *silhouette* 美腰型轮廓
sheath *silhouette* 鞘型轮廓(细长紧身廓型)

silk [sɪlk] 丝;蚕丝;丝绸,丝织品;〔复〕丝绸衣服
    Adelis *silk* 艾德莱斯绸(又名和田绸、舒库拉绸)
    artificial *silk* 人造丝
    bark *silk* 树皮纹绸
    cashmere *silk* 开司米毛葛,开司米绸
    China *silk* 中国丝绸
    cire *silk* 蜡光绸;砑光绸
    cloth and *silk* 布帛(棉布和丝绸)
    colored *silk* 彩绸
    damask *silk* 绫
    doupion *silk* 双宫绸
    embroidery *silk* 绣花丝线
    fibre *silk* 人造丝
    figured *silk* 提花丝织物;花绸
    Fuji *silk* 绢丝纺
    glass *silk* 玻璃纱
    greige *silk* 生丝,坯绸
    Hangzhou *silk* 杭纺
    kimono *silk* 和服绸
    mulberry *silk* 桑蚕丝
    nanometer *silk* 纳米丝(超细纤维)
    Nishang *silk* 霓裳绸
    noil *silk* 绸丝绸,绵绸
    oil *silk* (防水)油绸
    ottoman *silk* 四维呢
    printed *silk* 印花丝绸
    pure *silk* 纯丝绸(未加增重剂)
    raw *silk* 生丝,厂丝
    rayon lining *silk* 人丝美丽绸
    real(natural) *silk* 真丝;真丝绸
    rib *silk* 桑绢罗
    sheer *silk* 绡(轻薄丝绸)
    shot *silk* 闪光绸
    silk tartan 格子绸
    silk twill 斜纹绸
    space dyed *silk* 间隔染色绸
    spun *silk* 绢丝纺绸
    stiff *silk* 绢绸
    suiting *silk* 呢
    tencel/cuprene mixed *silk* 天丝/铜氨交织仿真丝绸
    tie *silk* 领带绸
    tussah *silk* 柞蚕丝;柞蚕绸
    umbrella *silk* 伞绸
    washable *silk* 洗水丝绸,〔粤〕水洗丝
    watered *silk* 波纹绸
    white *silk* 素绢
    Xinhua suiting *silk* 新华呢

silkwear [ˈsɪlkweə] 丝绸服装
    sand-washed *silkwear* 砂洗丝绸服装

simplicity [sɪmˈplɪsəti] 〔设〕简约,简单

simplification [ˌsɪmplɪfɪˈkeɪʃən] 〔设〕单纯化;简化

singelringen [ˈsɪŋglrɪŋɡən] 〔瑞典语〕单身戒指

singlet [ˈsɪŋɡlɪt] 〔英〕男汗衫;背心;文化衫
    athletic rib *singlet* 罗纹运动背心
    cotton interlock(sport)*singlet* and trousers 棉毛(运动)衫裤
    cotton lycra *singlet* 全棉弹力背心

double layered *singlet* （错色）双层背心（看似穿两件，实为一件）
mesh *singlet* 网眼背心
shelf bra *singlet* 胸罩形吊带背心
shoestring *singlet* 吊带背心
sleep *singlet* 女背心睡衣
V-neck sleeve less *singlet* V 领无袖针织衫

singlesuit [ˈsɪŋglsjuːt] 婴儿背心连裤装
size [saɪz] 大小，尺寸，尺码，号码
 chest *size* 胸围大小
 cutting *size* 裁剪尺寸
 extra-large(small) *size* 特大（小）号，特大（小）码
 finished-*size* 成品尺寸
 free *size* 可调尺寸
 full colour/*size* 齐色齐码
 garment *size* designation 服装号型
 head *size* 头围
 just *size* 准确尺寸
 kid's(boy's, adult's) *size* 小童（中童、成人）尺码
 king *size* 特大尺码
 neck *size* 颈围；领围
 off(incorrect) *size* 尺寸不合
 plus *size* 宽松尺寸；（特体的）超大尺码，加肥尺码
 queen *size* 第二大码（仅小于特大码 king *size*）
 reducing(enlarging) of *sizes* 缩小（放大）尺寸
 regular *size* 基准尺码
 series of apparel *sizes* and styles 服装号型系列
 short in *size* 〔检〕断码（尺码不全）
 size stamp 尺码印戳
 size tariff 尺码表
 small(medium, large) *size* 小（中、大）号，小（中、大）码
 solid(assorted) *size* 独（混）码
 special measurement *size* 特体尺码
 standard *size* 标准尺码
 thigh *size* 腿围
 wrong *size* indicated 〔检〕错码（尺码搞错）
 yarn *size* 纱支

skate [skeɪt] 溜冰鞋；（四轮）旱冰鞋
 figure *skates* 花样滑冰鞋
 hockey *skates* 冰球鞋
 ice *skates* 滑冰鞋
 roller *skates* 旱冰鞋
 speed *skates* 速度滑冰鞋

sketch [sketʃ] 图形，草图，示意图；素描
 design *sketch* 设计图
 detail *sketch* 分解图
 fashion *sketch* 时装画
 pattern *sketch* （服装）式样设计图；（布料）花纹图样
 sketch map 示意图
 working *sketch* 款式设计图

sketcher [ˈsketʃə] 描图员，描画师
skimmer [ˈskɪmə] 〔美〕（无袖圆领）裁剪简洁的合身女裙；平顶宽边草帽；女式平跟（或低跟）无带浅口皮鞋
skimp [skɪmp] 非常短的迷你装
skin [skɪn] 皮肤；毛皮；皮革
 crocodile *skin* 鳄鱼皮
 donkey *skin* 驴皮
 fabric-*skin* 护肤织物（具隔热功能）
 Persian lamb *skin* 波斯羔羊皮

python *skin* 蟒蛇皮
rabbit *skin* 兔毛皮
snake *skin* 蛇皮
walrus *skin* 海象皮
skintight [skɪn'taɪt] 紧身衣服；紧身内衣裤
skinwear ['skɪnweə] 紧身服装
skipping ['skɪpɪŋ] （缝纫中的）跳针；跳缝
skirt [skɜːt] 女裙，裙子；西服裙；（衣、裙的）裾，下摆

    accordion-pleat *skirt* 百褶裙
    A-line *skirt* A 型裙
    all-around pleated *skirt* 百褶裙
    apron *skirt* 围裙式裙
    asymmetrical *skirt* 不对称裙
    back pleated *skirt* 后襞褶裙
    ballet *skirt* 芭蕾短裙
    balloon *skirt* 气球裙
    bare-back *skirt* 露背裙
    barrel *skirt* 筒形裙，桶裙
    bell *skirt* 钟形裙
    bell type *skirt* 喇叭裙
    bias-cut *skirt* 斜裁裙
    bias *skirt* 斜裙
    Bohemian *skirt* 波希米亚裙（一种宽摆长裙，常饰以印花、绣花、蕾丝、流苏等）
    bouffant *skirt* 蓬松裙
    box pleat *skirt* 箱褶裙
    braces *skirt* 背带裙，吊带裙
    bubble *skirt* 泡泡裙
    button-down *skirt* 纽扣开襟裙
    button through *skirt* （部分或全部）开门襟裙
    cargo *skirt* 贴袋裙（裙身两侧有大贴袋）
    chemise *skirt* 马甲连衣裙，无袖衬衫裙
    cheongsam *skirt* 旗袍裙
    children's *skirt* 童裙
    circular *skirt* 圆裙
    coat and *skirt* （一套）上衣和裙子，衣裙套装
    cone *skirt* 圆锥裙
    costume *skirt* 西式裙
    culotte *skirt* 裤裙，裙裤
    cummerband *skirt* 宽腰带裙
    denim *skirt* 牛仔裙
    dervish *skirt* 伊斯兰教托钵僧裙
    detachable *skirt* 可拆卸的裙
    dirndl *skirt* 旦多尔裙（腰围打褶的宽裙）
    divided *skirt* 分衩裙；裙裤
    double *skirt* 双层裙
    draped *skirt* 垂饰裙，自然皱裙
    embroidered *skirt* 绣花裙
    empire *skirt* 高腰裙（高腰直线条宽裙）
    escargot *skirt* 蜗牛裙（拼缝呈蜗牛壳纹样）
    expansion *skirt* 大摆裙
    figured *skirt* 图案花裙
    fish-tail *skirt* 鱼尾裙
    fitted *skirt* 合身裙，贴身裙
    flare *skirt* 喇叭裙，宽摆裙，波浪裙
    flare tiers *skirt* 多层波浪裙
    flippy *skirt* 翘摆裙（下摆翘起的展宽裙）
    fluted hem *skirt* 波边下摆裙
    flounced *skirt* 荷叶边裙，摆饰裙
    formal *skirt* 正装裙，礼服裙
    full *skirt* 宽摆长裙

full length *skirt* 齐地长裙
gathered *skirt* 碎褶裙；腰褶裙；皱褶裙
gored *skirt* 片裙
grass *skirt* 草裙
gypsy *skirt* 吉卜赛裙（大摆裙）
half-circle *skirt* 半圆裙
half *skirt* 半截裙
harem *skirt* 闺阁裙，后宫裙（伊斯兰教女裙）
high waist *skirt* 高腰裙
hip bone *skirt* 腰骨裙，低腰裙
hip pleated *skirt* 襞褶裙
hipster *skirt* （低及臀部的）低腰裙
hobble *skirt* 窄摆裙，蹒跚裙（20世纪初流行，走路步幅受限）
hourglass-shape *skirt* 沙漏形裙
inverted pleat *skirt* 阴裥裙，倒褶裙
jersey *skirt* 针织裙
joint *skirt* 连结裙
jumper *skirt* 无袖连衣裙，背心裙，马甲裙，女学生裙
kick pleat *skirt* 倒褶裙（裙底开阴褶）
kilt *skirt* 褶叠短裙（苏格兰式男用）
knee *skirt* 齐膝裙
knitted *skirt* 针织裙
lace *skirt* 花边裙
leather *skirt* 皮裙
low-slung *skirt* 低吊裙
midi *skirt* 迷地裙；中庸裙，半长裙
mid length *skirt* 中长裙，半长裙
mini-mini-(micro-mini) *skirt* 超迷你裙（比迷你裙更短）
one-piece *skirt* 一片裙，单页裙

over *skirt* 短罩裙，两页裙
panel *skirt* （裙上缀缝布块的）掩块裙，拼幅裙
panne *skirt* 平绒裙
pant *skirt* 裤裙
parachute *skirt* 伞裙
peasant *skirt* 农妇裙（有绣花腰带的皱褶裙），村姑裙
pegged *skirt* （上宽下窄）陀螺裙
peg-top *skirt* 陀螺裙
pencil *skirt* （直筒紧身的）笔杆裙
permanently pleated *skirt* 永久褶裥裙
petal *skirt* 花瓣裙
pleated *skirt* 褶裥裙；对褶裙；百褶裙
pleat-gore *skirt* 片褶裙
print *skirt* 印花布裙
Qipao style *skirt* 旗袍裙
quilted *skirt* 绗缝裙，填棉裙
riding *skirt* 女骑装裙
rock *skirt* 摇滚裙（高腰大摆裙）
ruffle *skirt* 皱褶裙
salopette *skirt* 沙罗佩工装裙
sarong *skirt* 纱笼裙
semi-circular *skirt* 半圆裙
semi-flared *skirt* 小喇叭裙，半圆裙
separate *skirt* 裤裙，裙裤
sheath *skirt* 紧身裙
short(long) *skirt* 短（长）裙
short trapezium *skirt* 短梯形裙
side-draped *skirt* 边垂饰裙
side pleat *skirt* 边褶裙
silk noil *skirt* 绵绸裙
silk *skirt* 绸裙
skirt flare is uneven 〔检〕裙浪不匀

skirt hem line rides up 〔检〕裙摆起吊
skirt with laces 镶花边裙
slim *skirt* 窄摆裙；旗袍裙
spiral *skirt* 螺旋裙
split at the lower part of *skirt* 〔检〕裙裥豁开
square *skirt* 细襞裙，细褶裙
squaw *skirt* 印第安女裙
standard *skirt* 标准裙
straight *skirt* 直筒裙
suspender *skirt* 吊带裙，背带裙
swing *skirt* 摇曳裙
swirl *skirt* 旋涡裙，蜗牛裙
tailored *skirt* 西服裙
tennis *skirt* 网球（短）裙
tie-dyed *skirt* 扎染裙
tiered flounce *skirt* 宝塔荷叶边裙
tiered *skirt* 多层裙，宝塔裙，节式裙
tight *skirt* 紧身裙，窄裙
tiny *skirt* 超短裙
torso *skirt* 低腰裙
towelling *skirt* 毛巾裙
trumpet *skirt* 喇叭裙
tube *skirt* 筒型裙
tucked *skirt* 褶叠裙
tunic *skirt* 塔尼克裙（音译，紧身裙）
tutu *skirt* 芭蕾短裙
tuxido *skirt* 塔士多礼服裙
tweed *skirt* 粗花呢裙
two (four, six, eight)-gore *skirt* 两（四、六、八）幅裙
umbrella *skirt* 伞裙，伞裥裙
woolen *skirt* 毛呢裙
wrap-around *skirt* 围裹裙，卷裙，宽搭门裙
wrapped *skirt* 围裹裙，包裙
yoke *skirt* 拼腰裙，剪接裙

skirting [ˈskɜːtɪŋ] 裙料

skivvy [ˈskɪvi] 针织（立领）套头衫；（复）内衣（包括汗衫和短裤）

skullcap [ˈskʌlkæp] 室内便帽；瓜皮帽

skunk [skʌŋk] 臭鼬毛皮

slacks [slæks] 〔复〕长裤，宽松长裤，便裤
 bell bottom *slacks* 喇叭裤
 beltless *slacks* 不用腰带的裤子，无腰带裤
 Chinese-style *slacks* （旧时）中式长裤（大腰头，无前后档之分）
 dress *slacks* 礼服裤子
 high waist tapered *slacks* 高腰窄脚裤
 Ivy *slacks* 常春藤风格长裤（美国足球联盟服式）
 slim *slacks* 细脚口裤
 straight-leg *slacks* 直脚裤
 women's (ladies') *slacks* 女装长裤

slash [slæʃ] （衣服上的）长缝，剪口；衣衩

sleek [sliːk] （男装用）棉里布；口袋布

sleepbag [ˈsliːpbæg] 睡袋
 baby fleece *sleepbag* 一种有领和袖的婴幼绒布睡袋

sleepcoat [ˈsliːpkəʊt] 有腰带的长外套；门襟有纽扣的睡衣；宽松睡袍

sleepers [ˈsliːpəz] 〔复〕童睡衣裤

sleepshirt [ˈsliːpʃɜːt] 衬衣式睡袍

sleepsuit [ˈsliːpsuːt] 睡衣裤
 newborn *sleepsuit* 新生婴儿睡衣裤

sleepwear [ˈsliːpweə] （总称）睡衣裤
sleeve [ˈsliːv] 袖子；袖套
 batwing *sleeve* 蝙蝠袖
 bell *sleeve* 钟形袖；喇叭袖
 bishop *sleeve* 主教袖（袖口收紧呈蓬状的长袖）；泡泡袖
 bracelet *sleeve* 九分袖
 bubble *sleeve* 泡泡袖
 butterfly *sleeve* 蝴蝶袖
 cap *sleeve* 帽形袖
 cape *sleeve* （连肩）盖袖，喇叭袖，披肩袖
 circular cape *sleeve* （波浪褶）披肩短袖
 closed *sleeve* 封闭型袖
 cowl *sleeve* 垂褶袖
 cuffed *sleeve* 带袖头衣袖
 dolman *sleeve* 德尔曼袖（音译，袖口窄，袖窿宽），斗篷袖
 double *sleeve* 双层袖
 drawstring *sleeve* （袖口）束带袖；伸缩袖
 drop shoulder *sleeve* 落肩袖，低肩袖
 elbow *sleeve* 中袖，五分袖，肘长袖
 epaulet *sleeve* 肩章袖（过肩与袖连为一片）
 flare *sleeve* 喇叭袖
 flounced *sleeve* 荷叶边袖口袖，摆饰袖
 French *sleeve* 法式袖，连肩袖，和服袖；超短袖（袖长约为长袖的八分之一）
 front (back) *sleeve* 前（后）袖
 ful *sleeve* 宽大袖
 gigot *sleeve* 羊腿袖（袖窿松，袖口紧）
 half *sleeve* 半袖，中袖
 hanging *sleeve* 吊袖，挂袖
 high shoulder *sleeve* 高肩袖，耸肩袖
 horse-hoof (folded back) *sleeve* 马蹄袖
 Juliet *sleeve* 朱丽叶袖（泡状短袖）
 kimono *sleeve* 和服袖，连袖
 kite *sleeve* 风筝袖
 lamp *sleeve* 灯泡袖
 lantern *sleeve* 灯笼袖（两头小中间大）
 Magyar *sleeve* 匈牙利式袖（连身出袖，类似和服袖）
 melon (balloon) *sleeve* 气球袖
 normal *sleeve* 普通袖
 one-piece *sleeve* 整片袖，大裁袖，一片袖
 open (closed) *sleeve* 开放（封闭）式袖
 outer (inner) *sleeve* 外（内）袖
 over *sleeve* 罩袖，套袖
 pagoda *sleeve* 宝塔袖，佛塔袖
 petal *sleeve* 花瓣（短）袖（两袖片交叠）
 puff *sleeve* 泡泡袖（上膨下紧），膨胀袖
 push-up *sleeve* 推上袖（肘部以下紧贴，肘部以上宽松）
 raglan *sleeve* 套袖，连肩袖，插肩袖，斜肩袖，包肩袖，拔肩袖，〔粤〕牛角袖
 roll-up *sleeve* 翻折袖（袖口翻折）
 rufled *sleeve* 褶饰边袖；荷叶边袖
 semi-raglan *sleeve* 半连肩袖，半插肩袖

set-in sleeve 装袖，圆袖，接袖，普通袖
set-in sleeve with centre seam 中缝圆袖
shirred sleeves 抽褶袖
shirt sleeve 衬衫袖
short(long) sleeve 短（长）袖
sleeve leans to front(back) 〔检〕袖子偏前（后）
sleeve opening is uneven 〔检〕袖口不齐
slit sleeve 开衩袖
split raglan sleeve 前圆后连袖
square sleeve 方形直袖
strap shoulder sleeve 肩章袖
styled armhole sleeve 变形袖
three quarter sleeve 七分袖，半长袖
tiered sleeve 多层袖
tight(loose) sleeve 紧身（宽松）袖
top sleeve 大袖，外袖
trumpet sleeve 喇叭袖
two-piece sleeve 两片袖
under sleeve 小袖，内袖
unmatched sleeves 〔检〕（左右）袖长不一，〔粤〕长短袖
wedge sleeve 楔形袖
yoke sleeve 与过肩连裁的袖
zip-off sleeve 拉链脱卸袖

slicker [ˈslɪkə] 〔美〕宽长油布雨衣
sliders [ˈslaɪdəz] 〔复〕（平底或有跟）拖鞋
    one band sliders 宽带襻拖鞋
    sliders with knot 花结拖鞋
    wedge sliders 坡跟拖鞋
slimming [ˈslɪmɪŋ] （用减食、运动等方法）减重疗法，减肥

slingback [ˈslɪŋbæk] 后襻式女装鞋（带襻式鞋后帮）
slingbag [ˈslɪŋbæg] 吊带挎袋
slip [slɪp] 女无袖衬裙；套裙；童外衣；〔复〕男游泳裤
    bouffant slip 向外蓬起的衬裙；蓬松套裙
    bra slip 带乳罩的衬裙
    camisole slip 背心式衬裙
    flared slip 宽摆衬裙
    full slip 长衬裙
    semi-(half) slip 短衬裙
    shaping slip 塑形衬裙；紧身衬裙；紧身套裙
    sleeve slip 有袖衬裙
    slim slip 窄身衬裙
    slip ons 易穿戴的鞋，手套等
    strapless slip 无吊带衬裙（或套裙）
    tight slip 紧身衬裙（或套裙）
    waist slip 短衬裙
slip-on [ˈslɪpɒn] 套领衫；套裙；无扣手套；无带（扣）鞋
    built-up shoulder slip-on 挂肩套裙
    kimono slip-on 和服式（易穿脱）外套（衣身连袖未裁断）
    metal slip-on 有金属装饰的套穿鞋
    round neckline slip-on 圆领套裙
slipover [ˈslɪpˌəʊvə] 套领衫，套领运动衣
    sleeveless slipover 无袖套领衫
slipper [ˈslɪpə] 〔常用复〕拖鞋；便鞋
    ballet slippers 女低跟软底便鞋；芭蕾舞鞋
    bath slippers 浴鞋
    bedroom slippers 卧室用拖鞋
    boot slippers 靴式便鞋，便靴（室内防寒穿）

collar slippers 翻口便鞋（鞋口外翻似衣领）
embroidered slippers 绣花拖鞋
embroidered slippers with beads 珠绣拖鞋
fleece-lined slippers 绒里拖鞋
foamed plastic slippers 泡沫塑料拖鞋
fur slippers 毛皮拖鞋
health care slippers 保健拖鞋
indoor(house) slippers 室内拖鞋
leather slippers 皮拖鞋
paper slippers 纸拖鞋
pile slippers 剪绒毛拖鞋
plastic slippers 塑料拖鞋
printed slippers 印花拖鞋
scuff slippers 平底拖鞋
sock top slippers （婴幼童穿）袜口鞋
sponge rubber slippers 海绵拖鞋
straw slippers 草编拖鞋
suede slippers 绒面革拖鞋
toe slippers 芭蕾鞋
vinyl slippers 塑料拖鞋
winter slippers 冬用拖鞋
wooden slippers 木拖鞋

slipstick [ˈslɪpstɪk] 计算尺

slit [slɪt] 细长的切口；（服装上无搭接位的）衩口
   dizzying slit （服装）高开衩
   horizontal slit （牛仔裤前面）水平袋口
   side slit 摆缝衩；边衩
   sleeve slit 袖衩

slop [slɒp] （宽松）罩衣，外衣，工作服；〔复〕廉价成衣

Sloppy Joe (=sloppy joe) [ˈslɒpɪˈdʒəʊ] 肥大的女毛衫；宽松套头衫

smartphone [ˈsmɑːtfəʊn] 智能手机

smartwear [ˈsmɑːtweə] 智能服装

smock [smɒk] 工作罩衫，工作服；童罩衣；司马克（音译，装饰性缩缝褶）
   dress smock 披风
   painter smock 画家工作罩衫；油漆工罩衫
   raglan smock 套袖工作服；连肩袖罩衫
   smock frock 长罩衫；工作服

smocking [ˈsmɒkɪŋ] 装饰性缩缝，〔粤〕打揽；缩绣缝纫

smoker [ˈsməʊkə] 〔英〕略式晚礼服

snap [snæp] 四合纽，撳扣，按扣，子母扣，钩扣，〔粤〕唵纽
   diamond snap 宝石按扣
   insecure snap KG2.5mm]〔检〕四合纽钉得不牢
   pearl snap 珠光四合纽
   plastic(metal) snap 塑料（金属）四合纽
   prong snap 五爪按扣，五爪纽
   ring(cap) snap 圈（帽）形四合纽
   snap attacher 钉（四合）扣机

sneakers [ˈsniːkəz] (=sneaks [sniːks])〔复〕〔美〕（胶底帆布面）轻便运动鞋；旅游鞋；软底鞋
   athletic slip-on sneakers 无带扣运动鞋
   yacht sneakers 快艇运动鞋（鞋面有条纹的帆布鞋）

snippers [ˈsnɪpəz] 〔复〕手剪，线剪

snood [snuːd] （女用）束发带；发网

snowshoe [ˈsnəʊʃuː] 雪鞋
snowsuit [ˈsnəʊsuːt] 童风雪大衣
snugglesuit [ˈsnʌglsuːt] 一种婴儿舒适连身装
sock [sɒk] 〔常用复〕袜子，短袜；（古希腊、罗马喜剧演员穿）轻软鞋

 anklet socks 短袜；袜套
 anklet pocket socks 口袋袜（袜筒上有拉链袋）
 antibacterial deodorant socks 抗菌防臭袜
 baby socks 宝宝袜
 baby moccasin socks 宝宝莫卡辛袜（软袜鞋）
 baseball socks 棒球袜
 bed socks 睡觉用袜，睡袜
 business socks 上班袜，工作袜
 casual socks 休闲袜
 children's socks 童袜
 computer pattern socks 计算机花袜
 cotton socks （棉）纱袜
 cotton thread socks （棉）线袜
 crew socks 针织罗口短袜；水手短袜
 cycling socks 骑自行车袜
 dress socks 男用短丝袜；礼服袜
 electronic pattern socks 电子提花短袜
 embroidered socks 绣花袜
 fancy socks 花袜
 fashion socks 时装短袜
 girl's socks 少女袜
 hiking socks 徒步旅行袜，远足袜
 home socks 家用袜鞋（室内穿用）
 hunting socks 狩猎袜
 jacquard socks 提花袜
 knee-high socks 及膝短袜
 low cut socks 齐踝短袜
 lycra socks 莱卡袜，弹力袜
 men's dress socks 绅士袜
 midway socks 短袜
 moisture wicking socks 吸湿排汗袜，快干袜
 non skip socks 连续线迹袜
 nylon stretch socks 弹力锦纶袜
 open toe socks 露趾袜
 polyamide socks 锦纶袜，尼龙袜
 reversible socks 双面穿短袜
 room socks （室内用）袜鞋
 running socks 跑步袜
 seamless socks 无缝袜
 sheer toe socks 脚尖透明短袜
 silk socks 丝袜
 ski socks 滑雪袜
 slipper socks （室内用）皮底袜鞋
 smart socks 智能袜（采用特殊的传感器纤维面料，可以检测穿着者的运动数据，包括体重变化等）
 snowboard socks 滑雪板袜
 sport socks 运动短袜
 sport shoe socks 运动袜鞋
 tall socks 高筒短袜，高勒袜（长及腿肚）
 toe socks 脚趾分开的短袜，分趾袜
 towelling socks 毛巾袜
 trainer socks 训练短袜，运动短袜
 turnover socks 翻（袜）口短袜
 wood(bamboo)fibre socks 木（竹）纤维袜
 wool socks 毛袜
 work socks 工作袜
 worsted(knitting wool) socks 毛线

袜，绒线袜
socklet ['sɒklet] 翻口短袜；套袜
socquette [sɒkɛt] 〔法〕短袜
softness ['sɒftnəs] （布料的）柔软度
soil [sɔil] 〔检〕污渍
sole [səʊl] 脚底；鞋底；袜底
    cleated *sole* 防滑鞋底
    cloth *sole* 布鞋底
    inner *sole* 鞋垫
    leather *sole* 皮鞋底
    moulded *sole* 模压底
    plastic *sole* 塑料底
    rubber *sole* 胶底
solitaire [ˌsɒlɪ'teə] 单颗宝石；镶嵌单颗宝石的饰物（如耳环、戒指等）
    1/2 carat diamond *solitaire* 镶嵌半克拉钻石的戒指（或耳环）
soutane [suː'tɑːn] （天主教）祭司法衣
southwester [saʊθ'westə] （水手用）雨帽
sox(=**socks**) [sɒks] 〔复〕短袜
spacer ['speɪsə] 间隔器（确定纽扣、扣眼位置）
spacesuit ['speɪssuːt] 宇航服，太空服
    EVA *spacesuit* 舱外航天服
    IVA *spacesuit* 舱内航天服
spacewear ['speɪsweə] 宇航服，太空服
Spacie ['speɪsɪ] 斯佩西针织布（日本产运动服装面料）
spandex ['spændeks] 氨纶；氨纶织物；斯潘德克斯（弹性纤维）
spangle ['spæŋgl] （装饰于晚装、戏装上的）亮片，珠片
spanner ['spænə] 扳手；螺丝扳钳
    solid (adjustable) *spanner* 固定（活动）扳手

specification [ˌspesɪfɪ'keɪʃən] 规格；说明书
    machine *specification* 机种
    size *specifications* 尺码表
spectacles ['spektəklz] 〔复〕眼镜；护目镜
spencer ['spensə] 短上衣；短外套；短夹克；女针织套头衫；针织短外衣
    spencer croisé 〔法〕双排扣短上衣
spinning ['spɪnɪŋ] 纺纱
split [splɪt] （横剖的一层）兽皮；（衣服的）衩口
    cow *split* 牛二层革
    pig *split* 猪二层革
    side *splits* 边衩
sponge [spʌndʒ] 海绵
    ironing *sponge* 熨烫海绵垫（烫布的衬垫）
    sponge shoulder pad 海绵垫肩
sportscoat ['spɔːtskəʊt] 轻便上衣，运动上衣；休闲外套
sportswear ['spɔːtsweə] 运动服装
spot [spɒt] 〔检〕（服装上的）污迹，脏迹
    spot remover 去污剂
sprayer ['spreɪə] （熨烫用）喷水壶，水枪
spreading ['spredɪŋ] （裁剪前）铺布，平布，拉布
    cloth *spreading* 铺布，平布
    intelligent fabric *spreading* 智能拉布
square [skweə] 直角尺，矩尺；正方形；平方
    L-*square* L形直角尺
    set *square* 三角尺，三角板
    T *square* 丁字尺
squirrel ['skwɪrəl] 松鼠毛皮

stain [steɪn] 污渍
 oily *stains* 油渍，油污
 stain release 防污
 water *stain* 〔检〕（熨烫的）水渍
stand [stænd] （机器）架；台；人体模型；位置
 body *stand* 人体模型
 button *stand* 扣位（钉扣处）；（衬衫等）纽扣前襟
 clothes *stand* 衣架
 combined pressing *stands* 组合烫台
 machine *stand* 机座，机台
 press *stand* （裁剪、缝制用）工作台板
 steam pressing *stand* 蒸汽烫台
 thread *stand* 线架
 top collar *stand* 领里口
 uneven collar *stand* 〔检〕领座不均匀
standard [ˈstændəd] 标准，规格
 appearance *standard* 外观标准
 ministerial *standard* 部颁标准
 national *standard* 国家标准
 off *standard* 不合标准，等外级
 quality *standard* 质量标准
 test *standard* 测试标准
 women's wear *standard* 女装标准
staple [ˈsteɪpl] （棉、麻、毛、化纤）纤维，纤维（平均）长度；（人造）短纤维；订书钉
 long *staple* 长绒棉纤维，长纤维
 short *staple* 短纤维
stapler [ˈsteɪplə] 订书机
stature [ˈstætʃə] 身材；身高
stay [steɪ] 撑条，拉条；加固布条（块）；（领尖）插角片；绲边窄带；〔复〕女用束腹带；紧身胸衣
 armhole *stay* 袖窿牵条
 button *stay* 门襟
 collar *stay* 领插竹，插角片
 heel *stay* 贴脚条（毛呢裤脚口的贴条布）
 pocket *stay* 垫袋条料
 sleeve placket *stay* 袖衩条
 special *stay* 特殊支撑物
steamer [ˈstimə] 蒸汽机；汽蒸器
 crepe-smoothing *steamer* 蒸汽除皱熨机
stencil [ˈstensl] （镂空）模版；唛版；唛头版
step-ins [ˈstepɪnz] 〔复〕女内裤；船鞋；易穿便鞋，无带扣鞋
stereotype [ˈsterɪətaɪp] 模式化形象；板型
 clothes *stereotype* 服装板型
sticker [ˈstɪkə] 背面有粘胶的标签，胶贴，贴纸
 bar code *sticker* 条形码贴纸
 carton *sticker* 纸箱贴纸，箱唛
 hangtag *sticker* 吊牌贴纸
 polybag *sticker* 胶袋贴纸
 price *sticker* 价格贴纸
 QC *sticker* （查货用来标明缺陷处的）QC贴纸，〔粤〕鸡纸
 size *sticker* 尺码贴纸
stick-on [ˈstɪkɒn] 有背胶的标签；一种粘在脚板上的隐形鞋
stickpin [ˈstɪkpɪn] （领带）装饰别针
stitch [stɪtʃ] 线迹，针迹，针脚，〔粤〕线步；针步；针法；装饰线
 abutting *stitch* 拼合线迹
 angle *stitch* 角形线迹
 arched *stitch* 弓形线迹
 back *stitch* 回针，倒针，倒钩针

bartack *stitch* 倒针法
basic *stitch* 基本线迹；基本针法
basket *stitch* 网眼线迹
basting *stitch* 疏缝线迹；㩜针法
binding *stitch* 绲边线迹
blind hemming *stitch* 暗卷缝线迹；暗针法
blind *stitch* （缝纫中）暗针；暗缝线迹
broken *stitch* 间断线迹；〔检〕断线
buttonhole *stitch* 纽孔线迹；锁针法
Byzantine *stitch* 斜 Z 形线迹
catch *stitch* Z 形线迹；环针法
chain lock *stitch* 链锁式线迹；双线链式线迹
chain *stitch* 链式线迹，链式针迹，〔粤〕锁链底针步
change *stitch* 变形线迹
circle *stitch* 圆形线迹
coarse(close) *stitch* 疏（密）缝
combination *stitch* 复合线迹
cord *stitch* 嵌缝线迹，绳纹线迹
counter *stitch* 对称线迹
covering chain *stitch* 覆盖链式线迹，绷缝线迹
crochet *stitch* 钩编线迹
cross *stitch* 十字（交叉）线迹，挑缝；十字绣针迹
darning *stitch* 织补线迹
dot dash *stitch* 点画线
double action *stitch* 复式线迹
double cross *stitch* 双十字线迹
edge *stitch* 止口线迹；压边线
embroidery(crewel) *stitch* 刺绣线迹
face *stitch* 面缝线迹
fancy *stitch* 花式线迹

fastening *stitch* 加固线迹，保险线迹
feather *stitch* 羽状线迹；杨树花针法
fell *stitch* 折缝线迹；明缲针法
flat-lock(covering) *stitch* 覆盖线迹，绷缝线迹，〔粤〕拉冚线步
fly *stitch* 比翼线迹
functional *stitch* 功能线迹
hand *stitch* 手缝线迹；手缝针法；珠边
hemming *stitch* 卷边线迹
herringbone *stitch* 人字线迹；三角针法
honeycomb *stitch* 蜂窝线迹
interlock *stitch* 联锁线迹，多线链式线迹
invisible *stitch* 暗缝线迹
irregular top *stitch* 〔检〕面线不良
knotting *stitch* 打结线迹
lock *stitch* 锁式线迹
loose floating *stitch* 〔检〕浮线（底线不紧所致）
missed *stitch* 〔检〕（缝纫中）漏针
number of *stitch* 针（脚）数
ornamental(decorative) *stitch* 装饰线迹
outline *stitch* 轮廓线迹
overcasting *stitch* 包边线迹
overedge *stitch* 包缝线迹，锁（拷）边线迹
overedge chain *stitch* 包边链式线迹
overlock *stitch* 包缝线迹
pad *stitch* （衬垫的）疏缝线迹，扎缚线迹；扎缚针法

pattern *stitch* 花样线迹
pearl *stitch* 珠式线迹
picot *stitch* 锯齿线迹
plain *stitch* 平缝线迹，普通线迹，平缝针迹，〔粤〕平车针步
prick *stitch* 拱针法
quilted *stitch* 绗缝线迹
reversible(backward) *stitch* 倒缝线迹
running *stitch* 缲缝线迹，撩针线迹；撩针法
safety *stitch* 安全线迹
seam *stitch* 缝式线迹
shell *stitch* 贝壳形线迹
shirring *stitch* 抽褶线迹
single(multi-)chain *stitch* 单（多）线链式线迹
skipping *stitch* 跳针线迹；跳针法；〔检〕（缝纫中）跳针漏线
slant *stitch* 斜缝线迹；明针
slip *stitch* 短和松的暗缝线迹，缲缝线迹；挑针，漏针；撩针法（缲针）
slip *stitch* armhole 〔工〕缲袖窿
slip *stitch* button loop 〔工〕缲扣襻
slip *stitch* buttonhole 〔工〕缲扣眼
slip *stitch* collar to bodice 〔工〕缲领下口
slip *stitch* facing 〔工〕缲暗门襟
slip *stitch* gorge line 〔工〕缲领串口
slip *stitch* patch pocket 〔工〕缲口袋
slip *stitch* reinforcement for knee 〔工〕缲（裤）膝盖绸
slip *stitch* shoulder seam 〔工〕缲肩缝，叠肩缝
slip *stitch* sleeve slit 〔工〕缲袖衩
small(long) *stitch* 密（疏）缝
special *stitch* 特殊缝迹，专用线迹
split *stitch* 复合线迹，双重线迹
staggering *stitch* 〔检〕线迹歪斜
stay *stitch* 定位线迹
stitch indicator plate （缝纫机）针距指示牌
stitch jamming （缝纫中）卡线
stitch overlapping 〔检〕线迹重叠，〔粤〕驳线
stitch per inch(SPI) 每英寸针数
stitch per minute(SPM) 每分钟针数
stitch plate （缝纫机的）针板
stitch regulating dial （缝纫机）针距旋钮，针脚调节盘
stitch size spacing 针距
stitch width(length) regulate 线迹宽（长）度调节
straight *stitch* 直形线迹
stretch(elastic) *stitch* 伸缩线迹，弹性线迹
tacking *stitch* 加固线迹；假缝线迹
top *stitch* 面缝线迹，明线迹
twice(triple) *stitch* 双（三）重线迹
twin needle *stitch* 双针线迹
wave *stitch* 波形线迹
zigzag *stitch* 曲折线迹，锯齿形线迹，人字形线迹，Z形线迹；花针法
stitcher ['stɪtʃə] 缝纫机
straight lock *stitcher* with automatic thread trimmer 自动剪线平缝机
stitching [stɪtʃɪŋ] 缝纫；刺绣；线缝

abutted *stitching* 拼合线缝
back *stitching* 回针（缝纫）
bad join *stitching* 〔检〕（缝纫中）接线不准
blind *stitching* 暗缲针缝，暗缝，缲边，〔粤〕挑脚
contrast *stitching* 镶（拼）色线缝
double *stitching* 双线缝纫；双线缝
double lock *stitching* 双线锁式线缝
double whip *stitching* 双搭接线缝，包边线缝
edge *stitching* 缉边线，车边〔粤〕间边线
elasticity *stitching* 弹性线缝，伸缩线缝
fancy *stitching* 花式线缝
feather *stitching* 羽状线缝
heavy *stitching* 粗线缝纫；跳针（缝纫），珠边
hem *stitching* 卷边缝纫；卷边线缝
lock *stitching* 锁式线缝
ornamental (decorative) *stitching* 装饰缝纫；装饰线缝
overlap *stitching* 搭接线缝；搭接缝纫
pleated *stitching* 〔检〕线缝起褶
poor *stitching*〔检〕缝制不良
puck-free *stitching* 无皱缩线缝
run off *stitching* 缝纫偏离滑落，缉线上、下坑，〔粤〕落坑
safety *stitching* 安全线缝
shirring *stitching* 抽褶线缝
single *stitching* 单线缝纫；单线缝
space *stitching* 间隔线缝
stay *stitching* 稳定缝纫
stitching accuracy 缝纫精度
stitching area 缝纫部位

stitching as stripes (checks) matching 对条（格）缝纫
stitching horse 压脚
stitching speed 缝纫速度，缝速
straight-line *stitching* 直线线缝
stretch *stitching* 伸缩线缝
through *stitching* 全缝
top *stitching* 面缝，明线缝纫
triple *stitching* 三线缝纫；三线线缝
under *stitching* 贴底缝纫
whip (catch) *stitching* 曲折线缝
stitchwork ['stɪtʃwɜːk] 缝纫；刺绣
stockinet [ˌstɒkɪ'net] 松紧织物；针织衣料
stocking ['stɒkɪŋ] 长袜
air *stocking* 空气丝袜，隐形丝袜（用喷雾式粉霜薄薄涂在腿上，具柔滑丝袜效果，事后可洗去）
body stocking 紧身连袜内衣，（性感）连衣丝袜，连体袜衣，连体衣
cloth *stockings* 布袜
elastic *stockings* 弹力袜
fence net *stockings* （大网眼）围栏网袜
figured *stockings* 花袜
fishnet *stockings* 网眼丝袜
flesh (black) *stockings* 肉（黑）色丝袜
gauze *stockings* 纱布袜
hold up *stockings* 紧口长袜
knee (half) *stockings* 中筒袜
lace *stockings* 网眼袜
Lisle thread *stockings* 莱尔线袜
mercerized *stockings* 丝光袜
mesh *stockings* 网眼袜

  net *stockings* 网眼袜
  non stretch *stockings* 无弹力袜
  nylon *stockings* 尼龙袜
  over knee *stockings* 过膝长袜
  panty-*stockings* 连裤袜
  patterned *stockings* 花纹长袜
  quilted *stockings* （间）棉袜
  rayon *stockings* 黏胶纤维袜
  service *stockings* 厚袜，耐穿袜
  sheer silk *stockings* 透明丝袜
  woolen *stockings* 毛袜
  yarn *stockings* 纱袜
stole [stəʊl] 女用披巾；长围巾
stomach ['stʌmək] 腹部，肚
stomacher ['stʌməkə] 兜肚衣
stonewash ['stəʊnwɔːʃ] （服装的）石磨水洗
stopper ['stɒpə] 拉绳（弹簧）扣，绳索扣，绳扣，〔粤〕绳制
  end *stopper* 吊钟（拉绳末端的钟状扣）
  nylon *stopper* 尼龙绳索扣
stormbreaker ['stɔːmbreɪkə] 防风暴衣，风雪衣
strap [stræp] 带子；皮带；布带；吊带；搭扣鞋
  back *strap* （可调节的）后腰襻带（早年牛仔裤以此调节腰围）
  button *strap* 扣襻
  collar *strap* 领带襻
  cuff *strap* 假翻袖头
  elastic bottom *strap* 下摆（裤脚）松紧带
  facing *strap* 贴门襟
  foot *strap* （踩脚裤的）脚口踩带
  front *strap* 明门襟，〔粤〕明筒
  shoe *strap* 鞋搭扣带
  shoulder *strap* （军服）肩带；肩章；肩饰；（西裤的）吊裤带
  wrist *strap* 腕部扣带
strawboard ['strɔːbɔːd] 草纸板，马粪纸
streamline ['striːmlaɪn] 流线型
streetwear ['striːtweə] 街头服饰；时尚休闲装
strength [streŋθ] （织物）强度
  tear *strength* 撕裂强度
  tensile *strength* 拉伸强度
stretcher ['stretʃə] （撑开帽、鞋的）撑具
  shoe（cap）*stretcher* 鞋（帽）撑
string [strɪŋ] 线；细绳；带子
  interior *string* 内置绳带
  nylon(cotton)*string* 尼龙（棉）绳
  plastic *string* 塑胶绳带；（打吊牌用）胶针
  shoe *string* 鞋带
strip [strɪp] 条；带
  bias *strip* 斜布，绳条；帮胸衬
  collar *strip* 领条
  hanging *strip* 吊带
  narrow *strip* 狭带（用于衣服某部位起功能或装饰作用）
  reinforcing *strip* 加固带
  strip for location 定位带
stripe ['straɪp] 条纹，色条；条纹布，柳条布，〔粤〕间条布；〔复〕条纹囚衣
  block *stripe* 块状条纹，宽条纹
  cascade *stripe* 粗细相间条纹
  mismatched *stripes* 〔检〕对条不准
  multicoloured *stripe* 多色相间条纹
  pencil *stripe* 细条纹
  shadow *stripe* 阴影条纹

vertical(horizontal) *stripe* 直（横）条纹

yarn-dyed *stripe* 色织条纹布

stud [stʌd] 领扣；袖口饰纽；衬衫前胸饰纽；饰钉，窝钉

 collar *stud* 领扣

 multicolored stone *stud* （装饰女皮鞋头的）彩石花鞋扣

 press *stud* 按扣，揿纽

 shoe *stud* 鞋扣

studding [ˈstʌdɪŋ] 钉饰钉，打窝钉

stuff [stʌf] 织品（尤指毛织品）；呢绒；填塞料

 flared skirt in *stuff* 宽摆呢裙

 silk *stuff* 丝织品

stuffing [ˈstʌfɪŋ] 填塞；（衣、被）填料（棉花、羽绒等）

style [staɪl] 款式，样式；型；风格

 Academic *style* 学院派风格

 ancient *style* 古代风格

 apparel size and *style* 服装号型

 artistic *style* 艺术风格

 authentic *style* 正宗风格，地道风格

 avant-garde *style* 前卫风格

 basic *style* 基本款式

 Bohemian *style* 波希米亚风格（一种游牧民族的服饰风格）

 cargo *style* 大贴袋款式（通常在裤腿或裙身上缝有大贴袋）

 chic *style* 新潮款式

 Chinese *style* 中国风格；中式

 classical *style* 古典风格

 day-to-day *style* 日常款式

 dress *style*-book 服装款式书，服装样本

 Empire *style* 帝国风格（大领口短上衣，高腰裙）

 European and American *style* 欧美风格，欧美型

 excellent *style* 极好的款式，漂亮款式

 fancy *style* 奇特款式

 fashionable *style* 流行款式；流行风格

 Gothic *style* 哥特风格（源自一种建筑风格，也深深影响着服饰审美和服饰创造）

 grunge *style* 垃圾（摇滚）风格

 hair *style* 发型

 hippie *style* 嬉皮士款式

 hot *style* 热销款式，爆款，爆板

 idyllic *style* 田园风格

 innovation of *style* 款式翻新

 masculine(feminine) *style* 男（女）式

 national *style* 民族风格

 Nehru *style* 尼赫鲁款式

 neuter *style* 中性风格

 organic *style* 天然模拟风格

 pair *style* 成双款式（亲人或情人为增强亲切感，统一着装的款式）

 revival *style* 复古型；复兴风格

 romantic *style* 浪漫风格

 sleeve(collar) *style* 袖（领）型

 sports *style* 运动风格

 stable *style* 固定款式

 style chase 〔法〕狩猎款式

 style corsaire 〔法〕海盗款式

 style description 款式描述

 style lingerie 〔法〕女内衣款式

 style number(No.) 款号

 style of clothing 服装样式

  style petite fille 〔法〕少女服款式
  the latest *style* 最新款式
  up-to-date *style* 时新式样
  vintage *style* 古老风格，怀旧风格
  Western *style* 西方风格，西式
stylist [ˈstaɪlɪst] （服装、发型等）设计新款式，新花样的人；设计造型师；服饰搭配师
  apparel *stylist* 服装设计师
  hair-*stylist* 发型师
suede [sweɪd] 小山羊皮；人造麂皮；绒面皮革（光面革的反面，又称反绒皮，猄皮）
  cow(goat) *suede* 牛（羊）绒面革，牛（羊）反绒皮，牛（羊）猄皮
  embossed knitted *suede* 针织拷花麂皮绒
  lycra knitted *suede* 针织弹力麂皮绒
  micro *suede* 仿麂皮
  pig *suede* 猪绒面革，猪反绒皮，猪猄皮
  spandex micro *suede* 弹力仿麂皮
  T/R micro *suede* 涤（人）棉仿麂皮
  warp knitted *suede* 经编麂皮绒
suedette [sweɪˈdet] 仿麂皮（绒），仿绒面革
suédine [ˈsuedin] 〔法〕仿麂皮织物
suit [suːt] （一套）衣服，套装；（一套）西装；常服
  afternoon *suit* 日常套装
  anti-G *suit* 抗超重飞行服
  anti-radiation *suit* 防辐射服
  archery *suit* 射箭运动服
  athlete *suit* 田径服

baby's *suit* 婴儿套装，〔粤〕BB装
badminton *suit* 羽毛球服
battle *suit* 战斗套装
beach *suit* 沙滩装
beefy *suit* （呢绒制）厚实套装
bespoke *suit* 定做的衣服；定制西装
black *suit* 黑色西装（一般礼服）
blazer *suit* 户外运动套装
blouson *suit* 短夹克套装（宽松短夹克与裙或裤组合）
body *suit*(B.S.) （女）紧身套装；健美服；紧身连衣裤，B.S.装
boiler *suit* 连衫裤工作服
bolero *suit* 白来罗套装（短上衣配裙）
British-style *suit* 英式西装（上身长，收腰围，下摆宽）
business *suit* 商务套装，办公服，套装西服，男式常服
camouflage pattern(camo) *suit* 伪装迷彩服
cardigan *suit* 卡帝冈套装（音译，无领开襟毛衣裙套装）
cat *suit* 猫式套装，黑色跳跃套装，（衣裤相连，上松下紧的女装，黑色光亮柔软衣料制）
chameleon *suit* 变色龙套装，变色龙军服（采用高新技术和材料制作，能依照环境不断改变颜色的迷彩伪装服）
Chanel *suit* 香奈儿套装，仙奴服（音译，开襟上衣和直线型裙的组合）
Charlie Chaplin *suit* 卓别林套装（便礼服上衣加宽肥裤）
Chinese style *suit* 中式套装，中装，

〔粤〕唐装
Chinese tunic suit 中山装（套服）
classic model suit 古典型西装
classic suit 正式套装，正统男西装
coat suit 外套套装（外套与西裤同料）
cocktail suit 酒会礼服
continental suit 欧洲大陆型套装（自然肩线的上衣配上宽下窄的裤子，无腰带）
coordinate suit 调和套装（衣裤用不同衣料，但在色彩、花样上搭配协调）
country suit 郊外套装
cowboy suit 牛仔套装
culotte suit 裙裤套装
cycling suit 骑自行车（运动）套装
dark suit （办公制服多采用）暗色套装
denim suit 牛仔（布）套装
diving suit 潜水服
dress suit 礼服套装，男夜礼服，燕尾服
dry suit 干式潜水服；干式防寒服
easy suit 简便套装
Eton suit 伊顿服（美国伊顿公学学生制服）
evening suit （男）晚礼服
fancy dinner suit 变化式礼服
fitness suit 健身套装，健身服
fitting suit 贴身套装
flying suit （连体）飞行服
four-piece suit 四件式套装（衬衫、背心、夹克、裤或裙）
career suit 作业服（工作服、制服等）
G(gravity) suit 抗超重飞行服
gentlemen's suit 绅士服
girl's(boy's) suit 女（男）童套装
gymnastic(gym) suit 体操服
Gongfu suit （中国）功夫服（音译，武术服装）
Hakka suit 客家装（音译，中国香港早期女性套装）
hazmat suit 危险品防护服
hockey suit 曲棍球服
infant's(baby's) suit 婴儿套装
infant's grow suit （可放长的）婴儿成长套装
interview suit 面试服装
Italian Continental suit 意大利欧式西装
jacket suit 短夹克套装
jazz suit 爵士套装
jogging suit 慢跑运动套装，跑步装
judo suit 柔道服
jumper suit 女短外衣套装；女紧身连衫裤
jump suit 连裤工作服；伞兵服；跳跃套装
karate suit 空手道格斗服
knicker suit 尼卡套装（音译，用相同衣料做成的上衣、背心、裤三件套）
knitted suit 针织套装
lady's suit 女西装，女套装
leisure(casual) suit 休闲套装，便套装；休闲西装
light suit 夏季套装
lounge suit 〔英〕男普通西装；男式常服

L-peak lapel *suit*　L形尖驳领西装
luncheon *suit*　午餐套装（上下不同衣料轻松套装）
maid's(amah's) *suit*　妈祖装（音译，中国香港早期女佣穿的白衫黑裤套装）
Mao *suit*　（毛式）中山装
men's *suit*　男服，男套装
mixed *suit*　混合套装
monkey *suit*　制服；礼服；军服
mosquito prevent *suit*　防蚊服
Nehru *suit*　尼赫鲁套装（包括尼赫鲁外衣和紧身长裤）
night *suit*　（一套）睡衣裤
non *suit*　变化套装（不受限制，可组合变化的套装）
Norfolk *suit*　诺福克套装（音译，中国上衣背部打褶套装）
ocean survival *suit*　海洋救生服
one-button *suit*　单扣西装
one-piece dress *suit*　连衣裙套装
over *suit*　（穿在衣上的）外套装，罩衣
package *suit*　组合套装（几种套装互换组合的新套装）
pantaloons *suit*　裤子套装
penguin *suit*　宇航服，太空服
peplum *suit*　褶襞短裙套装
pressure *suit*　（高空飞行用）增压服
racing *suit*　竞赛套装（上下相连的服装）
radiation-proof *suit*　防辐射装
rah-rah *suit*　（学生）拉拉队套装
rainproof *suit*　防雨西装
riding *suit*　骑马套装
sack *suit*　〔美〕普通西装
sailor *suit*　水手套装，海员服
setup *suit*　组合套装
shirt *suit*　衬衫套装
shooting(safari) *suit*　狩猎套装
single（double）-breasted *suit*　单（双）排扣西服（套装）
ski *suit*　滑雪套装
skirt *suit* with knitted collar and sleeves　针织领、袖裙套装
slack *suit*　宽松便套装
sleeping *suit*　睡衣裤
smock *suit*　罩衫套装（宽松罩衫和裙）
snow *suit*　防寒服
softball *suit*　垒球运动套装
space *suit*　宇航服，航天服
Spencer *suit*　史本塞套装（短上衣与裙组合）
split *suit*　（衣裤背心不同衣料）分开套装
spring or autumn *suit*　春秋套装
suit of clothes　套装，西装
suit-suits　（相同衣料）正统套装；正统西装
summer *suit*　夏季套装
sun *suit*　夏季套装，日光服
surfing *suit*　冲浪服
sweater *suit*　毛线套装
sweat(sports) *suit*　运动套装，运动服
swimming(bathing) *suit*　泳装
table tennis *suit*　乒乓球服
taekwondo *suit*　跆拳道格斗套装
tailored skirt *suit*　西服裙套装
tailored *suit*　西装；西装式女套装
Tang-style(Tang) *suit*　唐装（中式服装）

tank *suit* 坦克套装（针织短袖衫与短裤组合）；坦克车型泳装；（无袖无领）一件泳装；有肩垫浴袍
teddy *suit* 女连裤内衣
toreador *suit* 斗牛士装
town *suit* 上街套装
track *suit* 田径运动套装
trouser(pant) *suit* 裤套装
tunic *suit* 束腰外衣套装
tuxido *suit* 塔士多（礼服）套装
tweed *suit* 粗花呢西装
two(three)-piece *suit* 两（三）件套装；两（三）件套西服
ultraman *suit* 奥特曼套装（一种动力盔甲）
ultraviolet resistance *suit* 防紫外线服装
unconstructed *suit* 简易套装，无构造套装（尽量简略衬、里等）
union *suit* 连裤内衣
un-matched *suit* 异料套装（衣、裤不同衣料和花色）
vest *suit* 背心套装（背心与裙或背心与西裤）；（含背心的）三件套西装
walking *suit* 散步套装
weekend *suit* 周末套装（洋溢着舒畅轻松感）
Western *suit* 西装
wet *suit* （用海绵胶制）保暖紧身潜水衣
wide(small) lapel *suit* 宽（窄）驳领西装
windbreaker *suit* 风衣套装
women's *suit* 女式常服，女套装，女西装
wrestling *suit* 摔跤运动套装
year-round *suit* 全年通用套装
yoga *suit* 瑜伽（练功）服
Y-shaped *suit* Y型西装
zip *suit* 拉链套装（衣裤用拉链相连，亦可分开）
zoot *suit* 阻特套装（音译，上衣齐膝，宽肩，裤子肥大而脚口狭窄）

suitcase ['suːtkeɪs] 衣箱
suiting ['suːtɪŋ] 衣料；套装料；西服料
 acrylic fancy *suiting* 腈纶花呢
 cotton *suiting* 线呢
 fancy *suiting* 花呢
 herringbone fancy *suiting* 人字花呢
 imitation wool *suiting* 仿毛西服料
 ladies' *suiting* （精纺）女衣呢
 midfibre fancy *suiting* 中长花呢
 navy *suiting* 海军呢
 one-side fancy *suiting* 单面花呢
 overcoat *suiting* 大衣呢
 polyester *suiting* 涤纶西服料
 polyester/viscose fancy *suiting* 涤/黏花呢
 polyester woolly-tex (spun) *suiting* 涤纶仿毛料
 pure wool fancy *suiting* 全毛花呢
 rayon *suiting* 黏胶纤维西服料
 striped(checked) fancy *suiting* 条（格）子花呢
 tropical *suiting* 热季套装料，轻薄精纺呢
 wool/polyester/viscose *suiting* （毛、涤、黏）三合一衣料
 wool *suiting* 呢子，毛料
 wool/viscose(polyester) fancy *suiting* 毛/黏（涤）花呢
 worsted(woolen) *suiting* 精（粗）

纺毛料
yarn-dyed fancy *suiting* 色织花呢
sunbonnet [ˈsʌnˌbɒnɪt] （女用）阔边太阳帽
sunday best [ˈsʌndɪˌbest] 星期日服装，礼拜服
sundown [ˈsʌndaʊn] 〔美〕阔边女帽
sundress [ˈsʌndres] 太阳（连衣）裙，（吊带）背心裙
sunglasses [ˈsʌnˌglɑːsɪz] 〔复〕太阳镜，墨镜
    bold framed *sunglasses* 宽框边太阳镜
    rimless *sunglasses* 无框边太阳镜
sunnies [ˈsʌnɪz] 〔澳，非正式〕太阳镜，墨镜
sun-shade [ˈsʌnʃeɪd] （女用）阳伞，〔复〕太阳镜
sunsuit [ˈsʌnsuːt] 太阳装，日光装
sunwear [ˈsʌnweə] 太阳装，日光服
supercrease [ˈsjuːpəkriːs] 犀牛褶（一种永久性定型的裤线折痕）
suppression [səˈpreʃn] 压制；压褶塑形（用各种方法使平面衣料变形为立体廓型）
surcoat [ˈsɜːkəʊt] （旧时）女上衣，罩衣
surfwear [ˈsɜːfweə] 冲浪（运动）服装
surplice [ˈsɜːpləs] 牧师法袍；和尚袍
surveste [syrvɛst] 〔法〕（披在夹克上）夹克宽外衣
suso [θˈsəʊ] 〔日〕下摆
susomawari [θˈsəʊmʌˌwʌri] 〔日〕摆围
suspenders [səsˈpendəz] 〔复〕〔美〕吊裤带；背带；〔英〕吊袜带

stocking- *suspenders* 吊袜带
swallowtail [ˈswɒləʊteɪl] （衣服的）燕尾式后摆
swatch [swɒtʃ] 样片，样品；小块布样，〔粤〕布碎；样本
    fabric *swatch* 布样，〔粤〕布办
sweatband [ˈswetbænd] （防汗用）帽内皮圈
sweater [ˈswetə] 厚运动衫；针织套衫；卫生衫；毛衣，毛线衫，〔粤〕冷衫
    acrylic *sweater* 腈纶衫
    angora *sweater* 兔毛衫
    Aran *sweater* 艾伦毛衫（爱尔兰传统的渔人毛衣）
    Argyle *sweater* 阿盖尔毛衫（菱形图案的提花毛衣）
    beaded embroidery *sweater* 珠绣毛衫
    beaded *sweater* 珠饰毛衫
    blended *sweater* 混纺毛衫
    bolero *sweater* 开胸短毛衫
    brocade *sweater* 提花毛衫
    button-shoulder *sweater* 开肩毛衫
    camel wool *sweater* 驼毛衫
    cardigan *sweater* 开襟毛线衫
    cashmere *sweater* 羊绒衫
    cotton *sweater* 棉织毛衫，棉纱衫
    cotton *sweater* and trousers 卫生衫裤
    cotton yarn *sweater* 棉线衫
    crew *sweater* 水手毛衫
    cropped *sweater* 短毛衫（长至腰围）
    dolman *sweater* 德尔曼毛衫（连肩蝙蝠袖，袖口、下摆收窄）
    drawstring *sweater* （下摆）束带毛衫

embossed *sweater* 拷花毛衫
Fair Isle *sweater* 费尔（岛式）毛衫（早年的渔夫毛衣，现代以多色几何图案设计，有圆形抵肩和袖饰）
fancy knitted *sweater* 织花毛衫
fisherman *sweater* 渔人毛衣（早年以未脱脂羊毛织成，有独特图案，是当时英国渔夫们的工作服，如费尔毛衫、艾伦毛衫和耿西毛衫）
form-fitting *sweater* 紧身运动衫
full fashioned *sweater* 全成形毛衫
Guernsey *sweater* 耿西毛衫（早年的渔夫毛衣）
hand crocheted *sweater* 钩编毛衫
hand embroidery *sweater* 手绣毛衫
jacquard *sweater* 提花毛衫
knitted *sweater* 毛线衫
lace-collared *sweater* 系带领毛衫
lamb's wool *sweater* 羊仔毛衫
layered *sweater* 多层级毛衫
letter *sweater* （编织或贴绣的）字母毛衫
middy *sweater* 水手领毛衫
mini *sweater* 超短毛衫
mohair *sweater* 马海绒衫
naked *sweater* 紧身毛衫
Nordic *sweater* 北欧毛衫
nylon *sweater* 锦纶毛衫
polo *sweater* 马球毛衫（按马球衬衫款制作）
polo neck *sweater* 高翻领套头衫
poor-boy *sweater* 罗纹紧身毛线衫
ramie/cotton yarn *sweater* 麻棉混纺线衫
reversible *sweater* 两面穿毛衫
ribbed *sweater* 罗纹带运动衫

round-neck *sweater* 圆领毛衫
Scandinavian *sweater* （滑雪用）斯堪的纳维亚羊毛衫
shetland wool *sweater* 雪特兰毛衫
silk *sweater* 丝织毛衫
Siwash *sweater* 北美印第安人厚毛衫
skate *sweater* 溜冰套头卫衣
ski *sweater* 滑雪毛衫
sleeveless woolen *sweater* 毛背心
sports *sweater* and trousers 运动衫裤
sweater made of cotton (ramie) yarn 棉（麻）线衫
sweater on *sweater* 重穿毛衫（套头毛衣上再穿一件开襟毛衣）
sweater with embroideries (patches, beadings) 绣花（补花、珠片）毛衫
target-shooting *sweater* 打靶用毛衫
tennis *sweater* 网球毛衫（V领，套头或开襟）
turtle neck *sweater* 高翻领毛线衫，高领衫
twin *sweater* set 套装毛衫
vest *sweater* 背心式毛衫，针织毛背心
V-neck *sweater* 尖领毛衫，V形领毛衫
woolen hand-knitted *sweater* 棒针毛衫
woolen *sweater* 纯毛毛衫，羊毛衫
woolen *sweater* and trousers 羊毛衫裤
wrap *sweater* 无扣开襟衫，卷裹毛衫

sweatpants [ˈswetpænts] 宽松长运动

裤；运动裤；卫裤

sweatshirt ['swetʃɜːt] 圆领长袖汗衫（或运动衫）
 cotton print *sweatshirt* 全棉印花圆领衫
 sloopy *sweatshirt* 宽松圆领长袖衫
 sweatshirt with elongated tail 长后摆衫
 sweatshirt with hood 带帽卫衣

sweep [swiːp] （衬衫、裙、大衣的）下摆，后摆
 bottom *sweep* 下摆

swifttuck ['swɪftʌk] （挂吊牌的）胶针，胶圈

swimsuit ['swɪmsuːt] 游泳衣，泳装
 apron *swimsuit* （极短小的）两件式女泳装，比基尼泳装
 bloomer *swimsuit* 灯笼裤女泳装
 jet-concept *swimsuit* 喷气概念泳装（一种高新技术的赛用泳装）
 LZR Racer *swimsuit* 鲨鱼皮泳装（一种高科技面料制的赛用连衣泳装）
 one-piece underwire *swimsuit* 内衬钢丝胸托的一件头女泳装
 one(two)-piece *swimsuit* 单（两）件式女泳装
 racer *swimsuit* （一件式）比赛用女泳衣
 surfer *swimsuit* 冲浪泳装
 tank *swimsuit* 大圆领口连身女泳装

swimwear ['swɪmweə] （总称）游泳衣，泳装
 silk spandex *swimwear* 丝弹力泳装

sword [sɔːd] 剑；刀
 ceremonial *sword* 礼仪佩剑

symmetry ['sɪmətri] 〔设〕对称；匀称

system ['sɪstəm] 系统；体系；方式；分类
 automated *system* for clothing production 服装生产自动化系统
 body measurement *system* (BMS) 人体尺寸测量系统
 colour *system* 颜色体系，色系
 computer aided design *system* 计算机辅助设计系统
 computer aided machineshop operation *system* (CAMOS) 计算机辅助车间操作系统
 computer automatic cutting *system* 计算机自动裁剪系统
 computer embroidery design and punching *system* 计算机绣花设计打板系统
 computer grading and marking *system* 计算机放码及排唛架系统
 computer integrated clothing *system* 计算机综合成衣系统
 computer management/information *system* for apparel industry 制衣业电脑管理资讯系统
 cost control *system* 成本控制系统
 3D somatometry *system* 三维人体测量系统
 3D virtual fitting *system* 三维虚拟试衣系统
 fabric/accessory inventory *system* 面辅料存货系统
 financial management *system* 财务管理系统
 intelligent custom-made clothing *system* 智能定制服装系统
 intelligent garment design *system* 智能服装设计系统

intelligent manufacturing *system* 智能制造系统
logistics information *system* 物流信息系统
new manufacturing *system* 新制造系统
order management *system* 订单管理系统
Pantone Matching System(PMS) 潘通配色系统
personnel management *system* 人事管理系统
pressing department conveyor *system* 熨烫部输送带系统
process template design *system* 工艺模板设计系统
production control *system* 生产管理系统
project file *system* 跟单系统
quality control *system* 质量管理系统
quota management *system* 配额管理系统
quotation *system* 报价系统
retail P.O.S. *system* 零售系统
sampling *system* 打样系统
shipping *system* 船务系统
size *system* and designation for costume 服装号型系列
visual control *system*(VCS) 视觉控制（管理）系统

T

tab [tæb] （衣服上的）垂片；挂襻；（鞋）拉襻；（帽）护耳；领角衬片
 adjustable *tab* 活动扣襻；裤腰襻
 bottom *tab* 下摆（脚口）扣襻
 button *tab* at right fly 裤里襟尖嘴，里襟宝剑头
 collar band *tab* 领舌
 collar *tab* 领襻
 extended *tab* 活动扣襻；裤腰宝剑头
 facing *tab* 贴边襻
 neck *tab* 领口襻
 shoe *tab* 鞋拉襻
 shoulder *tab* 肩襻
 sleeve *tab* 袖襻
 storm *tab* 防风袖襻
 waist *tab* （上衣、背心的）腰扣襻
tabard [ˈtæbəd] （古代）武士穿在铠甲上的绣有纹章的外衣；无袖（短袖）外衣
table [ˈteɪbl] 桌；台；工作台；案板；表格
 air flotation/vacuum cutting *table* 气垫/真空裁床
 cleaning *table* 清除衣物污垢台
 computerized automatic cutting *table* 计算机自动裁床
 dressing *table* 化妆台
 ironing *table* 熨烫台板
 sectional cutting *table* 组合裁床
 sewing *table* 缝纫机台板
 shirt folding *table* 叠衫台
 size *table* 尺码表
 steam *table* （整烫）蒸汽台
 toilet *table* 梳妆台
 vacuum blowing *table* 吹吸熨烫台

vacuum ironing table （熨烫）抽湿台

tack [tæk] （作临时固定用的）粗缝，疏缝，假缝，线丁；加固缝；平头钉
 arrowhead tack 三角形加固缝，三角结
 back tack 回针加固缝
 bar tack 打结缝
 French tack 线襻（用粗线做的襻，多用在夏季衫、裙上）
 shoe tack 鞋钉
 tailor's tacks 疏缝线做记号，打线丁
 X-shape tack X形加固缝，X形结

tacker ['tækə] 打结机，加固机；用粗针脚缝纫的人
 bar tacker 打结机
 blind tacker 暗缝机

tacking ['tækɪŋ] 粗缝，假缝；加固缝纫，打结，〔粤〕打枣
 slant tacking 斜针假缝

tactel ['tæktl] 塔克特（音译，又译特达，系美国杜邦公司研发的一种新的高品质锦纶纤维）

taffeta ['tæfɪtə]（= taffety ['tæfɪti]）塔夫绸
 changeable taffeta 闪光塔夫绸
 cotton taffeta 棉塔夫绸
 doupion silk taffeta 双宫丝塔夫绸
 lining taffeta 里子塔夫绸
 nylon taffeta 锦纶塔夫绸，锦纶绸（俗称尼龙绸）
 nylon taffeta water-proof and cire 防水轧光锦纶塔夫绸
 nylon taffeta with PVC (PU) coating PVC (PU) 涂层锦纶布（风雨衣、羽绒衣面料）
 polyester taffeta 涤纶塔夫绸
 quadrille taffeta 格子塔夫绸
 rayon taffeta 黏胶纤维塔夫绸
 silk taffeta 蚕丝塔夫绸
 spun silk taffeta 绢丝塔夫绸
 taffeta Fuchun rayon 富春纺
 taffeta glacé 〔法〕闪光塔夫绸

tag [tæg] （衣服）挂襻；（鞋）拉襻；标签
 back tag 后襻；（牛仔裤）后腰牌
 hang-tag 吊牌，挂牌
 price tag 价格吊牌
 self-adhesive tag 尼龙搭扣，魔术贴
 shipping tag 货运标签
 shoe tag 鞋拉襻
 swing tag 吊牌
 UPC tag 条形码吊牌
 waist tag 腰卡（裤腰上的纸牌）

tail [teɪl] （衣服）后摆；〔复〕燕尾服
 tail of shirt 衬衫后摆

tailcoat ['teɪlkəʊt] 燕尾服，男子夜礼服

tailleur [tajcer] 〔法〕定制的衣服（指男装套装、西式女套装）；男装裁缝
 tailleur Broadway 〔法〕百老汇套装（有大剧院风格套装）
 tailleur masculin 〔法〕男式套装

tailor ['teɪlə] 成衣工人；（指男服）裁缝师
 tailor's clapper 拱型烫木
 tailor's clippings 裁剪碎料；衣料小样
 tailor's ham 布馍头（烫衣胸部、裤臀部的托垫）；烫垫

tailor shop 裁缝铺，成衣店
tailor's twist 缝纫用粗丝线
tailoring [ˈteɪlərɪŋ] 裁制（衣服），缝制；裁缝行业；成衣
tailorship [ˈteɪləʃɪp] （裁缝）手工；裁缝业
talma [ˈtælmə] 宽大短外衣；（旧时）宽大披肩
tam-o'-shanter [ˌtæməˈʃæntə] 一种顶无檐圆帽；（饰羽毛的）苏格兰大黑帽
tanga [ˈtæŋgə] 女三角内裤，丁字裤
tangle [ˈtæŋgl] （饰于服装或头发上的）缠结
tank [ˈtæŋk] 针织圆领背心衫；运动背心
    active bra *tank* 带胸罩的运动背心
    longline *tank* 长身线形背心衫
    racer back *tank* T形后身背心衫
    sloppy *tank* 宽松背心衫
    swing *tank* 宽摆背心衫
    toning *tank* 弹力紧身背心衫
tankini [ˈtæŋkiːnɪ] 背心式女泳装（吊带背心加比基尼泳裤）
tanktop [ˈtæŋktɒp] （女）大圆领背心衫
tap [tæp] 鞋掌
    shoe *tap* 鞋掌；鞋跟铁片
tape [teɪp] 线带；贴边；牵条；卷尺，带尺
    adhesive *tape* 胶粘带
    bias *tape* 斜布条
    binding *tape* 绲边带
    bottom *tape* （裤）贴脚条
    braided *tape* 编织带
    elastic *tape* 松紧带
    fashion *tape* 装饰带

fluorescent *tape* 荧光带
hot air sealing *tape* 热风胶贴带（为防水在衣内缝上热粘此带）
magic *tape* 魔术贴，尼龙搭扣带，尼龙粘扣带
measuring *tape* 皮尺，带尺
MOBILON *tape* "无比耐"松紧带（弹性透明胶带，日本研制）
neck *tape* 领圈压条；领圈绳条
nylon fastener *tape* 尼龙搭扣带
printed *tape* 印带，（装饰）吊带
reinforced *tape* 加固带；（制鞋）补强带
sealing *tape* （热风）胶贴带
sleeve placket *tape* 袖衩条
stay *tape* 牵条；绲边带；定位带；胸衬条
steel(cloth) *tape* 钢（布）卷尺
tailor's *tape* 量身软尺
twill *tape* 人字带
Velcro *tape* 尼龙搭扣带
waistband *tape* 腰头牵带，腰头丝里带（缝在腰头上口内以控制尺寸）
woven *tape* 织带
zipper *tape* 拉链（齿）带，拉链（齿）布边
tapemeasure [ˈteɪpˌmeʒə] 卷尺；带尺
    steel *tapemeasure* 钢卷尺
tapestry [ˈtæpəstri] 绒绣，织锦
tarboosh [tɑːˈbuːʃ] 土耳其帽（无边圆塔型）
taping [ˈteɪpɪŋ] 以带子捆扎；用带子贴边；敷（覆）牵条用胶带粘上
    seam *taping* 〔工〕热压胶贴带（粘在衣内缝上以防水）

  taping armhole 〔工〕敷袖窿牵条
  taping back vent 〔工〕敷背衩牵条
  taping front edge 〔工〕敷止口牵条
  taping lapel roll line 〔工〕敷驳口牵条
  taping pocket opening 〔工〕敷袋口牵条
  taping waist line 〔工〕敷（裤、裙）腰口牵条
tarpaulin [tɑːˈpɔːlɪn] 防水布，油布；油布雨衣；油布雨帽
tartan [ˈtɑːtən] 格子花呢；格子织物；格子服装
  Scotch *tartan* 苏格兰格子呢
taslan (= taslon) [ˈtɑːslən] 塔斯纶（运动套装料）
  polyester honeycomb *taslan* (*taslon*) 涤纶蜂窝（状）塔斯纶
  (printed) nylon *taslan* (*taslon*) （印花）锦纶塔斯纶
  taslan (*taslon*) rip-stop 塔斯纶格布
tassel [ˈtæsəl] （装饰用）流苏
  cap *tassel* 帽流苏
taste [teɪst] 情趣；品位；鉴赏力；审美力
  good *taste* 良好的审美力，高品位
  oriental *taste* 东方风味
  poor *taste* 审美力差；风度差
technics [ˈteknɪks] 工艺学；工艺
technology [tekˈnɒlədʒɪ] 工艺学；工艺；技术
  clothing *technology* 服装工艺
  clothing template *technology* 服装模板技术（一种基于服装工艺工程，机械工程及 CAD 数字化原理并将之相互结合的新型缝制技术）
  digital *technology* 数字化技术
  3D human scanning *technology* 三维人体扫描技术
  3D *technology* 三维技术
  information *technology*(IT) 信息技术，资讯科技
  process of *technology* 工艺流程
  requirements in *technology* 工艺要求
  technology change 工艺变更
  technology for making suit 西装制作技术
  technology innovation 工艺革新
  textile and apparel *technology* 纺织服装技术
  traditional *technology* 传统工艺
  virtual stitching and fitting *technology* 虚拟缝制试穿技术
teddy [ˈtedɪ] 〔常用复 teddies〕〔主美〕女无袖连衫衬裤；无袖内衣
tee(s) [tiː(s)] 针织圆领短袖衫，T 恤衫
  active dry *tee* 快干运动 T 恤衫
  cropped *tee* 截筒 T 恤衫，短身 T 恤衫（宽松露腰）
  double layer *tee* （错色）双层 T 恤衫（在领口、袖口及下摆处露出两层，看似穿两件实为一件）
  functional *tee* 功能 T 恤衫
  graphic *tee* 图形印花 T 恤衫（印字、图、漫画、照片）等
  Henley *tee* 半开襟 T 恤衫
  individual *tee* 个性 T 恤衫
  label *tee* 标签 T 恤衫（以若干商

标装饰 T 恤）
layered sleeve tee 层叠袖 T 恤衫
logo tee 标识 T 恤衫
longline tee 长身 T 恤衫
moisture wicking（absorbing）tee 吸湿排汗 T 恤衫
oversize tee 阔身 T 恤衫，宽松 T 恤衫
panel V-neck tee 拼色（或拼块）V 领 T 恤衫
printed tee 印花 T 恤衫
raglan tee 连肩袖 T 恤衫
rugby tee 橄榄球衫
scoop gathered tee 勺形收褶领口 T 恤衫
sequined tee 珠片 T 恤衫
sleep tee T 恤式睡衣
slogan tee 标语 T 恤衫，广告衫
studded tee 窝钉 T 恤衫
swallowtail tee 燕尾 T 恤衫
sweethearts（couple）tee 情侣 T 恤衫
tie-dyed tee 扎染 T 恤衫
woven fabric tee 机织面料 T 恤衫

telephone [ˈtelɪfəʊn] 电话（缩写为 Tel. tel.）

telextra [ˈtelekstrə] 雪花波拉呢

template [ˈtempleɪt] 模板；样板；型板
clothing template 服装模板

temple [ˈtempl] 太阳穴；鬓角

tencel [ˈtensl] 天丝（一种以木浆为原料的全新人造纤维素纤维）

terry [ˈteri] 毛圈织物；厚绒布

terylene [ˈterɪliːn] 〔英〕涤纶，特丽绫（音译）
terylene / cotton 涤棉（混纺或交织，通常缩写为 T/C）

test [test] 试验；测试
colour fastness test 色牢度试验
dimensional stability and related test 尺寸稳定性及相关测试
fabric construction（composition）test 织物结构（成分）测试
fabric performance test 织物性能测试
fibre test 纤维测试
garment accessory test 成衣辅料测试（如拉链、纽扣、衬布等）
physical and chemical test 〔工〕（布料）理化试验
shrinkage test 缩水率测试
test of new products 新产品测试
test working （设备）试车
wash test （服装和布料）耐洗牢度测试
wearing test （服装和布料）耐磨测试

textile [ˈtekstaɪl] 纺织品；纺织原料
clothing（apparel）textile 服装纺织品
digital textile 数字化纺织
environmentally friendly textile 环保纺织品
functional textiles 功能性纺织材料
healthcare textiles 保健纺织材料
household textile 家用纺织品
smart（intelligent）textile 智能纺织品
textile industry 纺织工业
textile mill 纺织厂
woolen（cotton, linen, silk）textile 毛（棉、麻、丝）纺织品

texture [ˈtekstʃə] （织物）结构，质地；织物，织品

theory [ˈθɪəri] 理论；学说；意见

garments basic *theory* 服装基础理论

garments mark *theory* 服装符号学

thermal(s) [ˈθɜːməl(z)] 保暖衣服；保暖内衣

    cotton lycra *thermal(s)* 全棉弹力保暖内衣

thermoboots [ˈθɜːməubuːts] 保暖靴，冬靴

thermowear [ˈθɜːməuweə] 保暖衣服，防寒服

thickness [ˈθɪknɪs] 厚度，密度

thigh [θaɪ] 大腿，股；裤横裆，〔粤〕胜围

thimble [ˈθɪmbl] （缝纫用）顶针，针箍，针抵子

thong [θɒŋ] 狭长的皮带；〔复〕夹（带式）人字平底拖鞋（或凉鞋）

thread [θred] 线；〔复〕（美国俚语）衣服

    asbestos *thread* 石棉线

    ball of *thread* 线团

    basting *thread* 假缝用线；扎线

    blind hemming *thread* 暗缝缲边线

    card of *thread* 纸板线

    combed yarn cotton white *thread* 精梳全棉白线

    cone of *thread* 宝塔线

    cop of *thread* 纸芯线

    cotton *thread* 棉线

    cotton wrapped core spun *thread* 混合包芯线

    covering *thread* 绷缝线

    embroidery *thread* 绣花线

    eyelet buttonholing *thread* 锁凤眼线

    fancy *thread* 花色线，花式线

    fire-retarded *thread* （防火服装用）阻燃线

    flax *thread* 麻线

    floating *thread* ends 〔检〕活线头

    glacé *thread* 〔法〕蜡光线

    gold and silver *thread* 金银线

    leather article *thread* 皮件线

    lisle *thread* （制袜、手套用）莱尔线

    luminous *thread* 夜光线（蓄光材料制，黑暗中可发光）

    matching *thread* 配色线

    merceried *thread* 丝光线

    metallic *thread* 金属线；金银线

    mismatching of *thread* 〔检〕线色不配

    nylon *thread* 锦纶线

    ornamental *thread* 装饰线

    overlocking *thread* 包缝线，锁边线

    plyprism *thread* 特彩线（异色纱拼成的彩线）

    polyamide *thread* 锦纶线

    polyester high tenacity *thread* 涤纶高强线

    polyester *thread* 涤纶线

    reel of *thread* 木芯线

    sewing *thread* 缝纫线

    shuttle *thread* （绕在缝纫机梭芯上的）梭线，底线

    silk *thread* 丝线

    sole *thread* 鞋底线

    spool of *thread* 木纱团线

    spun polyester *thread* 纺纱涤纶线，PP 线

    stretch(elastic) *thread* 弹力线

    synthetic *thread* 合成纤维线

    tacking *thread* 粗缝（疏缝、假缝）用线

textured polyester thread　涤纶变形线

thin(thick) thread　细（粗）线

thread breakage　断线

thread drawing finger　松线器

thread end　线头

thread end left inside　〔检〕（浅色）衣里内透线头

thread feeder　（缝纫机）梭头

thread flaying　〔检〕缝线起毛（缝纫机过线件不光滑所致）

thread guard　（缝纫机）护线器

thread release　松线

thread residue　〔检〕线头线尾（服装成品上残留短线）

thread slipping　脱线

thread take-up　（缝纫机头上的）挑线杆

thread tone in tone　配色线

thread wiper　拨线器

top(bottom) thread　面（底）线

trueran(T/C) thread　涤/棉线

upper thread tension regulator　（缝纫机头上的）面线调节器，夹线器

vinylon thread　维纶线

waxed thread　（制鞋）蜡线

threader [θredə]　穿线器

threading [θredɪŋ]　穿线

　lower threading　穿底线

tiara [tɪˈɑːrə]　罗马教皇法冠；女用冠状头饰

tie [taɪ]　领带；领结，〔粤〕呔；绳；鞋带

　Albert tie　艾伯特领结（蝴蝶领结）

　ascot tie　阿斯可领带（围巾式宽领带）

　black tie　黑领带（正式场合用）；黑领结

　boater tie　便装领带

　Bohemian tie　波希米亚时尚领带

　bolo(bola) tie　（用饰针扣住）流星式领带

　bow tie　领结；蝴蝶领结，〔粤〕煲呔

　crossover tie　交叉领带（在脖前交叉扣住）

　derby tie　德贝领带（前端如剑尖）

　ivy tie　常春藤领带（一种窄领带）

　polyester jacquard woven tie　涤纶提花领带

　polyester knitted tie　涤纶针织领带

　polyester tie hand-printed　涤纶手工印花领带

　pure silk tie　真丝领带

　ribbon tie　缎带领带

　rouleau tie　筒型领带

　string tie　丝带领带

　teflon tie　防污领带（布料经teflon技术处理）

　tie chain　领带饰链

　tie holder　领带夹

　Western tie　（美国）西部领带

　white tie　白领带；白领结

　wide tie　宽领带

　wool knitted tie　毛针织领带

　zipper tie　拉链领带

tielocken [ˈtaɪlɒken]　门襟叠合无纽束腰男雨衣

tiepin [ˈtaɪpɪn]　领带扣针，领带夹；领结别针

tights [ˈtaɪts]　〔复〕紧身衣裤，紧身裙；紧身连袜裤

　active tights　紧身运动衣裤

　ballet tights　芭蕾舞紧身衣

nylon stretch *tights* 锦纶弹力紧身连袜裤
    skin *tights* 紧身内衣裤
tile [taɪl] 丝质高顶礼帽
tint [tɪnt] 色泽，色彩；色度；颜色的浓淡；浅色
    flat *tint* 制服
    middle *tint* 中间色
tinting [ˈtɪntɪŋ] 染色；吊色（洗水工艺之一）
tip [tɪp] 末端；鞋尖；尖状鞋掌
    collar *tip* 领尖
    shoulder-*tip* to *tip* 总肩宽
tippet [ˈtɪpɪt] （法官、教士用）披肩；女式披肩
tire [ˈtaɪə] （妇女）头饰；（古称）服装
    head *tire* 头饰
tissue [ˈtɪsjuː] 薄绢；纱；织物；薄纸
    cleaning *tissue* 洁面纸
    face *tissue* 化妆纸
to adjust [tu əˈdʒʌst] 调节；调整；整理
    adjust armhole 调节袖窿
    adjust hat and dress 整衣冠
    adjust needle pitch 调针距
    adjust sleeve and waist line 调整袖围线与腰围线
    adjust stitch 调整线迹
to adorn [tu əˈdɔːn] 装饰
    adorn skirt with flounce 给裙子饰荷叶边
to alter [tu ˈɔːltə] 改做（衣服）；改动（设计）
    alter blouse collar 改衬衫领
    alter design 修改设计
to appear [tu əˈpɪə] 呈现，显得

collar edge *appears* tight (loose) 〔检〕领外口紧（松）
lapel edge *appears* tight (loose) 〔检〕驳头外口紧（松）
top collar *appears* tight 〔检〕领面紧
top flap *appears* tight 〔检〕袋盖反翘
top lapel *appears* tight 〔检〕驳头反翘
to appliqué [tu æˈpliːkeɪ] 〔法〕缝饰，缝贴花
to attach [tu əˈtætʃ] 缝绱；覆，贴；钉；系
    attach back crotch slay 〔工〕绱大裤底
    attach belt loop 〔工〕绱裤带襻
    attach chest interlining to front piece 〔工〕敷衬（衣前片敷胸衬）
    attach collar stay 〔工〕领角薄膜定位
    attach collar to band 〔工〕夹翻领（翻领夹进底领缝合）
    attach cuff to sleeve 〔工〕装袖头，绱袖头
    attach elastic 〔工〕绱橡筋
    attach eye to waistband 〔工〕钉裤钩襻
    attach eyelet 〔工〕钉鸡眼
    attach facing to fly 〔工〕敷挂面
    attach front band 〔工〕绱明门襟
    attach heel stay 〔工〕绱贴脚条
    attach hook to band 〔工〕手工缲领钩
    attach hook to waistband 〔工〕钉裤钩
    attach labels 〔工〕钉标签；绱商

标

attach lining to bodice 〔工〕敷大身里子

attach rib 〔工〕绱罗纹

attach right (left) fly 〔工〕绱裤里（门）襟

attach rivet 〔工〕钉撞钉

attach snap 〔工〕钉四合纽

attach tape to armhole 〔工〕敷袖窿牵条

attach tape to back vent 〔工〕敷背衩牵条

attach tape to front edge 〔工〕敷止口牵条

attach tape to hood brim 〔工〕绱风帽檐

attach tape to lapel roll line 〔工〕敷驳口牵条

attach tape to pocket opening 〔工〕敷袋口牵条

attach tape to waistband of skirt 〔工〕敷裙腰口牵条

attach trouser curtain 〔工〕绱雨水布（缝在裤腰里下口）

attach waistband 〔工〕绱腰头

attach zipper 〔工〕绱拉链

to attire [tu əˈtaɪə] 装饰，打扮

to backstitch [tu ˈbækstɪtʃ] 倒针，回针

backstitch armhole 〔工〕倒钩袖窿

backstitch at the end of opening 剪口末端处回针

backstitch back rise 〔工〕倒钩后裆缝

backstitch neck line 〔工〕倒扎领窝

to bartack [tu ˈbɑːtæk] 加固缝纫，打套结

bartack back vent end 〔工〕封背衩

bartack the ends of pocket mouth 〔工〕封袋口

bartack front rise 〔工〕封小裆

bartack sleeve slit end 〔工〕封袖衩

to baste [tu beɪst] 粗缝，疏缝，假缝

baste front edge 〔工〕疏缝止口，擦止口

baste hem 〔工〕疏缝底边，擦底边

to bind [tu baɪnd] 给……绲边

bind armhole 〔工〕绲袖窿边

bind buttonhole 〔工〕绲扣眼边

bind off 绲边；锁边

to box [tu bɒks] 把……装箱（盒）

to buckle [tu ˈbʌkl] 把……扣住（扣紧）

buckle belt up 扣紧腰带

to bushel [tu ˈbʊʃl] 修改（衣服）；翻新（衣服）

bushel woolen coat 翻新呢外套

to button [tu bʌtn] 钉纽扣（于）；扣纽扣

to buttonhole [tu ˈbʌtnhəʊl] 开扣眼

to case [tu keɪs] 把……装箱

to chalk [tu tʃɔːk] 用粉笔写、画；打……的图样

to chalk out 打……的图样；设计

to change [tu tʃeɪndʒ] 更换；换衣

change bad quality cutted pieces 〔工〕换（裁）片

change a collar style 更换领型

change into one's overall 换上工作裤

change one's clothes 换衣服
change the style of summer dresses 变更夏装款式
to check [tu tʃek] 检查，查验；核对
  check accounts 查账
  check button and snap 〔工〕查纽扣（质量）
  check cutted pieces 〔工〕验裁片
  check cutting edge 〔工〕查裁片刀口
  check flaw 〔工〕查（布料）疵点
  check grain 〔工〕查（布料）纬斜
  check lining and interlining colour and lustre 〔工〕查衬布的色泽
  check marker 〔工〕检查唛架，查排料图
  check number 对号码
  check off 查讫，验毕
  check off colour 〔工〕验（布料）色差
  check out 检验
  check pattern 〔工〕核对划样；查纸样
  check spot 〔工〕查（布料）污渍
  check zipper 〔工〕查拉链（质量）
to choose [tu tʃu:z] 选择，挑选
  choose dress materials 挑选衣料
  choose samples 选样
to clip [tu klɪp] 剪；修剪
  clip cutted pieces 修剪裁片
  clip thread residue 剪线头
  clip underarm's seam allowance 〔工〕抬根缝剪口
to close [tu kləʊz] 关闭；缝合
  close centre seam of collar interlining 〔工〕缝合领衬
  close centre seam of top collar 〔工〕缝合领面
  close crotch 〔工〕缝合裤裆
  close hood seam 〔工〕缝合风帽缝
  close side seam 〔工〕缝合侧缝，〔粤〕埋侧骨
to clothe [tu kləʊð] 给……穿衣
  be clothed in pink 穿着粉红色衣服
  be clothed in suit 穿着套装
  be clothed in wool 穿着毛料衣服
to clout [tu klaʊt] 给（衣服）打补丁；给（鞋底）钉铁掌
to crease [tu kri:s] 打折痕；弄皱；起皱
to cuff [tu kʌf] 给……装袖头（翻折边）
  cuff jacket with rib 给夹克衫绱罗纹袖头
to cut [tu kʌt] 裁、剪、切
  cut along this line 裁开线
  cut away 剪掉
  cut buttonhole 〔工〕（剪）开扣眼
  cut dart 〔工〕剪省缝，开省线
  cut details 〔工〕（裁剪）配零料
  cut edge 切边
  cut from end 从布头上裁零件
  cut into 裁（剪）成
  cut off 裁（剪）下
  cut open 裁（剪）开
  cut pocket opening (mouth) 〔工〕开袋口
  cut out 裁（剪）出；镂空
  cut out back 镂空后身，露背
  cut out neckline 〔工〕挖领圈
  cut out template 切割模板
  cut up 裁得开
to darn [tu dɑ:n] 织补

to dart [tu dɑːt] 在……短缝，开省（俗称开省道）

to decorate [tu ˈdekəreɪt] 装饰
    decorate cuffs and flaps with plush 用长毛绒装饰袖头和袋盖

to design [tu dɪˈzaɪn] 设计，构思；绘制
    design dress in smart shape 设计时新女装
    design package 设计包装
    design trademark 设计商标

to draft [tu drɑːft] 制图；打样

to draw [tu drɔː] 划，画，绘制
    draw design(blue print) （裁剪）打样，划样
    draw pattern 〔工〕（裁剪）表层划样
    draw quilting 〔工〕画绗棉线

to dress [tu dres] 给……穿衣，穿衣；穿夜礼服
    dress up 穿上盛装，打扮

to drill [tu drɪl] 钻（孔）；在（裁片上）钻孔

to ease [tu iːz] 放宽（尺寸）；容位
    ease bottom 放宽下摆
    ease neckline 放领线
    ease waistband 放腰头

to edge [tu edʒ] 给（衣服）镶边
    edge skirt with lace 给裙镶花边

to embroider [tu ɪmˈbrɔɪdə] 在……绣花；刺绣
    embroider cheongsam in pattern of "two dragons playing with a pearl" 在旗袍上绣"二龙戏珠"图

to fade [tu feɪd] （颜色）褪去；使褪色

to fell [tu fel] （缝纫）平式接缝，平缝
    fell seam 〔工〕折边叠缝，折缝，合缝，〔粤〕埋夹

to fit [tu fɪt] 适合，使（服装）合身
    fit on 试穿
    fit the dress to the figure 量体裁衣
    fit well （衣服）很合身

to fold [tu fəʊld] 折叠，打折
    fold hem 折叠底边
    fold here 折叠处
    fold inside 往里折
    fold seam allowance 折缝份
    fold three 折三褶
    fold up(down) 向上（下）折

to fuse [tu fjuːz] 熔（化）；熔合，黏合
    fuse fusible interlining 〔工〕热熔黏合衬
    fuse interlining to top collar 〔工〕热粘领衬、面

to gather [tu ˈgæðə] 给（衣、裙）打褶裥

to hem [tu hem] 给……卷边（镶边）
    hem bottom 卷底（脚口）边
    hem pocket 卷口袋边
    hem sleeve 卷衣袖边

to insert [tu ɪnˈsɜːt] （缝纫）镶，嵌，补
    insert cord 嵌绳带
    insert elastic 嵌橡筋
    insert fur 镶毛皮
    insert lace 镶花边
    insert leather 嵌饰皮革
    insert patch 补（布）片
    insert pocket 挖口袋
    insert sleeve 装袖，绱袖
    insert strip 嵌条带

to inspect [tu ɪnˈspekt] 检查，检验
    inspect exported clothes 检验出口

服装
inspect imported fabrics 检验进口布料

to iron [tu ˈaɪən] 熨烫，烫平
  iron fabrics 熨烫衣料
  iron out wrinkles 熨平皱襞
  iron shirts 熨烫衬衫
  iron top collar for preshrinking 〔工〕热缩领面

to join [tu dʒɔɪn] 缝合；连接；加入
  join back centre seam 〔工〕合背缝
  join crotch 〔工〕缝合裤裆，〔粤〕埋浪
  join front crotch 〔工〕缝合小裆，〔粤〕埋小浪
  join inside seam 〔工〕合下裆缝
  join shoulder seam 〔工〕合肩缝
  join side seam 〔工〕合摆缝；合裤侧缝，合裙摆缝
  join sleeve seam 〔工〕合袖缝

to knit [tu nɪt] 把……编织，编（针）织
  knit up 织补
  knit wool into sweater (stockings) 用毛线织成毛衫（毛线袜）

to launder [tu ˈlɔːndə] 浆洗；（洗后）熨烫

to lay [tu leɪ] 放，搁；铺；排
  lay marker 〔工〕排唛架，排料

to let [tu let] 让，放
  let down 放长（衣服）
  let out 放大，放宽，（衣服）；放出（吊边）
  let out round waist （衣、裤）放腰

to line [tu laɪn] 给（衣服）装衬里；划线于
  line jacket with nylon taffeta 用锦纶绸做夹克衬里

to make [tu meɪk] 做（衣），缝制；制作；制造
  make an opening for pen on flap 〔工〕做（胸袋）插笔口
  make belt loop 〔工〕缝裤带襻
  make Chinese frog 〔工〕盘花纽
  make cloth 织布
  make clothing template 制作服装模板
  make coat and skirt 缝制衣裙
  make collar 制领
  make collar band tab 〔工〕做领舌
  make cuff 缝制袖头
  make down 改小（衣服）
  make flap 〔工〕做装盖
  make French tack at cuff 〔工〕叠裤卷脚
  make into 把……制成……
  make marker (lay) 裁剪排板，制图
  make of 用……制造……
  make over 把（衣服）改制（或翻新）
  make pleats 缝制褶裥
  make pocket flap 〔工〕做袋盖
  make pocket welt 缝袋牙
  make sample 做样品，打样
  make shoes (hat) 制鞋（帽）
  make shorter (longer) 改短（长）些
  make shoulder pads 〔工〕制肩垫
  make tailor's tack 〔工〕打线丁
  make tuck 打褶
  make up 缝制，制成，合成
  make waistband 缝制腰带（腰头）

to mark [tu mɑːk] 标记于，标明；（裁

片上）打号

    mark button and buttonhole position 〔工〕标扣位和扣眼位

    mark pocket opening line 标记开袋口处

    mark position 标出位置，〔粤〕点位

to measure [tu ˈmeʒə] 量，度量，测量

    measure and cut fabric 量裁布料

    measure sb for new suit 给某人量身做衣

to mend [tu mend] 修补；缝补；织补；打补丁

    mend clothes 补衣

    mend shoes 修鞋

    mend woolens 织补毛料衣服

to model [tu ˈmɒdl] 做（服装）模特

to ornament [tu ˈɔːnəmənt] 装饰

    ornament neckline with shirring 用抽褶装饰领口

to overcast [tu ˈəʊvəkɑːst] 包缝，锁边，拷边

    overcast armhole 锁袖窿边

    overcast bottom 锁缝底边

to overedge [tu ˈəʊvedʒ] 包缝，锁（拷）边

to overlock [tu ˈəʊvəlɒk] 包缝，锁（拷）边

    overlock with three (five)-thread 锁三（五）线

to pack [tu pæk] 包装，捆扎；打包，装箱

    pack clothes after ironing 熨烫后包装衣服

    pack goods into box 将货物装箱

to pad [tu pæd] 填塞；衬垫

    pad hood with down 将羽绒填入风帽

    pad shoulder with sponge lumps 衬海绵垫肩

to patch [tu pætʃ] 补缀，缝补

to perforate [tu ˈpɜːfəreɪt] 打眼

to piece [tu piːs] 修理，拼合

    piece together flange 〔工〕拼接耳朵皮

    piece together gore to sleeve 〔工〕拼袖角

    piece together under collar 〔工〕拼领里

    piece up 修补

to pin [tu pɪn] （用针）别住；钉住

    pin top collar to under collar together 〔工〕覆领面（领面上覆领里）

to pipe [tu paɪp] 为（衣服）绲边，拷边

    pipe buttonhole 〔工〕绲扣眼边

    pipe edge 镶绲（衣、布）边

    pipe inside line of facing 〔工〕绲挂面

    pipe pocket mouth 〔工〕绲袋口

to pleat [tu pliːt] 使打褶；编织

to press [tu pres] 压，按；熨烫，烫平

    fold and *press* back yoke 〔工〕扣烫过肩

    fold and *press* crotch reinforcement patch 〔工〕扣烫裤底

    fold and *press* heel stay 〔工〕扣烫贴脚条

    fold and *press* hem 〔工〕扣烫底边

    fold and *press* reinforcement for knee 〔工〕扣烫膝盖绸

    press chest interlining 〔工〕烫胸

衬
press collar interlining 〔工〕压领衬
press collar point 〔工〕热压领角定型
press open collar seam 〔工〕分烫领缝
press open dart 〔工〕分烫省缝
press open French dart 〔工〕分烫刀背缝
press open gorge line seam 〔工〕分烫领串口
press open lining seam 〔工〕分烫里子缝
press open seam 烫开缝
press open shoulder seam 〔工〕分烫肩缝
press open side seam 〔工〕分烫摆缝
press open sleeve seam 〔工〕分烫袖缝
press seam opening 烫开缝
press skirt(clothes) 熨平裙子（衣服）
press trouser crease 烫裤挺缝线
to prink [tu prɪŋk] 给……梳妆打扮，化妆
prink up 打扮，化妆
to print [tu prɪnt] （在织物、衣服上）印花
to pucker [tu pʌkə] 使起皱，皱缩；折叠
pucker up round neckline 领口皱缩
to punch [tu pʌntʃ] 冲压；冲孔
punch collar interlining 〔工〕冲领衬
punch collar stay 〔工〕冲领角薄膜
punch cuff interlining 〔工〕冲袖头衬
to put [tu pʊt] 放；装；使穿进；写上
put off one's coat 脱掉外衣
put on 穿上；戴上
put one's signature to a contract 在合同上签名
put together 使构成整体；新组合技术（以西装为中心，逐次添加与其调和的服装）
to quilt [tu kwɪlt] 绗缝（衣服等）；用垫料填塞
to repair [tu rɪˈpeə] 修补；维修
repair boots and shoes 修鞋，补鞋
repair sewing machine 修缝纫机
to rework [tu rɪˈwɜːk] （对不合格产品）返工
to ruffle [tu ˈrʌfl] 将（布料等）打褶裥；给……装褶边
to sample [tu sɑːmpl] 抽样检验；取样；采样
to seam [tu siːm] 缝合，缝纫
seam up a dress 缝制一件衣服
to set [tu set] 装置；缝、绱（in）；（衣服）合身
set in back yoke 〔工〕绱过肩
set in collar 〔工〕绱领子
set in hood 〔工〕绱风帽
set in pocket 〔工〕绱口袋
set in shoulder pad 〔工〕装肩垫
set in sleeve 〔工〕绱袖
set in sleeve tab 〔工〕绱袖襻
to sequin [tu ˈsiːkwɪn] 用闪光饰片装饰，给……饰亮片
to sew [tu səʊ] 缝；缝制；缝叠（合）；

缝补
 sew front and back rise together 〔工〕合前后裆缝
 sew hook and eye 〔工〕装订领钩
 sew in 缝入
 sew on 缝上，钉上
 sew one-way pleat 〔工〕缝叠顺裥
 sew placket to sleeve 〔工〕绱袖衩条
 sew pocket welt 〔工〕缉口袋嵌线
 sew to here 缝至此
 sew together bodice and its lining 〔工〕合大身里子
 sew together front and back rise 〔工〕合前后裆缝
 sew together hood and its lining 〔工〕合帽面里
 sew together sleeve and its lining 〔工〕合袖面里
 sew together waist band and its lining 〔工〕合腰头
 sew twice 缝两道线
 sew up 缝拢，缝合；把……缝入

to shir [tu ʃɜ:] 使成抽（多行碎）褶

to shrink [tu ʃrɪŋk] 使收（皱）缩；收（皱）缩（指布料缩水）
 shrink sleeve cap 〔工〕收袖山

to snip [tu snɪp] 剪，剪断；剪去
 snip off thread residue 剪线头
 snip paper (cloth) 剪纸（布）

to stitch [tu stɪtʃ] 缝合；缝缉；缝缀；缝纫
 stitch back centre seam 〔工〕合背缝
 stitch belt loop 〔工〕缝裤带襻
 stitch chest interlining 〔工〕缉胸衬
 stitch collars together 〔工〕合领子（领面、里缝合）
 stitch dart 〔工〕缉省缝
 stitch dart in interlining 〔工〕缉衬省
 stitch down piping lip 〔工〕绲边
 stitch French dart 〔工〕合刀背缝
 stitch front edge 〔工〕合止口
 stitch inside seam 〔工〕合下裆缝
 stitch out 绗缝
 stitch pleat 〔工〕缉裥
 stitch shoulder seam 〔工〕合肩缝
 stitch side seam 〔工〕合衣摆缝；合裤侧缝；合裙缝
 stitch sleeve seam 〔工〕合袖缝
 stitch up 缝补
 stitch waist band 〔工〕缝合腰带
 stitch waist pleat 〔工〕缉裤腰裥

to stretch [tu stretʃ] 伸展；拉伸；推；拔
 shrink and *stretch* back piece 〔工〕归拔后背
 shrink and *stretch* sleeve inseam 〔工〕归拔偏袖
 shrink and *stretch* under (top) collar 〔工〕归拔领里（面）
 stretch crotch 〔工〕拔裆
 stretch front piece 〔工〕推门（将平面衣片推烫为立体衣片）

to stud [tu stʌd] 用饰钉（或嵌钉）装饰；给……钉饰纽；打窝钉
 stud jeans with brass nails 在牛仔裤上饰铜钉

to tack [tu tæk] 粗缝，疏缝，假缝；加固缝纫，打结

to tailor [tu 'teɪlə] 裁制衣服；做裁缝；剪裁，制作；按男装式样裁制（女服）

to take [tu teɪk] 拿，取；（衣服）改动
    take in 改小（衣服）；归拢；打褶
    take off 脱下（衣帽）；改瘦（衣服）
    take on （服装式样）流行，风行
    take size 量尺码
    take up 改短（衣服）
to tape [tu teɪp] 用带子捆扎；用胶布把……粘牢；用卷尺（或带尺）量
    tape seam 胶贴线缝（用热风机将胶带粘贴在衣缝底以防水）
to thread [tu θred] 穿线于……；把（珠子）穿成串
    thread a needle 穿针
to tie [tu taɪ] （用带、绳、线）系，扎，束紧；系结（丝带、绳线）；打（结）
    tie bonnet 束紧帽带
    tie label 系标签
    tie on 系上
    tie ribbon in bow 把丝带打成蝴蝶结
    tie scarf 系围巾
    tie shoelaces 系鞋带
to topstitch [tu ˈtɒpstɪtʃ] 正面缝纫，缉明线
    topstitch collar band bottom 〔工〕正面缝领座底
    topstitch cuff 〔工〕缉袖口
    topstitch hem 〔工〕正面缝底边
    topstitch under collar 〔工〕缉领里
to trim [tu trɪm] 修剪；装饰
    trim cutted pieces 〔工〕撇片（修剪毛坯裁片）
    trim edge 〔工〕镶边、修边
    trim front edge 〔工〕修剪止口
    trim sleeve with leather 给衣袖装饰皮革
    trim thread 〔工〕剪线，修剪线头
to try [tu traɪ] 试，尝试
    try on 试穿（衣、鞋）
to tuck [tu tʌk] 折短；卷起；在（衣服上）打横褶（up）；塞
    tuck in 塞进，塞入
    tuck up skirt 提起裙子
    tuck up sleeve 卷起袖子
to turn [tu tɜːn] 翻；翻折；翻新
    turn belt loop 〔工〕翻带襻
    turn collar 〔工〕翻领子（领正面翻出）
    turn down 翻下
    turn hood 〔工〕翻风帽
    turn in collar band edge and stitch it 〔工〕包底领
    turn in top collar edge and stitch it 〔工〕包领面
    turn lining 〔工〕翻里子
    turn out 翻出
    turn pocket flap 〔工〕翻袋盖
    turn right (left) fly 〔工〕翻里（门）襟
    turn sleeve 〔工〕翻袖子
    turn tab 〔工〕翻小襻
    turn up 卷起；折起
    turn waistband 〔工〕翻腰头（或腰带）
    turn woolens 翻新毛料衣服
    turn wristband inside 袖口往里折
to unbutton [tu ˌʌnˈbʌtn] 解开……纽扣；解开纽扣
to undress [tu ˌʌnˈdres] 给……脱衣；脱衣服

to unpick [tu ˌʌnˈpɪk] 拆开（衣服等的）针脚；拆开线缝

to untie [tu ˌʌnˈtaɪ] 解开，松开
- *untie* one's tie 解开领带
- *untie* scarf 松开围巾

to wad [tu wɒd] （用填料）填塞，填衬

to wash [tu wɒʃ] 洗，洗涤，耐洗；（布料、服装）水洗
- *wash* and wear(use) 免烫
- *wash* clothes 洗衣
- *wash* off (out) 洗去，洗掉
- *wash* well 耐洗

to wear [tu weə] 穿；戴；佩；磨损，穿破；耐穿
- *wear* away （使）磨损，磨薄
- *wear* black 穿丧服
- *wear* denim 穿牛仔服装
- *wear* down （使）磨损
- *wear* face mask 戴口罩
- *wear* flower in buttonhole 在驳领纽孔中佩花
- *wear* gloves 戴手套
- *wear* grey 穿灰衣服
- *wear* leather shoes 穿皮鞋
- *wear* off 磨去，磨损掉
- *wear* out 穿破，穿坏
- *wear* ring(necklace) 戴戒指（项链）
- *wear* spectacles(glasses) 戴眼镜
- *wear* stockings(socks) 穿袜子
- *wear* sunbonnet 戴太阳帽
- *wear* uniform 穿制服
- *wear* watch 戴手表
- *wear* well 耐穿，耐用
- *wear* wool 穿毛料衣服

to whip [tu wɪp] 包缝，锁边；织补

to zip [tu zɪp] 拉开（拉上）……拉链；拉拉链
- *zip* jacket up 拉上夹克拉链
- *zip* pocket open 拉开口袋拉链

toe [təʊ] 脚趾，脚尖；鞋（袜）头
- duckbilled *toe* 鸭嘴式鞋头
- egg *toe* 蛋形鞋头
- knot *toe* 花结式鞋头
- medallion *toe* 孔饰鞋尖
- plain *toe* 普通鞋头；平鞋头
- pointed *toe* 尖鞋头
- round *toe* 圆鞋头
- shoe *toe* 鞋头
- sock(hosiery)*toe* 袜头
- square(sharp) *toe* leather shoes 方（尖）头皮鞋
- *toe* cap 鞋头；鞋尖饰皮
- *toe* puff 鞋头衬垫
- wing *toe* 翼形鞋头

toga [ˈtəʊgə] （古罗马）宽松外袍；（某一职业、官职的）专用袍裾

toggery [ˈtɒgəri] 衣服；特殊衣服；〔英〕服装店，服装用品店

toggle [ˈtɒgl] 套索纽，绳结纽，棒形纽，牛角纽（各种形状的短栓，套进襻索里作纽扣用）

togs [tɒgz] 〔复〕衣服，服装
- football *togs* 足球队服
- long *togs* （海员）上岸穿的衣服
- tech *togs* 科技服装
- tennis togs 网球衣

toile [twɑːl] 一种棉质印花布；薄麻布

toilet [ˈtɔɪlət] 梳妆，化妆，女服；服饰
- elaborate *toilet* 盛服；艳妆

toilette [twɑːˈlet] 梳妆；正式礼服；时

髦装束
    grande *toilette* 〔法〕大礼服
tolerance [ˈtɒlərəns] 公差；宽松量，〔粤〕抛位
    measurement out of *tolerance* 〔检〕尺寸超出公差
tone [təʊn] 色调
    medium *tone* 中间色调
    subtle *tone* 淡色调
    tender(pastel)*tone* 柔和色调
    tone in *tone*(TIT) 同色调配色
    tone on *tone*(TOT) 同色系配色
    warm(cold) *tone* 暖（冷）色调
tongue [tʌŋ] 舌状物；鞋舌；扣环搭针
tool [tuːl] 工具；器具
    packing *tool* 打包工具
    sewing *tool* 缝纫工具
top [tɒp] 上面，顶部；（套装）上装；上衣（如T恤衫等）祖胸衣服；表面镀过的金属纽扣；（羊毛）毛条；化学纤维条
    acrylic *top* 腈纶毛条
    asymmetric *top* 不对称衫
    babydoll *top* 娃娃上衣，娃娃衫
    bandeau *top* 狭带式胸衣
    bare *top* 露上装（指露肩、背、胸晚礼服，太阳装）
    batwing *top* 蝙蝠衫
    camisole *top* 吊带胸衫；短袖女衬衫
    coolmax bra *top* （背心式）凉爽胸衣
    crop *top* 极短的短上衣；（针织）吊带短衫；吊带胸衣
    cross front shirred *top* 门襟交叠式抽褶衫

    double layer *top* （错色）双层衫（看似穿两件，实为一件）
    drape neck slinky *top* 垂坠领口紧身衫
    fitness *top* 健身衫，健美衫
    gathered neck *top* 褶裥领口衫
    halter *top* 挂脖式露背衫
    lace cami *top* 网眼花边胸衣
    leopard print *top* 豹纹衫
    maternity *top* 孕妇衫
    mesh knit *top* 针织网眼衫
    midriff *top* 露腰衫
    muscle *top* 无袖圆领衫；健美衫
    nylon *top* 锦纶毛条
    off the shoulder *top* （女）露肩衫，吊带衫
    one shoulder stretch rib *top* （女）单肩带弹力罗纹衫
    peasant *top* 农妇衫，村姑衫
    polo *top* 马球衫
    polyester *top* 涤纶毛条
    polypropylene *top* 丙纶毛条
    ramie *top* 苎麻条
    round neck striped *top* 圆领间条衫
    rugby *top* 橄榄球衫
    shell *top* 无袖紧身套头衫
    shoestring cami *top* 女吊带背心
    silk *top* 绸面
    ski fleece *top* 滑雪绒衫
    sleeve *top* 袖山
    sleeveless V-neck *top* V领无袖衫
    sun-*top* 无肩带的胸衣
    sweat *top* 厚针织衫；毛线衣；卫生衣
    swing *top* 高尔夫短夹克
    tank *top* 无袖圆领衫；大圆领女背心；吊带式女背心

T-back sport *top*　T型运动衫
top and bottom　〔设〕上下调和（指上装和下装）
top of hat　帽顶
top of shoes　鞋面
tube *top*　（弹性）套筒胸衣
turtleneck *top*　高领上衣
twofer *top*　（错色）双层衫（看似穿两件，实为一件）
viscose *top*　黏胶毛条
voluminous (tight) *top*　蓬松（紧身）上衣
wool *top*　毛条
wrap lace *top*　（针织）网眼叠合衫
wrap frill *top*　饰边叠合衫
topcoat [ˈtɒpkəʊt]　轻便大衣；风衣
　covert *topcoat*　轻便大衣
topi(=topee) [ˈtəʊpi]　遮阳帽盔；兜帽
topknot [ˈtɒpnɒt]　顶髻；头饰
topless [ˈtɒplɪs]　露胸服，上空装，无上装
toplesssuit [ˈtɒplɪssuːt]　无上装比基尼，无胸罩泳装
topline [ˈtɒplaɪn]　上口线；鞋口
topper [ˈtɒpə]　（女用）轻便短大衣；单上衣；礼帽；（立体熨烫用）裤像机
　belted *topper*　束带短大衣
　topper-one　（立体熨烫用）人身机
　T-*topper*　圆领衫
　warp *topper*　宽松短大衣
topside [ˈtɒpsaɪd]　裤腰侧
　right (left) *topside*　右（左）腰头
topstitching [ˈtɒpstɪtʃɪŋ]　正面缝；〔工〕缉明线，缉面线
　uneven *topstitching*　〔检〕面线车得不均匀

toque [tɒk]　〔法〕（女用）小圆帽；无边女帽；（法官，厨师等戴的）直筒无边高帽
torchon [ˈtɔːʃən]　饰带花边；镶边花边
torque [tɔːk]　项链；项圈
torso [ˈtɔːsəʊ]　（人体）躯干
tote [təʊt]　大手提包（袋），箱形手提包
totem [ˈtəʊtəm]　〔设〕图腾；图腾形象
tow [təʊ]　丝束，纤维束
　acetate *tow*　醋酸丝束
　acrylic *tow*　腈纶丝束
　nylon *tow*　锦纶丝束
　polyester *tow*　涤纶丝束
　polypropylene *tow*　丙纶丝束
　ramie *tow*　苎麻束
　viscose rayon *tow*　黏胶丝束
　wool *tow*　毛条
towel [ˈtaʊəl]　毛巾
　bath *towel*　浴巾
　beach *towel*　沙滩巾
　dry hooded *towel*　带兜帽保暖毛巾（婴儿浴后用）
towelling [ˈtaʊəlɪŋ]　毛巾布料
　stretch *towelling*　弹力毛巾布
tracer [ˈtreɪsə]　（制图用）描绘工具；点线器
trackie [ˈtræki]　宽松式运动衣裤
trackpants [ˈtrækpænts]　〔复〕运动裤；田径裤
　fleece *trackpants*　绒布运动裤
　jersey *trackpants*　针织运动裤
　side stripe *trackpants*　两侧拼条运动裤
tracksuit [ˈtræksuːt]　（由宽松毛巾上

衣和长至足踝的长裤组合的）轻快套装；运动套装
  polyester tricot tracksuit　涤纶经编料运动套装
trademark [ˈtreɪdmɑːk]　商标
  registered trademark　注册商标
train [treɪn]　（结婚礼服）拖裙；长裙；裙长摆
  apron train　围裙长裙（礼服上的拖地蝴蝶结）
  fish tail train　鱼尾形拖裙
trainer [ˈtreɪnə]　（运动员穿的）训练服；训练鞋
  hooded trainer　带风帽运动衣
transfer [trænsˈfɜː]　转移；烫画（转移图案的工艺）
transition [trænˈzɪʃən]　〔设〕渐变
trappings [ˈtræpɪŋz]　〔复〕服饰；礼服
treadle [ˈtredl]　（缝纫机）踏板
tree [triː]　（柱式）木架；鞋楦
  clothes tree　（柱式）衣帽架
  hat tree　帽架
  shoe tree　鞋楦
trend [trend]　（流行）趋势，倾向；时尚
  apparel market trend　服装市场趋势
  fashion trend　时装潮流
  top trend　顶级时尚，最新流行
triacetate [traɪˈæsɪteɪt]　三醋酸酯，三醋酯纤维
triangle [ˈtraɪæŋgl]　三角板；三角裤
trichologist [trɪˈkɒlədʒɪst]　美发师
tricot [ˈtrɪkəʊ]　〔法〕经编针织物；斜纹毛织物；毛衫
tricotine [trɪkəˈtiːn]　斜纹织物；巧克丁（音译，精纺呢绒）

wool/polyester tricotine　毛涤巧克丁
trilby [ˈtrɪlbi]　〔主英〕一种软毡帽（= trilby hat）
trim [trɪm]　修剪；镶绲；装饰物；（华丽）服装
trimmer [ˈtrɪmə]　修边器；〔复〕剪刀；剪线工
  collar contour trimmer　修领器，修领脚机
  dressmaker's straight (bent) trimmers　裁缝用直（弯）形剪刀
  inlaid trimmers　嵌花剪
  pieces trimming　〔工〕（裁剪后）修片
  thread trimmer　剪线器
trimming [ˈtrɪmɪŋ]　装饰；修饰；镶绲；修剪，剪线；〔复〕服饰件（织带、饰带、花边）
  beaded trimming　珠饰带
  edge trimming　（衣服）饰边，修边
  lace trimming　花边装饰
  thread trimming　〔工〕修剪线头
trinket [ˈtrɪŋkɪt]　小件饰物
trios [ˈtrɪəʊz]　〔复〕三件套装（衣、裤、裙不同衣料但十分协调）
trousering [ˈtraʊzərɪŋ]　裤料；西裤料
  trousering in wool serge　毛哔叽裤料
trouser [ˈtraʊzə]　〔常用复〕〔英〕西裤；长裤
  acrylic trousers　腈纶裤
  ankle-length trousers　及踝长裤，八分之七裤，九分裤
  body fitting trousers　紧身裤
  British trousers　英式（直筒）西裤
  casual trousers　便裤，休闲裤
  cotton interlock trousers　（针织）棉

毛裤
cotton twill *trousers* 纯棉斜纹裤
cotton wadded (padded) *trousers* 棉裤
denim *trousers* 牛仔裤
down wadded *trousers* 羽绒裤
dress *trousers* 礼服裤
easy care *trousers* 便裤
end (bottom) of *trousers* 裤脚
flared *trousers* 宽脚裤，喇叭裤
formal *trousers* 礼服裤
fur cloth *trousers* 人造毛皮裤
harem *trousers* （伊斯兰）闺阁裤（宽松扎口长裤）
hose bottom *trousers* 窄裤管裤
knitted *trousers* 针织裤；毛线裤，〔粤〕冷裤
leather *trousers* 皮裤
lined *trousers* 夹裤（有衬里的裤）
oxford bag *trousers* 袋型裤
pegged *trousers* （上宽下窄）木钉裤，楔形裤
pencil *trousers* （直筒形）铅笔裤，笔杆裤
pleated *trousers* 腰头前面打褶裥的西裤
plush *trousers* 长毛绒裤
polyester *trousers* 涤纶裤
pyjama *trousers* 睡裤
rain *trousers* 雨裤
spare *trousers* （不成套西裤的）备用裤
stem-pipe *trousers* 直筒裤
straight *trousers* 直筒裤
striped *trousers* 礼服条纹西裤
tailored *trousers* 西服长裤，西裤
tapered *trousers* 窄脚裤，小裤脚裤

top of *trousers* 裤腰头
trousers curtain （裤）雨水布（用于遮盖腰头衬布）
trousers without belt 无腰带西裤
truearan *trousers* 涤/棉裤
tunic *trousers* 贴身裤
wedge *trousers* 楔形裤（马裤）
woolen *trousers* 毛呢裤，料子裤
working *trousers* 工作裤
wrinkle resistant *trousers* 防皱裤

trousseau [ˈtruːsəʊ] 出嫁时的衣物，嫁妆

try-on [ˈtraɪɒn] （衣服）试穿
3D virtual *try-on* 三维虚拟（服装）试穿

truearan [ˈtruːrən] 棉涤纶，涤/棉布，棉的确良

trunk [trʌŋk] （人体）躯干；衣箱，皮箱；〔复〕男运动短裤
boxing *trunks* 拳击短裤
guy (fly) front *trunks* （针织男式）前开门襟短裤
hipster *trunks* 低腰运动短裤
long leg *trunks* 长脚运动短裤（脚口至大腿中）
pouch front *trunks* 前中膨鼓形短裤
seamfree *trunks* 无缝运动短裤
swimming *trunks* 游泳裤
wardrobe trunk 柜式衣箱
Y-generation ribbed *trunks* 前开门襟罗纹运动短裤

tuck [tʌk] 褶裥；横褶；缝褶；塔克（音译，意即缝褶）
dart *tuck* 半活褶；塔克省
decorative *tuck* 装饰性褶裥
invisible *tuck* 暗褶
pin *tuck* 细褶，窄褶；针纹褶饰

place of *tuck* 打褶处
  two(one)-*tuck* 双（单）褶
  visible *tuck* 明裥
  wide *tuck* 宽褶
tucker [ˈtʌkə] （缝纫机）打褶裥装置；活动衣领
tucking [ˈtʌkɪŋ] 打横褶，打裥，〔粤〕打排褶
tulle [tjuːl] （女服和面纱用）的绢网；网眼布；（丝质或尼龙的）薄纱
tumbler [ˈtʌmblə] （衣服）干燥机，干衣机
  industrial drying *tumbler* 工业干衣机
tunic [ˈtjuːnɪk] （长至膝盖）束腰外衣；紧身短上衣；（女）罩衫；法衣，祭服；（古希腊、罗马人穿的）短袖或无袖的齐膝长袍
  camisole *tunic* 吊带束腰外衣
  tunic with V-neckline V 领束腰外衣
  Turkish *tunic* 土耳其式束腰外衣
  turtle neck *tunic* （针织）高圆领外衣
tunicle [ˈtjuːnɪkl] （天主教主教、助祭穿的）法衣，祭服
tunique [tynɪk] 〔法〕（古希腊、罗马人穿的）内长衣；宽长女裙；制服上衣；紧身女服；（主教等穿的）祭服
turban [ˈtɜːbən] 穆斯林头巾；印度头巾帽；无檐小帽
turnup [ˈtɜːnʌp] 衣服卷起部分；裤脚卷边（反吊边）
turtleneck [ˈtɜːtlnek] 高而紧的衣领；高领绒衣
tussore (=tussah) [ˈtʌsɔː] 柞蚕丝
tussores [ˈtʌsəz] 罗缎；线绢
  T/C dyed *tussores* 染色涤棉线绢

tutu [ˈtuːtuː] 芭蕾舞短裙；蓬松短裙
tuxedo [tʌkˈsiːdəʊ] (= tux [tʌks]) 〔美〕（男用）夜小礼服；无尾夜常礼服
  fancy *tuxedo* 变化小礼服
tweed [twiːd] （粗）花呢，〔粤〕粗绒；〔复〕花呢衣服
  check *tweed* 格子粗花呢
  connemara *tweed* 科纳马拉粗花呢
  costume *tweed* 粗毛呢，粗花呢
  Harris *tweed* 海力斯粗呢
  hemp/wool *tweed* 大麻/毛混纺粗花呢
  herringbone *tweed* 人字粗花呢
  micro *tweed* 超细斜纹布
  livery *tweed* 制服呢
  nub *tweed* 结子粗花呢
  pashmina *tweed* 羊绒粗呢
tweezers [ˈtwiːzəz] 〔复〕（拔除线头用的）镊子
twill [twɪl] 斜纹织物，斜纹布，〔粤〕斜布；〔复〕哔叽；绫（斜纹丝绸）
  cavalry *twill* 马裤呢
  cotton *twill* 全棉斜纹布
  herringbone *twill* 人字斜纹布
  nylon *twill* 锦纶斜纹布
  peach *twill* 卡丹绒（一种新型涤纶斜纹面料）
  ramie *twill* 苎麻斜纹布
  ramie/cotton blended *twill* 苎麻/棉混纺斜纹布
  rayon *twill* 黏胶纤维斜纹绸
  single (double) sided *twill* 单（双）面斜纹织物
  stretch *twill* 弹力斜纹布
  zigzag *twill* 人字呢
twine [twaɪn] 合股线（两股或两股

以上）；麻线；细绳
twist [twɪst] 缝纫用丝线
tying ['taɪŋ] 结，结子；系结；捆扎
type [taɪp] 型；类型；式
 basic type 基本类型
 big neckline type 粗脖子型
 classical type 古典型
 different types of bodies 各种体型
 high collarbone type 高锁骨型
 high shoulder blade type 高肩胛骨型
 hunchbacked type 驼背型

machine type 机型，机种
natural type 自然型
new type cutting (sewing) method 新式裁剪（缝制）方法
personality type 彰显个性型
pigeon-breasted type 鸡胸型
popularity type 流行前卫型
romantic type 浪漫型
seam type 缝型，缝式
shoulder type 肩型
stitching type 针迹类型

U

Uggs/uggs [ʌgs] 〔复〕羊毛皮靴；雪地靴
ulster ['ʌlstə] 阿尔斯特大衣（音译，有腰带的宽大长外套）；阿尔斯特长毛大衣呢；风衣
umanori [wʊ'mʌnɒrɪ] 〔日〕开缝；（和式）短外褂、贴身衬衫的襟开衩；西服的背开衩
umbrella [ʌm'brelə] 伞，雨伞，阳伞，〔粤〕遮
 cloth umbrella 布伞
 folding umbrella 折叠伞
 hand (auto)-open umbrella 手开（自动）伞
 nylon umbrella 尼龙伞
 oiled paper umbrella 油纸伞
 plaid (lattice) umbrella 格子伞
 polka dot umbrella 圆点花伞
 silk umbrella 绸伞
 splash umbrella 散点花伞
 stick umbrella 手杖伞

 topless umbrella 折叠伞，〔粤〕缩骨遮
 trifolding umbrella 三折伞
 variegated umbrella 花伞
unbalance [ˌʌn'bæləns] 〔设〕不均衡，失衡
underclothes ['ʌndəˌkləʊðz] (= underclothing) ['ʌndəˌkləʊðɪŋ] 内衣裤，衬衣裤（指贴身衣裤）〔粤〕底衫
 leopard print underclothes 豹纹内衣裤
undercoat ['ʌndəkəʊt] 大衣内的上衣，外套里衣，衬裙
undercollar ['ʌndəkɒlə] 底领
underdress ['ʌndədres] （女）内衣；衬裙
undergarment ['ʌndəˌgɑːmənt] （一件）内衣（衬衣）
underlap [ˌʌndə'læp] （衣）里襟；小袖衩边
underlinen ['ʌndəˌlɪnɪn] （麻布或其

他薄料）内衣，衬衣
underlining [ˌʌndəˈlaɪnɪŋ] 衬布
underpants [ˈʌndəpænts]〔复〕衬裤，内裤，汗裤，〔粤〕底裤
undershirt [ˈʌndəʃɜːt] 贴身内衣；汗衫；汗背心，〔粤〕笠衫
 elastic *undershirt* 弹力汗衫
 mesh *undershirt* 网眼汗衫
undershorts [ˈʌndəʃɔːts]〔复〕〔美〕男短衬裤，三角裤
underskirt [ˈʌndəskɜːt] 衬裙
undersleeve [ˈʌndəsliːv] 小袖，内袖，衬袖
underthings [ˈʌndəθɪŋz]〔复〕女内衣裤
undervest [ˈʌndəvest]〔英〕贴身内衣，汗衫，汗背心
underwaist [ˈʌndəweɪst] 穿在里面的背心；童内衣
underwear [ˈʌndəweə]（总称）内衣，衬衣，贴身衣
 aloe fibre *underwear* 芦荟纤维内衣（芦荟精华渗入面料滋养皮肤）
 collagen *underwear* 胶原蛋白内衣（胶原蛋白润肤护肤）
 cotton interlock *underwear* 棉毛类衫裤
 dralon *underwear* 德绒（保暖）内衣
 ecological *underwear* 环保内衣
 elastic *underwear* 弹力内衣
 far-infrared *underwear* 远红外内衣（具有磁疗保健作用）
 flannelette *underwear* 绒布类衫裤
 fragrance *underwear* 香味内衣
 functional *underwear* 功能内衣（具有矫型、保暖、保健等功能）
 health care *underwear* 保健内衣
 knitted *underwear* 针织内衣
 lovers *underwear* 情侣内衣
 moisture wicking (absorbing) *underwear* 吸湿排汗内衣
 seamless *underwear* 无缝内衣
 sexy *underwear* 性感内衣，情趣内衣
 shaping *underwear* 塑形内衣，美体内衣
 skin care *underwear* 护肤内衣
 smart *underwear* 智能内衣（可随时监测人体生理指标）
 space *underwear* 航天内衣（具有抗静电，阻燃，抗菌防臭，吸湿排汗等功能）
 sports *underwear* 运动内衣
 thermal *underwear* 保暖内衣
 vitamin E *underwear* 维生素 E 内衣（维 E 营养皮肤）
 wood (bamboo) fibre *underwear* 木（竹）纤维内衣
 X-ray proof *underwear* 防 X 射线内衣
undies [ˈʌndɪz]〔复〕内衣；女内衣
undress [ʌnˈdres] 便服；军便服；晨衣
uniform [ˈjuːnɪfɔːm] 制服；职业（工作）服；军服
 aircrew *uniform* 机组人员制服
 antistatic *uniform* 防静电制服
 army combat *uniform* (ACU) 陆军作战服，陆战服
 army service *uniform* (ASU) 服役军装（非战时穿）
 bank *uniform* 银行制服
 civil aviation *uniform* 民航制服

conductors' *uniform* 列车员制服
cooking *uniform* 炊事服
customs *uniform* 海关制服
dress *uniform* 军礼服（美国陆、海军制服）；仪仗队制服
dustmen's *uniform* 环卫工人服
family *uniform* 家庭常服
field *uniform* 野战军服
field training *uniform* 作（战）训（练）服
firemen's *uniform* 消防服
flight attendants' *uniform* 空姐服
full-dress *uniform* 大礼服
hotel stuff *uniform* 酒店制服
industrial and commercial managers' *uniform* 工商管理人员制服，工商服
kindergarten *uniform* 幼儿园服
medical (hospital) *uniform* 医务服
military (army) *uniform* 军服
nano-protective *uniform* 纳米防护服
nurse *uniform* 护士服
office *uniform* 办公服
official *uniform* 礼服
oil workers' *uniform* 石油工人服
out of *uniform* 穿着便服
policemen's *uniform* 警官服，警服
post and telecom *uniform* 邮电服
railwaymen's *uniform* 铁路制服
road maintenance worker's *uniform* 养路工人服
school *uniform* 校服，学生制服
security personnel's *uniform* 保安制服
special working *uniform* 特种工作服
Sun Yat Sen's *uniform* 中山服
super-excellent clean *uniform* 超净防尘服
tax collectors' *uniform* 税务服
team *uniform* 团队制服，队服
undress *uniform* 军便服；便装
waiters' *uniform* 服务员制服
working *uniform* 工作服

union [ˈjuːniən] 混纺（交织）织物
unisex [ˈjuːnɪseks] 不分男女的服装，无性别服装，中性服装
unit [ˈjuːnɪt] （机械）部件，零件；装置；（计数）单位
unitard [ˈjuːnɪtɑːd] （体操运动员、舞蹈演员等穿着）紧身套装或连衣裤，婴幼连衫裤
unpicker [ʌnˈpɪkə] 拆线刀
unsuits [ʌnˈsuːts] 无构造服，简易套装；便服
uplift [ʌpˈlɪft] （女服）胸衬
upper [ˈʌpə] 鞋帮；腰带；〔复〕绑腿，腿套
ushanka [wuˈʃʌnkʌ] 一种俄罗斯护耳毛皮帽

V

valitin [ˈvælɪtɪn] 凡立丁（音译，精纺呢绒）
 polyester/viscose *valitin* 涤/黏凡立丁
 wool/polyester *valitin* 毛/涤凡立丁
value [ˈvæljuː] 色明度（色彩明暗的

配合)

vamp [væmp] 靴(鞋)面;靴(鞋)面补片

    whole *vamp* 鞋全帮

veil [veɪl] 面纱,面罩

    nose *veil* (挂在女帽前能遮住眼鼻)短面纱

    wedding(bridal) *veil* 婚礼面纱

velcro [ˈvelkrəʊ] 维可牢(音译,一种尼龙搭扣的商品名),尼龙搭扣,〔粤〕魔术贴

velour [vəˈlʊə] 丝绒,天鹅绒;棉绒;拉绒织物;(制帽用)兔绒皮(海狸绒皮)

    knitted *velour* 针织丝绒;针织天鹅绒

    velours crepe-de-chine 双绉丝绒

    velours glacé 〔法〕闪光丝绒

velvet [ˈvelvɪt] 丝绒,天鹅绒,立绒,经绒

    antique *velvet* 仿古丝绒

    bright *velvet* 光明绒

    brocaded *velvet* 缎花丝绒

    chiffon *velvet* 薄天鹅绒,雪纺丝绒

    coral *velvet* 珊瑚绒

    cotton *velvet* 棉天鹅绒

    cut *velvet* 配饰丝绒;立绒呢

    embossed *velvet* 拷花丝绒

    faconné *velvet* 〔法〕烂花丝绒

    figured *velvet* 花丝绒

    georgette *velvet* 乔其绒

    ginning *velvet* 扎花天鹅绒

    melange *velvet* 混色雪花天鹅绒

    nylon *velvet* 锦纶丝绒

    pellet fleece *velvet* 粒粒绒布

    plain(panne) *velvet* 平绒

    rayon *velvet* 黏胶纤维丝绒

    rib *velvet* 棱条丝绒

    rib fleece *velvet* 抽条磨毛天鹅绒

    sculptured *velvet* 凹凸绒

    shoe-top *velvet* 鞋面丝绒

    silk *velvet* 丝天鹅绒,丝绒

    solid *velvet* 素色天鹅绒

    transparent *velvet* 透明丝绒,乔其绒

    velvet for toy(garments) 玩具(服装)绒

    Zhangzhou *velvet* 漳绒

velveteen [ˈvelvɪtiːn] 棉绒,平绒,纬绒;〔复〕棉绒衣服(尤指裤)

    both-side-raised *velveteen* 双面绒

    printed (dyed) *velveteen* 印花(染色)平绒

    shu *velveteen* 舒棉绒(人造绵绒的,一种涤纶新面料;类似摇粒绒)

venetian [vɪˈniːʃən] 直贡呢(精纺呢绒);棉直贡;礼服呢;威尼斯缩绒呢;威尼斯精纺呢

    cotton *venetian* 元贡呢,泰西缎

    wool *venetian* 威尼斯毛呢,毛直贡呢

    wool/polyester *venetian* 毛涤直贡呢

vent [vent] (西装上衣的)衩口,开衩,骑马缝

    back *vent* 背衩

    box *vent* 工字衩

    centre *vent* 中心开衩,单衩,后衩

    crossing at the back *vent* 〔检〕背衩搅(衩下部搭叠过多)

    cuff *vent* 袖(头)衩

    hook *vent* 明单衩,明后衩,钩形衩

    inverted *vent* 暗衩,阴衩

    mock *vent* 假衩

side vent 双开衩,边衩,摆衩
sleeve vent 袖衩
split at the back vent 〔检〕背衩豁（衩下部豁开）
top vent 面衩
uneven vents length 〔检〕左右衩长不一样
unmeet(closed) vent 〔检〕背衩豁（搅）

vest [vest] 〔美〕背心、马甲（=〔英〕waistcoat）；（作战时穿的）防护衣；〔英〕汗衫；内衣,衬衣；（女服前胸）V形饰布；（古语）外衣；法衣；长袍；衣服
 angler's vest 钓鱼背心（多口袋的背心）
 anti-radiation maternity vest 防辐射孕妇背心
 argyle vest （针织）菱形花格背心
 armoured vest 防弹背心
 baby's vest 婴儿汗衫
 backless vest （配男夜礼服的）无背背心
 buckskin vest 鹿皮背心
 bulky vest 膨体纱背心
 bullet proof vest 防弹背心
 cable vest 绞花编织背心
 chenille vest 绳绒线背心
 collar vest 有领背心
 cotton-padded vest 填棉背心,间棉背心
 cowboy vest 牛仔背心
 crocheted vest 钩编背心
 denim vest 牛仔（布）背心
 down vest 羽绒背心
 empire vest 帝国风格女吊带背心
 evening(dress) vest 礼服背心
 fancy vest 替换背心,变化背心
 far-infrared heating vest 远红外加热背心
 field training vest 作（战）训（练）背心
 fishing vest 钓鱼背心
 flannelet vest 绒布背心
 formal vest 礼服背心
 fur vest 毛皮背心
 golf pullover vest （针织）高尔夫套头背心
 knitted vest 针织背心,毛线背心
 lapeled vest 有驳领背心
 leather vest 皮背心
 leisure vest 休闲背心
 life vest 救生背心
 longline vest 长身线形背心
 mesh(cellular) vest 网眼背心
 multi-functional vest 多功能背心
 odd vest （套装中）单零背心,替换背心
 outer vest 户外背心
 padded vest 填棉背心,间棉背心
 photographer's vest 摄影师背心
 pig suede(split) vest 猪绒面（二层）皮背心
 polar fleece vest 摇粒绒背心
 police vest 警察背心
 postboy vest （旧时）邮递员背心
 puff vest 蓬松背心（绗缝衣面成泡泡状）,间棉背心
 pullover vest 套头背心
 rash vest 插肩袖圆领短袖衫
 reflective PVC/nylon vest （缝有荧光条带的）单胶安全背心
 reversible vest 两面穿背心
 shearling vest 毛羊皮衬里背心

shorn sheepskin vest 羊剪绒背心
single（double）breasted vest 单（双）襟背心
smart vest 智能背心（可实时监测心脏情况；可自动加热控温；可矫正坐姿，减轻背痛，等等。不同品种的智能背心具有不同功能）
sport(gym) vest 运动背心
stretched vest 弹力背心
suede vest 仿麂皮背心，反绒皮（獐皮）背心，绒面革背心
suit vest 套装背心
survival vest 救生背心
swallowtail vest 燕尾背心
sweater vest 针织V领套头背心；针织开襟背心
tactical vest 战术背心（缝有各种型号口袋，能将军人、警察大部分个人战斗所需装备整合穿戴于身）
texture vest 花式背心，织花背心
thermal vest 保暖（内衣）背心
velvet vest 丝绒背心
vest de chasse 〔法〕猎装夹克
vest de judo-ka 〔法〕（日式）柔道上衣
vest with hood 带帽背心
wool vest 羊毛背心
zipper（snap）front vest 拉链（按扣）背心

vestee [vesˈtiː] 背心形衣着；（女服前胸）V形饰布

vestiary [ˈvestɪərɪ] 衣帽间
vesting [ˈvestɪŋ] 背心料，马甲料
vestment [ˈvestmənt] 外衣；制服；法衣，祭服；〔复〕衣服
vesture [ˈvestʃə] 衣服；罩衣，罩袍
vicuna（= vicugna）[vɪˈkjuːnə] 骆马绒；骆马毛呢料
view [vjuː] 观点，视点；视角，视野；（制图的）视图
　　new view 新视点，新观点
vinylon [ˈvaɪnɪlɒn] 维纶
viscose [ˈvɪskəʊs] 黏胶纤维，黏纤
vision [ˈvɪʒn] 视野；想象力；构想；愿景
visor [ˈvaɪzə] （古时）脸盔，面甲；假面具；帽舌，帽檐
　　sun visor 遮阳帽舌（仅有前帽檐的帽子），空顶帽
vogue [vəʊg] 时尚；（服饰的）流行，时髦；流行物
　　best vogue matches 最佳时尚搭配
voile [vɔɪl] 巴里纱（透明薄纱，纯棉或涤/棉制，又名玻璃纱）
　　printed voile 印花巴里纱
V-string [ˈviːstrɪŋ] V形布窄带女短内裤（两侧为弹力窄带）
　　embroidered V-string 绣花窄带女内裤
　　lace V-string 花边窄带女内裤
　　mini V-string 仅有一小块V形遮布的窄带女内裤

W

wad [wɒd] 软填料（棉花、絮片）
wadding [ˈwɒdɪŋ] 软填料；衬料；棉胎；絮片
 armhole *wadding* 袖窿衬垫（细棉条或布条，俗称弹袖棉）
 bamboo fibre *wadding* 竹纤维絮片
 breast *wadding* （绒布）胸衬
 corn fibre *wadding* 玉米纤维絮片
 quilt *wadding* 被胎，絮被
 silk *wadding* 丝绵
 staple rayon *wadding* 喷胶棉絮片
 thermofusible cotton *wadding* 热熔棉絮片
waders [ˈweɪdəz] 〔复〕（涉水、捕鱼用）高筒防水靴
 fishing *waders* 钓鱼靴
wafuku [wʌfʊgə] 〔日〕和服
waist [weɪst] 腰，腰部；腰围；（衣裤的）腰身；背心；紧身胸衣；童内衣
 centre *waist* 中腰位
 fork to *waist* 裤直裆
 low(high) *waist* 低（高）腰身
 nape to *waist*(N.to W.) 背长，腰直（后颈点至腰距）
 nipped *waist* 掐腰，紧腰
 small(large) *waist* 细（粗）腰
 waist banding 缩裤腰头，〔粤〕拉裤头
 waist nipper （女用）腰封，束腰紧身衣；（妇女）紧身腹带
 waist relaxed(extended) 腰围收缩（拉开）量（尺寸）
waistband [ˈweɪstbænd] 腰带；裤带；裙带；裤（裙）腰头
 canvas *waistband* 帆布裤带
 draped *waistband* 褶裥腰头；垂饰腰头
 elastic *waistband* 松紧腰头；弹性腰带
 end of *waistband* is uneven 〔检〕腰头探出（前口不齐）
 high/low *waistband* ends 〔检〕腰头高低（不对称）
 knitted *waistband* （裤，裙）针织罗纹腰头；（夹克衫）罗纹下摆
 twisted *waistband* 〔检〕腰头扭曲
 upper(under) end of *waistband* 腰带上（下）口
 waistband extension 腰带嘴；〔检〕腰头探出
 waistband for pyjamas 睡衣腰带
 waistband is extension of body 原身裤头，连身裤头（即裤头连裤身未裁断）
 waistband neckline 腰头下口线
waistbelt [ˈweɪstbelt] 腰带；裤带
waistcloth [ˈweɪstklɒθ] 围腰布
waistcoat [ˈweɪskəʊt] 〔英〕背心，马甲（=vest〔美〕）
 dress *waistcoat* 礼服背心
 postboy *waistcoat* （旧时）邮递员背心
 two (three)-button *waistcoat* 两（三）扣背心
 waistcoat in jacket style 夹克式背心
waistline [ˈweɪstlaɪn] 腰围，腰身，腰

节

waistpack ['weɪstpæk] （系在腰上的）腰袋

walkers ['wɔːkəz] 〔复〕散步鞋，轻便鞋

wallet ['wɒlɪt] 钱包；皮夹子（=〔美〕pocket-book）

wardrobe ['wɔːdrəub] 衣柜；衣橱；藏衣室；（挂衣的）大衣箱；（个人的）全部服装；（某季节或某活动穿着的）全套服装；（剧团）全部戏装
    wardrobe mirror 衣柜镜，穿衣镜
    winter *wardrobe* 冬装

warm [wɔːm] 保暖的东西；〔英〕军用短大衣（= British warm）

warmer ['wɔːmə] 保暖衣物
    arm *warmer* （保暖）臂套
    bench *warmer* 替补球员大衣
    British *warmer* 英式厚大衣
    leg *warmer* （保暖）袜套，暖腿套

warp [wɔːp] （织物的）经纱

wash [wɒʃ] 洗涤；（服装）洗水；洗的衣物；洗衣房；洗涤剂
    acid *wash* 酸洗（加草酸等洗，使衣服颜色更艳丽）
    antique *wash* 仿古洗，怀旧洗（洗出古旧效果）
    berkeley *wash* 伯克利洗水
    bio *wash* 酵素洗
    black *wash* 黑色洗
    black over black *wash* 黑染黑
    bleach *wash* 漂洗（加过氧化氢或次氯酸钠洗，使衣服褪色变白且手感柔软）
    bleach stone *wash* 石磨加漂洗，石漂洗（漂白后石磨）
    blue & black *wash* 蓝黑洗（混合洗出新效果）
    broken edge *wash* 烂边洗（用电动砂轮有意磨破衣边后洗水）
    chemical *wash* 化学洗（加入强碱助剂洗，使衣服褪色具有陈旧感）
    chemical stone *wash* 化石洗（在化学洗中加浮石，使衣服更具有残旧感）
    colour fixed *wash* 固色洗
    crystal *wash* 水晶洗
    dark(light) *wash* 深（浅）色洗
    dark(medium, light) blue *wash* 深（中，浅）蓝洗
    destarch *wash* 退浆洗
    destruction(destroy) *wash* 破坏洗，破损洗（经浮石打磨及化学助剂处理，使衣服某些部位破损，具有残旧效果）
    dirty *wash* 怀旧洗
    dissolved rubber ball *wash* 胶球洗（塑胶球，作用与浮石相似）
    double enzyme *wash* 重酵素洗
    dull *wash* 钝色洗，浊色洗（洗出暗浊效果）
    dye *wash* 染色洗
    enzyme *wash* 酵素洗，酵洗（加纤维素酶洗，使衣服褪毛褪色）
    enzyme bleach *wash* 酵素漂洗，酵漂洗
    enzyme stone *wash* 酵素石磨洗，酵石洗
    enzyme stone bleach *wash* 酵素石磨漂洗，酵石漂洗
    fabric *wash* 布料洗水
    ferment *wash* 酵素洗
    heavy worn *wash* 重残旧洗
    garment *wash* 成衣洗水，普通洗

水，普洗

grinding wash 磨烂洗（将衣服某些部位磨薄或磨烂后洗）

hand brush and rinse wash 手擦加漂（清）洗

hand brush enzyme rubber ball wash 手擦酵素胶球洗

hand brush enzyme stone bleach wash 手擦酵石漂洗

hand brush enzyme stone wash 手擦酵石洗

hand brush enzyme stone wash and tinting 手擦酵石洗加吊色

hand brush whisker, broken edges and enzyme stone wash 手擦猫须酵石洗加烂边

hang bleach wash 吊漂洗

heavy acid wash 重酸洗

light(heavy) enzyme stone wash 轻（重）酵石洗

light(heavy) garment wash 轻（重）普洗

machine(hand) wash 机（手）洗

mid stone(bleach) wash 中度石磨（漂）洗，中磨（漂）洗

monkey wash 喷马骝（将高锰酸钾喷在衣服的某些部位使之褪色），马骝洗

normal wash 普通洗水，普洗

overdyed wash 套色洗（洗后可改变牛仔服的颜色）

pigment(pigment dyed)wash 碧纹洗（碧纹是 pigment 的音译。将涂料染色的衣服洗水，使之褪色变软）

pinching and creasing wash 收皱洗，抓皱洗（洗出皱褶效果，如牛仔裤膝内弯处）

powder milling wash 粉磨洗

PP（potassium permanganate）spray wash 喷灰锰氧（即高锰酸钾），喷马骝

PP stone vibration wash 雪花洗，炒雪花（用浸透了高锰酸钾的浮石在转缸内与衣服打磨，形成雪花点）

resin wrinkle effect wash 树脂压皱洗（衣服经树脂浸泡后，压皱效果更佳）

retro wash 怀旧洗

rinse wash 退浆保色洗

sand wash 砂洗（加碱性或氧化性助剂，使衣服褪色具有陈旧感）

sandblast wash 喷砂，打砂（高压喷砂在衣服某些部位上，造成局部磨损，使表层褪色）

sandblast enzyme bleach wash 喷砂酵漂洗

sandblast enzyme stone wash 喷砂酵石洗

scratch wash 猫须洗

silicone wash 硅油洗（硅油为柔软剂）

snow wash 雪花洗

soft wash 轻洗水，柔洗

spray colour wash 喷色

spray gold(silver) powder wash 喷金（银）粉

spray sand wash 喷砂，打砂

stone wash 石磨洗水，石洗（加浮石洗，与衣服摩擦出效果）

tie wash 扎洗（将衣服某些部位扎紧，洗后产生差异），扎花

tie bleach wash 扎漂洗

  underside garment *wash* （衣服）翻底洗
  vintage *wash* 怀旧洗（洗出古旧效果）
  vision *wash* 幻影洗
  warm(hot,cold) 温（热，冷）水洗
  wrinkle effect *wash* 压皱洗，折皱洗（将衣服某些部位压折定形后洗水）
  wrinkle free *wash* 防皱洗水（洗后衣服不易起皱）
washboard [ˈwɒʃbɔːd] 洗衣板，搓板
washer [ˈwɒʃə] 洗衣机；洗衣工；洗水布；垫圈
  bulk *washer* 大货洗衣机
washing [ˈwɒʃɪŋ] 洗水；洗涤；洗涤物；〔复〕洗涤剂
  over *washing* 〔检〕洗水过度，洗水太重
  poor *washing* 〔检〕洗水不良
  uneven *washing* effect 〔检〕洗水欠佳
  washing instruction 洗涤说明，洗涤法
  washing powder 洗衣粉
  washing streak 〔检〕（洗水不好所致）洗水痕迹
watch [wɒtʃ] 手表，挂表
  bracelet *watch* 手镯表
  brand *watch* 品牌表
  chain *watch* 手链表
  children's cartoon *watch* 儿童卡通表
  curio style *watch* 古董款式表
  dress *watch* 礼服表
  fashion *watch* 流行手表，时装表

  pocket *watch* 怀表
  revival(retro) *watch* 复古表
  smart *watch* 智能手表（具有提醒，导航，校准，监测等额外功能）
  souvenir *watch* 纪念表
  space *watch* 太空手表，航天表
  sports *watch* 运动手表，休闲表
  sweethearts *watches* 情侣表
  water resistant *watch* 防水表
watering [ˈwɔːtərɪŋ] 衣料下水（预缩水）
waterproof [ˈwɔːtəpruːf] 防水；防水布；〔英〕雨衣
wax [wæks] 蜡
  cobbler's *wax* （制鞋用）线蜡
  tailor's *wax* 裁缝用蜡
wear [weə] 穿，戴；佩；衣服，服装；时装；流行样式
  A B *wear* 两面穿服装
  acrobatic *wear* 杂技服
  active *wear* 运动服装
  artificial leather(fur) *wear* 人造革（毛皮）服装
  athlete's *wear* 田径服
  baby's *wear* 婴儿服装
  ballet *wear* 芭蕾舞服
  beach *wear* 海滨服，沙滩服
  between season *wear* 换季服装，春秋服
  business *wear* 职业服，上班服装，办公服装
  campus *wear* 校园服，学生装
  casual *wear* 便服，常服，生活服装
  cattle hide *wear* 牛皮服装
  ceremonial *wear* 礼仪服装，礼服
  children's *wear* 童服，童装
  Chinese *wear* 中国服装

cooking *wear* 炊事服
country *wear* 郊游服
crochet *wear* 钩编服装
cruising *wear* 巡航服
cycle *wear* 骑自行车服
down *wear* 羽绒服装
everyday *wear* 便服
executive *wear* 行政服
exercise *wear* 训练服；健身服
fencing *wear* 击剑服
flying *wear* 飞行服
formal *wear* 正式礼服，礼服
formal day *wear* 昼间礼服
formal evening *wear* 夜间礼服
goat fur *wear* 羊毛皮服装
golf *wear* 高尔夫球服
home *wear* 家居服，便服
hunting *wear* 猎装
infant's *wear* 婴儿服装，幼儿服装
informal *wear* 准礼服；便服，日常服装
in general *wear* （服装、衣着）流行一时
inner *wear* 穿在里面的衣服；内衣
insulated *wear* 绝缘服
judo *wear* 柔道服
kid's *wear* 儿童服装，童装
kindergarten *wear* 幼儿园服
leather *wear* 皮革服装
leisure *wear* 休闲服，便服
lounge *wear* 家居服，休闲服
maternity *wear* 孕妇服装
medical *wear* 医务服
men's *wear* 男服，男装
middle and old aged people's *wear* 中老年服装
miner's *wear* 矿工服

most formal *wear* 正式礼服
mountaineering *wear* 登山服
occasional *wear* 应时服装
office *wear* 办公服装
official *wear* 正式场合穿的服装总称，正式服装
pigskin (goatskin) *wear* 猪（羊）皮服装
play *wear* 游戏服；运动装
preschooler's *wear* 小童服（学龄前儿童服装）
prisoner's *wear* 囚服
rabbit *wear* 兔毛皮服装
rain *wear* 雨衣，防雨服装
ready-to-*wear* (R.T.W.) 现成服装，成衣
referee's *wear* 裁判服
resort *wear* 休闲服，旅游服
riding *wear* 骑马服，骑装
rodeo *wear* 竞技表演服
safety *wear* 安全服，劳保服
sailing *wear* 航海服
school-age's *wear* 少年服装，学生装
school *wear* 学生服，校服
search and rescue *wear* 搜救服
semi-formal *wear* 一般礼服，半正式礼服，简略礼服
shaped *wear* 塑形服装，紧身装
silk *wear* 丝绸服装
skating *wear* 溜冰服
ski *wear* 滑雪服
social *wear* 社交服装
spectator's *wear* 参观服
sport *wear* 运动服；两用衫；休闲装
spring (summer, autumn, winter) *wear*

春（夏、秋、冬）装
steelworker's wear 炼钢服
street wear 街头服装；时尚休闲服饰
student's(pupil's) wear 学生服
suburban wear 郊外服
teens' wear 青少年服装
thermal wear 防寒保暖服
toddler's wear 幼儿服装
town wear 街市服，外出服
training wear 训练服，运动服
urban wear 都市服
waiter's wear 服务员服装
warm wear 防寒服
wash and wear 洗即穿，免熨烫
wear property （衣服）服用性能
weight lifter's wear 举重服
women's(ladies') wear 女服，女装
working wear 工作服，劳保服
woven(knitted) wear 机（针）织服装

wearability [ˌweərəˈbɪlɪti] （衣服）服用性能

wearables [ˈweərəblz] 〔复〕衣服，服装

wearings [ˈweərɪŋz] 〔复〕（古语）衣服，服装

weave [wiːv] 织法，编法；织物
textile weave 织物组织
tight(loose) weave 紧（松）织
twill(plain, satin) weave 斜（平，缎）纹

weaving [ˈwiːvɪŋ] 机织，织造，织布

web [web] 织物；一匹布，一卷布；网；网织品；网络
information web 信息网
sales web 销售网络
web address 网址
web side 网站，网页

wedge [wedʒ] （鞋的）坡跟；坡跟鞋；（鞋）中插件，楔形发式
gladiator wedge 罗马风情坡跟鞋
platform wedge 厚底坡跟鞋
strappy wedge 带扣坡跟鞋

wedgies [ˈwedʒɪz] 女式坡跟鞋（商标名）

weeds [wiːdz] 〔复〕丧服（尤指寡妇所穿的）

weeper [ˈwiːpə] （丧服）黑纱

weft [weft] （织物）纬纱

weight [weɪt] （布料的）重量；〔复〕（压布用的）压铁
net(gross) weight 净（毛）重
weight per square meter （布料）每平方米（克）重
weight per yard （布料某个幅宽的）每码重量

welt [welt] （衣服，鞋袜的）贴边；（鞋）沿条；牙边；绲边；嵌条；嵌革
pocket welt 口袋开线；袋牙边；袋嵌条
reverse welt 翻口袜口
single(double) welt 单（双）贴边

welting [ˈweltɪŋ] （给衣、鞋等）贴（绲）边；嵌革；扎袜口

weskit [ˈweskɪt] 紧身背心；男西装背心

wetshirt [ˈwetʃəːt] 冲浪圆领短袖衫，冲浪T恤衫

wheel [wiːl] 轮子
balance wheel （缝纫机头上的）手

　　　　轮
　　tracing *wheel* （制图用）点线器，摇盘，描迹轮
whetstone [ˈwetstəʊn] （磨刀、剪用）磨石，油石
whipcord [ˈwɪpkɔːd] 马裤呢（精纺呢绒）
　　wool/polyester *whipcord* 毛/涤马裤呢
　　wool/viscose *whipcord* 毛/黏马裤呢
whipping [ˈwɪpɪŋ] 包缝，锁边
whisker [ˈwɪskə] （动物的）须；猫须（洗水手擦工艺名词之一，缩写为WHK）；〔复〕连鬓胡子，颊须，髯
whizzer [ˈwɪzə] 离心干燥机
　　industrial *whizzer* 工业脱水机（甩干衣服）
width [wɪdθ] 宽度；幅宽，〔粤〕封度；（某宽度的）一块布料
　　arm *width* 袖宽，中袖
　　armhole *width* 袖窿宽
　　back *width* 背宽
　　chest（breast, bust）*width* 胸宽，〔粤〕胸阔
　　collar point *width* 领尖宽
　　cuff *width* 袖头宽；裤卷脚宽
　　cuttable *width* （可裁用的）内门幅，有效幅宽
　　fabric *width* 布料幅宽，门幅
　　fly *width* 门襟宽，〔粤〕门筒阔；裤门襟宽，〔粤〕钮牌阔
　　forearm(elbow) *width* 肘宽
　　front *width* 前胸宽；门襟宽
　　grey（finished）*width* 坯布（成品）幅宽

　　hip *width* 臀宽
　　lapel *width* 驳头宽，〔粤〕襟宽位
　　narrow *width* 〔检〕布料幅宽不足
　　neck *width* 领（圈）宽，〔粤〕领横
　　placket *width* 门襟宽，〔粤〕筒阔
　　pocket *width* 口袋宽
　　pocket welt *width* 袋口（嵌边）高，袋唇高
　　point *width* 胸点间距，乳间宽
　　saddle *width* 臀宽
　　seam *width* 缝份宽
　　shoulder *width* 肩宽，〔粤〕膊阔
　　single(double) *width* 单（双）幅宽
　　sleeve *width* 袖宽
　　thigh *width* 大腿围；（裤）横裆宽
　　tuck *width* 打裥幅度，褶宽
　　yoke *width* 过肩宽
wig [wɪg] 假发
windbreaker [ˈwɪndˌbreɪkə] （皮或呢制）防风外衣，风衣，〔粤〕风褛
windcheater [ˈwɪndˌtʃiːtə] 〔英〕防风上衣，防风夹克
winder [ˈwɪndə] （缝纫机上的）绕线器
wolverene（= wolverine）[ˈwʊlvəriːn] 貂熊毛皮，狼獾毛皮
womenswear [ˈwɪmɪnsweə] 女服，女装
wonderbra [ˈwʌndəbrɑː] 神奇胸罩，魔术胸罩（一种上托胸罩，加拿大著名品牌，1990年后逐渐流行于全世界）
wondersuit [ˈwʌndəsuːt] 一种婴幼连衫裤装
woof [wʊf] （织物的）纬纱；布，织

物

wool [wʊl] 羊毛；驼毛；毛线；呢绒；毛织品；毛料衣服

    alpaca *wool* 羊驼毛，羊驼绒
    artificial *wool* 人造毛
    Australian *wool* 澳大利亚羊毛，澳毛（主要是美利奴优质细羊毛）
    Berlin *wool* 柏林绒线，细毛线
    blended *wool* 混纺毛
    brushed *wool* 拉绒毛织品
    cashmere *wool* 山羊绒
    combing (carding) *wool* 精（粗）梳毛
    greasy *wool* 原毛
    guanaco *wool* 原驼毛
    improved *wool* 改良毛
    knitting *wool* 毛线
    lamb's *wool* 羔羊毛
    llama *wool* 美洲驼毛，美洲驼绒
    merino *wool* 美利奴羊毛
    native (imported) *wool* 国产（进口）羊毛
    New Zealand *wool* 新西兰羊毛
    pure (all) *wool* 全毛，纯毛
    raw *wool* 原毛
    raw *wool* of goat 山羊毛
    raw *wool* of sheep 绵羊毛
    recovered *wool* 回毛，再生毛
    scoured *wool* 洗净毛
    shetland *wool* 雪特兰羊毛
    short (long) *wool* 粗（精）梳毛
    special *wool* 特种羊毛
    Tasmanian *wool* （澳大利亚）塔斯马尼亚羊毛（顶级的美利奴羊毛）
    vicuna *wool* 骆马毛

woolenet (= **woolenette**) [ˈwʊlənet] 轻薄毛呢，薄呢

woolens [ˈwʊlənz] 〔复〕粗纺毛织物，呢绒，〔粤〕绒布；毛织品；毛料衣服

    chequered *woolens* 格花呢
    Chinese *woolens* 中国呢绒
    fancy *woolens* 粗花呢
    herringbone *woolens* 人字呢绒

woolfell [ˈwʊlfel] 〔英〕羊毛皮

woolly [ˈwʊli] （常用复数）毛线衣；羊毛内衣

woolskin [ˈwʊlskɪn] 〔美〕羊毛皮

woolwork [ˈwʊlˌwɜːk] 绒绣

work [wɜːk] 工作；产品；工艺；工艺品；刺绣品；针线活

    bead-*work* 珠绣
    darning *work* 织补
    drawn *work* 抽绣，抽纱；抽绣品，抽纱品
    eyelet *work* 网眼刺绣
    finish *work* 后整理工作（如熨烫等）
    hand *work* 〔工〕做手工（如缝珠片，钉钩眼扣等）
    patch *work* （将布片）拼缝；补花
    pattern *work* 提花织品
    punch *work* 抽绣

workbag [ˈwɜːkbæg] 针线袋；工具袋

workbasket [ˈwɜːkˌbɑːskɪt] 针线筐

workboots [ˈwɜːkbuːts] 〔复〕工作靴

workbox [ˈwɜːkbɒks] 针线盒；工具箱

workmanship [ˈwɜːkmənʃɪp] 手艺，做工，工艺；工艺品

    cutter's *workmanship* 裁剪师的手艺
    quality *workmanship* 做工精湛
    traditional *workmanship* 传统工艺

workwear [ˈwɜːkweə] 工作服
worsted [ˈwʊstɪd] 毛线；精纺毛纱；精纺毛织物
    fancy *worsted* 精纺花呢
wrap [ræp] 包卷；叠盖；（常用复数）外套；罩衫；披风；围巾；头巾
    bathing *wrap* 浴衣，浴袍
    button *wrap* 纽扣搭门
    evening *wrap* 晚会外套（披在晚礼服外）
    wrap over front 暗门襟
wrapover [ˈræpəʊvə] 裹身叠合式衣裙
wrappage [ˈræpeɪdʒ] 女便服；化装衣；围巾
wrapper [ˈræpə] 女晨衣；睡衣；童披风；宽袍
    tennis *wrapper* 网球外套
wrapping [ˈræpɪŋ] （常用复数）用于包裹材料
    leg *wrappings* 绑腿
wrapround [ˈræpraʊnd]（= wraparound [ˈræpəˌraʊnd]） 裹身叠合式衣裙（尤指裙）
wrapseam [ˈræpsiːm] 包缝，〔粤〕包骨
wrinkle [ˈrɪŋkl] 皱，皱褶
    diagonal *wrinkles* at sleeve cap 〔检〕袖山起皱
    diagonal *wrinkles* at sleeve lining 〔检〕袖里拧（袖里、面错位）
wrinkle resistance 抗皱性
wrinkles at collar band facing 〔检〕底领里起皱
wrinkles at front rise 〔检〕小裆不平
wrinkles at hem 〔检〕底边起绺
wrinkles at lapel 〔检〕驳头起皱
wrinkles at lower armhole 〔检〕抬根窝起皱
wrinkles at shoulder 〔检〕塌肩（衣服胸部起绺）
wrinkles at sleeve opening 〔检〕袖口起绺
wrinkles at top collar 〔检〕领面松
wrinkles at top fly 〔检〕门襟起皱
wrinkles at waistband facing 〔检〕腰缝起皱
wrinkles at zip fly 〔检〕绱拉链处起绺
wrist [rɪst] 手腕；（袖子、手套的）腕部
wristband [ˈrɪstbænd] （衬衫等的）袖口；护腕
wristlet [ˈrɪstlɪt] （保暖用）腕带；手镯；手环

X

XL(extra large 的略写) 特大号，加大码
X-leg [eksˈleg] 叉形腿
X-shape [eksˈʃeɪp] X 形；交叉形
X-shoulder [eksˈʃəʊldə] 肩宽（= across shoulder)
X-wool [eksˈwʊl] 64 支羊毛（高级美利奴羊毛）
XXL(extra extra large 的略写) 超特大号

Y

yard [jɑːd] 码(英美长度单位,1 码 = 3 英尺 = 36 英寸 = 0.9144 米);场地
    cloth *yard* 布码
    clothes *yard* 晒衣场
    yard goods 按码销售的布料

yardage [ˈjɑːdɪdʒ] 码数(以码测量的长度)
    piece *yardage* 单件耗料(码数)

yardstick [ˈjɑːdstɪk] 〔美〕码尺(指直尺)

yardwand [ˈjɑːdwɒnd] 〔主英〕码尺(指直尺)

yarmulke(= yarmulka) [ˈjɑːmʊlkə] 亚莫克便帽(犹太男人在祷告时所戴的圆形小便帽)

yarn [jɑːn] 纱,纱线,丝
    acetate rayon *yarn* 醋酸人造丝
    acrylic/cotton *yarn* 腈/棉纱
    acrylic *yarn* 腈纶纱
    air-jet texturing *yarn* 空气变形纱
    aloe fibre *yarn* 芦荟纤维纱
    alpaca *yarn* 羊驼毛纱
    angora *yarn* 兔毛纱
    asbestos *yarn* 石棉纱
    bamboo fibre *yarn* 竹纤维纱
    bamboo fibril *yarn* 竹原纤维纱
    black *yarn* 黑丝
    bleached *yarn* 漂白纱
    bulked *yarn* 膨体纱
    carded *yarn* 普梳纱
    cashmere *yarn* 羊绒纱,开司米纱
    chenille *yarn* 绳绒纱线,雪尼尔纱(音译);结子精纱毛纱
    coloured knops *yarn* 结子彩点纱
    combed *yarn* 精梳纱
    coolmax *yarn* 酷美丝纱(吸湿排汗)
    coolplus *yarn* 酷帛丝纱(吸湿排汗)
    cooltech *yarn* 新光合纤纱(吸湿排汗)
    corn fibre *yarn* 玉米纤维纱
    cotton/spandex *yarn* 棉/氨纶包芯纱
    cotton *yarn* 棉纱
    cotton, *yarn* and cloth 花、纱、布
    60 count *yarn* 60 支纱
    crepe *yarn* 绉纱
    CVC blended *yarn* CVC 混纺纱,棉/涤纱
    denim *yarn* 牛仔纱
    drill *yarn*-twisted 全线卡其,线卡
    duppion silk *yarn* 双宫丝
    dyed *yarn* 染色纱,色纱
    elastic *yarn* 弹力丝,弹力纱
    extra fine *yarn* 特细支纱,高支纱
    fashion(fancy) *yarn* 花式纱
    feather *yarn* 羽毛纱
    finished *yarn* 加工纱
    flax/cotton blended *yarn* (亚)麻/棉纱
    functional *yarn* 功能纱(具有阻燃、抗菌、保暖、吸湿排汗等功能)
    glass *yarn* 玻璃纱
    grey *yarn* 原纱;原丝;本色纱
    guanaco *yarn* 原驼毛纱

hand knitting yarn 毛线，绒线
hard twist yarn 强捻纱
hemp yarn 大麻纱
hemp/cotton blended yarn （大）麻/棉纱
hemp/soybean protein blended yarn 大麻大豆蛋白混纺纱
hemp/tencel blended yarn 大麻天丝混纺纱
hemp/yak hair blended yarn 大麻牦牛绒混纺纱
high count yarn 高支纱
hosiery yarn 针织纱
knitting yarn 针织纱
knot yarn 结子纱，粒子纱
lamb's wool yarn 羔羊毛纱，短毛混纺纱
lamé yarn 金银线
linen/cotton blended yarn （亚）麻/棉纱
linen(flax) yarn 亚麻纱
livery yarn 皱缩纱
llama yarn 美洲驼毛纱
luster yarn 有光毛纱
medium(coarse, fine) yarn 中（粗、细）支纱
melange yarn 混色纱
mercerized yarn 丝光纱线
metal yarn 金属纱线
mixed(blended) yarn 混纺纱
mohair yarn 马海毛纱
moisture wicking yarn 吸湿排汗纱，吸湿快干纱
nanofiber yarn 纳米纤维纱
necked yarn 竹节纱
noil silk yarn 䌷丝
nylon yarn 锦纶丝，锦纶纱

pearl yarn 珍珠纱
peppermint fibre yarn 薄荷纤维纱
plain yarn 素色纱，单色纱
ply yarn 合股纱
polyamide yarn 锦纶丝
polyester/viscose blended yarn 涤/黏纱
polyester yarn 涤纶丝，涤纶纱
rainbow yarn 彩虹纱，渐变纱
ramie/acrylic blended yarn （苎）麻/腈纱
ramie/cotton blended yarn （苎）麻/棉纱
ramie/polyester blended yarn （苎）麻/涤纱
ramie/silk blended yarn （苎）麻/丝纱
ramie/viscose blended yarn （苎）麻/黏纱
ramie/wool blended yarn （苎）麻/毛纱
ramie yarn 苎麻纱
R/C (rayon/cotton) blended yarn R/C 纱，黏/棉纱
rayon yarn 黏胶纤维纱
resist-dyed yarn 防染纱
seaweed(alginate)yarn 海藻纤维纱
Shetland yarn 雪特兰羊毛纱
silk yarn 丝线
silk/cotton blended yarn 丝/棉纱
single yarn 单股纱，单纱
slub yarn 竹节花式纱；粗节纱；粒纱
soybean fibre yarn 大豆纤维纱
space dyed yarn 间隔染色纱，多色纱，彩虹纱，花色纱
spandex yarn 氨纶纱；弹力纱；

弹力丝
spun rayon yarn 黏胶纤维纱；人造短纤纱
spun silk yarn 绢丝
spun yarn 精纺用纱，细纱；短纤纱
stretch yarn 弹力纱，弹力丝
synthetic yarn 合成纤维纱，合纤纱
T/C (polyester/cotton) blended yarn T/C 纱，涤/棉纱
tencel yarn 天丝纱
textured yarn 变形纱，变形丝
topcool yarn 远东合纤纱（吸湿排汗）
T/R (polyester/rayon) blended yarn T/R 纱，涤/黏纱，涤纶黏胶纤维混纺纱
triangle profile yarn 三角异形纱
truerun yarn 涤/棉纱
tussah noil yarn 柞䌷丝
tussah silk yarn 柞蚕丝
tussah spun silk yarn 柞绢丝
twist(filling) yarn 经（纬）纱
union yarn 混纺纱
vicuna yarn 骆马毛纱
vinylon/cotton blended yarn 维/棉纱
vinylon yarn 维纶纱
viscose/cotton blended yarn 黏/棉纱
viscose/rayon yarn 黏胶人造丝
voluminous yarn 膨体纱
vortex yarn 涡流纺纱（用涡流纺技术纺出的具有独特结构的新型纱线）

weft(warp) yarn 纬（经）纱
white steam filature yarn 白厂丝
wood fibre yarn 木纤维纱
wool/acrylic blended yarn 毛/腈纱
wool/nylon blended yarn 毛/锦纱
wool/polyester blended yarn 毛/涤纱
wool/silk blended yarn 毛/丝纱
wool/viscose blended yarn 毛/黏纱
wool yarn 毛纱
worsted (woolen) yarn 精（粗）纺毛纱
yak hair yarn 牦牛毛纱
yarn in hanks 绞纱
yarn number(count) 纱支
yarn on cones 筒子纱，筒纱
yashimagh [ˈjæʃmæg] 亚希玛戈浮纹布（伊拉克人所穿斗篷装用）
yashmak(=yashmac) [ˈjɑːʃmæk] 双重面纱（穆斯林妇女戴）
yé-yé [jeje] 〔法〕（20 世纪 60 年代流行于法国的）耶耶派服式
Y-fronts [waɪ frʌnts] （前缝呈倒 Y 形的）男内裤，三角裤
yoke [jəʊk] （衣服）过肩，抵肩，披肩，托肩，覆肩，覆势，约克（育克，音译），〔粤〕担干；裤或裙腰部的拼接布，〔粤〕机头
   front(back) yoke 前（后）过肩
   neck yoke 项圈，领座圈
   two(one) layer yoke 双（单）层过肩
   uneven back yoke 〔检〕（裤、裙）拼腰大小不匀
   yoke slide insert 过肩衬垫

Z

zamarra [zəˈmɑːrə] 羊皮袄（牧民穿用）
zardozi [zəˈdɒsɪ] 〔印度〕金银线刺绣
zari [ʒəˈriː] 〔印度〕（服装上的）装饰金线
zeck [zek] 条子斜纹衬里（外套，风雨衣）
zephyr [ˈzefə] 轻薄织物；轻罗；细毛线
 zephyr worsted 轻软精纺毛线
zibeline（=zibelline）[ˈzɪbəlaɪn] 黑貂皮；齐贝林有光长绒呢；绒结子毛纱
zigzag [ˈzɪɡzæɡ] 之字形，人字形，Z形，曲折形，锯齿形；跳针
zigzagger [ˈzɪɡzæɡə] 曲折缝缝纫机
zimarra [zɪˈmɑːrə] 教士长袍
zip（zipper 的略写）[zɪp] 拉链
zip-fastener [ˈzɪpˌfɑːsənə] 拉链
zipper [ˈzɪpə] 拉链
 airtight *zipper* 气密拉链（高度密封防水）
 aluminium *zipper* 铝拉链
 anti-flaming *zipper* 阻燃拉链（防火服用）
 autolock slider *zipper* 自动头拉链
 bias *zipper* 斜拉链
 brass *zipper* 黄铜拉链
 bronze *zipper* 青铜拉链，古铜拉链
 closed-end *zipper* 密尾拉链（尾部封死不分开），闭口拉链
 coil *zipper* 环扣（式）拉链
 copper-plated *zipper* 镀铜拉链
 decorative *zipper* 装饰拉链（仅起装饰作用）
 exposed *zipper* 外露拉链，明拉链
 front fly *zipper* （衣）门襟拉链；（裤）前裆拉链
 hidden(concealed) *zipper* 暗缝拉链，隐形拉链
 invisible *zipper* 隐形拉链，暗缝拉链
 light(heavy) *zipper* 细（粗）齿拉链，小（大）齿拉链
 long chain *zipper* 加长拉链（夹克活动内里用的拉链）
 magnetic *zipper* 磁吸拉链（有磁性的拉链头靠近后自动合上，可单手操作）
 metal *zipper* 金属拉链
 moveable open-end *zipper* 双开拉链，双头拉链
 nickel teeth *zipper* 镍齿拉链
 nylon *zipper* 尼龙拉链
 plastic(delrin) *zipper* 塑料拉链
 polyester *zipper* 涤纶拉链
 resin *zipper* 树脂拉链
 resin silver(gold) teeth *zipper* 树脂银（金）齿拉链
 resin transparent (translucent) *zipper* 树脂透明（半透明）拉链
 sealed *zipper* 密封拉链，闭口拉链
 split(opened-end) *zipper* （尾部）分开拉链，开口拉链，开尾拉链
 two way *zipper* 双向拉链，双头拉链
 underarm *zipper* （连衣裙等）腋下拉链

water-proof *zipper* 防水拉链
wave *zipper* 〔检〕拉链不平伏,〔粤〕拉链起蛇
YKK *zipper* （日本名牌）YKK 拉链
zinc alloy *zipper* 锌合金拉链
zipper assembler 缝拉链的人；缝拉链机
zipper changing 换拉链
zipper is not moveable 〔检〕拉链不能开合
zipper slider(puller) 拉链头，链头
zipper teeth 拉链齿，链齿
zipper with double sliders 双头拉链
zipper with semi-autolock slider 半自动头拉链
zone [zəʊn] （古语）腰带；带子

zoo-suit [ˈzuːˈsjuːt] 宽肥套装（衣宽裤肥，脚口收紧）
zootsuiter [ˈzuːtsuːtə] 穿阻特装的人（比喻衣着时髦的人，尤指廉价服装）
zori [ˈzɔːrɪ] （日本）草履，拖鞋（平底有带的拖鞋）
zoster [ˈzɒstə] （古希腊的）腰带，束带
Zouave [zuːˈɑːv] 佐阿夫女长裤（裤管上宽下窄，至踝部收紧）；短至膝上的灯笼裤（早年法国轻步兵穿用）；女式绣花短上衣
zucchetto [tsʊˈketəʊ] （天主教戴）室内圆形小便帽（神父用黑色，主教用紫色，红衣主教用红色，教皇用白色）

附 录
Appendix

一、颜色
Colour

1. 汉英顺序

①红色类

红色	red;blush
朱红	vermeil;vermilion;ponceau;cinnabar;framboise
朱砂色，丹色	minium;vermilion
粉红	pink;soft red;rose bloom;blossom;petal
淡粉红	pale pink;soft pink;pink tint
灰粉红	dusty pink
深粉红	deep pink;radiance;clementine;framboise
鲜粉红	shocking pink;hot pink;pop pink
浅粉红	baby pink;chalk pink
妃红，妃色	light pink
浮雕石粉红	cameo pink
晚霞粉红	sunset pink
朝霞粉红	aurora pink
朝霞红	aurora red
水红	hot red
梅红	plum;crimson;fuchsia red
深梅红	deep plum
鲜梅红	berry
玫瑰红	rose madder;rose;damask;blush
鲜玫瑰红	bright rose
浅玫瑰红	light rose;misty rose;rosebud;rose pink
褐玫瑰红	rosy brown
灰玫瑰红	old rose
水晶玫瑰红	crystal rose
奶油玫瑰红	cream rose
浮雕石玫瑰红	cameo rose
粉玫瑰红	misty rose
桃红	peach blossom;peach red;peach-puff;peach;carmine rose
樱桃红	cherry;cerise
橘红	orange red;reddish orange;tangerine;jacinth;salmon pink;salmon;coral haze
浅橘红	sunset
鲜橘红	nacarat
朝霞橘红	aurora orange
石榴红	garnet;garnet red;garnet rose;pomegranate red
石榴汁红	grenadine red
枣红	purplish red;jujube red;date red;bay
酸果蔓红	cranberry
杏红	apricot blush
柿子红	persimmon red
草莓红	strawberry red
木莓红	raspberry
浆果红	red berry;berry
莲红	lotus red
浅莲红	fuchsia pink
深莲红	fuchsia red
豆沙红	russet red
豇豆红	bean red
辣椒红	capsicum red;cayenne red;paprika;pimiento
番茄红	tomato red
高粱红	Kaoliang red

红色类

芙蓉红　hibiscus red; poppy red; poppy
牡丹红　peony red
海棠红　begonia red
夹竹桃红　phlox red; phloxine
天竺葵红　geranium
茶藨子浆果红　currant red; currant
丁香红　lilac rose
山茶红　camellia red; camellia
罂粟红　poppy red; poppy
胭脂树红　achiote
胭脂红　rouge red; carmine; cochineal; coccinellin; nacarat; lake red; lake
印度胭脂红　Indian lake
鲑肉红　salmon pink; salmon
暗鲑肉红　dark salmon
淡鲑肉红　light salmon
火烈鸟红　flamingo red; flamingo
龙虾红　lobster red
玳瑁红　hawksbill turtle red
宝石红　ruby red; rubine
玛瑙红　agate red
珍珠红　pearl blush
浅珍珠红　pearl pink
尖晶石红　spinel red
珊瑚红　coral
浅珊瑚红　light coral
祭祀红　sacrificial red
烛光红　candle light peach
火红　fire red; fiery red; flame scarlet
亮红　bright red
银红　silver red
铜红　bronze red; copper red
铁红　iron oxide red
铁锈红　rust red; ferruginous
铬红　chrome red
镉红　cadmium red
砖红　brick red

火砖红　firebrick red; firebrick
深砖红　dark brick
瓦红　tile red
土红　laterite; reddle; earth red
郎窑红　lang-kiln red
均红　Jun-kiln red
釉底红　underglaze red
陶嫣红　pottery red
威尼斯红　Venetian red
法国红　French vermilion; French rose
西班牙红　Spanish red
土耳其红　Turkey red
中国红　Chinese red
印度红　Indian red
（墨西哥）阿兹特克红　Aztec red
茜红，茜色　alizarin red; alizarin crimson; madder red
洋红　carmine; magenta; tyraline
品红　pinkish red; magenta; fuchsin(e); solferino; tyraline; rubin
浅品红　magenta haze
猩红　scarlet red; scarlet; blood red; crimson
油红　oil red
奶油红　cream red
（葡萄）酒红　wine red; mauve wine; bordeaux; shiraz; burgundy
紫檀红　rosewood red
紫红　purplish red; madder red; wine red; imperial red; plum purple; magenta pink; magenta; plum, wine; carmine; amaranth; claret; fuchsia; mauve; damask; murrey; raspberry; beetroot
玫瑰紫红　rose carmine; rose mauve
优品紫红　opera mauve
深紫红　deep claret; deep mauve; gridelin; prune; mulberry; berry

鲜紫红　mauve glow; solferino
浅紫红　light mauve; light berry; dahlia; heliotrope
暗紫红　old mauve; bordeaux; burgundy; dark magenta
淡紫红　pale mauve; bright mauve; heather; lavender blush
深藕红　conch shell
淡藕红　cloud pink
棕红　henna; oxblood red; oxblood
深棕红　Aztec red
赭红　burnt henna
褐红　maroon; redexide; brick red
黑红　black red; black currant
殷红　blackish red; black red; dark red
浅蓝红　bluish red
灰红　gray red
暗红　dark red; dull red; breton red; dirty red; alizarin crimson; garnet; burgundy; shiraz
鲜红　scarlet red; scarlet; bright red; fresh red; blood red; shocking red; strong red; madder; ruby; cerise; cherry; florid
大红　oriental red; jester red; light scarlet; bright red
嫣红　bright red
血红　blood red; incarnadine; sanguine
牛血红　oxblood red; oxblood
血牙红　shell pink; peach beige
绯红　scarlet; crimson; geranium pink
米红　silver pink
深红　deep red; dark red; royal red; crimson; cardinal; cranberry; garnet; sardonyx; siam
淡红　light red; subtle red; light siam carnation

② 橙色类

橙色　orange; orange-chestnut; China orange; mandarin orange
铬橙　chrome orange
红锆石橙　jacint
野苹果橙　crab-apple
南瓜橙　pumpkin
瓜瓢橙　melon-based orange
蜜露橙　honeydew
蜂蜜橙　honey orange
柿子橙　persimmon orange
辣椒橙　chili orange
藏红花橙　crocus
波斯橙　Persian orange
雅法橙　Jaffa orange
救援橙　rescue orange
全浓橙　full strength orange
焦橙　burnt orange; burnt pumpkin
热带橙　tropical orange
日光橙　sun orange
亮光橙　brilliant orange
红橙　aurora(orange); orange red
浅红橙　peach; melon; grenadine red
灰橙　dusty orange
深橙　deep orange
浅橙　clear orange; honeydew
淡橙　light orange; jacinthe
淡白橙　pale orange

③ 黄色类

黄色　yellow
橘黄　orange; crocus; gamboge; cadmium orange; saffron
深橘黄　deep orange
暗橘黄　dark orange
浅橘黄　clear orange; light orange; rattan

黄色类

柑橘黄	citrus
焦橘黄	burnt orange
酸橙黄	lime yellow
橙杏黄	warm apricot
香橼黄	citron
香蕉黄	banana
杧果黄	mango
蜂蜜黄	honey yellow
黄油黄	butter
酒黄	wine yellow
柠檬黄	lemon yellow; lemon; cadmium yellow lemon; citrine
浅柠檬黄	light lemon
柠檬皮黄	zest
玉米黄	maize
橄榄黄	olive yellow; golden olive
樱草黄	primrose yellow
稻草黄	straw yellow; golden straw
香草黄	vanilla
含羞草黄，缃色（黄）	mimosa yellow
万寿菊黄	marigold
毛茛黄	buttercup
茉莉黄	jasmine
竹黄	bamboo yellow
柳黄	willow yellow
黄栌黄	smoke tree yellow
洋艾黄	absinthe yellow
洋槐黄	acacia yellow
水仙黄	daffodil
郁金香黄	tulip yellow
甘草黄	liquorice; licorice
亚麻黄	flaxen
姜黄	ginger
芥末黄	mustard; mustard yellow; karashi
杏黄	apricot; apricot buff; bronze yellow; cracker khaki
蛋黄	vitelline; yolk yellow; egg yellow
藤黄	rattan yellow
鳝鱼黄	eel yellow
象牙黄	ivory; ivory yellow
象牙奶油色	ivory cream
象牙淡黄	ivory deep
珍珠黄	pearl yellow
蜡黄	wax yellow
曙光黄	aurora yellow
雾黄	misty yellow
日光黄	sunny yellow
月光黄	moon yellow
荧光黄	lucifer yellow
石黄	mineral yellow
赭石黄	ocher yellow
土黄	earth yellow; yellowish brown; yellow ocher; golden apricot; khaki; citrus
深土黄	oxford ocher
暗土黄	dark khaki
浅土黄	pale khaki; natural
砂黄	sand yellow
金黄	golden yellow; gold; titian; sunshine
淡金黄	pale goldenrod
深金黄	dark goldenrod
奶油金黄	cream gold
淡金黄	platinum blond
铁黄	iron oxide yellow; iron buff
镉黄	cadmium yellow
铬黄	chrome yellow
钴黄	cobalt yellow
巴黎黄	Paris yellow
帝国黄	empire yellow
中国黄	Chinese yellow
南京黄	nankeen yellow
明瓷黄	Ming yellow
深黄，暗黄	deep yellow
浓黄	reddish yellow; goldenrod
苍黄	sallow

棕黄　tan
浅棕黄　light tan
淡棕黄　colonial yellow;pebble
奶油棕黄　cream tan
卡其棕黄　khaki tan
印度棕黄　Indian tan
印度黄　Indian yellow
青黄　bluish yellow
灰黄　isabella;sallow;grey yellow;beige; bisque;lark;maple;wheaten
米黄　apricot cream;cream;beige;onix; poloere
浅米黄　golden fleece
嫩黄　yellow cream;tender yellow
粉黄　lemonchiffon
鲜黄　cadmium yellow; canary; canary yellow
鹅黄　light yellow
雄黄　cock yellow
雌黄　king's gold; king's yellow; royal yellow
硫黄黄　sulfur yellow
中黄　medium yellow;yellow mid
浅黄　light yellow;pale yellow;buff;wheat
秋香黄　deep yellow
淡黄　pastel yellow; soft yellow; cream; nankeen;jasmin(e);primrose;cornsilk

④绿色类

绿色　green
豆绿　pea green;bean green;mistletoe
浅豆绿　light bean green;asparagus green
橄榄绿　olive green;olive;surplus green
淡橄榄绿　olive green clear
暗橄榄绿　olive green deep
墨橄榄绿　olive green black
中橄榄绿　mid-olive
深橄榄绿　dark olive
浅橄榄绿　light olive green
焦橄榄色　burnt olive
橄榄军服绿　olive drab
青梅绿　plum green
茉莉绿　jasmin(e) green
莴苣绿　lettuce green
韭葱绿　leek green
芹菜绿　celery green
欧芹绿　parsley green
查特酒绿　chartrellse
茶绿　tea green;celandine green;plantation
葱绿　onion green;pale green
苹果绿　apple green;granny smith
酸橙绿　lime green;lime
柑橘绿　citrus green
香蕉绿　green banana
梨绿　pear green;pear
春天绿　spring green
冬天绿　winter green
雾绿　fog green;misty green;green mist
原野绿　field green
森林绿　forest green
丛林绿　jungle green
松树绿　pine green
雪松绿　cedar green;cedar
柳树绿　willow green
榆树绿　elm green
柏树绿　cypress green
桉树绿　eucalyptus
常青树绿　evergreen
常青藤绿　ivy green
洋蓟绿　artichoke green
苔藓绿　moss green;moss;bracken green
海藻绿　seaweed green
马尾藻绿　sargasso green

绿色类

草地绿，草绿 grass green; meadow green; lawn green; olive drab; blue grass
水草绿 water grass green
深草绿 jungle green
三叶草绿 clover
草皮绿 turf
芽绿 bud green
叶绿 leaf green; foliage green
嫩叶绿 fresh leaves
竹绿 bamboo green
芦苇绿 reed green
芦荟绿 aloe green; aloe
仙人掌绿 cactus green; cactus
薄荷绿 mint green; mint
洋艾绿 absinthe green
灰湖绿 agate green
水绿 aqua green; aqua; water green
雾水绿 aqua haze
海水绿 marine green; sea green
海沫绿 seafoam green; seafoam
海洋绿 ocean green; neptune green
大西洋绿 atlantic green
浅大西洋绿 light atlantic green
冰河淡蓝绿 glacier
尼罗河绿 Nile green
湖绿 lake green
水塘绿 pool green
酸性绿 acid green
水晶绿 crystal green
玛瑙绿 agate green
翡翠绿 imperial green
玉绿 jade green; jade
深玉绿 dark jade
石绿 mineral green
松石绿 spearmint; viridis; turquoise
萤石绿 fluorite green
浮雕石绿 cameo green

石板绿 slate green
青瓷绿 celadon green; celadon
铜绿 copper green; blue verditer; verdigris
古铜绿 bronze green
铜锈绿 patina green
镍绿 nickel green
镉绿 cadmium green
铬绿 chrome green; viridian
钴绿 cobalt green
孔雀绿 peacock green
鹦鹉绿 parrot green
龟绿 turtle green
威尼斯绿 Venetian green
巴黎绿 Paris green; king's green
西班牙绿 Spanish green
爱尔兰绿 Irish green
凯利绿 Kelly green; Kelly
澳门绿 Macao green
中国绿 Chinese green
明瓷绿 Ming green
墨绿 blackish green; green black; jasper; dark green; deep green
黑玉绿 emerald black
黛绿 dark green
青黛绿 chateau green
深绿 dark green; Chinese green; bottle green; invisible green; August green
暗绿 sap green; dark green; deep green; dull green; dirty green
油绿 glossy dark green
青绿，青碧 blue green; bluish green; turquoise blue; turquoise; aeruginous; verdure; aquamarine
中青绿 medium turquoise
绀青绿 ultramarine green
暗青绿 gobelin blue; dark turquoise

碧绿，碧色　azure green; turquoise green; viridity; aquamarine
淡碧绿　light turquoise green
翠绿　emerald green; jade green; bright green; aloe green; verdancy; viridity; kingfisher
深翠绿　viridian
蓝绿　blue green; aquamarine; aquamarine blue; teal blue; pacific blue; pacific; harbour blue
浅蓝绿　aqua; light aqua
黄绿　yellow green; chartreuse; pea green; green almond
浅黄绿　light chartreuse; aqua green; off olive
鲜黄绿　Kelly green; Kelly
暗黄绿　lime green; lime
淡黄绿　pistachio
金黄绿　golden green
灰绿　grey green; sage green; dusky green; hedge green; slate green; mignonette; sea spray; celadon; reseda
银灰绿　silver green; silver sage
石墨灰绿　graphite blue
淡灰绿　petrol
褐绿　breen; sargasso green
土绿　terra verde〔拉〕
蟹绿　balsam
乳白绿　milky green
品绿　light green; malachite green
鲜绿　green clear; emerald green; emerald; vivid green
嫩绿　pomona green; verdancy; spring green; shoot green; tender green
中绿　medium green; golf green
浅绿　light green; mint green; green tint; minize; minze; bud green; aqua

淡绿　pale green; pastel green; soft green; viridescent

⑤青色类

青色　cerulean blue; blue; green; cyan
豆青　pea green; bean green
花青　flower blue
茶青　tea green; olive green; olive
竹青　blue green
梅青　plum green
葱青　onion green
翠青　bright green
天青　celeste; azure; reddish black
霁青　sky-clearing blue
石青　stone blue; mineral blue
铁青　electric blue; river blue
钢青　steel blue
蟹青　turquoise; ink blue; greenish-grey; teal blue
鳝鱼青　eel green; eel bluish
蛋青　egg blue
鸭蛋青　duck egg blue
影青　misty blue; white blue; bluish white
烟青　smoky blue
黛青　dark blue
群青，佛青　ultramarine
暗青　dark blue; deep cerulean; dark cyan
玄青　dark cyan
藏青　navy blue; dark blue; Ming blue
靛青　indigo
绿青　academy blue
红青，绀青　prune purple; dark purple
灰青　dark bluish grey
大青　smalt
粉青　light greenish blue
鲜青　clear cerulean
浅青　light blue; light cerulean

淡青　pale cerulean; light greenish blue; light cyan

⑥蓝色类

蓝色　blue
天蓝　sky blue; azure; celeste; azure cerulean blue; cerulean blue; Parisian blue; bristol
深天蓝　deep sky blue
墨天蓝　azure black
暗天蓝　azure deep
亮天蓝　light sky blue
天青蓝　azure blue; cerulean blue
蔚蓝　azure blue; azure; sky blue
黎明蓝　dawn blue
月光蓝　moon blue
午夜蓝　midnight blue
太空蓝　ethereal blue
海洋蓝　ocean blue
大西洋蓝　atlantic blue
海蓝　sea blue; marine blue
爱琴海蓝　Aegean blue
加勒比海蓝　Caribbean blue; Caribbean
河蓝　river blue
尼罗河蓝　Nile blue
湖蓝　acid blue; lake blue
深湖蓝　vivid blue
中湖蓝　bright blue
浅湖蓝　canal blue
礁湖蓝　lagoon
水塘蓝　pool blue
清水蓝　water blue; aqua blue
淡水蓝　aqua
深水蓝　bluebird; turquoise
月桂树蓝　daphne blue
风铃草蓝　campanula blue
鼠尾草蓝　salvia blue
薰衣草蓝　lavender blue
矢车菊蓝　cornflower blue; cornflower
冰蓝　ice blue
冰雪蓝　ice-snow blue
冰山蓝　iceberg
云蓝　cloud blue
雾蓝　fog blue; misty blue; blue haze
汽雾蓝　vapour blue
苍蓝　horizon blue
孔雀蓝　peacock blue
青鸟蓝　bluebird
水鸭蓝　teal blue
珐琅蓝　enamel blue
景泰蓝　imperial blue
瓷蓝　porcelain blue
韦奇伍德陶瓷蓝　Wedgwood blue
宝石蓝　sapphire; jewelry blue
鲜宝石蓝　bright sapphire
水晶蓝　crystal blue
浮雕石蓝　cameo blue
松石蓝　turquoise blue
天青石蓝　lapis lazuli blue
石板蓝　slate blue
粉末蓝　powder blue
青铜蓝　bronze blue
铜蓝　indigo copper
钢蓝　steel blue
淡钢蓝　light steel blue
铁蓝　iron blue
钴蓝　cobalt blue; king's blue; zaffer blue
鲜钴蓝　bright cobalt
锰蓝　manganese blue
皇家蓝　Royal blue
帝国蓝　empire blue
道奇蓝　Dodger blue
国际奇连蓝　international klein blue
波斯蓝　Persian blue

萨克斯蓝	Saxe blue
爱丽丝蓝	Alice blue
普鲁士蓝	Prussian blue
牛津蓝	Oxford blue
耶鲁蓝	Yale blue
卡普里蓝	Capri blue
巴哈马蓝	Bahama blue
威尼斯蓝	Venetian blue
土耳其蓝	Turkey blue
柏林蓝	Berlin blue
法国蓝	French blue
中国蓝	Chinese blue
明瓷蓝	Ming blue
北京蓝	Beijing blue;Peking blue
士林蓝	indanthrene blue
古典蓝	classic blue
公主蓝	princess blue
品蓝	reddish blue;royal blue;king's blue;blue purple
靛蓝	indigo;indigo blue;benzo blue
深(中、浅)靛蓝	dark(mid,light) indigo
古旧靛蓝	antique indigo
赭色靛蓝	rust indigo
菘蓝	woaded blue
石磨蓝	stone-washed indigo
原色蓝	cyanine blue
阴影蓝	blue shadow
清澈蓝	serene blue
龙胆蓝	qentian blue
藏蓝	purplish blue;navy blue;navy
海军蓝	navy blue;navy
法国海军蓝	French navy
空军蓝	airforce blue;airforce
水兵蓝	marine blue
宝蓝	royal blue;vivid turquoise;sapphire blue;sapphire
浅宝蓝	bristol
墨水蓝	ink blue
墨蓝	blue black;midnight navy
黛蓝	dark blue
绿蓝	turquoise blue;cyan;robin's egg blue
紫蓝	hyacinth;purplish blue;Lyons blue;enamel blue;cornflower blue;cornflower
射光紫蓝	bronze violet
淡紫蓝	lavender blue
浅紫蓝	Dutch blue;mauve orchid;periwinkle
青蓝	ultramarine;cyan blue
翠蓝	bright blue;kingfisher
灰蓝,军校蓝	cadet blue
深灰蓝	blue ashes;charcoal blue
暗灰蓝	slate blue;metropolitan
中灰蓝	midnight blue
淡灰蓝	glaucous;chambray
银灰蓝	silver blue
深蓝	deep blue;dark blue;navy blue;mandarin blue;midnight blue;cyan blue;Antwerp blue;mazarine;smalt;ultramarine
暗蓝	deep blue;dark blue;dull blue;dirty blue;oriental blue;wan blue
鲜蓝	blue clear;vivid blue
浓蓝	strong blue
粉蓝	lily sky
纯蓝	true blue
乳白蓝	milky blue;opal blue
中蓝	medium blue;azure blue;jay
浅蓝	light blue;chalk blue;baby blue;calamine blue;beryl
淡蓝	pale blue;baby blue;calamine blue;pastel blue

⑦ 紫色类

紫色	purple;purpl;violet

紫罗兰色　violet
三色紫罗兰　pansy
中国紫罗兰　Chinese violet
印度紫　Indian purple
紫藤色　lilac;wistaria
深紫藤色　deep wistaria
浅紫藤色　lilac clear
淡紫藤色　lilac pale
紫晶色　rose amethyst;amethyst
深紫晶色　dark amethyst
葡萄紫　grape
茄皮紫　aubergine
玫瑰紫　rose violet
丁香紫　lilac
淡丁香紫　pail lilac
风铃草紫　campanula purple
薰衣草紫　lavender
兰花紫　orchid
深兰花紫　dark orchid
中兰花紫　medium orchid
丁香花紫　clove
木槿紫　mauve
铁线莲紫　clematis
缬草紫　heliotrope
锦葵紫　mallow purple;mallow
古紫色　antique violet
龙胆紫　gentian violet
钴紫　cobalt violet
矿紫,石紫　mineral violet
墨紫　violet black
绛紫,酱紫　dark reddish purple
黛紫　dark purple
暗紫　violet deep;dull purple;dirty purple;dusky orchid;imperial purple;damson;admiral;petunia
乌紫　raisin
青紫　bruised purple;cyanosis

红紫　red violet;red purple
洋红紫　magenta purple
泰尔红紫　Tyrian purple
蓝紫　blue violet;royal purple
鲜紫　violet light
深紫　dark purple;deep claret;amaranth;modena;ultraviolet
中紫　medium purple
浅紫　grey violet;chalk violet;light purple;dry rose
淡紫　pale purple;pastel lilac;lavender;lilac;orchid;palatinate;lias;heliotrope;thistle
淡白紫　violet ash
淡灰紫　dusty lavender;mineral violet;grayish purple
红灰紫　dusty mauve
紫莲　violet lotus
青莲　pale purple;heliotrope
深青莲　amaranth purple
雪青　lilac;purple drab
浅雪青　chalk violet;pink lavender
墨绛红　purple black
暗绛红　purple deep
浅绛红　purple light
朦胧紫　twilight purple

⑧黑色类

黑色　black
土黑　black earth;terra nera〔拉〕
煤黑　coal black;coal
煤烟黑　soot black
煤玉黑　jet
炭黑　carbon black;charcoal black
石板黑　slate black
古铜黑　bronze black
铁黑　iron oxide black;iron black

铂黑　platinum black
洋苏木黑　logwood black
乌木黑　ebony
骨黑　bone black
橄榄黑　olive black
栗黑　chestnut black
蓝黑　blue black
紫黑　purple black; violet black; blueberry; admiral
灰黑　grey black; slate black
棕黑　sepia; brown black
青黑，黛色　lividity
漆黑　pitch-black; pitch-dark; inkiness; lake black; midnight black
乌黑　jet black; ebony
深黑　deep black; dark black; pitch black
暗黑　black dull; dull black; dark black
玄色　dark black

⑨ 白色类

白色　white
象牙白　ivory white; ivory
牡蛎白　oyster white
鱼肚白　fishbelly white
珍珠白　pearl white; gray lily
奶油珍珠白　cream pearl
奶油白　cream white; crystal cream
杏仁白　blanched almond
玉石白　jade white
雪花石膏白　alabaster white; alabaster
花白　flower white
骨白　bone
银白　silver white; silver
铅白　flake white; lead white; ceruse white
锌白　zinc white
锌钡白　lithopone
羊毛白　wool white
米白　off-white; shell
乳白　milk white; cream white; crystal cream; ivory white; ivory
薄荷乳白　mint cream
雪白　snow-white; snowy white
灰白　greyish white; off-white; chalky; canescence; hoar; lyart; pallar; pale
烟白　white smoke
花白　flower white
青白　bluish white; misty blue
(印第安) 那伐鹤白　Navaho white
月光白　moon white
幽灵白　ghost white
云白　cloud cream
冬天白　winter white
淡白　whitish
浊白　milkiness
纯白　clear white; crisp-white; pure white
精白　pure white; crisp white
本白；缟色　raw white; off-white
暖白色　warm white
蜡白色　white wax
老白色　old lace
粉红白　pinky white
淡紫白　lilac white
荧光白　fluorescent white
亮白　bright white
古典白　antique white
中国白　Chinese white
巴黎白　Paris white

⑩ 灰色类

灰色　grey; gray; grizzle
银灰　silver grey; chinchilla; grey morn; platinum
铁灰　iron grey; gunmetal grey
铅灰　lividity; leaden grey; lead grey

灰色类

锡镴灰	pewter grey; pewter
镍灰	nickel
锌灰	zinc
金属灰	metallic grey
玛瑙灰	agate grey
炭灰	charcoal grey; charcoal marl
驼灰	doe; doe marl
豆灰	rose dust
藕色	pale pinkish grey
藕灰	zephyr
莲灰	elderberry
浅莲灰	pale lilac
鸽子灰	dove grey; pigeon grey
海鸥灰	gull grey
雀灰	sparrow
羽灰	feather grey
大象灰	elephant grey
松鼠灰	squirrel grey
鼠灰	mouse grey; mouse
蟹灰	storm blue
天灰	sky grey
黎明灰	dawn grey
水灰	aqua grey
风景灰	storm grey
雨云灰	nimbus grey
云灰	cloud grey
石墨灰	graphite
石板灰，瓦灰	slate grey
暗瓦灰	dark slate grey
淡瓦灰	light slate grey
花岗石灰	granite grey
矿石灰	mineral grey
云石灰	sierra grey
土灰	dusty grey
泥灰	marl; grey marl
水泥灰	concrete grey
烟灰	smoky grey; ash
伦敦烟灰	London smoke
雾灰	misty grey; gray mist
玉灰	jade grey
水晶灰	crystal grey
黑灰	grey black; charcoal grey; black ash; gunmetal grey; gunmetal
墨灰	graphite
棕灰	beige
红光灰	reddish grey
黄光灰	yellowish grey
紫灰	purple grey; purple marl; cadet; dove grey; lavender grey
深紫灰	heron
淡紫灰	lilac grey
浅绿灰	eucalyptus
米灰	oyster grey; oatmeal
深米灰	dark oatmeal
浅米灰	moon light; moon beam
卡其灰	khaki grey
蓝灰	blue grey; slate; steel grey; pike grey; cadet grey; Russian blue; pearl
青灰	lividity; steel grey; powder grey; pewter grey; stale grey; balsam green
深青灰	dusky green blue
白灰	pale grey
乳白灰	opal grey
军装灰	field grey
橄榄灰	olive grey
贝壳灰	shell grey
珠光灰	pearl grey
深灰	dark grey; deep grey; dim grey; clerical grey; Oxford grey
暗灰	dull grey; dark grey; dirty grey; gray deep; sombre grey; dim grey
冷灰	cool grey
中灰	medium grey
浅灰	light grey; ash grey; French grey

淡灰　ash; ashes; gray pale; soft grey; gainsboro
暖灰　warm grey

⑪棕色类及其他

棕色，褐色　brown
红棕　umber; chili; sorrel; garnet brown; ginger spice
淡红棕　rose tan
黄棕　cocoa brown; cocoa; maple sugar; rame; spice
墨棕　roman sepia
深青棕　bark
玛瑙棕　agate
金棕　auburn
铁棕　iron brown; vandyke brown
铁锈棕　rustic brown
橘棕　orange brown
橄榄棕　olive brown
芥末棕　mustard brown
栗棕　ginger snap
驼棕　camel tan
土狼棕　coyote brown
无花果棕　fig brown
可可棕　cocoa brown
椰棕　coconut brown
秋叶棕　autumn brown
烟叶棕　tobacco brown
哈瓦那雪茄棕　Havana brown
胡桃棕　walnut brown
玫瑰棕　rosy brown
马鞍棕　saddle brown
页岩棕　shale
浮雕石棕　cameo brown
石棕　brown stone
陨石棕　meteorite
沙棕，沙褐　sandy brown; sand beige

骨棕　bone brown
羊皮纸浅棕　parchment
湿沙浅棕　wetsand
赤褐　sorrel; chocolate; mahogany; maroon; terra-cotta; hazel nut; ferruginous; rubiginous; rufous; russet
棕褐　summer tan; tan; burnt almond; swamp
茶褐　auburn; umber; dark brown
黑褐　black brown; cola
炭褐　charcoal brown
紫褐　puce
黄褐　drab; fulvous; cinnamon; ocher; tawny; russet brown; tan; cocoa; filemot; fulvous; mulatto; khaki
淡黄褐　fawn; palomino
金黄褐　gold brown; golden ocher
玳瑁黄褐　tortoise shell
栗褐　chestnut brown
灰褐　taupe; mouse; greige beige; rose beige; dust-colour; honeycomb
浅灰褐　putty; light taupe
橙褐　orange brown
土褐　soil brown; clay
深褐　dark brown; deep brown; nigger-brown; chocolate; bistre; burnt sienna; mocha; carob; husk; mink
暗褐　dull brown; fuscous; dun; sad brown; fuscous; seal
重褐　saddle brown
中褐　toffee brown; mid-brown
浅褐　sandy beige; mushroom
淡褐　light brown; caramel; hazel; biscuit; bisque; drab; ecru
咖啡色　coffee; fudge; espresso; global brown
摩卡咖啡色　mocha; moka
浓咖啡　espresso
深咖啡色　autumn mink

棕色类及其他

拿铁咖啡色　latte
可可色　cocoa;cacao brown
巧克力色　chocolate
奶油巧克力色　milk chocolate
酱色　caramel;reddish brown
紫酱色　marron;mauve wine
猪肝色　hepatic;auburn;liver brown
茶色　tea; umber; dun; dark brown; fulvous;tawny
浓茶色　umber
焦茶色　dun
淡茶色　sandy
赭色　ocher; auburn; chocolate; sienna; umber;rust;reddish brown;ruddle
深赭色　burnt umber
琥珀色　amber;succinite
栗色　chestnut; sorrel; marron; spadiceous; bark
深栗色　liver chestnut
鹿皮色　moccasin
金色　gold;gilt
古金色　old gold
黯金色　matte gold
暗金色　dark blonde;dark gold
深金色　dark gold;dark blonde
浅金色　light gold
珠光金色　pearlized gold;pearl gold
玫瑰金色　rose gold
绿金色　green gold
紫金色　purple gold
黑金色　black gold
白金色　white gold
铂金色　platinum
银色　silver;argent
珍珠银色　pearlized silver
雾银色　silver mist
哑银色　matte silver;dull silver

黑银色　black silver
墨灰银　graphite silver
铅色　lividity;lead
锌色　zinc
铬色　chrome
铁锈色　rust;rust brown
铜色　copper
青铜色　bronze;aeneous;gilt
青古铜色　antique bronze;bronze
黑古铜色　dark bronze
暗古铜色　dull antique
紫铜色　purple bronze
黄铜色　brassiness;brazen yellow
炮铜色　gunmetal
水银色　mercury
暖钴色　warm cobalt
木色　wood
实木色　burlywood
藤色　rattan
沙色　sand;sandiness
沙漠色　desert sand;desert
沙漠雾色　desert mist
沙漠玫瑰　desert rose
浓黄土色　sienna
泥色　mud
石头色　stone
卵石色　pebble
电石色　flint
驼色　camel;light tan
米色　beige;French beige;buff;cream;grey sand;bone beige
粉红米色　pink beige
卡其色　khaki
亚麻色　linen;flaxen
小麦色　wheat
奶油色　cream
金黄奶油色　golden cream

香槟酒色	champagne
蜂蜜色	honeydew
红糖色	brown sugar
棒糖色	lollipop
椰冰色	coconut ice
燕麦色	oatmeal
豆沙色	cameo brown; rose wine; cork
浅豆沙色	pale mauve
藕荷色	pale pinkish purple; bisque
薄荷色	mint; mint cream
番木瓜色	papayawhip; papaya; pawpaw
玉米穗丝色	cornsilk
杏仁色	almond
白杏色	blanched almond
山楂色	haw
香辛料色	spice
肉桂色	cinnamon
肉色	flesh; nude; carnation; incarnadine; pastel peach; yellowish pink
裸色	nude
浅灰棕裸色	beige nude
骨色	bone; bone white
肤色	skin
烟色	smoke
鼻烟色	nigger; snuff colour; snuff
绿玉色	jade; beryl
石墨色	graphite
海贝色	seashell
水晶色	crystal; crystaline
荧光色	iridescence; fluorescence
彩虹色	iris; iridescence; rainbow
霓虹色	neon
天然色	natural

2. 英汉顺序

-A-

absinthe green	洋艾绿（暗黄绿）
absinthe yellow	洋艾黄（暗绿光黄）
acacia yellow	洋槐黄
academy blue	绿青色
acajou	赤褐色
achiote	胭脂树红
acid blue	湖蓝
acid green	酸性绿
admiral	紫黑，暗紫
Aegean blue	爱琴海蓝
aeneous	青铜色
aeruginous	青绿；铜绿
agate	玛瑙棕（深棕）
agate green	玛瑙绿，灰湖绿
agate grey	玛瑙灰
agate red	玛瑙红
airforce(blue)	空军蓝（深蓝）
alabaster(white)	雪花石膏白（淡黄白）
Alice blue	爱丽丝蓝（浅灰蓝）
alizarin crimson	茜红，暗红
alizarin red	茜红，茜色
almond	杏仁色
aloe(green)	芦荟绿，翠绿
amaranth	深紫，紫红
amaranth purple	深青莲
amber	琥珀色
amethyst	紫晶色
antique bronze	青古铜色
antique indigo	古旧靛蓝
antique violet	古紫色
antique white	古典白，怀古白
Antwerp blue	深蓝
apple green	苹果绿

apricot 杏黄
apricot blush 杏红
apricot buff 杏黄
apricot cream 米黄
aqua 浅蓝绿，淡水蓝，水绿，浅绿
aqua blue 水蓝
aqua green 水绿，浅黄绿
aqua grey 水灰
aqua haze 雾水绿
aquamarine 碧绿，蓝绿，青绿
aquamarine blue 蓝绿
artichoke green 洋蓟绿
ash 烟灰，淡灰
ash grey 浅灰
ashes 淡灰
asparagus green 浅豆绿
atlantic blue 大西洋蓝
atlantic green 大西洋绿
aubergine 茄皮紫，紫红
auburn 金棕，茶褐，赭色，猪肝色
August green 深绿
aurora(orange) 朝霞橘红，红橙
aurora pink 朝霞粉红
aurora red 朝霞红
aurora yellow 曙光黄
autumn brown 秋天棕，秋叶棕
autumn mink 深咖啡色
Aztec red （墨西哥）阿兹特克红（深棕红）
azure 天蓝，天青，蔚蓝
azure black 墨天蓝
azure blue 天青蓝，蔚蓝，中蓝
azure cerulean blue 天蓝
azure deep 暗天蓝
azure green 碧绿，碧色

-B-

baby blue 浅蓝；淡蓝
baby pink 浅粉红
Bahama blue 巴哈马蓝（宝蓝）
balsam 蟹绿
balsam green 青灰
bamboo green 竹绿
bamboo yellow 竹黄
banana 香蕉黄，草黄
bark 深青褐，栗色，深青棕
bay 枣红
bean green 豆绿，豆青
bean red 豇豆红
beetroot 紫红
begonia red 海棠红
beige 米黄，灰黄，棕灰
beige nude 浅灰棕裸色
Beijing blue 北京蓝
benzo blue 靛蓝
Berlin blue 柏林蓝（深蓝，同普鲁士蓝）
berry 浆果色（深紫红），鲜梅红
beryl 绿玉色，浅蓝
biscuit 淡褐，浅棕
bisque 淡褐，灰黄，藕荷色
bistre, bister 深褐
bitter chocolate 深咖啡色
black 黑色
black ash 黑灰
black brown 黑褐
black currant 黑红
black dull 暗黑色
black earth 土黑
blackish green 墨绿
blackish red 殷红
black red 黑红，殷红
black silver 黑银色
blanched almond 杏仁白，白杏色
blaze orange 亮光橙，艳橙

blond 淡黄；浅黄棕	bright blue 鲜蓝，翠蓝，中湖蓝
blood red 血红，猩红，鲜红	bright cobalt 鲜钴蓝
blossom 粉红	bright green 翠绿，翠青
blue 蓝色，青色	bright mauve 淡紫红
blue ashes 深灰蓝	bright red 鲜红，大红，亮红，嫣红
blueberry 紫黑	bright rose 鲜玫瑰红
bluebird 青鸟蓝，深水蓝	bright sapphire 鲜宝石蓝
blue black 蓝黑，墨蓝	bright white 亮白色
blue clear 鲜蓝	brilliant orange 亮光橙
blue deep 深蓝	brimstone 硫黄（石）色
blue grass 青草色，草地绿	bristol 浅宝蓝，天蓝
blue green 蓝绿，青绿，竹青	bronze 青铜色，古铜色
blue grey 蓝灰	bronze black 古铜黑
blue haze 雾蓝	bronze blue 青铜蓝，深蓝
blue pale 淡蓝	bronze green 古铜绿，暗铜绿，灰橄榄绿
blue purple 品蓝	
blue shadow 阴影蓝	bronze red 铜红，橘棕
blue verditer 铜绿	bronze violet 射光紫蓝
blue violet 蓝紫	bronze yellow 杏黄
bluish green 青碧，青绿	brown 褐色，棕色
bluish red 浅蓝红	brown black 棕黑
bluish white 蓝白，青白，影青	brown deep 深褐
bluish yellow 蓝黄，青黄	brown stone 石棕
blush 红色；玫瑰红	brown sugar 红糖棕
bone 骨色，骨白色，淡灰黄	bruised purple 青紫
bone beige 米色	bud green 芽绿（浅草绿，淡绿）
bone black 骨黑色	buff 浅黄，米色
bone brown 骨棕色（暗棕）	burgundy 勃艮第葡萄酒红，暗紫红
bone white 骨色，骨白	burlywood 实木色
bordeaux （葡萄）酒红，暗紫红	burnt almond 棕褐
bottle green 深绿	burnt henna 赭红
bracken green 苔藓绿	burnt olive 焦橄榄色
brassiness 黄铜色	burnt orange 焦橘黄，焦橙
brazen yellow 黄铜色	burnt pumpkin 焦橙色
breen 褐绿，棕绿	burnt sienna 深褐
breton red 暗红	burnt umber 深赭色
brick red 砖红，褐红	butter 黄油黄，油黄色

buttercup 毛茛黄（浅黄）
butternut 灰胡桃色，淡褐

-C-

cacao brown 可可棕
cactus 仙人掌绿
cactus green 仙人掌绿
cadet 灰蓝，紫灰
cadet blue 军校蓝，灰蓝
cadet grey 蓝灰
cadmium green 镉绿
cadmium orange 镉橙；橙黄
cadmium red 镉红
cadmium yellow 镉黄，鲜黄
cadmium yellow lemon 柠檬黄
cadmium yellow pale 浅镉黄
calamine blue 浅蓝
camel 驼色
camel tan 驼棕色
cameo blue 浮雕石蓝
camellia(red) 山茶红
cameo brown 浮雕石棕，豆沙色
cameo green 浮雕石绿
cameo pink 浮雕石粉红
cameo rose 浮雕石玫瑰红
campanula blue 风铃草蓝（浅紫蓝）
campanula purple 风铃草紫（浅红紫）
canal blue 浅湖蓝
canamine blue 淡蓝
canary(yellow) 鲜黄
candle light peach 烛光红
canescence 灰白
Capri blue 卡普里蓝（绿蓝）
capsicum red 辣椒红
caramel 淡褐，酱色
carbon black 炭黑
cardinal 深红，鲜红（红衣主教服色）

cardinal red 深红，鲜红
Caribbean(blue) 加勒比海蓝
carmine 胭脂红，紫红，洋红
carmine rose 胭脂玫瑰红，桃红
carnation 肉色，淡红
carob 深褐，暗褐
cayenne red 辣椒红
cedar(green) 雪松绿，杉木绿
celadon(green) 青瓷绿，灰绿
celandine green 茶绿
celery green 芹菜绿
celeste 天蓝，天青
cement grey 水泥灰
cerise 樱桃红，鲜红
cerulean blue 天蓝，蔚蓝，天青蓝，青色
ceruse white 铅白
chalk blue 浅蓝
chalk pink 浅粉红
chalk violet 淡紫，浅雪青
chalky 灰白
chambray 青年布蓝（淡灰蓝）
champagne 香槟酒色（淡橘黄，浅灰褐）
charcoal black 炭黑
charcoal blue 深灰蓝
charcoal brown 炭褐
charcoal grey 炭灰，黑灰
charcoal marl 炭土色，炭灰
chartreuse 查特酒绿（黄绿）
chateau green 青黛绿（浓绿）
cherry 樱桃红，鲜红
chestnut 栗色
chestnut black 栗黑
chestnut brown 栗褐
chili (干)辣椒色，红棕
chili orange 辣椒橙

China orange 橙色
chinchilla 银灰
Chinese blue 中国蓝（深蓝）
Chinese green 中国绿（深绿）
Chinese red 中国红（大红）
Chinese violet 中国紫罗兰色
Chinese white 中国白（锌白）
Chinese yellow 中国黄（老黄色）
chocolate 巧克力色,深褐,红褐,赭
chrome 铬色（银色）,铬黄
chrome green 铬绿
chrome orange 铬橙
chrome red 铬红
chrome yellow 铬黄
cinnabar 朱砂色,朱红
cinnamon 肉桂色,黄褐,黄棕
citrine 柠檬黄
citron 香橼黄
citrus 柑橘黄（嫩黄）,土黄
citrus green 柑橘绿（浅绿）
claret 紫红
classic blue 古典蓝
clay 土褐,浅棕
clear cerulean 鲜青
clear white 纯白
clematis 铁线莲紫（浅紫）
clementine 深粉红
clerical grey 深灰
cloud blue 云蓝
cloud cream 云白
cloud grey 云灰
cloud pink 淡藕红,淡妃色
clove 丁香花色（多为淡紫,蓝紫）
clover 三叶草绿
coal (black) 煤黑
cobalt blue 钴蓝
cobalt green 钴绿

cobalt violet 钴紫
cobalt yellow 钴黄
coccinellin 卡红,胭脂红
cochineal 胭脂红
cock yellow 雄黄
cocoa 可可色,黄棕,黄褐
cocoa brown 可可棕（黄棕）
coconut brown 椰棕,椰褐
coconut ice 椰冰色（浅粉红）
coffee 咖啡色
cola 深褐,黑褐
colonial yellow 淡棕黄
conch shell 海螺壳红,深藕红
concrete grey 混凝土灰,水泥灰
cool grey 冷灰色
copper 铜色
copper brown (紫)铜棕
copper green 铜绿
copper red 铜红
coquelicot 虞美人花橙红
coral 珊瑚色,珊瑚红
coral blush 珊瑚红
coral haze 橘红
coral pink 浅珊瑚红
coral rose 珊瑚玫瑰红
cork 软木浅棕,豆沙色
cornflower(blue) 矢车菊蓝（紫蓝）
cornsilk 淡黄,玉米穗丝色
coyote brown 土狼棕,狼棕色
crab-apple 野生苹果色,野苹果橙
cracker khaki 杏黄
cranberry 酸果蔓红,深红,红褐
cream 奶油色,淡黄,米色
cream blush 奶油红
cream gold 奶油金黄
cream pearl 奶油珍珠白
cream rose 奶油玫瑰红

cream tan 奶油棕黄
cream white 奶油白，乳白
crimson 深红，梅红，绯红
crisp white 纯白，精白
crocus 藏红花橙，橘黄
crystal 水晶色
crystal blue 水晶蓝
crystal cream 乳白，奶油白
crystal green 水晶绿
crystal grey 水晶灰
crystaline 水晶色
crystal rose 水晶玫瑰红
currant(red) 茶藨子浆果红（深红）
cyan 青色，绿蓝
cyan blue 青蓝，深蓝，天蓝
cyanine blue 原色蓝
cyanosis 青紫
cypress green 柏树绿

-D-

daffodil 水仙黄，淡黄
dahlia 火丽花红（浅紫红）
damask 玫瑰红，紫红
damson 暗紫
daphne blue 月桂树蓝
dark amethyst 深紫晶色
dark black 深黑，暗黑，玄色
dark blonde 暗金色，深金色
dark blue 深蓝，暗蓝，黛蓝，暗青，藏青，黛青
dark bluish grey 灰青
dark brick 深砖红
dark bronze 黑古铜色
dark brown 深棕，暗棕，深褐，暗褐，茶褐
dark camel 暗驼色
dark cyan 暗青，玄青（有光泽的黑青色）
dark earth 暗土色
dark gold 深金色，暗金色
dark goldenrod 深金黄
dark green 深绿，暗绿，黛绿
dark grey 深灰，暗灰
dark khaki 暗土黄
dark magenta 暗洋红，暗紫红
dark indigo 深靛蓝
dark jade 深玉绿
dark navy 深藏青
dark oatmeal 深米灰
dark olive 深橄榄色
dark orange 暗橘黄
dark orchid 深兰花紫
dark purple 深紫，黛紫，红青，绀青
dark red 深红，暗红，殷红
dark reddish brown 酱色
dark reddish purple 酱紫
dark rose taupe 古铜色
dark salmon 暗鲑肉红
dark sea green 暗海水绿
dark slate grey 暗瓦灰
dark turqoise 暗青绿
date red 枣红
dawn blue 黎明蓝
dawn grey 黎明灰
deep black 深黑
deep blue 深蓝，暗蓝
deep brown 深褐，深棕
deep cerulean 暗青
deep claret 深紫红
deep green 墨绿，暗绿
deep grey 深灰
deep mauve 深紫红
deep olive 深橄榄色
deep orange 深橙，深橘黄

deep pink 深粉红
deep plum 深梅红
deep red 深红
deep sky blue 深天蓝
deep wistaria 深紫藤色
deep yellow 深黄，暗黄，秋香色
desert 沙漠色
desert mist 沙漠雾色
desert rose 沙漠玫瑰色
desert sand 沙漠色
dim grey 暗灰，深灰
dirty blue 暗蓝
dirty green 暗绿
dirty grey 暗灰
dirty purple 暗紫
dirty red 暗红
Dodger blue 道奇蓝（类似宝蓝）
doe 驼灰
doe marl 驼灰
dove grey 鸽子灰，紫灰
drab 黄褐，淡褐
dry rose 浅紫
duck egg blue 鸭蛋青
dull antique 暗古铜色
dull black 暗黑
dull blue 暗蓝
dull brown 暗褐
dull green 暗绿
dull grey 暗灰
dull purple 暗紫
dull red 暗红
dull silver 暗银色，哑银色，雾银色
dun 暗褐，焦茶色
dusk 黄昏暗色，微黑色
dusky green 灰绿
dusky green blue 深青灰
dusky orchid 暗紫

dust-colour 灰褐
dusty grey 土灰
dusty lavender 朦胧紫，淡灰紫
dusky mauve 红灰紫
dusty orange 灰橙色
dusty pink 灰粉红
dusty rose 朦胧玫瑰色，灰玫瑰红
Dutch blue 浅紫蓝

–E–

earth red 土红
earth yellow 土黄
ebony 乌木黑，乌黑
ecru 淡褐
ecru drab 淡褐
eel bluish 鳝鱼青
eel green 鳝鱼绿
eel yellow 鳝鱼黄
egg blue 蛋青
egg yellow 蛋黄
elderberry 莲灰
electric blue 铁青
elephant grey 象灰色，红灰
elm green 榆树绿
emerald 艳绿，鲜绿，翠绿
emerald black 黑玉绿
emerald green 翠玉绿，翠绿，鲜绿
empire blue 帝国蓝，灰蓝，青绿
empire yellow 帝国黄
enamel blue 珐琅蓝（紫蓝）
espresso 浓咖啡色
ethereal blue 太空蓝，纯蓝
eucalyptus 桉树绿，浅绿灰
evergreen 常青树绿

–F–

fallow 淡棕

fawn 淡黄褐
feather grey 羽灰
ferruginous 铁锈色，赤褐
field green 原野绿
field grey 军装绿（深灰）
fiery red 火焰红
fig brown 无花果棕
filemot 枯叶色，黄褐
firebrick(red) 火砖红
fire red 火红
fishbelly white 鱼肚白
flake white 铅白
flame orange 火橙色
flame scarlet 火红，大红
flamingo(red) 火烈鸟红，火鹤红
flaxen 亚麻色，淡黄
flesh 肉色
flesh blond 肉棕
flint 打火石色，电石色（浅褐）
flower blue 花青
flower white 花白
fluorescence 荧光色
fluorescent white 荧光白
fluorite green 萤石绿
fog blue 雾蓝（浅灰蓝）
fog green 雾绿
foliage green 叶绿
forest green 森林绿
framboise 朱红，深粉红
French beige 米色，灰棕
French blue 法国蓝（红光鲜蓝）
French grey 浅灰
French navy 法国海军蓝（无光深蓝）
French rose 法国玫瑰红
French vermilion 法国红
fresh leaves 嫩（叶）绿
fresh red 鲜红

frorid 鲜红
frost white 霜白，霜色
fuchsia 紫红
fuchsia pink 浅莲红
fuchsia red 梅红，深莲红
fuchsin(e) 品红
fudge 咖啡色
full strength orange 全浓橙色
fulvous 黄褐，茶色
fuscous 暗褐

-G-

gainsboro 淡灰
gamboge 橙黄
garnet 石榴（石）红，深红
garnet brown 红棕
garnet red 石榴（石）红
garnet rose 石榴（石）红
gentian blue 龙胆蓝
gentian violet 龙胆紫
geranium 天竺葵红，原色红
geranium pink 绯红
ghost white 幽灵白
gilt 金色，青铜色
ginger 姜黄
ginger snap 栗棕
ginger spice 红棕
glacier 冰河色（淡蓝绿）
glaucous 淡灰蓝，淡灰绿
global brown 咖啡色
glossy dark green 油绿
gobelin blue 暗青绿
gold 金色，金黄
gold brown 金褐
golden apricot 金杏色，土黄
golden cream 金黄奶油色
golden fleece 浅米黄

golden green 金黄绿	grenadine red 石榴汁红，浅红橙
golden ocher 金黄褐	grey 灰色（同 gray）
golden olive 橄榄黄	grey black 黑灰，墨灰
goldenrod 金麒麟色，金菊黄，浓黄	grey green 灰绿
golden straw 稻草黄	grey marl 泥灰
golden yellow 金黄	grey sand 米色
golf green 中绿	grey violet 浅紫
granite grey 花岗石灰	grey yellow 灰黄
granny smith 绿苹果色，青绿	greyish white 灰白
grape 葡萄紫	gridelin 深紫红
graphite 石墨灰，墨灰	grizzle 灰色，灰白
graphite blue 石墨绿（灰绿）	gull grey 海鸥灰
graphite silver 墨灰银	gunmetal 炮铜色，黑灰
grass green 草绿	gunmetal grey 铁灰，黑灰；暗灰
gray 灰色（同 grey）	

–H–

gray dark 深灰	
gray deep 暗灰	harbour blue 蓝绿
gray dull 暗灰，深灰	Havana brown 哈瓦那雪茄棕，黄棕
gray green 灰绿	haw 山楂色
grayish purple 浅灰紫	hawksbill turtle red 玳瑁红
gray lily 珍珠白	hazel 淡褐
gray mist 雾灰	hazel nut 赤褐
gray morn 晨灰，银灰	heather 欧石蓝色（淡紫红）
gray pale 白灰，淡灰	hedge green 灰绿
gray red 灰红	heliotrope 缬草紫（浅紫红），淡紫
green 绿色，青色	henna 棕红
green almond 绿杏仁色，黄绿	hepatic 猪肝色
green banana 香蕉绿	heron 深紫灰
green black 墨绿	hibiscus red 芙蓉红
green clear 鲜绿	hoar 灰白
green crystal 绿晶色	honeycomb 蜂巢色，灰褐，灰白
green gold 绿金色	honeydew 蜂蜜色，蜜露橙，浅橙
green mist 雾绿	honey orange 蜂蜜橙
green smalt 大青	honey yellow 蜂蜜黄
green tint 浅绿	horizon blue 苍蓝
greenish-grey 蟹青	hot pink 鲜粉红；水红
greige beige 灰褐	husk 深褐

hyacinth 紫蓝

-I-

iceberg 冰山蓝（淡绿光蓝）
iceberg green 冰山绿
ice blue 冰蓝，淡绿光蓝
ice cream 乳白
ice-snow blue 冰雪蓝
imperial blue 景泰蓝
imperial green 翡翠绿
imperial purple 暗紫
imperial red 紫红
incarnadine 肉色，血红
indathrene blue 士林蓝
Indian lake 印度胭脂红
Indian purple 印度紫
Indian red 印度红（红棕）
Indian tan 印度棕黄
Indian yellow 印度黄（老黄色）
indigo 靛蓝，靛青
indigo blue 靛蓝
indigo copper 铜蓝
ink blue 墨水蓝，蟹青
inkiness 墨黑，漆黑
international klein blue 国际奇连蓝（宝蓝）
invisible green 深绿
iridescence 彩虹色，荧光色
iridescent 荧光色
iris 彩虹色
Irish green 爱尔兰绿
iron black 铁黑
iron blue 铁蓝
iron brown 铁棕
iron buff 铁黄
iron grey 铁灰
iron oxide black 铁黑

iron oxide red 铁红
iron oxide yellow 铁黄
isabel 灰黄
ivory 象牙色，乳白，象牙黄
ivory cream 象牙奶油色
ivory deep 象牙淡黄
ivory white 象牙白，乳白
ivory yellow 象牙黄，乳白
ivy green 常春藤绿（暗橄榄绿）

-J-

jacinth 橘红，红锆石橙
jacinthe 淡橘红，淡橙
jade 绿玉色，浅绿
jade green 玉绿，翠绿
jade grey 玉灰
jade lime 翡翠橙色
jade white 玉石白
Jaffa orange 雅法橙色
jasmin(e) 茉莉黄（淡黄）
jasmin(e) green 茉莉绿
jasper 墨绿
jay 中蓝
jester red 大红
jet 煤玉黑，黑玉色
jet black 乌黑，深黑
jewelry blue 宝石蓝
jujube red 枣红
jungle green 丛林绿，深草绿
Jun-kiln red 均红

-K-

Kaoliang red 高粱红
Kelly 凯利绿，鲜黄绿
Kelly green 凯利绿，鲜黄绿
khaki 卡其色，土黄，黄褐
khaki grey 卡其灰

khaki tan　卡其棕黄
kingfisher　翠鸟色（翠蓝；翠绿）
king's blue　钴蓝，品蓝
king's gold　雌黄
king's green　巴黎绿
king's yellow　雌黄

−L−

lagoon　礁湖蓝（浅绿蓝）
lake　胭脂红
lake black　漆黑
lake blue　湖蓝
lake green　湖绿
lake red　胭脂红
lapis lazuli blue　天青石蓝
lang-kiln red　郎窑红
lark　云雀色，灰黄
laterite　土红
laurel green　月桂树绿，浅绿
laurel pink　月桂树粉红
latte　拿铁咖啡色（牛奶与咖啡混合色）
laurel　月桂花淡黄
lavender　薰衣草紫（淡紫）
lavender blue　淡紫蓝
lavender blush　淡紫红
lavender grey　紫灰
lawn green　草绿
lead　铅色，铅灰
lead grey　铅灰
lead white　铅白
leaf green　叶绿
leek green　韭葱绿（暗黄）
lemon　柠檬色，淡黄
lemonchiffon　粉黄
lemon yellow　柠檬黄
lettuce green　莴苣绿（黄绿）
lias　淡紫

licorice　甘草黄
light aqua　浅蓝绿
light atlantic　浅大西洋绿
light bean green　浅豆绿
light berry　浅紫红
light blue　浅蓝，淡蓝，浅青
light brown　淡褐
light cerulean　浅青
light chartreuse　浅黄绿
light chestnut　浅栗色
light coral　浅珊瑚红
light cyan　淡青
light gold　淡金色
light green　浅绿，品绿
light greenish blue　粉青，淡青
light grey　浅灰
light indigo　浅靛蓝
light lemon　浅柠檬黄
light lilac　浅丁香色
light mauve　浅紫红
light olive green　浅橄榄绿
light orange　浅橙色，浅橘黄
light pink　浅粉红，妃红，妃色
light purple　浅紫，浅绛红
light red　淡红
light rose　浅玫瑰红
light salmon　淡鲑肉红
light sapphire　浅宝石蓝
light scarlet　大红
light siam　浅红，淡红
light sky blue　亮天蓝
light slate grey　淡瓦灰
light steel blue　淡钢蓝
light tan　浅棕黄，驼色
light taupe　浅灰褐
light turquoise green　淡碧绿
light yellow　浅黄，鹅黄

lilac 丁香紫，淡紫，紫藤色，雪青
lilac clear 浅紫藤色
lilac grey 淡紫灰
lilac pale 淡紫藤色
lilac rose 丁香红
lilac snow 淡紫
lilac white 淡紫白
lily sky 粉蓝
lime 酸橙绿，暗黄绿
lime green 酸橙绿，暗黄绿
lime yellow 酸橙黄
linen 亚麻色
lipstick red 唇膏红，口红色
liquorice 甘草黄
lithopone 锌钡白
liver brown 肝棕色，红褐
liver chestnut 深栗色
lividity 铅色，铅灰，青灰，青黑，黛色
lobster red 龙虾红
loden green 深灰绿
logwood black 洋苏木黑
lollipop 棒糖色（粉红）
London smoke 伦敦烟灰色，暗灰
lotus red 莲红
lucifer yellow 荧光黄
lyart 灰白
Lyons blue 里昂蓝，紫蓝

-M-

Macao green 澳门绿（澳门特别行政区区旗底色）
madder 茜草色，鲜红
madder brown 鲜红棕
madder red 茜红，茜色，紫红，鲜红
magenta 洋红，品红，紫红
magenta haze 淡品红

magenta pink 紫红
magenta purple 洋红紫
mahogany 红木色，赤褐
maize 玉米黄
malachite green 孔雀石绿，品绿
mallow(purple) 锦葵紫（红紫）
mandarin blue 深蓝
mandarin orange 橙色
manganese blue 锰蓝
mandarin red 橙红
mango 杧果黄
maple 淡棕，灰黄
maple sugar 黄棕
marigold 万寿菊黄（橙黄）
marine blue 海水蓝
marine green 海水绿
marl 泥灰
maroon(=marron) 栗色，赤褐，褐红，紫酱色
matte gold 黯金色
matte silver 哑银色
mauve 紫红，木槿紫
mauve glow 鲜紫红
mauve orchid 浅紫蓝
mauve wine 酒红，紫酱色
mazarine 深蓝
meadow green 草地绿
medium blue 中蓝
medium green 中绿
medium grey 中灰
medium orchid 中兰花紫
medium purple 中紫
medium turquoise 中青绿
medium yellow 中黄
melon 瓜瓤红（浅红橙）
melon based orange 瓜瓤橙
mercury 水银色

metallic grey　金属灰（浅红灰）
meteorite　陨石棕
metropolitan　暗灰蓝
mid-brown　中褐
mid-indigo　中靛蓝
midnight black　午夜黑，深黑
midnight blue　午夜蓝，深蓝，中灰蓝
midnight navy　黑蓝
mid-olive　中橄榄绿
mignonette　灰绿
milk chocolate　奶油巧克力色，浅棕褐色
milk white　乳白
milkiness　浊白
milky blue　乳白蓝
milky green　乳白绿
mimosa yellow　含羞草黄，缃色（黄）
mineral blue　石青
mineral green　石绿
mineral grey　石灰
mineral violet　矿紫，石紫（淡灰紫）
mineral yellow　石黄
Ming blue　明瓷蓝，藏蓝，藏青
Ming green　明瓷绿，艳绿
Ming yellow　明瓷黄
minium　朱砂色，丹色
minize(=minze)　浅绿
mink　深褐
mint　薄荷色，浅绿
mint cream　薄荷乳白，薄荷色
mint green　薄荷绿，浅绿
mistletoe　豆绿
misty blue　雾蓝，影青，青白
misty green　雾绿
misty grey　雾灰
misty rose　雾玫瑰红，粉玫瑰红，浅玫瑰红

misty yellow　雾黄
moccasin　鹿皮色
mocha(=moka)　摩卡咖啡色（深褐）
modena　深紫
moonbeam　月光色（浅米灰）
moon blue　月光蓝
moon light　浅米灰
moon purple　月光紫红
moon white　月光白
moon yellow　月光黄
moss(green)　苔藓绿，黄绿
mouse　鼠灰，灰褐
mouse grey　鼠灰，灰褐
mud　泥色
mulatto　黄褐
mulberry　深紫红
murrey　桑果色，紫红
mushroom　浅褐
mustard　芥末黄，深黄，暗黄
mustard brown　芥末棕（黄棕）
mustard yellow　芥末黄

-N-

nacarat　鲜橘红
nankeen　本色，淡黄
nankeen yellow　南京黄，浅灰黄
nattier blue　淡蓝
natural　自然色（浅土黄，浅褐黄）
Navaho white　（印第安）那伐鹤白
navy　深蓝，藏蓝，海军蓝
navy blue　藏蓝，藏青，深蓝，海军蓝
neon　霓虹色
neptune green　海王星绿，海洋绿，浅粉绿
neutral tints　（各种）灰色
nickel　镍灰（红光浅灰）
nickel green　镍绿（浅暗绿）

nigger　鼻烟色，深棕
nigger-brown　深褐
nigritude　黑色
Nile blue　尼罗河蓝（浅绿蓝）
Nile green　尼罗河绿（浅青绿）
nimbus grey　雨云灰
nude　肉色，裸色

–O–

oatmeal　燕麦色，米灰
ocean blue　海洋蓝
ocean green　海洋绿
ocher(=ochre)　赭色，黄褐
ocher yellow　赭石黄
off-olive　浅黄绿
off-white　米白，灰白，本白
oil blue　油蓝
oil red　油红
old gold　古金色
oldlace　老白色，浅米色
old mauve　暗紫红
old rose　灰玫瑰红
olive　橄榄色，橄榄绿，茶青
olive black　橄榄黑
olive brown　橄榄棕，黄褐
olive drab　橄榄军服绿，草绿，草黄
olive green　橄榄绿，茶青
olive green black　墨橄榄绿
olive green clear　淡橄榄绿
olive green deep　暗橄榄绿
olive grey　橄榄灰
olive yellow　橄榄黄
onion green　洋葱绿，洋葱青
onix　米黄
opal blue　乳白蓝
opal grey　乳白灰
opera mauve　优品紫红

orange　橙色，橘黄
orange brown　橘棕，橙褐
orange-chestnut　橙色
orange clear　浅橙，浅橘黄
orange light　淡橙
orange pale　淡白橙
orange red　橘红，红橙
orchid　兰花紫，淡紫，粉紫
orchid bloom　兰花莲色
orchid pink　兰花粉红
oriental blue　暗蓝
oriental red　大红
oxblood(red)　牛血红，棕红
Oxford blue　牛津蓝，深蓝
Oxford grey　牛津灰，深灰
oxford ocher　深土黄
oyster grey　牡蛎灰，米灰，浅灰
oyster white　牡蛎白（灰黄白）

–P–

pacific(blue)　蓝绿
pail lilac　淡丁香紫
palatinate　淡紫
pale　苍白，灰白；淡色，浅色
pale blue　淡蓝
pale cerulean　淡青
pale goldenrod　淡金黄
pale green　淡绿，葱绿，苍绿
pale grey　白灰
pale khaki　浅土黄
pale lilac　浅紫藤，浅莲灰
pale mauve　淡紫红
pale pink　淡粉红
pale pinkish grey　藕色
pale pinkish purple　藕荷色
pale purple　淡紫，青莲色
pale yellow　浅黄，苍黄

pallar	灰白	pebble	卵石色（米黄）
palomino	淡黄褐	Peking blue	北京蓝
pansy	三色紫罗兰（深紫）	peony red	牡丹红
papaya(=pawpaw)	番木瓜酒色（浅朱红）	periwinkle	长春花蓝（浅紫蓝）
papayawhip	番木瓜色	Persian blue	波斯蓝（紫蓝）
paprika(=paprica)	（匈牙利）辣椒红	Persian orange	波斯橙
parchment	羊皮纸色（浅棕）	persimmon orange	柿子橙
Paris green	巴黎绿（黄光绿）	persimmon red	柿子红
Paris white	巴黎白	peru	秘鲁褐（浅褐）
Paris yellow	巴黎黄	petal	粉红
Parisian blue	巴黎蓝，天蓝	petrol	淡灰绿
parrot green	鹦鹉绿（黄绿）	petunia	牵牛花色，暗紫
parsley green	欧芹绿	pewter(grey)	锡镴灰，青灰
pastel blue	淡蓝	phlox	夹竹桃红
pastel green	淡绿	phlox red	夹竹桃红
pastel lilac	淡紫	pigeon grey	鸽子灰（紫光灰）
pastel yellow	淡黄	pike grey	蓝灰
patina green	铜锈绿	pimiento	辣椒红
pea green	豆绿，豆青，黄绿	pine green	松树绿
peach	桃色，桃红，浅红橙	pink	粉红
peach beige	血牙红	pink beige	粉红米色，米妃色
peach blossom	桃红	pink lavender	浅雪青
peachpuff	桃红	pink tint	淡粉红
peach red	桃红	pinkish red	品红
peacock blue	孔雀蓝（暗绿蓝）	pinky white	粉红白
peacock green	孔雀绿（暗蓝绿）	pistachio	淡黄绿
pear(green)	梨绿	pitch-black	漆黑，深黑
pearl	珍珠色，蓝灰	pitch-dark	漆黑，深黑
pearl blue	珍珠蓝，淡灰蓝	plantation	茶绿
pearl blush	珍珠红	platinum	铂金色，银灰
pearl gold	珠光金色	platinum black	铂黑
pearl grey	珠光灰	platinum blond	淡金黄
pearl pink	浅珍珠红	plum	梅红，紫红，青紫
pearlized gold	珍珠金色，珠光金色	plum green	青梅绿，梅青
pearlized silver	珍珠银色，珠光银色	plum purple	紫红
pearl white	珍珠白，珠光白	poloere	米黄
		pomegranate red	石榴红

ponceau 朱红，深红
pool blue 水塘蓝
pool green 水塘绿
pop pink 鲜粉红
poppy 芙蓉红，罂粟红（深红）
poppy red 芙蓉红，罂粟红（深红）
porcelain blue 瓷蓝
pottery red 陶器红，陶嫣红
powder blue 粉末蓝
powder grey 青灰
primrose 樱草色，樱草黄，淡黄
primrose yellow 樱草黄
princess blue 公主蓝
prune 深紫红
prune purple 红青，绀青
Prussian blue 普鲁士蓝（深蓝）
puce 紫褐
pumpkin 南瓜色，南瓜橙
pure white 纯白，精白
purpl 紫色
purple 紫色
purple black 墨绛红，紫黑色
purple bronze 紫铜色
purple deep 暗绛红
purple drab 雪青
purple gold 紫金色
purple grey 紫灰
purple haze 紫雾色
purple light 浅绛红
purple marl 紫
purplish blue 紫光蓝，藏蓝
purplish red 紫光红，枣红
putty 浅灰褐

—R—

radiance(=radiancy) 深粉红
rainbow 彩虹色

raisin 葡萄干色，乌紫
rame 黄棕，黄褐
raspberry 木莓色，紫红
raspberry jelly 木莓果冰色，紫红
rattan 藤色，浅橘黄
rattan yellow 藤黄
raw white 本白，缟色
red 红色，赤色，彤色
red berry 浆果红（深紫红）
red purple 红紫
red violet 红紫
reddish black 红黑，紫棠色，天青
reddish blue 品蓝
reddish brown 赭色
reddish grey 红光灰
reddish orange 橘红（红光橙）
reddish yellow 浓黄（红光黄）
reddle 土红
redexide 褐红
reed green 芦苇绿
rescue orange 救援橙色
reseda(green) 木犀草绿，灰绿
river blue 河蓝
robin's egg blue 绿蓝
roman sepia 墨棕
rose 玫瑰色，玫瑰红
rose amethyst 紫晶色
rose beige 灰褐
rose bloom 粉红
rosebud 玫瑰花苞色，浅玫瑰红
rose carmine 玫瑰紫红
rose dust 豆灰
rose gold 玫瑰金色
rose madder 玫瑰红
rose mauve 玫瑰紫红
rose pink 玫瑰粉红，浅玫瑰红
rose tan 淡红棕

rose violet	玫瑰紫
rose wine	豆沙色
rosewood red	紫檀红，紫檀色（紫红）
rosy brown	玫瑰棕，褐玫瑰红
rouge red	胭脂红
Royal blue	皇家蓝
royal blue	宝蓝，品蓝
royal purple	蓝紫
royal red	深红
royal yellow	雌黄
rubiginous	赤褐
rubin	品红
rubine	宝石红，暗红
ruby	红宝石色，鲜红
ruby red	宝石红
ruddle	赭色
rufous	赤褐
russet	赤褐
russet brown	黄褐
russet red	豆沙红
rust	铁锈色，赭色
rust brown	铁锈色
rust indigo	赭色靛蓝
rust red	铁锈红
rustic brown	铁锈棕

-S-

sacrificial red	祭祀红
sad brown	暗褐
saddle brown	马鞍棕，重褐色
saffron	藏红花色，橘黄
sage green	灰绿
sallow	苍黄，灰黄
salmon	橙红，鲑肉红
salmon pink	橙红，鲑肉红
salvia blue	鼠尾草蓝
sand	沙色，沙灰，黄灰，浅棕
sand beige	沙棕
sand yellow	沙黄
sandiness	沙色，浅茶色
sandy	淡茶色
sandy beige	浅褐
sandy brown	沙棕，沙黄
sanguine	血红
sap green	暗绿
sapphire(blue)	宝石蓝，天蓝，青玉色（深蓝）
sardonyx(玛瑙)	深红
sargasso green	马尾藻绿，褐绿
saxe blue	萨克斯蓝（浅孔雀蓝）
scarlet	猩红，鲜红，绯红
scarlet red	猩红，鲜红
sea blue	海水蓝，海蓝
seafoam(green)	海沫绿（浅蓝绿）
sea green	海水绿，海绿
sea spray	灰绿
seal	深褐，暗褐
seashell	海贝色
seaweed green	海藻绿（浅灰绿）
sepia	乌贼墨色（棕黑）
serene blue	清澈蓝
shale	页岩棕
shell	米白
shell grey	贝壳灰，淡黄灰
shell pink	贝壳粉红，血牙红
shiraz	深枣红，深赤褐，暗红，葡萄酒红
shocking pink	鲜粉红
shocking red	憬红，鲜艳红
shoot green	嫩绿
siam	深红
sienna	浓黄土色，赭色
sierra grey	云石灰
silver	银色，银白，银灰

silver blue	银灰蓝
silver green	银灰绿
silver grey	银灰
silver mist	雾银色
silver pink	银光粉红，米红
silver red	银红色
silver sage	银灰绿
silver white	银白
skin	肤色
sky blue	天蓝，蔚蓝
sky-clearing blue	雾青
sky grey	天灰
slate	石板灰，暗蓝灰
slate black	石板黑，灰黑
slate blue	石板蓝，暗灰蓝
slate green	石板绿，灰绿
slate grey	石板灰，青灰，瓦灰
smalt	深蓝，大青
smoke	烟色，淡蓝
smoke tree yellow	黄栌黄（近似土黄，棕黄）
smoky blue	烟青色，青灰
smoky grey	烟灰
snow-white	雪白
snowy white	雪白
snuff(colour)	鼻烟色，黄褐
soft green	柔绿，淡绿
soft grey	柔灰，淡灰
soft peach	淡桃红
soft pink	淡粉红
soft yellow	淡黄
soil brown	土褐
solferino	鲜紫红，品红
sombre grey	暗灰
soot black	煤烟黑
soot brown	煤烟褐
sorrel	红棕，赤褐，栗色
spadiceous	浅褐，栗色
Spanish green	西班牙绿（浅暗绿）
Spanish red	西班牙红（朱红）
sparrow	雀灰（浅灰褐）
spearmint	松石绿
spice	香辛料色（黄棕）
spinel red	尖晶石红
spring green	春天绿，嫩绿
squirrel grey	松鼠灰
steel blue	钢蓝，钢青
steel grey	蓝灰，青灰
stone	石头色，石青，褐灰，暗青灰
stone blue	石青，灰蓝
stone grey	石灰，青灰
stone-washed indigo	石磨蓝
storm blue	蟹灰
storm grey	风暴灰
straw yellow	稻草黄
strawberry red	草莓红，紫红
strong blue	浓蓝
strong red	鲜红
succinite	琥珀色
sulfur yellow	硫黄黄，嫩黄
summer tan	棕褐
sunny yellow	日光黄
sun orange	日光橙
sunset	晚霞色，浅橘红
sunshine	金黄
surplus green	橄榄绿
swamp	棕褐

— T —

tan	棕黄，黄褐，棕褐
tangerine	橘红
taupe	灰褐
tawny	黄褐，茶色
tea	茶色

tea green 茶绿，茶青
tea rose 茶玫瑰红（淡橙红）
teal(blue) 水鸭蓝（绿光暗蓝），蓝绿，蟹青
tender green 嫩绿，新绿
tender yellow 嫩黄
terra-cotta 赤褐，土红
terra nera 〔拉〕土黑
terra verde 〔拉〕土绿
thirsky blue 浅蓝
thistle 蓟色，淡紫
tile blue 瓦蓝（浅灰蓝）
tile red 瓦红（橙色）
titian 金黄，橙红
tobacco brown 烟叶棕
toffee brown 中褐
tomato red 番茄红
tortoise shell 玳瑁壳色（黄褐）
tropical orange 热带橙
true blue 纯蓝
tulip yellow 郁金香黄
turf 草皮绿，草绿
Turkey blue 土耳其蓝
Turkey red 土耳其红（鲜红）
turquoise 松石绿，蓝绿，青绿，蟹青，深水蓝
turquoise blue 松石蓝，蓝绿，青绿
turquoise green 碧绿
turtle green 龟绿色
twilight purple 朦胧紫
tyraline 品红，洋红
Tyrian purple 泰尔红紫

– U –

ultramarine 深蓝，青蓝，佛青，群青，绀青
ultramarine blue 深蓝，群青，饱和蓝

ultramarine green 绀青绿
ultraviolet 深紫
umber 红棕，浓茶色，焦茶色，赭色
underglaze red 釉底红

– V –

vandyke brown 铁棕
vanilla 香草黄，淡杏黄
vapour blue 汽雾蓝，烟灰
Venetian blue 威尼斯蓝
Venetian green 威尼斯绿
Venetian red 威尼斯红
verdancy 翠绿，嫩绿
verdant green 嫩绿
verdigris 铜绿
verdure 青绿，青翠
vermeil 朱红
vermilion 朱红，朱砂红，丹色
vert （纹章的）绿色
violet 紫罗兰色，紫色
violet ash 淡白紫
violet black 墨紫，紫黑色
violet deep 暗紫
violet light 鲜紫
violet lotus 紫莲色
viridescent 淡绿
viridian 铬绿，深翠绿
viridis 松石绿
viridity 碧绿，翠绿
vitelline 蛋黄
vivid blue 鲜蓝，碧蓝，深湖蓝
vivid chartreuse 嫩黄绿
vivid green 鲜绿，嫩绿
vivid turquoise 宝蓝

– W –

walnut brown 胡桃棕（浅棕）

wan blue　暗蓝
warm apricot　橙杏色
warm cobalt　暖钴色
warm grey　暖灰色
warm white　暖白色
water blue　清水蓝（浅暗绿蓝）
water grass green　水草绿
water green　清水绿（浅暗黄绿）
wax yellow　蜡黄
Wedgwood blue　韦奇伍德陶瓷蓝
wetsand　湿沙浅棕
wheat　小麦色（淡黄）
wheaten　灰黄
white　白色
white blue　影青
white gold　白金色
white smoke　烟白
white wax　蜡白色
whitish　淡白
willow green　柳树绿，柳绿
willow yellow　柳树黄，柳黄
wine　红葡萄酒色，深红，紫红
wine red　（葡萄）酒红，深红，紫红
wine yellow　酒黄（浅灰黄）
winter green　冬天绿
winter white　冬天白（灰白，粉白）
wistaria(=wisteria)　紫藤色，浅紫
woaded blue　菘蓝
wood　木色
wool white　羊毛白

-X-

xanthic　黄色的，带黄色的
xanthous　黄色的，浅黄色的

-Y-

Yale blue　耶鲁蓝
yellow　黄色
yellow cream　嫩黄
yellow green　黄绿
yellow mid　中黄
yellow ocher　赭石黄，土黄
yellowish brown　黄光棕，土黄
yellowish green　黄光绿
yellowish grey　黄光灰
yellowish orange　黄光橙
yellowish pink　肉色
yellowish red　黄光红
yolk yellow　蛋黄，深黄

-Z-

zaffer blue　钴蓝
zenith blue　天顶蓝（淡紫光蓝）
zephyr　藕灰，红莲灰
zest　柠檬皮黄
zinc　锌色，锌灰
zinc white　锌（氧粉）白
zinc yellow　（铬酸）锌黄
zircon blue　锆石蓝，浅蓝

二、常用缩写词
Common Abbreviation

1. 制图缩写词

A.A.P.	(Anterior Armpit Point)	腋窝前点
AH.	(Armhole)	袖窿
A.S.	(Arm Size)	肘围
B.	(Bust)	胸围
	(Back)	背；后身
	(Bottom)	脚口；底边
B.C.	(Biceps Circumference)	上臂围；袖宽
B.D.	(Bust Depth)	胸高
	(Back Depth)	后腋深
B.H.	(Button Hole)	扣眼（位）
BK.	(Back)	背；后身
B.L.	(Back Length)	背长
	(Bust Line)	胸围线
B.N.	(Back Neck)	后领围
B.N.L.	(Back Neck Line)	后领圈线
B.N.P	(Back Neck point)	后颈点
B.P.	(Bust Point)	胸高点，乳峰点
B.R.	(Back Rise)	后（直）裆，〔粤〕后浪
	(Body Rise)	股上
B.S.L.	(Back Shoulder Line)	后肩线
B.S.P.	(Back Shoulder Point)	后肩（颈）点
B.T.	(Bust Top)	乳围
BTM	(Bottom)	底边；脚口
BTN	(Button)	纽扣（位）
B.W.	(Back Width)	背宽
C.	(Chest)	胸围
C.B.	(Centre Back)	后中缝，后中长
C.B.F	(Centre Back Fold)	后中对折
C.B.L.	(Centre Back Length)	后中长
	(Centre Back Line)	后中线

缩写	英文	中文
C.B.N.-W.	(Centre Back Neck Point to Waist)	腰直（后颈点至腰距）
C.F.	(Centre Front)	前中长，前中缝
C.F.L.	(Centre Front Line)	前中线
	(Centre Front Length)	前中长
C.L.	(Coat Length)	衣长
	(Chest Line)	上胸围线
CLR	(Collar)	领子
C.P.L.	(Collar Point Length)	领尖长
C.P.W.	(Collar Point Width)	领尖宽
C.W.	(Cuff Width)	袖口宽
D.B.	(Double Breast)	双排纽，双襟
E.C.	(Elbow Circumference)	肘围
E.L.	(Elbow Length)	肘长
	(Elbow Line)	肘线
E.P.	(Elbow Point)	肘点
F.FRT	(Front)	前身，前胸
F.D.	(Front Depth)	前腋深
F.L.	(Front Length)	前长
F.N.	(Front Neck)	前领圈
F.N.P.	(Front Neck Point)	前颈点
F.R.	(Front Rise)	前（直）裆，〔粤〕前浪
F.S.	(Fist Size)	手头围
F.W.	(Front Width)	前胸宽
H.	(Hip)	臀围
H.L.	(Hip Line)	臀围线
	(Head Length)	头长
H.P.S.	(High Point of Shoulder)	肩顶点，肩高点
H.S.	(Head Size)	头围
I.	(Inseam)	内长
I.L.	(Inside Length)	股下，下裆长
K.L.	(Knee Line)	膝围线
L.	(Length)	衣（裤，裙等）长
	(Line)	线（条）
L/S	(Long Sleeve)	长袖
M.H.	(Middle Hip)	中臀围
M.H.L	(Middle Hip Line)	中臀围线
N. NK.	(Neck)	颈，领

N.H.	(Neck Hole)	领孔，领口，领圈
N.L.	(Neck Length)	领长
	(Neck Line)	领口线，领围线
N.P.	(Neck Point)	颈点；肩顶
N.R.	(Neck Rib)	领高
N.S.	(Neck Size)	颈围
N.S.P	(Neck Shoulder Point)	颈肩点
N.to W.	(Nape to Waist)	背长，腰直
N.W.L	(Neck Waist Length)	背长
O.S.	(Outside Seam)	外长
P.A.P.	(Posterior Armpit Point)	后腋窝点
PKT	(Pocket)	口袋（位）
PLKL	(Placket)	门襟
P. PNT	(Point)	点
P.S.	(Palm Size)	掌围
P.W.	(Point Width)	乳间宽；乳中
S.	(Sleeve)	袖长
	(Shoulder)	肩宽
S.A.	(Seam Allowance)	缝份，止口
S.B.	(Single Breast)	单排纽，单襟
	(Slack Bottom)	裤脚口
S.C.	(Stand Collar)	领座
S.D.	(Scye Depth)	腋深
SHLDER SH	(Shoulder)	肩宽
S.L.	(Sleeve Length)	袖长
	(Skirt Length)	裙长
SLV SV	(Sleeve)	袖子
S.N.P.	(Shoulder Neck Point)	肩颈点
	(Side of Neck Point)	旁颈点，颈侧点
S.P.	(Shoulder Point)	肩端点
S.S.	(Sleeve Slope)	肩斜
	(Side Seam)	侧缝
S/S	(Short Sleeve)	短袖
S.S.P	(Shoulder Sleeve Point)	肩袖点
S.T.	(Sleeve Top)	袖山
S.W.	(Shoulder Width)	肩宽
T.L.	(Trousers Length)	裤长

T.R.	(Trouser Rise)	裤(直)裆
T.S	(Thigh Size)	腿围
U.B.L.	(Under Bust Line)	下胸围线
W.	(Waist)	裤(裙)腰,腰节;腰围
WB W/B	(Waistband)	腰头
W.L.	(Waist Line)	腰围线,腰节线
Y	(Yoke)	过肩

2. 色相(hue)缩写词

R.	(Red)	红
O.	(Orange)	橙
Y.	(Yellow)	黄
G.	(Green)	绿
B.	(Blue)	蓝
P.	(Purple)	紫
YG.	(Yellow Green)	黄绿
BG.	(Blue Green)	蓝绿
PB.	(Purple Blue)	蓝紫
RP.	(Red Purple)	红紫
YR.	(Yellow Red)	红黄
pR.	(purplish Red)	紫调红
yR.	(yellowish Red)	黄调红
rO.	(reddish Orange)	红调橙
yO.	(yellowish Orange)	黄调橙
rY.	(reddish Yellow)	红调黄
gY.	(greenish Yellow)	绿调黄
yG.	(yellowish Green)	黄调绿
bG.	(blueish Green)	蓝调绿
gB.	(greenish Blue)	绿调蓝
pB.	(purplish Blue)	紫调蓝

3. 色调(tone)缩写词

l.	(light)	浅的
p.	(pale)	淡的
b.	(bright)	明亮的
d.	(dull)	浊的
s.	(strong)	强烈的

v.	(vivid)	鲜艳的
g.	(grayish)	灰调的
dk.	(dark)	暗的
dp.	(deep)	深的
lg.	(light grayish)	明灰调的
dg.	(dark grayish)	暗灰调的

4. 纤维（fibre）缩写词（或代码）

A AC	(acrylic)	腈纶
AC	(acetate)	醋酸酯纤维
AF	(anion fibre)	负离子纤维
	(aramid fibre)	芳纶纤维
AL	(albumen)	白蛋白（纤维）
	(alpaca)	羊驼毛
ALG	(alginate)	海藻纤维
AR	(aramid)	聚酰胺纤维，芳纶
BM	(bamboo)	竹（纤维）
C CO CTN	(cotton)	棉
CA	(cellulose acetate)	醋酯纤维
CF	(carbon fibre)	碳纤维
C.H.	(camel hair)	驼绒，驼毛
CL CLF	(chlorofibre)	含氯纤维
CLY	(lyocell)	莱赛尔纤维，天丝
CMD	(modal)	莫代尔纤维
CNF	(carbon nanofiber)	纳米碳纤维
CTA	(cellulose triacetate)	三醋酸酯纤维
CU CUP CUPRA	(cuprammonium)	铜氨纤维
CV	(viscose)	黏胶纤维
EL	(elastane)	弹性纤维
F	(flax)	亚麻
F fil.	(filament)	单纤维，长丝
GF	(glass fibre)	玻璃纤维
GNF	(graphene nanofiber)	石墨烯纳米纤维
HM HEM	(hemp)	大麻
JU	(jute)	黄麻
KP	(kapok)	木棉
L LI	(linen)	亚麻

LA	(lambswod)	羊羔毛
LC	(lyocell)	莱赛尔纤维，天丝
LY	(lycra)	莱卡（弹性纤维）
M	(mohair)	马海毛
m	(monofilament)	单纤维丝，单丝
MD	(modal)	莫代尔纤维
ME	(metallic)	金属纤维
MS	(mulberry silk)	桑蚕丝
MTF	(metal fibre)	金属纤维
MW	(merino wool)	美利奴羊毛
N	(nylon)	锦纶，尼龙
OP	(opelon)	奥佩纶（弹性纤维）
PA P.A.	(polyamide)	聚酰胺，锦纶
	(polyacrylate)	聚丙烯酸酯
PAN	(polyacrylonitrile)	聚丙烯腈纤维
PE	(polyethylene)	聚乙烯，乙纶
PLA	(polylactide)	聚乳酸纤维
P PE PES POLY	(polyester)	涤纶
PO	(polyolefine)	聚烯烃，环氧丙烷
PP Pp	(polypropylene)	聚丙烯，丙纶
PROT	(protein)	蛋白质纤维
PS	(polystyrene)	聚苯乙烯
PTT	(polytrimethylene terephthalate)	聚酯弹性纤维
PU	(polyurethane)	聚氨酯，氨纶
PV	(polyvinyl)	聚乙烯，维纶
PVA	(polyvinyl alcohol)	聚乙烯醇，维纶
PVC pvc P.V.C.	(polyvinyl chloride)	聚氯乙烯，氯纶
R RY RN	(rayon)	黏胶
RA	(ramie)	苎麻
RH	(rabbit hair)	兔毛
RY rn	(rayon)	人造丝
s	(staple)	短纤维，短丝
SB	(soybean)	大豆纤维
S SE	(silk)	丝
SP	(spandex)	氨纶
st.	(staple fibre)	人造短纤维
T	(terylene)	涤纶

TA	(tectel)	特达锦纶纤维
Tel	(tencel)	天丝
TPE	(thermo plastic polyethylene)	热塑聚乙烯
TPU	(thermo plastic polyurethane)	热塑聚氨酯
TS	(tencel)	天丝
	(tussah silk)	柞蚕丝
V	(vinylon)	维尼纶
VI v.	(viscose)	黏胶
w. wo.	(wool)	羊毛
WA	(angora)	安哥拉兔毛；安哥拉山羊毛
WK	(camel)	驼毛，驼绒
WL	(lambwool)	羔羊毛
	(llama)	美洲驼毛
WM	(mohair)	马海毛
WP	(alpaca)	羊驼毛
WS	(cashmere)	羊绒
WU	(guanaco)	原驼毛
WY	(yak)	牦牛毛
YH	(yak hair)	牦牛毛

5. 贸易和其他有关缩写词

@	(at)	在……（位置）
	(each)	每个，单价
a/c	(account)	账单；账户
acc.	(accessory)	辅料；配件
	(account)	账单；账户
ACU	(army combat uniform)	陆军作战服
ad. advt.	(advertisement)	广告
Add.add.	(address)	地址
adv.	(advice)	通知
A.F.B.	(Air Freight Bill)	航空提单
agt.	(agent)	代理商，代理人
a.m.	(ante meridiem)	上午
amt.	(amount)	总计，合计，总额
a.n.	(arrival note)	到货通知
a.o.p.	(all over print)	（图案）满印
app.	(approval)	确认

approx.	(approximately)	约计
Apr.	(April)	四月
AQL	(Acceptable Quality Level)	质量合格标准
AQSIQ	(General Administration of Quality Superrision, Ispection and Quarantine of the People's Republic of China)	国家质量监督检验检疫总局
a/r.	(all-round)	共计
art.	(article)	物品，商品；条款
AS as	(anti static)	防静电
asap	(as soon as possible)	尽快
asst.	(assortment)	品类；系列
ASU	(Army Service Uniform)	服役军装（非战时穿）
asym.	(asymmetric)	不对称的
att. attn.	(attention)	（请）注意
Aug.	(August)	八月
av.	(average)	平均
a/w.aw	(actual weight)	实际重量
a.w.	(all-wool)	全毛
A/W	(autumn/winter)	秋冬
B2C	(Business-to-Consumer)	商家对客户的电子商务
bh.	(buttonhole)	纽孔，扣眼
BK.	(black)	黑色
B/L b/l	(bill of lading)	提货单
BMS	(body measurement system)	人体尺寸测量系统
B.O.	(Branch Office)	分公司
BOC.	(Bank of China)	中国银行
BPM	(Basiness Process Management)	业务流程管理
B.RGDS.	(Best regards)	（信尾用语）谨致问候
B.S.	(body suit)	紧身套装，BS装
BT	(bartuck)	打结
B.T.	(bar tacker)	打结机
bx.	(boxes)	箱，盒（复数）
C/-	(case)	箱
CAD.	(computer aided design)	计算机辅助设计
C.A.D.	(cash against documents)	凭单据付款
CAL	(computer aided layout)	计算机辅助排料

CAM.	(computer aided manufacture)	计算机辅助生产
	(computer aided management)	计算机辅助管理
CAP	(computer aided pattern)	计算机辅助画样
CAT	(computer aided testing)	计算机辅助测试
cat.	(catalogue)	目录
CB.cb.	(cubic)	立方的
CBD.	(cash before delivery)	付款交货
CBM	(cubic meter)	立方米
C2C	(Consumer to Consumer)	个人对个人的电子商务
C.C.LBL.	(care content label)	洗水成分唛
CCM	(computer color matching)	计算机配色
C/D.	(certificate of delivery)	交货证明书
	(customs declaration)	报关单
C.& D.	(collected and delivered)	货款两清
CIEF	(China Import and Export Fair)	中国进出口商品交易会（广交会）
CELL.	(cellular phone)	手机
cert.	(certificate)	证书；执照
CFM	(confirm)	批准；确认
CFR C.F. c.f.C.&.F.	(cost and freight)	离岸加运费价格
CFW	(China Fashion Week)	中国国际时装周
CGFW	(China Graduate Fashion Week)	中国国际大学生时装周
C.H.	(Custom House)	海关
chg.	(change)	费用
CHIC	(China International Clothing & Accessories Fair)	中国国际服装服饰博览会
chq.	(cheque)	支票
C/I	(certificate of insurance)	保险证明书
CIF C.I.F.	(cost insurance and freight)	到岸价
CIM	(computer integrated manufacturing)	计算机综合制造
CIP	(carriage and insurance paid to)	运费和保险费付至（指定目的地）
Clthg.	(clothing)	服装
C/M	(certificate of manufacture)	制造商证明书
CM	(Change Management)	变革管理，应变管理
CM	(cutting and making)	（加工方负责）裁剪和缝制
cm.	(centimetre)	厘米，公分

C2M	(customer to manufacturer)	用户直连制造，客对厂
CMPT	(cutting, making, packing and trimmings)	（加工方负责）裁剪，缝制，包装及辅料
CMT	(cutting, making and trimmings)	（加工方负责）裁剪，缝制及辅料
CNC	(computer numerical control)	计算机数控
CNGA	(China National Garments Association)	中国服装协会
C/No.	(carton number)	箱号
CNTAC	(China National Textile & Apparel Council)	中国纺织工业联合会
CNY	(Chinese Yuan)	人民币
Co. co. coy.	(company)	公司
C/O	(certificate of origin)	产地证明书
c/o c.o.	(care of)	代收，转交
C.O.D. c.o.d.	(cash on delivery)	货到付款
col. clr.	(colour)	颜色，色彩
comm.	(commission)	佣金
cont.	(contract)	合同
c.o.o.	(country of origin)	原产地（国）
Corp.	(corporation)	公司
C.O.S.	(cash on shipment)	装船付款
cott.	(cotton)	棉花
CPT	(carriage paid to)	运费付至（指定目的地）
CQC	(China Quality Certification Centre)	中国质量认证中心
CRM	(Customer Relationship Management)	客户关系管理
CS	(commercial standard)	商业标准
c/s	(cases)	箱（复数）
CTN	(carton)	纸箱
CTR	(haute couture)	高级定制服装
CVC	(chief value of cotton)	以棉为主（50%或以上）的混纺织物
CXL	(cancel)	取消
D/A	(document against acceptance)	承兑交单
D. d.den.	(denier)	旦尼尔
	(density)	密度
	(diameter)	直径

3D	(three dimensions)	三维
DAF	(delivered at frontier)	边境交货
DBL	(double)	一对，双
D.D. D/D	(demand draft)	即期汇票
D/d	(document draft)	跟单汇票
DDP	(delivered duty paid)	进口国完税后交货
DDU	(delivered duty unpaid)	进口国未完税交货
Dec.	(December)	十二月
DEL	(delivery)	交付，交货
DELD	(delivered)	已交付
den. dens.	(density)	密度
dept.	(department)	部门
DEQ	(delivered ex quay)	目的港码头交货
DES	(delivered ex ship)	目的港船上交货
des.	(design)	设计
destn.	(destination)	抵达地
d.f.	(double fold)	双折
diam.	(diameter)	直径
diff.	(difference)	差异，差别
disc.	(discount)	折扣；贴现
DIY	(do it yourself)	自己动手做
DK. dk.	(dark)	深色（的）
dm.	(decimetre)	分米
DNTS	(double needle top stitch)	双针明线
d/o	(delivery order)	提货单
doc.	(document)	文件，单据
dol.	(dollar)	元（美国、加拿大、澳大利亚等的货币单位）
doz.	(dozen)	（一）打
DP	(durable press)	耐久压烫
D/P	(document against payment)	付款交单
DPC	(Dimensional Pattern Concept)	（服装）整体设计放码系统
Dr.	(Doctor)	博士
drt.	(draft)	汇票
ds	(detail sketch)	详图
d.t.	(delivery time)	交货时间
DTM	(dye to match)	配色

dto. do.	(ditto)	同上；同前
dup.	(duplicate)	副本
d/y	(delivery)	交货，交付
E.	(East)	东（方）
	(electron)	电子
EA ea	(each)	各个，每
eco.	(ecology)	生态学
E/D	(export declaration)	出口申报单
EDP	(electronic data processing)	电子数据处理
E.E.C.	(European Economic Community)	欧洲经济共同体（欧共体，欧盟的前身）
eg.	(拉丁文 exampli gratia)	例如
EMB embr	(embroidery)	绣花，刺绣
EMS	(express mail special)	特快专递
enc. encl.	(enclosure)	附件
E.& O.E.	(errors and omissions excepted)	差错待查
e.o.m.	(end of month)	月底
e.o.s.	(end of season)	季末
e.p.	(ever press)	永久熨烫
EQ	(excluding quota)	（报价时）不包配额
eq.	(equal)	等于
ERP	(Enterprise Resource Planning)	企业资源计划
ETA	(estimated time of arrival)	预计到达时间
etc.	(et cetera)	等等
ETD	(estimated time of departure)	预计发运时间
EU	(European Union)	欧洲联盟，欧盟
ex	(example)	例如
excl.	(excluding)	不包括
exp.	(export)	出口
exs.	(expenses)	费用
ext.	(extra)	特别的；额外的
EXW	(ex works)	工厂交货
F. f.	(foot)	英尺
	(front)	前面，前
fab.	(fabric)	织物，布
f.a.c.	(fast as can)	尽快
FAQ	(fair average quality)	中等品

FAS	(free alongside ship)	装运港船边交货
FAX. fax.	(facsimile)	传真
fb	(freight bill)	交货清单；运费单
FCA	(free carrier)	货交承运人
FCL	(full container load)	整（货）柜走货
f.e.	(for example)	例如
Feb.	(February)	二月
FFM	(fully fashion mark)	全成形标记
fib.	(fibre)	纤维
FLNG	(flange)	镶边
FLUO	(fluorescence)	荧光色
FM fm	(from)	从……起，来自
F.O.A.	(free on aircraft)	装运地飞机上交货
FOB f.o.b.	(free on board)	离岸价
f.o.c.	(free of change)	免费
FOR	(free on rail)	装运地火车上交货
FOS	(free on steamer)	装运地轮船上交货
FOT	(free on truck)	装运地卡车上交货
FP fp	(fire proof)	防火
FQC	(field quality control)	现场质量控制
FR fr	(flame retardant)	阻燃
fr. fc.	(franc)	法郎
	from	从……
	free	自由
Fri.	(Friday)	星期五
frt. frit. fgt.	(freight)	运费
FS	(fell seam)	对折缝
ft.	(foot, feet)	英尺
fty	(factory)	工厂
FW	(fabric weight)	织物重量
FX	(Foreign Exchange)	外汇
FYR	(for your reference)	供你参考
g. gm	[gram(s)]	克
GATT	(General Agreement on Tariffs & Trade)	关税及贸易总协定
GBP	(Great Britian Pound)	英磅
gds.	(goods)	货物

ggt.	(georgette)		乔其纱
GOH	(garment on hanger)		衣服挂装（走货）
gr.	(gross)		总额，总量；罗（=12打）
	[gram(s)]		克
GSM	(grams per square meter)		（布料）每平方米克重
GSP	(Generalised system of Preferences)		普遍优惠制（简称普惠制）
G.W. gr.wt.	(gross weight)		毛重
GWS	(garment wash sample)		成衣洗水样
H.O.	[head (home) office]		总公司，总部
HRM	(Human Resource Management)		人力资源管理
hr(s).	[hour (s)]		小时
H/S	(hand brush)		手砂，手擦
ht.	(height)		高度
	(hangtag)		吊牌
HTM	(How To Measure)		量身方法
IACD	(International Association of Clothing Designer)		国际服装设计师协会
ICA	(International Colour Authority)		国际色彩局
ICC	(International Chamber of Commerce)		国际商会
ICT	(information and communication technology)		信息通信技术
i.e.	(id est)		（拉丁语）那就是，即
IFC	(International Fashion Council)		国际流行（时装）评议会
IM	(Inventory Management)		库存管理
	(Intelligent Manufacturing)		智能制造
imp	(import)		进口
in.	[inch(es)]		英寸
ince. ins.	(insurance)		保险
incl.	(including)		包括
ind	(indent)		订单
inst.	(instant)		本月
int.	(interest)		利息
INTER-COLOR	(International Commission for Colour in Fashion and Textile)		国际流行色委员会
inv.	(invoice)		发票
I/P	(Insurance Policy)		保险单
IQ	(including quota)		（报价时）包配额

IQA	(integral quality assurance)	整体质量保证
irre.L/C	(irrevocable letter of credit)	不可撤销信用证
ISO, IOS	(International Organization for Standardization)	国际标准化组织
IT	(information technology)	信息技术
I.T.G.W.F.	(International Textile and Garment Worker's Federation)	国际纺织和服装工人联合会
ITMA	(International Textile Manufactures Association)	国际纺织业联合会
ITO	(International Trade Organization)	国际贸易组织
IWS	(International Wool Secretariat)	国际羊毛局
IWSFS	(International Wool Secretariat Fashion Studio)	国际羊毛局服装款式创作室
Jan.	(January)	一月
Jul.	(July)	七月
Jun.	(June)	六月
kg.	(kilogram)	千克,公斤
km.	(kilometer)	千米,公里
KM	(Knowledge Management)	知识管理
L.	(large)	大号
	(length)	长度
l.	(left)	左
lb.	(libra)	(一)磅
LBD	(little black dress)	小黑裙
lbl. lab.	(label)	唛头,商标
L/C l/c	(letter of credit)	信用证
LCL	(loose/less container loading)	拼(货)柜走货,走散货
L/D	(Lab Dip)	小块色样
Ld. Ltd.	(Limited)	有限股份(公司)
LG l/g	(letter of guarantee)	保函
LENG lg.lgth.	(Length)	长度
L/H	(label and hangtag)	唛头和吊牌
LM	(Logistics Management)	物流管理
LOA	(length over all)	全长,总长
l/p	(lab dip)	小块色样
LT. lt.	(light)	淡色(的)
L/y	(last year)	去年

m.	[metre(s)]	公尺，米
	(mark)	记号
	(month)	月
	(minute)	分钟
M. med.	(medium)	中号，中等
Mar.	(March)	三月
mat.	(material)	原料，布料
max.	(maximum)	最大
m/b	(must be)	必须
m/c	(machine)	机器
M2C	(manufacturer to customer)	生产厂家对消费者
meas.	(measurement)	尺寸；量身
meno.	(menorandum)	备忘录
MFN	(Most Favoured Nation)	最惠国
min.	(minimum)	最小
mk.	(mark)	商标；标志
mkt.	(market)	市场
mm.	(millimetre)	毫米
MNC	(multi-national corparation)	跨国公司
MO	(money order)	汇票
Mon.	(Monday)	星期一
mos.	(months)	月（复数）
Mr.	(Mister)	先生
Mrs.	(Mistress)	夫人，太太
Ms.	(Miss)	小姐
MT	(metric ton)	公吨
M/T	(mail transfer)	信汇
MTM	(made-to-measure)	量身定制
MTN	(Multilateral Trade Negotiations)	多边贸易谈判
MTO	(made-to-order)	按订单制作
MZ	(metal zipper)	金属拉链
N.	(North)	北（方）
n.c.v.	(no commercial value)	无商业价值
NDL	(needle)	针
NIL	(nothing)	无，没有
No. no.	[nomero(拉丁文), number]	数，号码
Nov.	(November)	十一月

n/p	(non payment)	拒付
N.W. nt. wt.	(net weight)	净重
NX	(next)	随后的；其次的
N.Y.	(New York)	（美）纽约
O/C	(outward collection)	出口托收
Oct.	(October)	十月
ODM	(Original Design Manufacturer)	贴牌设计生产
OEM	(original equipment manufacturing)	贴牌生产，来料（样）加工
OJT	(on the job training)	在职培训
O.K.	(agreed)	已获批准
o/l ovelk.	(overlock)	锁边
O/N	(Order No.)	订单号
O2O	(online to offline)	线上到线下（一体化）销售
OS	(oversize)	超大尺码
Oxf.	(Oxford)	（英）牛津
oz.	(ounce)	（一）盎司，英两，唡
p	(page)	页
PB	(Private Brand)	私人商标，自有品牌
pc.	(piece)	（一）件，（一）个
	[price's]	价格
P/C	(polyester/cotton)	涤棉混纺（织物）
PC. PCT.	(percent)	百分之
pcs. ps.	(pieces)	件，个（复数）
P.D. P/D	(piece-dyed)	匹染
pd	(paid)	付讫
P.I.	(preforma invoice)	形式发票
PJs PJ.	(pyjamas)	睡衣裤
pkg.	(package)	包装
P/L	(packing list)	装箱单
pls	(please)	请
p.m.	(past meridien)	下午
PM	(Project Management)	项目管理
PNT	(point)	点
P/O	(production order)	生产通知单
P.O.	(purchase order)	定购单

P.O.B.	(Post Office Box)	邮箱
p.o.d.	(payment on delivery)	交货时付款
pos	(position)	位置
pp	(pages)	页（复数）
p/p	(pre-production)	生产前
p.p.	(paper pattern)	纸样
PPS	(pre-production sample)	产前样
P.R.C.	(People's Republic of China)	中华人民共和国
Prof.	(Professor)	教授
prox.	(proximo)	下月
PRT	(print)	印花布
P.S. p.s.	(postscript)	再启，又及
PSI	(per square inch)	每平方英寸
pt. pnt.	(point)	点，尖
P.T.O. p.t.o.	(please turn over)	请看背面，见下页
PUR	(purchase)	购买
PX	(price)	价格
QA	(quality assurance)	品质保证
QC	(quality control)	质量管理，品质控制
	(quality controler)	质控员
QLY qlty.	(quality)	质量
qr.	(quarter)	四分之一；季度
QTY. qt.	(quantity)	数量
quotn.	(quotation)	报价
r.	(right)	右
®.	(registered trademark)	注册商标
RCPT.	(reciept)	收到；收据
rd.	(road)	路
REF	(reference)	参考
REJ.	(reject)	拒绝
RM	(Risk Management)	风险管理
rep.	(representative)	代表
RM. Rm.	(room)	房间，室
RN	(reference No.)	参考号
R.S.	(right side)	正面
R.T.W.	(ready to wear)	成衣
S.	(South)	南（方）

	(Small)	小号，小码
s	(count)	纱支
SA	(seam allowance)	缝份，止口
Sat.	(Saturday)	星期六
S/C	(sales contract)	销售合同
	(sales confirmation)	销售确认书
S.C.	(shopping center)	购物中心
SCM	(Supply Chain Management)	供应链管理
SD	(service dress)	军便服；制服
sec.	(second)	秒
Sep.	(September)	九月
SGS	(Societe Generale de Surveillance S.A.)	通用公证行
SHPG. SP.	(shipping)	装运；海运
SHPMT.	(shipment)	装船；装运
s.i.r.	(subject to immediate reply)	立即生效
S/M	(sewing machine)	缝纫机
smpl. spl.	(sample)	样品
S/N s/n	(shipping note)	装船通知
snl. sgl.	(single)	单（的），一个
SNTS	(single needle top stitch)	单针明线
SOHO	(small office/home office)	家居办公（自由职业）
S.P.	(sales promotion)	促销
SPA	(Speciality Retailer of Private Label Apparel)	自有品牌服饰专业零售商
	(Solus Por Aqua)	水疗健康美容
specs.	(specifications)	规格；说明书
SPI	(stitch per inch)	每英寸针数
SPM	(stitch per minute)	每分钟针数
sq.	(square)	平方
S/S	(spring/summer)	春/夏
S.S.	(steamship)	轮船（船名的前冠）
st.	(street)	街
std.	(standard)	标准
sty.	(style)	款式
Sun.	(Sunday)	星期天
s.w.	(shipping weight)	启运重量
sz.	(size)	尺码

t.	(ton)		（一）吨
tal.	(tailor)		裁缝，成衣工
TBA	(to be advised)		待复
TBC	(to be confirmed)		待确认
TBD	(to be decided)		待定
T/C	(terylene/cotton)		棉涤纶，涤棉
tech.	(technology)		技术；工艺
	(technician)		技术员
Tel. tel.	(telephone)		电话
tex. text.	(textile)		纺织品
tg.	(telegram)		电报
thd.	(thread)		线
thru.	(through)		穿过，通过
Thur(s)	(Thursday)		星期四
TIT	(tone in tone)		同色调配色
TOT	(tone on tone)		同色系配色
TQC	(total quality control)		全面质量管控
TQM	(total quality management)		全面质量管理
tr.	(tare)		皮重
T/R	(polyester/rayon)		涤黏（混纺）
T-S	(T-shirt)		T恤衫
TP/ST t/s	(top stitch)		明线迹
T/T	(telegraphic transfer)		电汇
Tu.Tues.	(Tuesday)		星期二
TV T.V.	(television)		电视
ult.	(ultimo)		上月
UPC	(universal product code)		通用产品代码，条形码
UV	(ultraviolet)		紫外线
via	(by way of)		经由
VM	(Value Management)		价值管理
vol.	(volume)		体积
W. w.	(west)		西（方）
	(weight)		重量
	(width)		宽度
Wed.	(Wednesday)		星期三
w.f.	(wrinkle-free)		无皱纹
WHK	(whisker)		猫须（洗水工艺）

W.I.P.	(work in process)	半成品
wk.	(week)	星期
WMSP	(workmanship)	手艺,做工
W/O	(without)	没有
w/p	(water proof)	防水
W/R	(warehouse receipt)	仓库
	(water repellent)	拒水,防水
W/S	(water shrinkage)	缩水
W.S.	(wrong side)	反面
wt.	(weight)	重量
WTO	(World Trade Organization)	世界贸易组织
W.W. W & W	(Wash and Wear)	洗可穿,免烫
XL	(extra large)	特大号
XOS	(extra outsize)	特大号
XXL	(extra extra large)	超特大号
y	(yard)	码
	(yuan)	元
	(year)	年
ya	(yard)	码
	(yarn)	纱线
Y.D. Y/D	(yarn-dyed)	纱线染色,色织
yd.	(yard)	码
yds.	(yards)	码(复数)
yr	(year)	年
	(your)	你(们)的

三、粤语、普通话、英语对照词 200 例
200 Clothing Words in Cantonese, Putonghua and English

粤语	普通话	英语
唐装	中式服装	Chinese-style costume
衫	衣服,衣裳	clothes
面衫	外衣	outer coat

粤语	普通话	英语
底衫	内衣	underclothes
底裤	内裤	underpants
笠衫	汗衫	undershirt
恤衫	衬衫	shirt
过头笠	套（头）衫	pullover
T恤	针织圆领衫，针织翻领衫	T-shirt
机恤	夹克衫	jacket
冷衫	针织毛衣，毛衣	sweater
冷裤	针织毛裤	knitted woolen trousers
褛	大衣	coat, overcoat
长褛	长大衣	long coat
中褛	半长大衣	midi coat
短褛	短大衣	short coat
太空褛	短外套	carcoat
晨褛	长晨衣	dressing gown
干湿褛	风雨衣	weather-all coat
风褛	风衣	windbreaker
间棉褛	棉大衣	cotton-padded coat
雪褛	风雪大衣	parka
皮褛	皮大衣	leather coat
皮草	毛皮大衣	fur coat
棉衲	棉袄	cotton quilted jacket
衫裙	连衣裙	dress
波裤	运动短裤	sport shorts
BB装	婴儿服装	babywear
高睁鞋	高跟鞋	high heeled shoes
船睁鞋	坡跟鞋	wedge shoes
波鞋	运动鞋	sport shoes
手袜	手套	gloves
手梆	手镯	bracelet
颈巾	围巾	scarf
颈链	项链	necklace

粤语	普通话	英语
枙	领带	tie
煲枙	蝴蝶领结	bow-tie
幼腰带	细窄腰带	narrow belt
手巾仔	手帕	handkerchief
遮	伞	umbrella
缩骨遮	折叠伞	topless umbrella
板	样衣,样品	sample
头板	初样,原样	initial sample
批板	确认样	approval sample
影相板	照相样	photo sample
行街板	推销样	salesman sample
船头板	船样	shipping sample
办房	样品间	sample room
衫身	上衣主要部分	body
上级领	领面	collar fall
下级领	底领	collar band
上下级领	翻领	turn-down collar
企领	立领	stand-up collar
樽领	瓶颈领	bottled collar
襟领	驳领,驳头	lapel
襟贴	驳领面	lapel facing
襟褶	驳头省	lapel dart
襟宽位	驳领宽	lapel step
襟纽门	驳头眼,假扣眼	lapel buttonhole
襟领口位	驳领串口	gorge
扼位	领嘴	notch
膊阔	肩宽	shoulder width
膊平	上平线	imaginary line
牛角袖	连肩袖	raglan sleeve
夹圈,夹围	袖窿	armhole
袖肶围	袖肥	biceps

粤语	普通话	英语
夹底	袖下；腋下	underarm
夹底褶	袖底省，胁省	underarm dart
领横	领宽	neck width
胸阔	胸宽	chest width
门筒，筒位	门襟	placket
明筒	明门襟	front strap
暗筒	暗门襟	cover placket
假筒	假门襟	false placket
底筒	里襟	under placket
孖襟	双门襟	double breast
筒阔	门襟宽	placket width
搭位	搭门	overlap
担干，擔干	过肩	yoke
龟背	半月形贴布	half-moon patch
鸡英，介英	袖口，袖头	cuff
摺脚	（下摆，脚口的）折边	hem
侧骨	侧缝，摆缝	side seam
包骨	包叠缝	wrap seam
皿骨	覆盖缝，绷缝	covered seam
膊头骨	肩缝	shoulder seam
后中骨	背中缝	centre back seam
胸底骨	高腰缝	empire seam
公主骨	公主缝，刀背缝	princess seam
衫脚	底边	bottom
脚围，脚阔	底边宽；脚口宽	bottom width
裤浪	裤（直）裆	rise
前浪	前（直）裆	front rise
后浪	后（直）裆	back rise
后浪中骨	后裆缝	seat seam
外肽骨	外侧缝	outseam
内肽骨	内侧缝，下裆缝	inseam

粤语	普通话	英语
十字骨	裤裆底	cross crotch
反脚	卷边裤脚	cuffed bottom
坐围	臀围	seat, hip
肶围	横裆	thigh
拉度（松紧部位）	拉开量（度）	stretched, extended
平度（松紧部位）	放松量（度）	relaxed
机头	裤裙腰部拼接布	yoke
肶袋	裤腿贴袋	thigh pocket
侧骨袋	外侧袋	outseam pocket
耳仔	裤带襻	belt loop
纽门	扣眼	buttonhole
（腰头）搭位	宝剑头	extended tab
纽牌	门襟	fly
纽牌里	门襟里	fly lining
纽子	扣位，里襟	button stand
贴	贴边，贴布	facing
袋贴	袋口贴边	pocket facing
襟贴	驳领面	lapel facing
原身出贴	原身布折边	fold back facing
裤脚增强位	贴脚条	bottom binding
贴布	贴花	appliqué
梳织	机织	weaving
斜布	斜纹布	twill, drill
格仔布	格子布	check
间条布	条纹布	stripe
色丁	缎子	satin
水洗丝	洗水绸	washed silk
抓毛布	起绒布	fleece, flannelette
绒布	呢绒，呢子	woolens
粗绒	粗花呢	tweed
噶唏	马海毛	mohair

粤语、普通话、英语对照词 200 例

粤语	普通话	英语
拉架	莱卡（弹性纤维商名）	lycra
布碎	小布块	swatch
布办	布样	fabric swatch
散口	毛边	fringed edge
布封，封度	幅宽	fabric width
朴，里朴	衬布	interlining
主朴	主衬布	main interlining
黏补	黏合衬	fusible interlining
生朴	非黏合衬	non-fusible interlining
纸朴	无纺衬	non-woven interlining
亚沙的	醋酸里子绸	acetate lining
纽	纽扣	button
士啤纽	备用扣	space button
噏纽	四合纽	snap fastener
乌蝇扣	钩眼扣	hook and eye
辘扣	旋筒带扣	roll buckle
叻色	（金属扣）有亮光	light
哑叻	（金属扣）无亮光	dark
克叻	（金属扣）黑色亮光	black & light
无叻	（金属扣）环保无毒	nonpoisonous
绳制	拉绳扣	stopper
魔术贴	尼龙搭扣	velcro
橡根，丈根	橡筋，松紧带	elastic
膊头棉	肩垫，垫肩	shoulder pad
厘士	花边	lace
胶蝴蝶	领口蝴蝶插片	collar fly
插竹	领插角片	collar stay
烟子，烟治	尺码唛	size label
挂咭	吊牌	hang-tag
平车	平缝机	flat sewing machine
拉冚车，冚车，虾苏网车	绷缝机	covering stitch machine

粤语	普通话	英语
钑骨车	包缝机，锁边机	overlock machine
埋夹车	折缝机	fell seaming machine
人字车	曲折缝缝纫机	zigzag sewing machine
筒车	门襟机	placket machine
切筒车	切门襟机	placket cutting machine
反鸡英车	翻袖头机	cuff turning machine
拉裤头车	腰头机	waistband machine
耳仔车	裤襻机	loop sewing machine
打枣车	打结机	bar tacker
钉纽车	钉扣机	buttoning machine
啤纽车	钉四合扣机	snap fixing machine
纽门车	锁扣眼机	button holing machine
挑脚车	暗缝机	blind stitching machine
辘脚车	脚口卷边机	bottom hemming machine
推波车	波边机	fluted hem machine
拉筒蝴蝶	折边器	folder
绲边蝴蝶	绲边器	binder
梭仔	梭芯	bobbin
狗牙	送布牙	feed dog
靴仔	压脚	foot
啤机	冲压裁剪机	die cutting machine
夹机	（开合式）熨烫机	pressing machine
辘烫机	滚动式熨烫机	rotary pressing machine
黏朴机	黏合机	fusing press machine
焗领机	热压领机	collar pressing machine
焗炉	（整熨用）烘箱	oven
唛架	排料图	marker
排唛架	排纸样，排料	marker laying
车工	缝纫工	sewer
车缝	机器缝纫	machine sewing
车花	机绣	machine embroidering

粤语	普通话	英语
切驳	裁剪和缝纫	cutting and sewing
执扎	（裁片）捆扎	bunding
返针	回针	back stitching
线步，针步	线迹，针迹	stitch
平车针步	平缝针迹	plain stitch
锁链底针步	链式针迹	chain stitch
拉冚线步	绷缝线迹	covering stitch
抛位	宽松量	tolerance
间边线	缝（衣）边	edge stitching
埋夹	合袖下缝或裤内缝	feed-off-the-arm felling
纳膊	合肩缝	joining shoulder seam
拉裤头	上腰头	waist banding
埋浪	合裤裆	join crotch
埋小浪	合小裆	join front crotch
埋侧骨	合侧缝	close side seam
打纽	钉扣	button(snap) attaching
打纽门	开扣眼	buttonholing
打枣	打结	tacking, bartack
钑骨	包缝，锁边	overlocking
挑脚	暗缝缲边	blind stitching
手针挑脚	手工缲边	hand hemming
打揽	装饰性缩缝	smocking
还口	卷边	hemming
暗枣	隐形套结	hidden bartack
襟着	耐穿	be endurable
克色	黑色	black
甩色	脱色，掉色	fading
起镜	（熨烫不良所致）亮光	iron-shine
针窿	针孔	needle hole
鸡	缺陷，疵点	defect
大鸡	大缺陷	major defect

粤语	普通话	英语
细鸡	小缺陷	minor defect
鸡纸	查货贴纸	QC sticker
执鸡，扎鸡	返工	rework
爆口	线缝断开	seam broken
扭骨	线缝扭曲	twisted seam
扭肶	裤腿扭曲	twisted leg
落坑	缉线上、下炕	run off stitching
驳线	线迹重叠	stitch overlapping
骨位皱	线缝皱缩	seam puckering
骨位起蛇	线缝起拱	wavy seam
筒起蛇	门襟起拱	wavy placket
拉链起蛇	拉链不平伏	wavy zipper
露底筒	里襟外露	exposed under placket
长短筒，高低脚	门、里襟长短不一	unmatched front fly
长短袖	袖长不一	unmatched sleeve
大小唇	袋唇宽窄不匀	uneven pocket lip

四、公、市、英制长度单位及其换算
The Units Of Length in Metric, Chinese, British & U.S. System and Their Conversion with Each Other

1. 长度单位

①公制　The Metric System

　　米　metre(m)
　　分米　decimetre(dm)
　　厘米　centimetre(cm)
　　毫米　millimetre(mm)

1 米 = 10 分米 = 100 厘米 = 1000 毫米

②市制　The Chinese System

　　丈　zhang
　　尺　chi
　　寸　cun
　　分　fen

1 丈 = 10 尺 = 100 寸 = 1000 分

③英制 The British and U.S. System
码 yard(yd)
英尺 foot(ft)
英寸 inch(in)
1 码 = 3 英尺 = 36 英寸

2. 长度换算

①公制折合市制或英制

公 制	市 制	英 制
1 米	3 尺	3.2806 英尺或 1.0936 码
1 分米	3 寸	3.9370 英寸
1 厘米	3 分	0.3937 英寸
2 厘米	6 分	0.7874 英寸
3 厘米	9 分	1.1811 英寸
4 厘米	1.2 寸	1.5748 英寸
5 厘米	1.5 寸	1.9685 英寸
6 厘米	1.8 寸	2.3622 英寸
7 厘米	2.1 寸	2.7559 英寸
8 厘米	2.4 寸	3.1496 英寸
9 厘米	2.7 寸	3.5433 英寸

②市制折合公制或英制

市 制	公 制	英 制
1 丈	$3\frac{1}{3}$ 米	10.936 英尺
1 尺	$\frac{1}{3}$ 米或 $33\frac{1}{3}$ 厘米	1.0936 英尺
1 寸	$3\frac{1}{3}$ 厘米	1.3123 英寸
2 寸	$6\frac{2}{3}$ 厘米	2.6246 英寸
3 寸	10 厘米	3.9370 英寸
4 寸	$13\frac{1}{3}$ 厘米	5.2493 英寸

续表

市制	公制	英制
5寸	$16\frac{2}{3}$ 厘米	6.5616 英寸
6寸	20 厘米	7.8739 英寸
7寸	$23\frac{1}{3}$ 厘米	9.1862 英寸
8寸	$26\frac{2}{3}$ 厘米	10.4986 英寸
9寸	30 厘米	11.8109 英寸

③英制折合公制或市制

英制	公制	市制
1 码	0.9144 米	2.7432 尺
1 英尺	0.3048 米或 30.48 厘米	0.9144 尺
1 英寸	2.54 厘米	0.762 寸
2 英寸	5.08 厘米	1.524 寸
3 英寸	7.62 厘米	2.286 寸
4 英寸	10.16 厘米	3.048 寸
5 英寸	12.70 厘米	3.810 寸
6 英寸	15.24 厘米	4.572 寸
7 英寸	17.78 厘米	5.334 寸
8 英寸	20.32 厘米	6.096 寸
9 英寸	22.86 厘米	6.858 寸
10 英寸	25.40 厘米	7.620 寸
11 英寸	27.94 厘米	8.382 寸

五、国际通用服装洗涤、熨烫符号及含义
International General Clothing Washing and Ironing Symbols and Meanings

为了帮助消费者对不同质地的纺织品、服装进行恰当的保养，国际标准化组织制定了国际标准 ISO3756，建立了一整套图形的符号体系，以此作为纺织品、服装保养的永久标记。保养符号及含义如下：

1. 基本符号及其附加符号

基本符号	⊔	洗涤符号
	△	氯漂符号
	⊿	熨烫符号
	○	干洗符号
	⊡	滚筒干燥符号
附加符号	×	禁止符号，若加在上述任何一种符号上，则表示该符号所代表的处理方法禁止采用
	—	轻度处理符号，若在洗涤或干洗的符号下加一条横线，则表示所做的处理程度要比无线的相同符号轻
	=	极轻度处理符号，若在洗涤符号下加两条横线，则表示温度在40℃下，非常轻柔的处理

2. 洗涤符号

⊔95	最高水温：95℃ 机械动作：常规	漂洗：常规 脱水：常规
⊔95	最高水温：95℃ 机械动作：减弱	漂洗：逐渐降温（至冷却） 脱水：减弱
⊔70	最高洗涤温度：70℃ 机械动作：常规	漂洗：常规 脱水：常规
⊔60	最高洗涤温度：60℃ 机械动作：常规	漂洗：常规 脱水：常规
⊔60	最高洗涤温度：60℃ 机械动作：减弱	漂洗：逐渐降温（至冷却） 脱水：减弱
⊔50	最高洗涤温度：50℃ 机械动作：减弱	漂洗：逐渐降温（至冷却） 脱水：减弱
⊔40	最高洗涤温度：40℃ 机械动作：常规	漂洗：常规 脱水：常规

☐40	最高洗涤温度：40℃ 机械动作：减弱	漂洗：逐渐降温（至冷却） 脱水：减弱
☐40	最高洗涤温度：40℃ 机械动作：极弱	漂洗：常规 脱水：常规
☐30	最高洗涤温度：30℃ 机械动作：减弱	漂洗：逐渐降温（至冷却） 脱水：减弱
☐	最高洗涤温度：40℃，只可手洗，禁止机洗，洗涤小心	
☒	禁止洗涤，潮湿时处理要小心	

3. 氯漂符号

△Cl	在洗涤前、后或过程中可以加放氯基漂白剂，但仅可用冷而稀的溶液漂白
⊠	禁止氯基漂白

4. 熨烫符号

⌒⋯	熨斗底板最高温度：200℃
⌒⋯	熨斗底板最高温度：150℃
⌒·	熨斗底板最高温度：110℃ 不宜蒸汽熨烫
⌀	禁止熨烫 不准汽烫和蒸汽处理

5. 干洗符号

Ⓐ	可以干洗，所有类型的干洗溶剂均可使用

Ⓟ	可以干洗，按常规洗涤程序洗涤，无需限制。用四氯乙烯、一氟三氯甲烷和Ⓕ栏中所有溶剂干洗
Ⓟ̱	严格限制干洗时的水分和机械运动，在漂洗和干燥时严格控制温度，禁止自动清洗，干洗剂同Ⓟ栏
Ⓕ	按正常漂洗程序洗涤，无需任何限制。用三氟三氯乙烷、石油精干洗
Ⓕ̱	严格限制干洗时的水分和机械运动，在漂洗和干燥时严格控制温度，禁止自动清洗，干洗剂同Ⓕ栏
⊠	禁止干洗。禁用溶剂除污

6. 滚筒干燥符号（不适于真皮和毛皮制品）

⊡⋅⋅	可用滚筒烘干，正常的烘干周期
⊡⋅	可用滚筒烘干，设定低温烘干
⊠	禁止采用滚筒烘干

六、部分词汇图释
Illustrations of Some Vocabularies

1. 上衣

① 领座　stand collar
② 领吊襻　hanging loop
③ 领面　top collar
④ 肩线　shoulder line
⑤ 袖山　sleeve top
⑥ 领嘴　lapel point (notch)
⑦ 假眼　mock buttonhole
⑧ 袖窿　armhole
⑨ 胸袋　breast pocket
⑩ 扣眼　buttonhole
⑪ 门襟　top fly (left front)
⑫ 前胸省　front dart
⑬ 胁省　underarm dart
⑭ 大袖　top sleeve
⑮ 袖扣　sleeve button
⑯ 袖口　sleeve opening
⑰ 底边，下摆　hem
⑱ 止口圆角　front cut

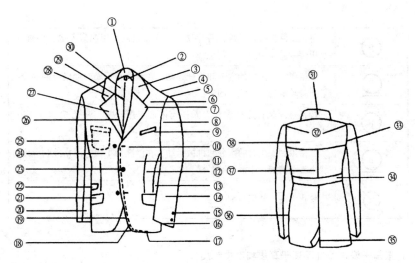

⑲ 门襟止口　front edge
⑳ 小袖　under sleeve
㉑ 袋盖　flap
㉒ 零钱袋　change pocket
㉓ 前扣　front button
㉔ 里襟　under fly（right front）
㉕ 里袋　inside breast pocket
㉖ 驳口　fold line for lapel
㉗ 驳头　lapel
㉘ 串口　gorge line

㉙ 后片里中缝　back lining centre pleat
㉚ 后片里　back lining
㉛ 领面　top collar
㉜ 总肩　across back shoulders
㉝ 后袖隆　back armhole
㉞ 半腰带　half back belt
㉟ 背衩　back vent
㊱ 侧缝　side seam
㊲ 背缝　centre back seam
㊳ 后过肩　back yoke

2. 裤

① 腰头　waistband
② 腰头纽　waistband button
③ 里襟尖嘴　button tab
④ 腰头里　waistband lining
⑤ 裤头纽　bearer button
⑥ 腰头宝剑头　extended tab
⑦ 门襟里　left fly lining
⑧ 扣眼　left fly buttonhole
⑨ 裤门襟　left fly
⑩ 裤裆垫布　crutch lining
⑪ 外侧缝　side seam
⑫ 下档缝　inside seam
⑬ 膝盖绸　reinforcement for knees

⑭ 脚口　leg opening
⑮ 卷脚　turn-up cuff
⑯ 贴脚条　heel stay
⑰ 裤中线　crease line
⑱ 纽扣　right fly button
⑲ 里襟　right fly
⑳ 侧斜袋　slant side pocket
㉑ 裤裥　waist pleat
㉒ 表袋　watch pocket
㉓ 裤带襻　belt loop
㉔ 后袋　hip pocket
㉕ 后档缝　seat seam
㉖ 后裥　back pleat

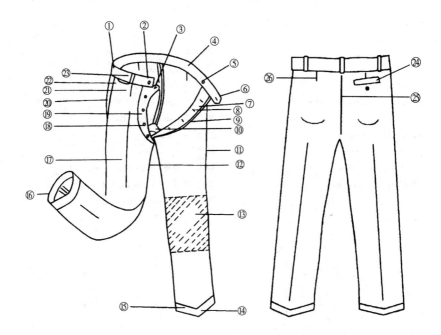

3. 裙

① 裙腰头　waistband
② 侧拉链口　side opening
③ 暗裥　inverted pleat
④ 中缝　centre seam
⑤ 裙腰省　waist dart
⑥ 裙腰省　waist dart
⑦ 侧开口　side opening
⑧ 前育克　front yoke
⑨ 裙褶　pleats
⑩ 褶脚　hem
⑪ 裙摆缝　side seam
⑫ 斜袋　slant welt pocket
⑬ 腰带襻　belt loop

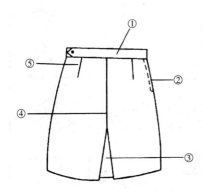
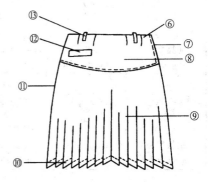

4. 背心

① 衬里　lining
② 肩缝　shoulder seam
③ 袖窿　armhole
④ 胸袋　breast welt pocket
⑤ 门襟　top fly
⑥ 扣眼　buttonhole
⑦ 边衩　side vent
⑧ 尖角　front point
⑨ 衬布　interlining
⑩ 里襟贴　right facing
⑪ 纽扣　button
⑫ 腰袋　waist welt pocket
⑬ 摆缝侧　seam
⑭ 胸省　breast dart
⑮ 里襟　under fly
⑯ 肩缝　shoulder seam
⑰ 袖窿　armhole
⑱ 摆缝侧　seam
⑲ 活动腰襻　adjustable waist tab
⑳ 后腰省　back waist dart
㉑ 背中缝　centre back seam

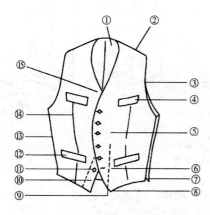
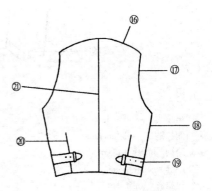

5. 领子

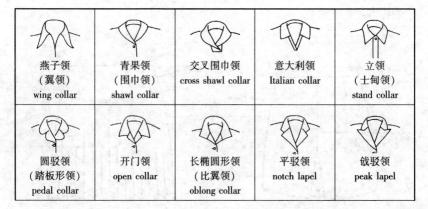

领子 459

续表

翻下领 turn collar	远离领 far-away collar	马蹄领 horse-hoof collar	小圆领（彼得·潘领）Peter Pan collar	伊顿领 Eton collar	
伊丽莎白女王领 Elizabethan collar	两用领（换形领）convertible collar	下扣领（扣贴领）button-down collar	别针扣领 pinhole collar	小方领（短领）short square collar	
帆形叠领 reefer collar	巴尔马干领 Balmacaan collar	拿破仑领 Napoleon collar	倒挂领（阿尔斯特领）ulster collar	中式领（旗袍领）mandarin collar	
军官领 officer collar	颚领 chin collar	偏侧领 sideway collar	农民领 peasant collar	带扣领 belt collar	
单折领 single collar	隧道领 tunnel collar	小斗篷领（清教徒领）puritanical collar	披肩领 cape collar	巴莎领 purser collar	
围兜领 bib collar	皱褶领 ruffled collar	棒形领 stick collar	直离立领 stand away collar	小丑领 pierrot collar	

续表

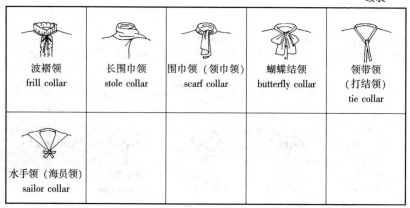

波裥领 frill collar	长围巾领 stole collar	围巾领（领巾领）scarf collar	蝴蝶结领 butterfly collar	领带领（打结领）tie collar
水手领（海员领）sailor collar				

6. 领口

玳瑁领口 turtle neckline	高领口 high neckline	中装领领口 Chinese neckline	农民领领口 peasant neckline	水手领领口 crew neckline
圆形领口 round neckline	直离领领口 stand away neckline	远开领领口 far-away neckline	船型领口 boat neckline	荷兰领领口 Dutch neckline
长椭圆形领口 oblong neckline	U形领口 U-neckline	马蹄形领口 horse-hoof neckline	蛋形领口 oval neckline	勺形领口 scoop neckline
宽开领口 off neckline	低胸领口 low neckline	低肩领口 off-shoulder neckline	鸡心领口 sweetheart neckline	扇贝形领口 scalloped neckline

袖

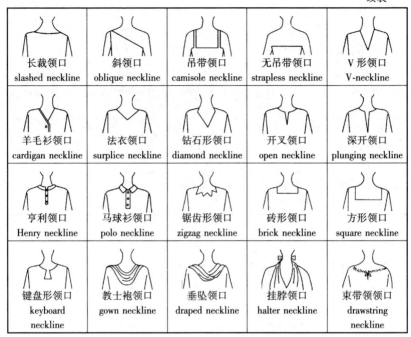

长裁领口 slashed neckline	斜领口 oblique neckline	吊带领口 camisole neckline	无吊带领口 strapless neckline	V形领口 V-neckline
羊毛衫领口 cardigan neckline	法衣领口 surplice neckline	钻石形领口 diamond neckline	开叉领口 open neckline	深开领口 plunging neckline
亨利领口 Henry neckline	马球衫领口 polo neckline	锯齿形领口 zigzag neckline	砖形领口 brick neckline	方形领口 square neckline
键盘形领口 keyboard neckline	教士袍领口 gown neckline	垂坠领口 draped neckline	挂脖领口 halter neckline	束带领口 drawstring neckline

7. 袖

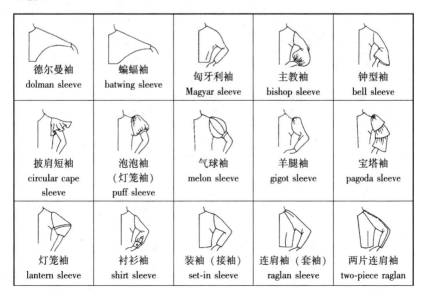

德尔曼袖 dolman sleeve	蝙蝠袖 batwing sleeve	匈牙利袖 Magyar sleeve	主教袖 bishop sleeve	钟型袖 bell sleeve
披肩短袖 circular cape sleeve	泡泡袖(灯笼袖) puff sleeve	气球袖 melon sleeve	羊腿袖 gigot sleeve	宝塔袖 pagoda sleeve
灯笼袖 lantern sleeve	衬衫袖 shirt sleeve	装袖(接袖) set-in sleeve	连肩袖(套袖) raglan sleeve	两片连肩袖 two-piece raglan

续表

两片袖 two-piece sleeve	落肩袖（肩章袖）epaulet sleeve	翻折袖 roll-up sleeve	花瓣袖 petal sleeve	垂褶袖 cowl sleeve
盖袖（喇叭袖）cape sleeve	变形袖 styled sleeve	和服袖（连袖）kimono sleeve		

8. 口袋

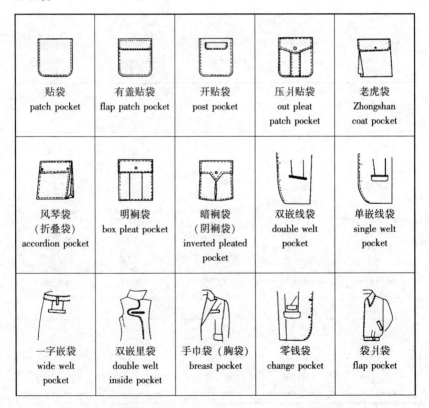

贴袋 patch pocket	有盖贴袋 flap patch pocket	开贴袋 post pocket	压爿贴袋 out pleat patch pocket	老虎袋 Zhongshan coat pocket
风琴袋（折叠袋）accordion pocket	明裥袋 box pleat pocket	暗裥袋（阴裥袋）inverted pleated pocket	双嵌线袋 double welt pocket	单嵌线袋 single welt pocket
一字嵌袋 wide welt pocket	双嵌里袋 double welt inside pocket	手巾袋（胸袋）breast pocket	零钱袋 change pocket	袋爿袋 flap pocket

续表

卡片袋 card pocket	眼镜袋 glasses pocket	侧直袋 vertical side pocket	侧斜袋 slant side pocket	侧横袋 horizontal side pocket
表口袋（一）watch pocket	表口袋（二）watch pocket	裤后袋（臀袋）hip pocket	锯齿形里袋 zigzag inside pocket	

9. 廓型

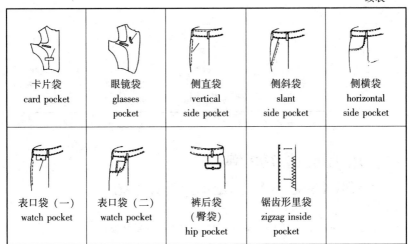

苗条线型（紧身线型） slim line	低腰线型 drop waist line	公主线型 princess line	丹尼尔线型 Daniel line	自然线型 natural line
直线型 straight line	长身线型 long torso line	喇叭线型 trumpet line	T恤线型 T-shirt line	宽松线型 ample line
矩形线型 rectangle line	锥形线型 tapered line	Y形线型 Y-line	沙漏线型 hourglass line	X形线型 X-line

| 盒形线型 | 帐篷线型 | 气球线型 | 桶形线型 | 宽大线型 |
| box line | tent line | balloon line | barrel line | big line |

七、服装英语 100 例
100 Cases of Clothing English

1. Clothing performs a range of social and culture functions, such as individual, occupational and gender difference, and social status. It may also function as a form of adornment and an expression of personal taste or style.

服装具有一系列社会和文化功能，如反映个人、职业和性别差异，以及社会地位。服装也可以作为装饰品和一种个人品位或风格的表达。

2. The garment industry involves many upstream and downstream enterprises, and employs millions of people. It's one of the biggest industries in the world.

服装业涉及众多上下游企业，从业人员有数百万人。这是世界上最大的产业之一。

3. Modern apparel engineering is composed of these parts as style design, structure design and process design. Clothing pattern drafting belongs to the structure design.

现代服装工程是由款式造型设计、结构设计和工艺设计三部分组成。服装制板属于结构设计。

4. Clothtech (clothing textile) is a segment of technical textile that includes all textile components used primarily in clothing and footwear. Clothtech adds functional properties to the product that improves specific and critical objectives.

服装纺织品是技术纺织品的一部分，包括所有主要用于服装和鞋类的纺织品类。它增加了产品的功能属性，以改善特定和关键的目标。

5. Fashion industry is facing new challenges. The intelligent textiles, smart clothing and fashion engineering are only a few of the keywords.

时尚产业正面临着新的挑战，智能纺织品、智能服装和时尚工程只是其中几个关键词。

6. The cities of New York, London, Paris and Milan, as the global "Big Four" fashion capitals of the 21st century, have major impact on international fashion trends. Meanwhile, some countries also hold their own Fashion Week.

纽约、伦敦、巴黎和米兰，作为21世纪四大时尚之都，对国际流行趋势有重大影响。同时，有些国家也在本土举办他们自己的时装周。

7. Every year in the fashion capitals, the design, production and retailing of fashion products, plus events such as fashion week, awards and trade fairs all generate significant economic output.

每年在时尚之都，时装产品的设计、生产和销售，加上时装周、颁奖典礼和交易会等活动，都产生了巨大的经济产出。

8. In the fashion industry, fashion modelling is often used as a type of advertising. Models are often featured in magazines and catalogues as well as on television and the internet.

在时尚界，时尚造型经常被用作广告的一种类型。模特经常出现在杂志和商品目录上，也出现在电视和互联网上。

9. Vegan clothing is considered an eco-fashion. It's free from any type of animal products and only made with plant and man-made fibers. These are typically sustainable as well.

素食服装被认为是一种生态时尚。它不含任何类型的动物产品，只用植物纤维和人造纤维制成。这些通常也是可持续的。

10. China Fashion Week, originated in 1997, is an international fashion event held twice a year in Beijing, China in March and October. It showcases fashion collections from various designers, including ready-to-wear, accessories, styling and other new designs.

起源于1997年的"中国国际时装周"是一项国际时装活动，每年3月和10月分两季在北京举行。这个活动展示了不同设计师的时装系列，包括成衣、配饰、造型和其他新设计。

11. Chinoiserie (China Chic) became the most popular element at Fashion Week in the Big Four cities on the eve of the 2008 Beijing Olympics.

在2008年北京奥运会前夕，在四大时尚之都的时装周上，中国风成为最受欢迎的元素。

12. There was a nice sartorial symbiosis playing out at the Beijing 2022 Olympic Winter Games. Fashion had never been more techwear-centric, and the Olympics had never been more capital-Fashion-minded. Many internationally clothing brands were displayed here. It really was the fashion Olympics.

在2022年北京冬奥会上，上演了一幕美好的服装共生关系。时尚界从未像现在这样以科技服装为中心，奥运会也从未像今天这样以时尚为中心。众多国际知名服装品牌在这里展示，这才是真正的时尚奥运会。

13. It is obvious that the cultural elements from Chinese traditional costume have impact on the European brand fashion.

中国传统服饰文化元素对欧洲品牌时装的影响显而易见。

14. The "catwalk" is a long stage along which models walk to display various of clothes created by the designers during the fashion shows.

T台是一个狭长的表演台，在时装秀中，模特走在上面展示设计师设计的各种服装。

15. Although many young girls dream of being a fashion model and think that model just involves wearing different fashions and looking pretty. However, the job is in fact rather demanding and stressful.

虽然很多女孩子都梦想当一名时装模特，认为模特只是穿着不同的时装，看起来很漂亮。然而事实上，这份工作要求很高，压力也很大。

16. In recent years, more and more Chinese fashion designers have appeared in the "Big Four" fashion weeks, showing the results of local apparel education in China.

近年来，越来越多的中国服装设计师出现在四大国际时装周上，显示了中国服装教育的成果。

17. More than 80 universities and colleges in China have established the fashion design majors. Some of them are very famous, such as Beijing Institute of Fashion Technology, Donghua University, Academy of Arts and Design, Tsinghua University, etc.

中国已开设服装设计专业的院校超过80所，其中著名的有北京服装学院、东华大学、清华大学美术学院等。

18. Clothing is the specific fruit of human labour. Costume culture, which is one of the symbols of social progress and civilization, has unique aesthetic value.

服饰是人类特有的劳动成果。服饰文化是社会进步和文明的标志之一，具有独特的审美价值。

19. The costume style and features of Dunhuang murals and coloured sculpture are undoubtedly a vital part of Chinese costume history and treasure of art.

敦煌壁画和彩塑的服饰风格无疑是中国服饰史的重要组成部分，是艺术瑰宝。

20. Dunhuang murals contain rich traditional clothing culture, among them the Tang Dynasty costume is considered the most gorgeous and elegant.

敦煌壁画蕴含着丰富的传统服饰文化，其中唐朝服饰最为绚丽多姿和雍容大度。

21. Cheongsam is a kind of typical national costume with internal and external harmony. It's praised as the representative of Chinese clothing culture.

旗袍是一种内与外和谐统一的典型民族服装，被誉为中华服饰文化的代表。

22. Streetwear, rooted in West Coast surf and skate culture, is a distinctive style of street fashion. It has grown to encompass elements of hip hop fashion, Japanese street fashion and modern haute couture fashion.

街头服饰根植于美国西海岸的冲浪和滑板文化，是一种有特色的街头时尚。它已经发展到包含嘻哈时尚、日本街头时尚和现代高级定制时装的元素。

23. Streetwear commonly centres on casual, comfortable pieces such as jeans, T-shirts, baseball caps and sneakers and exclusivity through intentional products scarcity.

街头服饰通常以休闲舒适的单品，

如牛仔裤、T恤、棒球帽和运动鞋为中心，并通过有意的产品稀缺来体现其专营权。

24. Functional clothing represents an area where clothing crosses the conventional boundaries and integrates with the domains of medicine, biotechnology, nanotech, physics and computing among others, to meet the multifaceted and complex requirements of the users.

功能服装代表着服装跨越传统边界，与医学、生物技术、纳米技术、物理学和计算机等领域相结合，以满足用户多方面和复杂的需求。

25. Functional clothing is mainly divided into four categories, including protective clothing, sports functional clothing, medical functional clothing and special need clothing.

功能服装主要分为四类，包括防护服、运动功能服、医疗功能服和特需服装。

26. Hazmat suit is a piece of personal protective equipment that consists of an impermeable whole-body garment worn as protection against hazardous materials. Such suits are often combined with self-contained breathing apparatus (SCBA) to ensure a supply of breathable air.

危险品防护服是一个个人防护设备，由一种防渗漏的全身服装组成，用来抵御有害物质。这类防护服通常与自给式呼吸器（缩写为SCBA）相结合，以保证呼吸空气的供应。

27. Intelligent clothing is the direction and future of the wearable industry. Keep an eye on the smart shirts that could give you super powers to your body, or smart sneakers that help you run better.

智能服装是可穿戴产业的方向和未来。留意那些可以给你超能力的智能衬衫，或能助你跑得更快的智能运动鞋。

28. Have you ever seen clothes that can change their colours based on ambient light or heat? This kind of clothes is becoming fairly popular as they use a technology to easily set the clothes apart from others through passive stimuli light or body heat.

你见过可以根据环境光线或温度改变颜色的衣服吗？这种衣服正在变得相当流行。因为他们使用了一种技术，通过被动刺激光线或体温，很容易地使变色衣区别于其他衣服。

29. The "motion response sportswear" uses themochromic printiny techniques to enable the fabric to change colour in response to the wearer's heart rate and body temperature.

这款"活动响应运动服"采用热致变色技术，使面料能够根据穿着者心率和体温的变化而改变颜色。

30. This is a type of high-tech field uniform. It was colloquially named the "chameleon suit" because the variable-camouflage system changed colour and pattern in a similar manner to a chameleon and assimilates it into the surrounding.

这是一种高科技野战军服，俗称"变色龙套装"，因为这种可变伪装系统以类似变色龙的方式改变服装的颜色和图案，使之融入周围的环境。

31. This tactical vest is a primarily protective gear for military and polices. It

features six M4mag pockets, two internal pockets and two multi-purpose pockets.

这件战术背心是军队和警察的主要防护装备。它有六个 M4 弹匣口袋，两个内置口袋和两个多用途口袋。

32. The jacket has specially designed pockets to hold an iPod flatly in place. It also has a special fabric touchpad to run, which allows you to turn the iPod on, control the volume, skip forward or back, and auto-lock the device.

这件夹克有专门设计的口袋，用来放置播放器，另配有一个特殊的布料触摸板，可以让你打开播放器，控制音量，前后滑动选歌，并自动锁定设备。

33. Sensors nowadays can be embedded directly into textile. With an array of sensors woven into the fabric, this high-tech underwear is intended to help users track their health.

现在的传感器可以直接嵌入纺织品。这款高科技内衣在面料上织入了一系列传感器，旨在帮助用户追踪自己的健康状况。

34. A simple T-shirt can emit vibrantly coloured messages, images and animations by uniting LED with it.

通过与 LED 照明技术结合，一件简单的 T 恤可以映发出彩色信息、图像和动画。

35. The Mao suit remained as the standard formal dress many years for the Chinese leaders. From the 1980s, more and more Chinese politicians and common people began wearing the European-style suits with neckties.

中山装一直是中国领导人的标准礼服。从 20 世纪 80 年代开始，越来越多的中国政治家和普通民众开始穿西装，打领带。

36. In the 1960s and 1970s, the Mao suit became fashionable among Western European, Australian and New Zealander socialists and intellectuals. It was sometimes worn over a turtleneck.

在 20 世纪 60 年代和 70 年代，中山装曾在西欧、澳大利亚、新西兰的社会主义者和知识分子中成为时尚，有时是穿在一件高领毛衣上。

37. Qipao, also known as cheongsam in Cantonese, is a one-piece Chinese traditional dress that has its origins in Manchu-ruled China back in the 17th century. Qipao has evolved over the decades and is still worn today.

旗袍在广东话中也叫长衫，是一件式的中国传统服装，起源于 17 世纪统治中国的满族。旗袍经过几十年的演变，今天仍在穿。

38. The original Qipao was very loose and covered most of the wearer's body. But the modern Qipao worn today is a form-fitting dress, that has a high slit on one or both sides, classic stand collar and short sleeves, bell sleeves or sleeveless, etc.

最初的旗袍非常宽松，罩住了穿着者的大部分身体。而现代的旗袍十分合体，具有一边或两边的高衩，经典立领、短袖、喇叭袖或无袖等。

39. Nowadays, some women wear the Qipao as everyday attire. Cheongsams are especially worn during formal occasions like weddings, parties and beauty pag-

eants. Qipao is also used as a uniform at restaurants, hotels and on airplanes in Asia.

如今，一些女性把旗袍作为日常服装，尤其在正式场合穿用，如婚礼、派对或选美比赛。在亚洲，旗袍还被用作餐厅、酒店和飞机上的制服。

40. A mandarin jacket is a short clothing worn over a gown. It reaches to the waist with sleeves covering only the elbows, which made it easy for riding horses. So, it is called Magua which was popular in the Qing Dynasty and the Republic of China.

马褂是一种穿在长袍外的短衣。其衣长及腰，袖仅遮肘。这种服装便于骑马，因而得名马褂。它流行于清代及民国时期。

41. Tibetan garments still keep with the original styles. Both men and women wear long-sleeved silk or cotton jackets topped with loose gowns tied with a band on the right. In addition, they also wear woolen or leather boots.

藏族服装仍然保持着原来的款式。藏族男女穿丝或棉的长袖上衣，配以右边有系带的宽松长袍。此外，他们还穿羊毛靴或皮靴。

42. Jeans are pants made from denim or dungaree cloth. They were invented by German-American Levi Strauss and his partner Jacob Davis in 1873. Jeans marked culture of the last 150 years probably more than we think.

牛仔裤是用粗斜纹布或粗布制成的裤子，是由德裔美国人李维·斯特劳斯和他的合伙人雅格布·戴维斯于1873年发明的。牛仔裤可能比我们想象的更能代表过去150年的文化。

43. Originally designed for miners, modern jeans were popularized as casual wear by Marlon Brando and James Dean in their 1950s films. Nowadays, jeans become one of the most popular types of practical clothes in the world.

现代牛仔裤最初是为矿工设计的。20世纪50年代，马龙·白兰度和詹姆斯·迪恩主演的几部电影让牛仔裤作为休闲服装流行起来。如今，它成了世界上最受欢迎的实用服装之一。

44. The little black dress, designed by Coco Chanel in the 1920s, is a black evening or cocktail dress, cut simply and often quite short. Back then it's considered essential to a complete wardrobe for every woman.

由可可·香奈儿在20世纪20年代设计的小黑裙，是一款黑色的晚礼服或鸡尾酒会礼服，剪裁简单，通常很短。当年，它被视为每个女人衣柜里必备的衣服。

45. In addition to the motifs of wildlife, this elegant girl's striped dress, featured with a Peter Pan collar and ruffled sleeves, also comprises a classic floral belt. It shows a beautiful summer scenery.

除了野生动物图案，这条优雅的条纹女童连衣裙还采用小圆领和荷叶边袖，并带有一条经典的花卉腰带，展示了美丽的夏日风景。

46. Gaucho pants are all about the cut-off. Keep the length above the ankle but only bare a bit of the shin. In the 1970s, they were a nice compromise be-

tween formal and informal clothing.

高乔裤都是截短式的，其长度应保持在踝部以上，只露出一段小腿。在20世纪70年代，高乔裤是正式和非正式服装之间的一种很好的折中选择。

47. Never considered to be appropriate office attire, these early gaucho pants never made it out of the realm of casual clothing. But for several years they were almost a necessity in many well-dressed women's wardrobes.

这些早期的高乔裤从未被认为是合适的办公室着装，也从未走出过休闲服装的领域。但几年下来，它们几乎成了很多衣着考究的女性衣柜里的必需品。

48. Ceremonial dress is a clothing worn for very special situations, such as state occasions, celebrations, graduations, weddings and other important events. It is often considered one of the most formal in all dress codes.

礼仪服是在特殊场合，如国事活动、庆祝典礼、毕业典礼、婚礼和其他重要活动中穿着的服装。它通常被认为是所有着装规范中最正式的一种。

49. Wearing a suit and tie or a dress or on skirted suit was once the norm in workplaces. Gradually, however, the norm has become business casual style. Many employees like to wear business casual attire at work.

穿西装打领带、穿连衣裙或裙套装曾经是职场的常态。然而，渐渐地，商务休闲已成为一种常态。很多员工在工作时喜欢穿商务休闲装。

50. The company has a strict dress code. All employees are expected to wear uniforms to work. Casual wear is only allowed on Fridays.

这家公司有严格的着装要求，所有员工都要穿制服上班，只有星期五可以穿休闲装。

51. The oldest knitted items have been found in Egypt and dated between 11th and 14th centuries AD. The first pieces of clothing made by the technique similar knitting were socks.

在埃及发现了最古老的针织品，它们的年代为11～14世纪。第一件用类似编织技术制作的服用品是袜子。

52. Knitwear did not become popular until the early 20th century. In 1950, sweaters and cardigans were both in fashion magazines and almost in every wardrobe. It was hard to imagine a party without a lady wearing a jersey cocktail dress.

直到20世纪初，针织服装才开始流行起来。1950年，针织套衫和开襟毛衫都出现在时尚杂志上和几乎每个衣柜里，很难想象酒会上没有一位女士穿针织礼服。

53. Modern apparel design is divided into two basic categories: haute couture and ready-to-wear. The haute couture collection is dedicated to certain customers to made-to-measure. And ready-to-wear is standard sized, not custom made, more suitable for large production runs.

现代服装设计分为两大类：高级时装定制和成衣。高级定制系列是专门为特定客户量身定制的，而成衣系列是标准尺寸，不是定制的，更适合大批量生产。

54. Smart clothing is being developed. Designers are increasingly marring technology with fashion in the hopes of making garments more intelligent and functional.

智能服装正在被开发。设计师们越来越多地将科技与时尚结合起来，希望让服装更加智能，更具功能性。

55. Fashion designers have to be very interested in learning new things and read magazines, journals and books on fashion design history and new trends. They should also be aware of the fashion market needs and have some knowledges and experiences of tailoring.

时装设计师必须非常喜欢学习新事物，阅读与时尚设计历史和新趋势有关的杂志和书籍。他们还应该了解时装市场的需求，并具有一些裁制衣服的知识和经验。

56. Brand garment enterprises are increasingly eager for excellent fashion design talents. Sometimes they go to the fashion universities and colleges to scout the outstanding fresh graduates.

品牌服装企业对优秀的服装设计人才的渴望日趋强烈，有时他们会去服装院校寻找出色的应届毕业生。

57. Chinese-style stand collar, also known as Zhongshan coat collar or mandarin collar, is a short unfolded stand-up collar style on a jacket or shirt. It originated from ancient Chinese culture and become one of the quintessences of Hanfu.

中式立领，又称中山装领或马褂领，是夹克或衬衫上的一种不折叠的立领。它源于中国古代文化，是汉服的精髓之一。

58. A detachable collar, also called false collar, is a shirt collar separate from the shirt, fastened to it by studs. The collar is made of a different fabric from the shirt. Usually, it is always white.

可拆卸领，也叫假领，是一种与衬衫分开，要用扣子固定上去的衬衫领。领子用不同的衬衫料制作，通常情况下是白色的。

59. The design of the raglan sleeve structure is more complicated than other sleeve types. Its reverse thinking design is different from traditional structural design. The analysis of the test datas must be fully utilized for the design.

插肩袖结构设计比其他袖型复杂，其逆向思维设计不同于传统结构设计。必须充分利用测试数据的分析来进行设计。

60. Trouser length is sometimes called "outseam". Measure it from the centre of side waist to the outer ankle. Adjust the measurement to the desired length when drafting.

裤子的长度有时被称为"外缝长"，其为从侧腰中位至外踝处的长度。打样时可根据所需长度调节尺寸。

61. Textile engineers and designers from all over the world have already worked on fabrics that are not only green, but capable of changing colours, adjusting the body's temperature and working like a touchscreen. The smart fabrics will be widely used in clothing and other textiles in the coming years.

来自世界各地的纺织工程师和设计

师都在研制不仅绿色环保，而且能够改变颜色，调节体温，像触摸屏一样工作的面料。未来几年，这种智能面料将广泛应用于服装和其他纺织品。

62. In this thesis, property, processing technology and market prospect of the new moisture wicking fibers were analyzed, and several methods for designing woven and knitted clothing fabrics with the fibers were proposed.

这篇论文，分析了新型吸湿排汗纤维的性能、加工技术和市场前景，提出了以这种纤维设计机织和针织服装面料的几种方法。

63. Natural fibers exist in nature and can be obtained directly. They are divided into plant, animal and mineral fibers according to their source. The four kinds of natural fibers used long in textile and garment industry are cotton, linen, silk and wool.

天然纤维存在于自然界，可以直接获取。根据来源，天然纤维分为植物纤维、动物纤维和矿物纤维。长期用于纺织服装工业的天然纤维有棉、麻、毛、丝四种。

64. Tencel is a form of rayon. It consists of cellulose fibre, made from dissolving pulp and then reconstituting it by dry jet-wet spinning. The typical green fibre is used to make clothing and other textiles.

天丝是一种人造纤维素纤维，经过干喷湿法纺成丝。这种典型的绿色纤维被用来制作服装和其他纺织品。

65. Seaweed fibre is produced entirely from sustainable raw materials of wood and seaweed. It uses methods that save both energy and resources. The fibre is completely biodegradable.

海藻纤维全部是由可持续的原料木材和海藻制成，采用既节约能源又节约资源的方法。这种纤维是完全可以生物降解的。

66. Aramid fibre is a man-made synthetic fibre, known as its strong heat-resistance and extreme durability. The high-performance material is now used for flame-retardant clothes, bulletproof vests, military helmets, even aeroplanes and spaceships.

芳纶纤维是一种人造合成纤维，因超强耐热性和极端耐用性而受到青睐。这种高性能材料现被用于阻燃服装、防弹背心、军用头盔，甚至飞机和宇宙飞船上。

67. Clothing accessories can be broadly categorized into two general areas: those that are carried and those that are worn. The former includes handbags, hand fans, parasols, wallets, canes, etc. And the latter variety contains more, such as headgear, footwear, hosiery, gloves, ties, scarves, belts, glasses, necklaces, bracelets and so on.

服装配饰大致可以分为两类：随身携带和穿戴在身上的。前者包括手袋、手扇、阳伞、钱包、手杖等。而后者包含的品种更多，如帽类、鞋类、袜类、手套、领带、围巾、腰带、眼镜、项链、手镯等。

68. A beret is a round, soft, brimless, flat-crowned cap typically made of felt that fits snugly. It probably originated as the cap worn by shepherds in southern

France in the 15th century. Berets now became a kind of signature military caps for special forces in some countries.

贝雷帽，是一种圆形、无檐、平顶软帽，通常用毡布制作，戴起来很舒服。贝雷帽可能起源于15世纪法国南部牧羊人戴的帽子，现在它成了一些国家特种部队的标志性军帽。

69. The picklhaube was worn in the 19th and 20th centuries by Prussian and German military, firemen and police. It's still worn today as part of ceremonial wear in the militaries of certain countries, such as Sweden, Chile and Colombia.

这种尖顶盔在19世纪和20世纪被普鲁士和德国的军人、消防员和警察使用。如今，在某些国家如瑞典、智利和哥伦比亚的军队里，它仍然是礼仪服装的组成部分。

70. The ballet boots, originally gained popularity in the 1980s, are a contemporary style of footwear, that uses the extreme design and merges the look of the point shoes with extremely long heels. They should not be confused with ballet shoes which are dancing wear.

芭蕾靴是一种现代风格的靴子，采用极端设计，融合了超高跟尖头鞋的外观，最早流行于20世纪80年代。这种靴不应该和芭蕾鞋混淆，芭蕾鞋是跳舞用的。

71. Mary Jane, also known as bar shoes or doll shoes, is a closed, round toe, low-cut leather shoes with one or more straps across the instep. Classic Mary Jane for children is typically made of black leather or patent leather.

玛丽珍鞋，又称带襻鞋或娃娃鞋，是一种包脚、圆头的浅口皮鞋，在脚背上有一条或多条带襻。传统的儿童版玛丽珍鞋通常是用黑色皮革或漆皮做的。

72. A lining is an inner layer of fabric, fur, or other material inserted into clothing, hats, shoes and similar items. It may reduce the wearing stain on clothes, extending the useful life of the lined clothes. It can also add warmth to cold weather wear.

衬里是缝在服装、帽子、鞋子等物品内层的布料、毛皮或其他材料。它可以减少对衣服的磨损，延长其使用寿命，还可以为寒衣增添温暖。

73. Clothing lining can be classified as full lined, half lined or partial lined. Jackets are mostly full or half lined. Pants are usually lined from waist to knees. In haute couture, the body and sleeves are lined separately before assembly.

服装衬里可分为全衬里、半衬里或部分衬里。夹克衫大多是全衬里或半衬里，裤子通常从腰部到膝部都有里布。在高级时装中，衣身和袖子是在缝合前分别上衬里的。

74. Invisible zippers are usually nylon coil zippers, using lighter lace-like fabrics on the zipper tapes. They are common in skirts and dresses, also seeing increased use by the military and emergence service.

隐形拉链一般是尼龙环扣拉链，拉链带使用的是轻质的类似花边布料。这种拉链在裙子和连衣裙上常见，也越来越多地用于军事和紧急服务。

75. The airtight zipper is built like a standard toothed zipper, but with a water-

proof sheeting wrapped around the outside of each row of zipper teeth. When the zipper is closed, the two facing sides of the waterproof sheeting are squeezed tightly forming a double seal.

气密拉链的构造与标准的齿形拉链类似，但在每一排拉链齿的外侧都包裹着防水布。当拉链闭合时，两边的防水布被紧紧地挤压，形成双重密封。

76. Button is one of the accessories for garments. The word, "button", roots in the Old French word "boton". From 14th century, the term was used for clothing fasteners, as it still is today. However, the modern buttons have made tremendous changes and progresses in their varieties, materials and properties.

纽扣是服装辅料的一种，纽扣这个词源于古法语单词"boton"。从14世纪开始，这个词就被用于服装扣件，直到今天。不过，现代纽扣在品种、材质和性能方面都有了巨大的变化和进步。

77. In costume designs, designers pay more and more attention to the fashion elements of accessories such as zippers, buttons, laces, etc. On the basis of the original functions, accessories are now endowed with the meaning of decorations.

在服装设计中，设计师们越来越注重辅料如拉链、纽扣、花边等的时尚元素。如今，辅料在原有功能的基础上，被赋予了装饰的含义。

78. Garment template technology is a new sewing technique integrating clothing process engineering, mechanical engineering and CAD digitalization principle, which brings new vigor and vitality into the apparel industry.

服装模板技术是一种集服装工艺工程、机械工程和CAD数字化原理于一体的新型缝制技术，给服装行业注入了新的生机和活力。

79. Clothing craft template, also known as clothing template, is a mould for fixing fabric and assisting sewing according to the technical requirements of the specified process. Using it makes some difficult crafts simple, fast and quality.

服装工艺模板，也称服装模板，是一种按照指定工序的工艺要求，来固定面料、辅助缝制的模具。运用它使一些有难度的工艺变得简单、快速和优质。

80. The washing is a way that is done in garments for increasing their added value. There are various washing processes for the denim clothes such as stonewash, bleach wash, enzyme wash, pigment wash, monkey wash, destroy wash, sand blast, hand brush and so on.

洗水是服装增加其附加值的一种方法。牛仔服装有多种洗水工艺，如石磨洗、漂洗、酵素洗、碧纹洗、马骝洗、破坏洗、喷砂、手擦等。

81. The steaming heat setting is the last process of the finishing for formed knit clothing. Its purpose is to make the clothing have durable and stable dimension, soft handfeel, smooth surface and beautiful appearance.

蒸烫定型是成型针织服装后整理的最后一道工序。其目的是使成型服装具有持久、稳定的尺寸，手感柔软，表面平整，外形美观。

82. There is a shortage of skilled workers in the clothing industry, and consumer preference changes faster than ever pushed by fast fashion thends. Therefore, the garment manufacturing industry is urgeed to strive for automation of their eqipment.

服装行业熟练工人短缺，而在快时尚的推动下，消费者的喜好比以往任何时候变化得都快。因此，迫切要求服装制造业努力实现其设备自动化。

83. In recent years, high and new technology at home and abroad has been widely used in clothing equipment with functional, intelligent, automatic and high-speed features. These new equipment for cutting, sewing, ironing and bonding promote the improvement of garment technology, quality and efficiency.

近年来，国内外高新技术已被广泛应用于服装设备，使具有功能性、智能化、自动高速的特征。这些裁剪、缝制、整烫和黏合设备，促进了服装技术、质量和效率的提高。

84. This full-automatic computerized cutting machine provides the most accurate possible cutting at high speed. Marker is not necessary to put over the fabrics lays during cutting. The technology has the advantage of being highly precise and fast, also simple operation.

这台全自动裁剪机能在高速下提供最精确的裁剪。在裁剪过程中，无需在布料上做记号。这种技术的优点是精度高、速度快、操作简单。

85. One of the best pros of computer-ized sewing machines is that users can update its software, which means that users can have a modern new machine as long as the manufacturer releases updates. Also, users can download new stitches and patterns.

计算机缝纫机最大的优点之一，是用户可以更新它的软件。这意味着，只要制造商发布更新，用户就可以拥有一台现代化新机器。此外，用户还可下载新的针法和花样。

86. Seamless clothing is a kind of tri-dimensional one-time forming clothes, produced by the seamless circular knitting machine, which complete whole clothes based on pre-programmed computer commands.

无缝服装是一种由无缝针织大圆头机生产的三维一次性成型衣服。大圆机根据预先编程的计算机命令完成整件衣服。

87. Please make sure the collar is balanced on both sides. Besides, the collar band must be not visible at back when wear.

请确保领子两边对称。此外，衣服穿着时，后身底领不能外露。

88. Left and right back neck patch length along raglan sleeve seam has 1/4" difference. It makes the back neck patch shape a little deformed.

左右后领贴在插肩袖缝处有1/4英寸长短，导致后领贴有点变形。

89. The front inner facing must be long enough in order to avoid hiking up at the placket bottom.

门襟内贴必须足够长，以避免门襟底部起吊。

90. Please check that the closure is tight when buttoned. There should not be puckering of fabrics around the fastened buttons. Such puckering would mean unleveled sewing of buttons and buttonholes.

请检查（衣服）纽扣是否扣紧，扣紧的纽扣周围不应有衣料起皱的情况。这种起皱说明纽扣和扣眼缝得高低不平。

91. Pay attention to the colour, which within one clothes must be the same colour shade. Also pay attention to the trimming and cleaning.

注意同一件衣服不能有色差，还要留意线头和污迹。

92. Founded in 1983, this fashion company is a large enterprise integrating design, production and sales. The listed company owns two high-end women's clothing brands.

这间时装公司成立于1983年，是一家集设计、生产和销售于一体的大型企业。这间上市公司拥有两个高端女装品牌。

93. We are a supplier of various kinds of brand clothes and accessories, such as women's fashion, sportwear, jeans, beach shorts, swimming suit, sun hat, sunglasses, etc.

我们是各类品牌服装和服饰品的供应商，品种有时款女装、运动服装、牛仔裤、沙滩短裤、泳装、遮阳帽、太阳镜等。

94. As an important achievement of the modern business management, virtual operation mode has gradually been recognized by many enterprises. Some apparel companies use the mode to increase their market competitiveness and build their strong brand.

虚拟经营模式作为现代企业管理的一项重要成果，逐渐被很多企业认可。一些服装公司采用这种模式提升市场竞争力，打造强势品牌。

95. The clothing business has become more and more difficult with the stiff competition and increased cost. However, our silk embroidered pyjamas still sell well both at home and abroad.

随着竞争的激烈和成本的上升，服装生意变得越来越难做。不过，我们的丝绸绣花睡衣依然畅销国内外。

96. If you want to know more, you could check our company's website. You are also welcome to our factory to visit the cutting, sewing and pressing workshop, especially the sample room.

如果你想了解更多，可以去我司的网站查看。也欢迎你来我们工厂，参观裁剪、缝制和整烫车间，特别是样品间。

97. We'll alter paper pattern, lay marker and arrange the production in large scale as soon as possible after we get the approval from you.

收到你们的确认样后，我们会尽快调纸样、排唛架和安排大货生产。

98. Quality is even more important than quantity. We always adhere to put the quality control of the exported garments as top priority.

质量比数量更重要。我们始终坚持将出口服装的质量管理放在首位。

99. Young people like to choose fashionable clothes, because they want to look

cool. For a party with friends, pick a well-tailored jumpsuit that flatters your figure.

年轻人喜欢选择新潮服装，因为他们想看上去很酷。与朋友聚会，选一条剪裁得体的连体裤会很显身材。

100. It seems that this tie in brown does not look to go well with your suit. Also, you should wear dress shoes to your friend's wedding.

这条棕色领带似乎与你的西装不太搭配。还有，你应该穿正装鞋出席朋友的婚礼。